PRACTICAL MULTISCALING

PRACTICAL MULTISCALING

Jacob Fish
Columbia University, USA

WILEY

This edition first published 2014

Registered Office
John Wiley & Sons, Ltd, The Atrium, Southern Gate, Chichester, West Sussex, PO19 8SQ, United Kingdom

For details of our global editorial offices, for customer services and for information about how to apply for permission to reuse the copyright material in this book please see our website at www.wiley.com.

Library of Congress Cataloging-in-Publication Data

Fish, J. (Jacob)
Practical multiscaling / Jacob Fish.
pages cm
Includes bibliographical references and index.
ISBN 978-1-118-41068-4 (hardback)
1. Mechanical engineering–Mathematical models. 2. Continuum mechanics–Mathematical models. 3. Materials–Mathematical models. 4. Multiscale modeling. 5. Scaling laws (Statistical physics) 6. Mechanical engineering–Computer simulation. 7. Continuum mechanics–Computer simulation. 8. Materials–Computer simulation. I. Title.
TJ153.F455 2013
620.001′51–dc23

2013027948

A catalogue record for this book is available from the British Library.

Set in 10/12pt Times by SPi Publisher Services, Pondicherry, India

1 2014

To my wife Ora and to my children Adam and Effie

Contents

Preface

This textbook covers fundamental modeling techniques aimed at bridging diverse temporal and spatial scales ranging from the atomic level to a full-scale product level. The focus is on *practical multiscale methods* that account for fine-scale (material) details but do not require their precise resolution. The text material evolved from over 20 years of teaching experience, which included the development of Multiscale Science and Engineering courses at Rensselaer Polytechnic Institute and Columbia University, as well as from practical experience gained in the application of multiscale software.

Due to a broad spectrum of application areas, this course is intended to be of interest and use to a varied audience, including:

- graduate students and researchers in academia and government laboratories who are interested in acquiring fundamental skills that will enable them to advance the state-of-the-art in the field;
- practitioners in civil, aerospace, pharmaceutical, electronics, and automotive industries who are interested in taking advantage of existing multiscale tools; and
- commercial software vendors who are interested in extending their product portfolios and tapping into new markets.

This textbook is unique in three respects:

- Theory and implementation. The text provides a detailed exposition of the state-of-the-art multiscale theories and their insertion into conventional (single-scale) finite element code architecture.
- Predictability and design. The text emphasizes the *robustness* and *design* aspects of multiscale methods. This is accomplished via four building blocks: *upscaling* of information, *systematic reduction* of information, *characterization* of information utilizing experimental data, and *material optimization* (Figure 1).

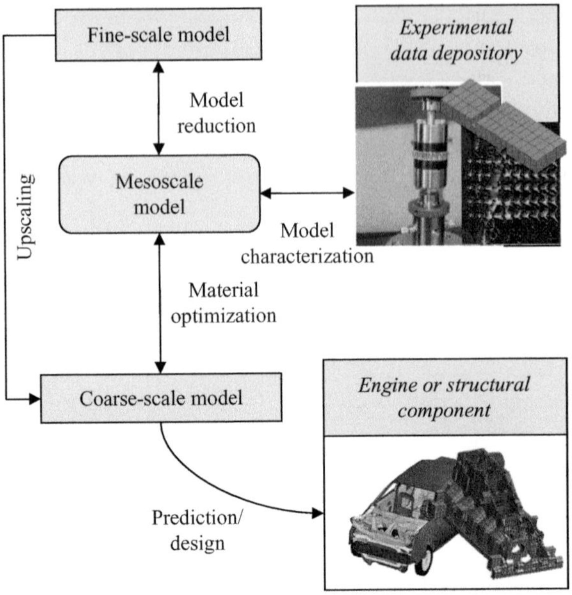

Figure 1 Building blocks of multiscale design. *Upscaling*: derivation of coarse-scale equations from fine-scale equations using homogenization-like theories. *Model reduction*: reducing the complexity of solving fine-scale problems. *Model characterization*: solving an inverse problem for reduced model parameters. *Material optimization*: optimizing microstructure based on design criteria

- Hands-on experience. Included with this textbook is an academic version of the multiscale design software (MDS-Lite) [1], which serves as a seamless plug-in to commercial software. A full integration with a built-in coarse-scale solver is also provided.

The material in this book can be covered in a single semester, and a meaningful course can be constructed from a subset of the chapters in this book for a one-quarter course. Following the Introduction to Multiscale Methods (Chapter 1), course material is organized in five chronological chapters: Upscaling/Downscaling of Continua (Chapter 2), Upscaling/Downscaling of Atomistic/Continuum Media (Chapter 3), Reduced Order Homogenization (Chapter 4), Scale-separation-free Upscaling/Downscaling of Continua (Chapter 5), and Multiscale Design Software (Chapter 6). Basic knowledge of continuum mechanics and finite elements is required. Chapters 2–4 focus on multiscale methods that take advantage of the scale separation hypothesis stemming from the infinitesimality of fine-scale features compared with the coarse-scale problem. The issue of how to systematically reduce fine-scale information and to characterize it against available experimental data is detailed in Chapter 4. Multiscale design software, which incorporates the aforementioned building blocks, including continua upscaling, model reduction, experimental characterization, and material optimization, is described in Chapter 6. The software can be used in conjunction with one of the commercial macroscopic solvers, ANSYS, ABAQUS, or LS-DYNA, or, alternatively, with the built-in coarse-scale solver, MDS-Macro. Use of this software provides a valuable hands-on experience to both students and practitioners. Chapters 2, 4, and 6 represent the core course material, which is recommended for one-quarter or full semester courses when supplementary material is used. A link between upscaling methods and the exact solution of the fine-scale problem is provided within the framework of the multigrid methods in Chapters 2 and 3 for continua and discrete media, respectively. Chapter 3, which details upscaling of atomistic media, is self-contained and can be taught independently of the core course material in Chapters 2, 4, and 6. Chapter 5, which is intended for an advanced audience, describes advanced multiscale and model reduction methods that are free of scale separation hypothesis.

Reference

[1] http://multiscale.biz.

Acknowledgments

I have many people to thank. I am grateful to Ted Belytschko for his constructive suggestions, which aided me in writing the text. I also appreciate the support of Zheng Yuan, my outstanding former PhD student and the current Chief Technological Officer of MDS, for "MDS-izing" the text and making it useful to both the academic community and practitioners. I received valuable assistance from Vasilina Filonova, an outstanding Research Associate at Columbia University, and a multitude of my PhD students including Sergey Kuznetsov, Nan Hu, Zifeng Yuan, Dimitrios Fafalis and Mahesh Bailakanavar. I express my deepest appreciation to Wendy Bickel for her assistance in editing the text. Finally, I wish to express my gratitude and apologies to my family, Adam, Effie, and Ora, to whom this book is dedicated, for enduring this past year when much of my time and energy should have been devoted to them rather than to this book.

1

Introduction to Multiscale Methods

1.1 The Rationale for Multiscale Computations

Consider a textbook boundary value problem that consists of equilibrium, kinematical, and constitutive equations together with essential and natural boundary conditions. These equations can be classified into two categories: those that directly follow from physical laws and those that do not. A constitutive equation demonstrates a relation between two physical quantities that is specific to a material or substance and does not follow directly from physical laws. It can be combined with other equations (equilibrium and kinematical equations, which do represent physical laws) to solve specific physical problems.

In other words, it is convenient to label all that we do not know about the boundary value problem as a *constitutive law* (a term originally coined by Walter Noll in 1954) and designate an experimentalist to quantify the constitutive law parameters. While this is a trivial exercise for linear elastic materials, this is not the case for anisotropic history-dependent materials well into their nonlinear regime. In theory, if a material response is history-dependent, an infinite number of experiments would be needed to quantify its response. In practice, however, a handful of constitutive law parameters are believed to "capture" the various failure mechanisms that have been observed experimentally. This is known as *phenomenological modeling*, which relates several different empirical observations of phenomena to each other in a way that is consistent with fundamental theory but is not directly derived from it.

An alternative to phenomenological modeling is to derive constitutive equations (or directly, field quantities) from finer scale(s) where established laws of physics are believed to be better understood. The enormous gains that can be accrued by this so-called multiscale approach have been reported in numerous articles [1,2,3,4,5,6]. Multiscale computations have been identified (see page 14 in [7]) as one of the areas critical to future nanotechnology advances.

Practical Multiscaling, First Edition. Jacob Fish.

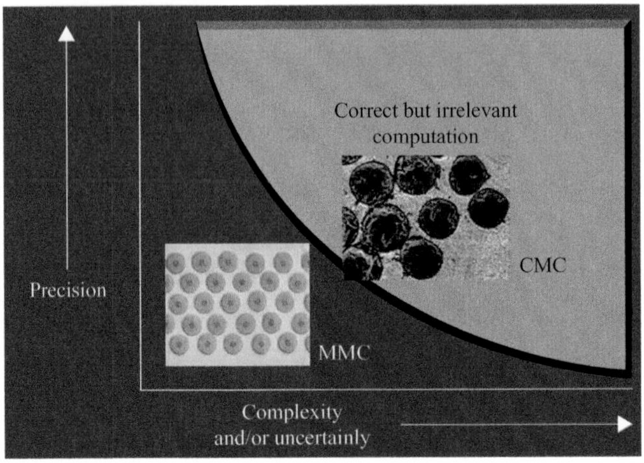

Figure 1.1 Reduced precision due to increase in uncertainty and/or complexity. CMC, ceramic matrix composite; MMC, metal matrix composite

For example, the FY2004 US$3.7 billion National Nanotechnology Bill (page 14 in [7]) states that "approaches that integrate more than one such technique (…molecular simulations, continuum-based models, etc.) will play an important role in this effort."

One of the main barriers to such a multiscale approach is the increased uncertainty and complexity introduced by finer scales, as illustrated in Figure 1.1. As a guiding principle for assessing the need for finer scales, it is appropriate to recall Einstein's statement that "the model used should be the simplest one possible, but not simpler." The use of any multiscale approach has to be carefully weighed on a case-by-case basis. For example, in the case of metal matrix composites (MMCs) with an almost periodic arrangement of fibers, introducing finer scales might be advantageous since the bulk material typically does not follow normality rules, and developing a phenomenological coarse-scale constitutive model might be challenging at best. The behavior of each phase is well understood, and obtaining the overall response of the material from its fine-scale constituents can be obtained using homogenization. On the other hand, in brittle ceramic matrix composites (CMCs), the microcracks are often randomly distributed and characterization of their interface properties is difficult. In this case, the use of a multiscale approach may not be the best choice.

1.2 The Hype and the Reality

Multiscale Science and Engineering is a relatively new field [8,9] and, as with most new technologies, began with a *naive euphoria* (Figure 1.2). During the euphoria stage of technology development, inventors can become immersed in the ideas themselves and may overpromise, in part to generate funds to continue their work. Hype is a natural handmaiden to overpromise, and most technologies build rapidly to a peak of hype [10].

For instance, early success in expert systems led to inflated claims and unrealistic expectations. The field did not grow as rapidly as investors had been led to expect, and this translated into disillusionment. In 1981 Feigenbaum *et al.* [11] reckoned that although artificial intelligence (AI) was already 25 years old, it "was a gangly and arrogant youth, yearning for

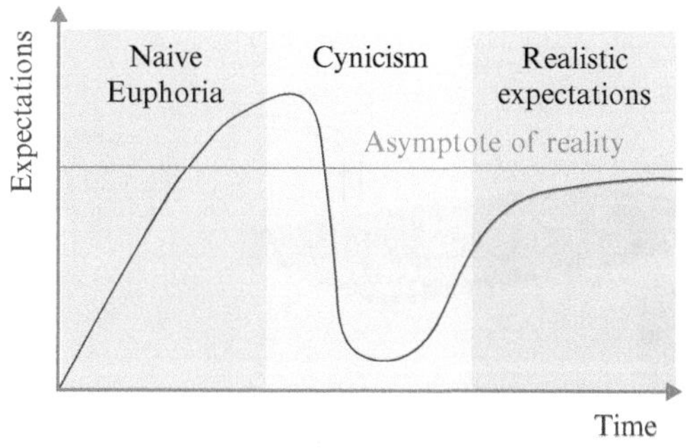

Figure 1.2 Evolution of new technology

a maturity that was nowhere evident." Interestingly, today you can purchase the hardcover AI handbook [11] for as little as US$0.73 on Amazon. Multiscale computations also had their share of overpromise, such as inflated claims of designing drugs atom by atom [12] or reliably designing the Boeing 787 from first principles, just to mention a few.

Following this naive euphoria (Figure 1.2), there is almost always an overreaction to ideas that are not fully developed, and this inevitably leads to a crash, followed by a period of wallowing in the depths of cynicism. Many new technologies evolve to this point and then fade away. The ones that survive do so because industry (or perhaps someone else) finds a "good use" (a true user benefit) for this new technology.

The author of this book believes that the state of the art today in multiscale science and engineering is sufficiently mature to take on the more than 50-year-old challenge [13] posed by Nobel Prize Laureate Richard Feynman: "What would the properties of materials be if we could really arrange the atoms the way we want them?" However, progress toward fulfilling the promise of multiscale science and engineering hinges not only on its development as a discipline concerned with the understanding and integration of mathematical, computational, and domain expertise sciences, but more so with its ability to meet broader societal needs beyond those of interest to the academic community. After all, as compelling as a finite element theory is, the future of that field might have been in doubt if practitioners had not embraced it.

Thus, the primary objective of this book is to focus not only on theory but also on practical utilization of multiscale methods.

1.3 Examples and Qualification of Multiscale Methods

Nature and man-made products are replete with multiple scales. Consider, for instance, the Airbus A380 depicted in Figure 1.3. It is 53 m long with a wingspan of 80 m and height of 24 m. The A380 consists of hundreds of thousands of structural components and many more structural details. Just in the fuselage alone there are more than 750,000 holes and cutouts. In addition to various structural scales, there are numerous material scales. At the coarsest material scales, the composites portion of the fuselage consists of laminate and woven/textile composite scales; at the intermediate scale is a tow or yarn, which consists of a bundle of fibers; and finally, there are one or more discrete scales, including atomistic and ab initio (quantum) scales. The metal portion of the airplane consists of a polycrystalline scale, a single crystal scale that considers dislocation density, a discrete dislocation scale, and finally, atomistic and ab initio scales.

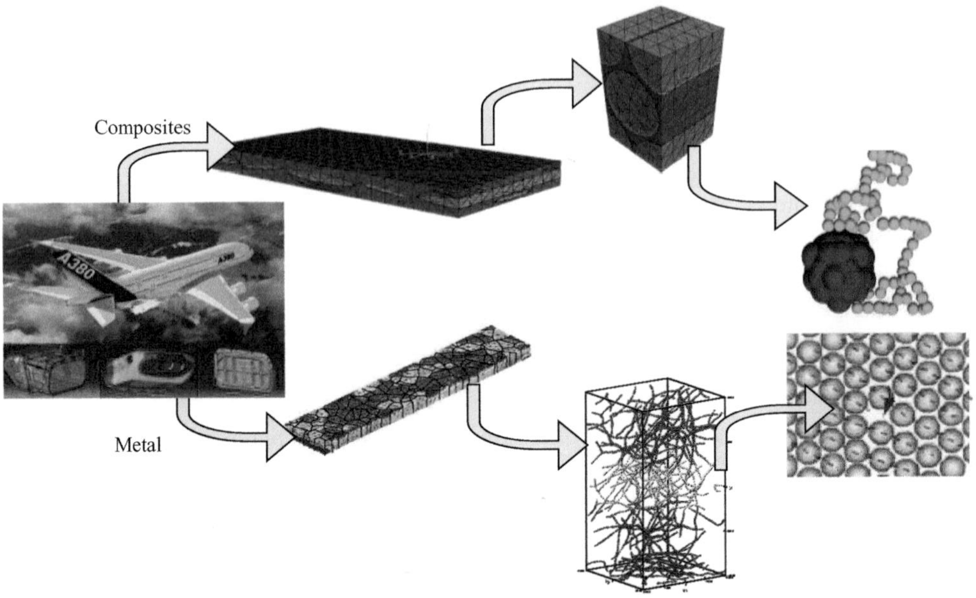

Figure 1.3 Multiple scales in the Airbus A380

It is tempting to start at the ab initio scale and to upscale, scale after scale, all the way to the product scale. This, unfortunately, is neither a realistic undertaking nor the goal of the present book. Our goal here is much more modest. We will focus on modeling and simulation approaches that can predict certain quantities of interest with significantly lower computational cost than solving the corresponding fine-scale system. The starting point for the fine-scale system of choice is not necessarily the ab initio scale; instead, the computational resources available and the accuracy requirement determine the starting point.

A modeling and simulation approach will be considered multiscale if it is capable of resolving certain quantities of interest with significantly lower cost than solving the corresponding fine-scale system. Schematically, a multiscale method has to satisfy the so-called Accuracy and Cost Requirements (ACR) test:

$$\boxed{\begin{array}{l} \textit{Error in quantities of interest} < tol \\ \dfrac{\textit{Cost of multiscale solver}}{\textit{Cost of fine scale solver}} \ll 1 \end{array}}$$

In general, multiscale approaches fall into one of two categories: information-passing (or hierarchical) or concurrent. In the information-passing multiscale approach, which is the main focus of this book, the fine-scale response is idealized (approximated or unresolved) and its overall (average) response is infused into the coarse scale. In the concurrent approaches, fine- and coarse-scale resolutions are simultaneously employed in different portions of the problem domain, and the exchange of information occurs through the interface. The subdomains where different scale resolutions are employed can be either disjoint or overlapping.

Information-passing multiscale methods are typically used to model the overall response of the fine scale, except for the hot spots in the vicinity of cutouts and boundary layers where concurrent multiscale methods are more appropriate.

To this end, we will focus on the qualification of multiscale methods. Loosely speaking, the information-passing multiscale approach is likely to pass the ACR test provided that:

(i) quantities of interest are limited to or defined only on the coarse scale (provided that these quantities are computable from the fine scale); and
(ii) special features of the fine-scale problem, such as scale separation and self-similarity, are taken advantage of.

On the other hand, for the concurrent multiscale approach to pass the ACR test, the following conditions must be satisfied:

(i) the interface (or interphase) between the fine and coarse scales should be properly engineered;
(ii) the fine-scale model should be limited to a small portion of the computational domain; and
(iii) the precise material microstructure should be known in the subdomain where the fine-scale model is considered.

It is important to note that even though the concurrent approach may pass the first two criteria in the ACR test, its computational cost will typically exceed that of the information-passing methods. Furthermore, the main hurdle to successful utilization of concurrent methods in practice is a lack of knowledge of precise material microstructure in the hot spots. In these locations, fine-scale resolution is required, as opposed to the information-passing multiscale methods where material microstructure in small representative windows is reconstructed from various test coupons.

1.4 Nomenclature and Definitions

Since various multiscale methods were conceived in different scientific communities, there has been a proliferation of definitions, some of which are contradictory or overlapping. For instance, various information-passing multiscale methods have been labeled by different names, including upscaling methods, coarse-graining methods, homogenization methods, or simply multiscale methods. There are also subcategories of the above definitions, such as systematic upscaling (with obvious implications), operator upscaling, variational multiscale, computational homogenization, multigrid homogenization, numerical homogenization, numerical upscaling, and computational coarse-graining, just to mention a few.

Some authors draw a distinction between upscaling and multiscale methods. According to one such definition, upscaling forms a coarse-scale model with an a priori defined mathematical structure; once the model is conceived, the fine-scale information is discarded, whereas in multiscale methods, the fine-scale information is retained throughout the simulation and the coarse-scale structure is generally not expressed analytically. However, there is no consensus on the above definition. For instance, the variational multiscale method (VMS) [14] is considered to fall into the category of (operator) upscaling methods. Yet, for nonlinear problems, fine-scale information is not discarded in VMS, suggesting that it belongs to the category of multiscale methods. Likewise, the homogenization method for linear problems provides effective properties, and this is obviously an upscaling method based on the aforementioned definition; and yet, for nonlinear problems, fine- and coarse-scale problems are fully coupled throughout the analysis. Another misconception is the supposition that upscaling is a form of homogenization that is free of the periodicity assumption. Homogenization, like most of the upscaling methods, assumes some form of scale separation, but it can be used to homogenize random heterogeneous media with either periodic, weakly periodic, essential, natural, or hybrid boundary conditions.

Hereafter, upscaling and downscaling will be understood as two building blocks of the information-passing multiscale method. For nonlinear processes, upscaling is a history-dependent process of constructing coarse-scale equations from well-defined fine-scale equations. *History dependence* means that the fine-scale information is retained and used throughout the simulation to update the coarse-scale problem. Downscaling, often called localization, is the second building block of the information-passing multiscale approach. Downscaling is a history-dependent process by which fine-scale information is continuously reconstructed in small windows using the information from the coarse-scale problem. The information-passing multiscale approach is a continuous process of upscaling and downscaling. The window in this information-passing process can be a point in the coarse-scale domain, in which case the information-passing multiscale approach is synonymous with nonlinear (computational) homogenization, a single coarse-scale element [14], or a patch of coarse-scale elements. In the former case, this small window is often referred to as a unit cell or representative volume element. For linear problems, the fine- and coarse-scale problems are one-way coupled, where upscaling provides coarse-scale (effective) properties, while downscaling plays the role of postprocessing of the fine-scale solution. Hereafter, the nested process of upscaling and downscaling will be termed as *upscaling/downscaling*.

Coarse-graining is a subclass of upscaling methods where a coarse-scale (or coarse-grained) model is constructed from the fine-scale information in the preprocessing stage prior to nonlinear analysis. Coarse-grained molecular dynamics is a typical example of such coarse-graining. The fact that fine-scale information is not revisited in these methods offers considerable computational advantages, but often at the expense of accuracy.

Different terminologies are used to indicate various scales. In the case of two scales, the fine scale is often referred to as a microscale, unresolvable scale, atomistic scale, or discrete scale; the coarse scale is often labeled as a macroscale, resolvable scale, component scale, or continuum scale. Here we will simply refer to the two scales as fine and coarse scales. For more than two scales, we will refer to the additional scales as mesoscales.

1.5 Notation

1.5.1 Index and Matrix Notation

Two types of notation will be used: (i) indicial notation; and (ii) matrix notation. All the derivations will be made in the indicial notation. The equations pertaining to the finite element implementation will be given in indicial or matrix notation.

In the indicial notation, the components of tensors or matrices are explicitly specified. Thus a vector, which is a first-order tensor, is denoted in indicial notation by a_i where the range of the index is the number of spatial dimensions n_{sd}. Indices repeated twice in a term are summed, in conformance with the rules of Einstein notation. Spatial tensor components are denoted by lowercase Latin subscripts, which are always on the right of the tensor. Spatial components of a second-order tensor are indicated by two Latin subscripts, and they always refer to the Cartesian coordinate system. For example, small strain tensor components are denoted by ε_{ij}.

We will alternate between two notations for finite element nodes and degrees of freedom. Nodal indices will always be indicated by uppercase Latin letters positioned at the bottom right of the tensor, vector, or matrix. For example, v_{iA} is the velocity of node A in the direction i. Indices representing finite element degrees of freedom will always be indicated by lowercase Greek letters positioned at the bottom right of the tensor or vector. For example, v_α is the velocity of degree-of-freedom α. The degrees of freedom are related to nodes by

$$\alpha = (A-1)n_{sd} + i$$

where n_{sd} denotes the number of spatial dimensions.

When nodal and degrees-of-freedom indices are repeated twice, they will be summed over their range, which depends on the context. When dealing with an element, the range is over the nodes or degrees of freedom of the element, whereas when dealing with a mesh, the range is over the nodes or degrees of freedom of the mesh.

In the finite element implementation, we will often use matrix notation. We will indicate matrices and vectors, which are the first-order matrices, in boldface. Second-order tensor components will often be converted to Voigt notation in the implementation phase. In Voigt notation, kinetic symmetric tensors, such as Cauchy stress σ_{ij}, and kinematic symmetric tensors, such as small strain ε_{ij}, are written as column matrices:

$$\sigma_{ij} \rightarrow \boldsymbol{\sigma} = \begin{bmatrix} \sigma_{11} \\ \sigma_{22} \\ \sigma_{33} \\ \sigma_{23} \\ \sigma_{13} \\ \sigma_{12} \end{bmatrix}; \quad \varepsilon_{ij} \rightarrow \boldsymbol{\varepsilon} = \begin{bmatrix} \varepsilon_{11} \\ \varepsilon_{22} \\ \varepsilon_{33} \\ 2\varepsilon_{23} \\ 2\varepsilon_{13} \\ 2\varepsilon_{12} \end{bmatrix}$$

Note that kinematic tensor components for which indices are not equal are multiplied by 2 in Voigt notation. The Voigt rule is particularly useful for converting fourth-order tensors. For example, the linear elastic constitutive tensor components L_{ijkl} are written in Voigt notation as

$$L_{ijkl} \rightarrow \boldsymbol{L} = \begin{bmatrix} L_{1111} & L_{1122} & L_{1133} & L_{1123} & L_{1113} & L_{1112} \\ L_{2211} & L_{2222} & L_{2233} & L_{2223} & L_{2213} & L_{2212} \\ L_{3311} & L_{3322} & L_{3333} & L_{3323} & L_{3313} & L_{3312} \\ L_{2311} & L_{2322} & L_{2333} & L_{2323} & L_{2313} & L_{2312} \\ L_{1311} & L_{1322} & L_{1333} & L_{1323} & L_{1313} & L_{1312} \\ L_{1211} & L_{1222} & L_{1233} & L_{1223} & L_{1213} & L_{1212} \end{bmatrix}$$

such that $\boldsymbol{\sigma} = \boldsymbol{L}\boldsymbol{\varepsilon}$ in matrix notation.

The fourth-order identity tensor in the indicial notation is given as

$$I_{ijkl} = \frac{1}{2}\left(\delta_{ik}\delta_{jl} + \delta_{il}\delta_{jk}\right)$$

where δ_{ik} is the Kronecker delta, which is equal to zero for $i \neq j$ and one for $i = j$. The nonzero components of I_{ijkl} are

$$I_{1111} = I_{2222} = I_{3333} = 1$$

$$I_{1212} = I_{1313} = I_{2323} = I_{2121} = I_{3131} = I_{3232} = \frac{1}{2}$$

$$I_{2112} = I_{3113} = I_{3223} = I_{1221} = I_{1331} = I_{2332} = \frac{1}{2}$$

Let $\boldsymbol{I}$ be a 6×6 diagonal matrix and $\boldsymbol{H}$ be any $6 \times n$ matrix. The identity matrix $\boldsymbol{I}$ is defined so that $\boldsymbol{IH} = \boldsymbol{H}$, which requires diagonal terms with unequal indices to be multiplied by two.

$$\boldsymbol{I} = \begin{bmatrix} 1 & 0 & 0 & 0 & 0 & 0 \\ 0 & 1 & 0 & 0 & 0 & 0 \\ 0 & 0 & 1 & 0 & 0 & 0 \\ 0 & 0 & 0 & 1 & 0 & 0 \\ 0 & 0 & 0 & 0 & 1 & 0 \\ 0 & 0 & 0 & 0 & 0 & 1 \end{bmatrix}$$

1.5.2 Multiple Spatial Scale Coordinates

The coordinates in the coarse-scale deformed (or current) and undeformed (initial) configurations will be denoted by $\boldsymbol{x}$ and $\boldsymbol{X}$, respectively. For small deformation problems, a single coarse-scale coordinate $\boldsymbol{x}$ will be used.

The focus of this book is on two-scale analysis. The fine-scale problems will be considered in a small representative window often referred to as a unit cell. The unit cell will be generally assumed to be much smaller than the coarse-scale domain, and therefore its deformed and undeformed coordinates, denoted by $\boldsymbol{y}$ and $\boldsymbol{Y}$, respectively, will be rescaled by a small positive parameter ζ as

$$\boldsymbol{y} = \boldsymbol{x}/\zeta; \quad \boldsymbol{Y} = \boldsymbol{X}/\zeta \quad 0 < \zeta \ll 1$$

For three-scale problems, $\boldsymbol{y}$ and $\boldsymbol{Y}$ will denote the intermediate scale (or mesoscale) configuration, whereas $\boldsymbol{z}$ and $\boldsymbol{Z}$ will denote the finest scale configuration, such that

$$\boldsymbol{z} = \boldsymbol{y}/\zeta; \quad \boldsymbol{Z} = \boldsymbol{Y}/\zeta$$

For the general case of n_{sc} scales, the left uppercase superscript in the brackets will denote the scale, with 0 denoting the coarsest scale and $n_{sc} - 1$ the finest scale. The position vector at scale I, ${}^{(I)}\boldsymbol{x}$, will be related to the position vector at scale I–1, ${}^{(I-1)}\boldsymbol{x}$, by

$$\left.\begin{aligned} {}^{(I)}\boldsymbol{X} &= {}^{(I-1)}\boldsymbol{X}/\zeta \\ {}^{(I)}\boldsymbol{x} &= {}^{(I-1)}\boldsymbol{x}/\zeta \end{aligned}\right. \quad \text{for } I = 1, \ldots, n_{sc} - 1$$

1.5.3 Domains and Boundaries

For two-scale problems, we will consider three types of problem domains: (i) the composite domain denoted by Ω^ζ; (ii) the coarse-scale domain denoted by Ω; and (iii) the unit cell domain denoted by Θ. The corresponding boundaries are denoted by $\partial\Omega^\zeta$, $\partial\Omega$, and $\partial\Theta$, respectively. $\boldsymbol{n}^\zeta$, $\boldsymbol{n}^c$, and $\boldsymbol{n}^\Theta$ denote unit normals to the boundaries $\partial\Omega^\zeta$, $\partial\Omega$, and $\partial\Theta$, respectively. The volumes of the three domains are denoted by $|\Omega^\zeta|$, $|\Omega|$, and $|\Theta|$.

Throughout this book, the right superscript ζ will denote the existence of fine-scale features. The source problem will be always stated on a domain that, in addition to heterogeneities, may include microstructural voids. The composite domain Ω^ζ is defined as a solid part of the coarse-scale domain that does not contain voids in the material microstructure. Furthermore, the boundary of the composite domain $\partial\Omega^\zeta$ may be rough due to the intersection of voids with the external boundary. On the other hand, the coarse-scale domain Ω and its boundary $\partial\Omega$ are free of fine-scale material features. In the absence of information about surface roughness, we will often assume that $\partial\Omega^\zeta=\partial\Omega$ and $\boldsymbol{n}^\zeta=\boldsymbol{n}^c$.

$\partial\Omega^{u\zeta}$, $\partial\Omega^{t\zeta}$ and $\partial\Omega^u$, $\partial\Omega^t$ denote the essential (displacement) and natural (traction) boundaries of the composite and coarse-scale domains, respectively, related by

$$\partial\Omega^{t\zeta}\cup\partial\Omega^{u\zeta}=\partial\Omega^\zeta \quad \text{and} \quad \partial\Omega^{t\zeta}\cap\partial\Omega^{u\zeta}=0$$

$$\partial\Omega^{t}\cup\partial\Omega^{u}=\partial\Omega \quad \text{and} \quad \partial\Omega^{t}\cap\partial\Omega^{u}=0$$

A unit cell may consist of two or more fine-scale phases. The internal boundary between the fine-scale phases will be denoted by S, with $\breve{\boldsymbol{n}}$ being the unit normal to the boundary.

For three-scale problems, Θ_z will denote the unit cell domain at the finest scale, and $\partial\Theta_z$ will denote its boundary. For more than three scales, ${}^{(I)}\Theta$ and $\partial^{(I)}\Theta$ will be the unit cell domain and its boundary at scale I, with indices $I=1$ and $I=n_{sc}-1$ denoting the coarsest and finest scale unit cell domains, respectively.

For large deformation problems, we will distinguish between deformed and undeformed configurations. Ω_X^ζ and Ω_X will denote initial (undeformed) composite and coarse-scale domains, whereas Ω_x^ζ and Ω_x are the corresponding current (deformed) configurations. $\partial\Omega_X^{u\zeta}, \partial\Omega_X^{t\zeta}$ and $\partial\Omega_x^{u\zeta}, \partial\Omega_x^{t\zeta}$ will denote the essential and natural boundaries of the initial and current composite domains. Similarly, $\partial\Omega_X^u, \partial\Omega_X^t$ and $\partial\Omega_x^u, \partial\Omega_x^t$ are the essential and natural boundaries of the initial and current coarse-scale domains, respectively. Unit normals to the initial and current composite, coarse-scale, and unit cell domains will be denoted by ($\boldsymbol{N}^\zeta$, $\boldsymbol{N}^c$, $\boldsymbol{N}^\Theta$) and ($\boldsymbol{n}^\zeta$, $\boldsymbol{n}^c$, $\boldsymbol{n}^\Theta$), respectively.

The unit cell domains Θ_Y and Θ_y will denote initial and current configurations, with $\partial\Theta_Y$ and $\partial\Theta_y$ being the corresponding boundaries.

1.5.4 Spatial and Temporal Derivatives

Upscaling methods will be predominantly derived from either the Hill–Mandel macrohomogeneity condition [15] or by using multiple-scale asymptotic methods. For two-scale problems, the various response fields $f^\zeta(\boldsymbol{x})$ will be assumed to depend on the fine- and coarse-scale coordinates

$$f^\zeta(\boldsymbol{x})=f(\boldsymbol{x},\boldsymbol{y})$$

Spatial derivatives of the response function $f^\zeta(\boldsymbol{x})$ can be calculated by the chain rule as

$$f^\zeta{}_{,i} = \frac{\partial f(\boldsymbol{x},\boldsymbol{y})}{\partial x_i} + \frac{1}{\zeta}\frac{\partial f(\boldsymbol{x},\boldsymbol{y})}{\partial y_i} = f_{,x_i} + \frac{1}{\zeta} f_{,y_i}$$

where a comma followed by a subscript variable denotes a partial derivative with respect to the subscript variable. Symmetric spatial derivatives are denoted as

$$f_{(i,x_j)} = \frac{1}{2}\left(\frac{\partial f_i}{\partial x_j} + \frac{\partial f_j}{\partial x_i}\right) \quad \text{and} \quad f_{(i,y_j)} = \frac{1}{2}\left(\frac{\partial f_i}{\partial y_j} + \frac{\partial f_j}{\partial y_i}\right)$$

For problems involving multiple temporal scales, such as fatigue in Chapter 4 and lattice vibration in Chapter 3, various response fields $f^\zeta(\boldsymbol{x},t)$ will be assumed to depend on multiple spatial and temporal coordinates

$$f^\zeta(\boldsymbol{x},t) = f(\boldsymbol{x},\boldsymbol{y},t,\tau)$$

where τ is the fast time coordinate related to the slow time coordinate t by

$$\tau = \frac{t}{\eta} \qquad 0 < \eta \ll 1$$

Time differentiation of response fields with respect to multiple temporal scales is given by the chain rule

$$\frac{df^\zeta(\boldsymbol{x},t)}{dt} = \frac{\partial f(\boldsymbol{x},\boldsymbol{y},t,\tau)}{\partial t} + \frac{1}{\eta}\frac{\partial f(\boldsymbol{x},\boldsymbol{y},t,\tau)}{\partial \tau} = \dot{f}(\boldsymbol{x},\boldsymbol{y},t,\tau) + \frac{1}{\eta} f'(\boldsymbol{x},\boldsymbol{y},t,\tau)$$

Most often, it will be assumed that spatial and temporal scaling parameters are identical, that is, $\zeta = \eta$.

1.5.5 Special Symbols

Throughout this book, special notations will denote certain attributes, as follows:

$\hat{\boldsymbol{x}}, \hat{\boldsymbol{X}}$ – coordinates of the unit cell centroid
$\bar{\boldsymbol{u}}, \bar{\boldsymbol{t}}$ – prescribed fields (displacements and tractions)
χ – local Cartesian coordinate in the physical domain placed at the unit cell centroid
$(\,)^{(k)}$ – right Latin superscript in parentheses denotes kth term in asymptotic expansion
$(\,)^f$ – right superscript f denotes fine-scale fields and properties
$(\,)^c$ – right superscript c denotes coarse-scale fields and properties
$(\,)^M$ – right superscript M denotes master (independent) nodes on the unit cell boundary
$(\,)^S$ – right superscript S denotes slave (dependent) nodes on the unit cell boundary

$(\,)^T$ – right superscript T denotes transpose
$(\,)^{-1}$ – right superscript -1 denotes inverse
${}^i(\,)$ – left superscript denotes iteration count
${}_k(\,)$ – left subscript denotes time increment or load parameter
$(\,)^{(\alpha)}$ – right Greek superscript in parentheses denotes phase or interface partition in the unit cell

References

[1] Curtin, W.A. and Miller, R.E. Atomistic/continuum coupling in computational materials science. Modeling and Simulation in Materials Science and Engineering 2003, 11(3), R33–R68.

[2] Fish, J. Bridging the scales in nano engineering and science. Journal of Nanoparticle Research 2006, 8, 577–594.

[3] Fish, J., ed. Bridging the Scales in Science and Engineering. Oxford University Press, 2007.

[4] Ghoniem, N.M. and Cho, K. The emerging role of multiscale modeling in nano- and micro-mechanics of materials. Modeling in Engineering and Sciences 2002, 3(2), 147–173.

[5] Liu, W.K., Karpov, E.G., Zhang, S. and Park, H.S. An introduction to computational nanomechanics and materials. Computer Methods in Applied Mechanics and Engineering 2004, 193, 1529–1578.

[6] Khare, R., Mielke, S.L., Paci, J.T., Zhang, S.L., Ballarini, R., Schatz, G.C. and Belytschko, T. Coupled quantum mechanical/molecular mechanical modeling of the fracture of defective carbon nanotubes and graphene sheets. Physical Review B 2007, 75(7), 075412.

[7] National Nanotechnology Initiative. Supplement to the President's FY 2004 Budget. National Science and Technology Council Committee on Technology, 2004.

[8] Horstemeyer, M.F. Multiscale modeling: a review. In Practical Aspects of Computational Chemistry, eds J. Leszczynski and M.K. Shukla. Springer Science Business Media, 2009, pp. 87–135.

[9] Belytschko, T. and de Borst, R. Multiscale methods in computational mechanics. International Journal for Numerical Methods in Engineering 2010, 89(8–9), 939–1271.

[10] Bezdek, J. Fuzzy models–what are they, and why? IEEE Transactions on Fuzzy Systems 1993, 1, 1–5.

[11] Barr, A., Cohen, P.R. and Feigenbaum, E.A., ed. The Handbook of Artificial Intelligence, Volume IV. Addison-Wesley, 1990.

[12] The Next Industrial Revolution: designing drugs by computer at Merck. Fortune Magazine, October 5, 1981.

[13] Feynman, R.P. There's plenty of room at the bottom. 29th Annual Meeting of the American Physical Society. California Institute of Technology, 1959.

[14] Hughes, T.J.R. Multiscale phenomena: Greens functions, the Dirichlet to Neumann formulation, subgrid scale models, bubbles and the origin of stabilized methods. Computer Methods in Applied Mechanics and Engineering 1995, 127, 387–401.

[15] Hill, R. Elastic properties of reinforced solids: some theoretical principles. Journal of the Mechanics and Physics of Solids 1963, 11, 357–372.

2

Upscaling/Downscaling of Continua

2.1 Introduction

Heterogeneous materials, such as composites, concrete, solid foams, soils, and polycrystals, consist of clearly distinguishable constituents (or phases) that show different mechanical and transport material properties. The premise of *homogenization of continua* is that the governing equations of individual phases, including their geometry and constitutive equations, are well understood at the fine-scale phases, or at least better understood than at the coarse-scale phases. Under this premise, homogenization provides a mathematical framework by which coarse-scale equations can be deduced from well-defined fine-scale equations. The benefit of homogenization is that the behavior of a heterogeneous material can be determined, at least in theory, without resorting to potentially expensive testing. Homogenization of continua can predict the full multiaxial properties and responses of heterogeneous materials, which are often anisotropic. Such properties are often difficult to measure experimentally or model phenomenologically. In addition to describing the overall behavior of heterogeneous materials, homogenization provides local fields (given coarse-scale fields, phase properties, and phase geometries) in a process known as *downscaling*, *localization*, or *postprocessing*. Such knowledge is especially important in understanding and describing material damage and failure.

Over the past two decades there has been a proliferation of methods aimed at improving the accuracy of the classical homogenization theory. These methods are often labeled as *upscaling* methods. Within this broader definition, homogenization is considered to be one of the upscaling methods.

We distinguish upscaling (or homogenization) of continua, which is based on the methods of continuum mechanics, from the *upscaling of discrete media* where fine-scale phases are governed by the motion of atoms as described by molecular dynamics. The latter is considered

Practical Multiscaling, First Edition. Jacob Fish.

in Chapter 3. In Chapter 2 we focus on the mechanical response of heterogeneous materials. A coupled vector-scalar field problem at multiple scales, such as environmental degradation of composites, is considered in Chapter 4.

From the mathematical point of view, homogenization of continua is the study of partial differential equations with rapidly oscillating periodic coefficients $L(x/\zeta)$ [1,2,3,4,5], such as

$$\frac{d}{dx}\left(L(x/\zeta)\frac{du}{dx}\right)+b=0 \tag{2.1}$$

where L denotes material property (Young's modulus for elasticity problems), u denotes the solution field (displacements for elasticity problems), b denotes the source term (body forces for elasticity problems), and ζ is a very small positive parameter ($0<\zeta\ll 1$). Heterogeneous materials are often subjected to loads that vary on a length-scale far bigger than the characteristic length-scale of the microstructure. In this situation, we can replace (2.1) with an equation of the form

$$\frac{d}{dx}\left(L^c\frac{du^c}{dx}\right)+b^c=0 \tag{2.2}$$

where L^c is the constant known as the effective property, b^c is the effective (smoothed out) source term, and u^c is the coarse-scale solution field. The process of replacing an equation having a highly oscillatory periodic coefficient (2.1) with one having a homogeneous (uniform) coefficient (2.2) is known as *upscaling of continua* or simply upscaling. The effective properties are calculated by solving a fine-scale problem over the so-called representative volume element (RVE) or unit cell (UC) domain, which is defined to contain enough statistical information about the heterogeneous medium to be representative of the material. The link between the coarse- and fine-scale equations can be established using one of the following: (i) asymptotic methods (section 2.2 and section 2.4); (ii) the so-called Hill–Mandel macrohomogeneity condition [6] (section 2.2.4); or (iii) the heterogeneous multiscale method (HMM) [7] (section 2.2.4.5).

As an alternative to finding effective properties, coarse-scale continuum equations can be constructed by enriching coarse-scale kinematical relations to capture the oscillatory response of a heterogeneous medium. This idea was originally introduced by Babuska [8,9] and later enhanced by several noteworthy contributions, including the multiscale finite element method (MsFEM) [10], the variational multiscale method (VMS) [11], the discontinuous enrichment method (DEM) [12], and the multiscale enrichment based on partition of unity method (MEPU) [13,14], just to mention a few. These enhancements differ in terms of: (i) the definition of the fine-scale window and boundary conditions from which the enriched kinematics is calculated; (ii) how the enriched kinematics is introduced into the coarse-scale equation; and (iii) how the coarse-scale equations are integrated and solved.

For the nonlinear problems with periodic oscillatory coefficients considered in section 2.3, we will consider a multidimensional counterpart of

$$\frac{d}{dx}\sigma(x/\zeta,x)+b=0 \tag{2.3}$$

The resulting coarse-scale problem is defined as

$$\frac{d}{dx}\sigma^c(x)+b^c=0 \tag{2.4}$$

with the UC or RVE problem schematically described by the following input–output relation

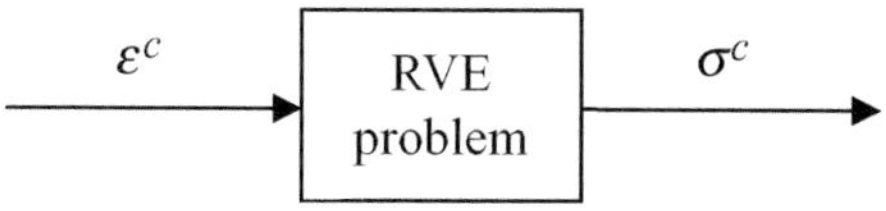

$$\varepsilon^c \longrightarrow \boxed{\begin{array}{c} \text{RVE} \\ \text{problem} \end{array}} \xrightarrow{\sigma^c} \tag{2.5}$$

where ε^c is a coarse-scale strain and σ^c is a coarse-scale stress.

For problems where the coarse-scale solution varies rapidly over the unit cell domain, ε^c is no longer constant over the unit cell domain and, consequently, the coarse-scale problem may include higher order coarse-scale stresses. This case is addressed by the higher order homogenization theory considered in section 2.4.

The unit cell problem can be solved either analytically or numerically. The simplest possible analytical methods are based on the assumption that either strains or stresses are constant within the unit cell. The constant strain approach was pioneered by Voigt [15] and is often referred to as the rule of mixtures for stiffness components. The constant stress approach is identified with Reuss [16] and is often referred to as the rule of mixtures for compliance components [17]. A more sophisticated analytical method is based on the Eshelby elasticity solution [18] of a boundary value problem for a spherical or ellipsoidal inclusion of one material in an infinite matrix of another material, which was further developed by Hashin [19] and Mori and Tanaka [20]. An extension of this method is the self-consistent approach, in which an inclusion is embedded into an effective material, the properties of which are not known a priori [21,22].

Over the past decade, numerical methods for solving the RVE problem have attracted considerable attention. These numerical methods are often referred to as computational upscaling or *computational homogenization* methods. A catchy label of FE[2] was suggested by Feyel and Chaboche [23] to emphasize the fact that finite elements are typically employed to discretize the boundary value problems on both the fine and coarse scales. Computational homogenization methods were employed by Suquet [24], Guedes and Kikuchi [25], Terada and Kikuchi [26,27], Matsui *et al.* [28], Gosh *et al.* [29,30], Miehe and Koch [31], Smit *et al.* [32], Kouznetsova *et al.* [33], Geers *et al.* [17], Michel *et al.* [34], and Yuan and Fish [35], among many others. For higher order computational homogenization methods, the interested reader is referred to the literature [17,33,36,37,38,39].

The outline of this chapter is as follows. Asymptotic homogenization theory for two-scale linear elastic periodic structures is detailed in section 2.2. Starting from the governing equations on the composite domain, the two-scale governing equations are derived using differential (strong) and variational (weak) forms. The Hill–Mandel macrohomogeneity condition, as well as Hill–Reuss–Voigt bounds, are derived from the asymptotic homogenization theory. Section 2.2 concludes with computational aspects of the linear homogenization theory, formulation for boundary layers, convergence estimates, and preliminary material design. Exposition of modern variants of upscaling methods [10,11,12] takes place in section 2.3. The two-scale nonlinear homogenization theory, encompassing both geometric and material nonlinearities, is detailed in section 2.4. Extension to higher order homogenization that accounts for variation of coarse-scale fields within a unit cell domain is considered in section 2.5. Extension to more than two scales is outlined in section 2.6. Chapter 2 concludes with homogenization-based multigrid approaches, which provide considerable solution quality improvement over homogenization theories.

2.2 Homogenizaton of Linear Heterogeneous Media

2.2.1 Two-Scale Formulation

In this section we focus on a heterogeneous linear elastic solid on a composite domain Ω^{ζ} with boundary $\partial\Omega^{\zeta}$. The superscript ζ denotes the existence of fine-scale features. We draw a distinction between the composite domain Ω^{ζ} and the coarse-scale domain Ω as follows. Ω^{ζ} is the solid part of the domain that does not contain voids in the material microstructure. Furthermore, the boundary of the composite domain $\partial\Omega^{\zeta}$ may be rough at the fine scale, whereas the coarse-scale domain Ω with smooth boundary $\partial\Omega$ is free of fine-scale features. $\boldsymbol{n}^{\zeta}$ and $\boldsymbol{n}^{c}$ will denote the unit normal to the boundaries $\partial\Omega^{\zeta}$ and $\partial\Omega$, respectively. In the absence of information about surface roughness, it will be assumed that $\partial\Omega^{\zeta}=\partial\Omega$, and thus $\boldsymbol{n}^{\zeta}=\boldsymbol{n}^{c}$. Note that the precise structure of the composite domain cases is often unknown (unless the component has been scanned). This fact will be taken advantage of in the homogenization process; for example, we will conveniently assume that a unit cell is centered at the quadrature point of the coarse-scale element.

We start with a strong form of the boundary value problem on a composite domain given as

$$\sigma_{ij,j}^{\zeta}+b_i^{\zeta}=0 \quad \text{on} \quad \Omega^{\zeta} \tag{2.6a}$$

$$\sigma_{ij}^{\zeta}=L_{ijkl}^{\zeta}\varepsilon_{kl}^{\zeta} \quad \text{on} \quad \Omega^{\zeta} \tag{2.6b}$$

$$\varepsilon_{ij}^{\zeta}=u_{(i,j)}^{\zeta}\equiv\frac{1}{2}\left(u_{i,j}^{\zeta}+u_{j,i}^{\zeta}\right) \quad \text{on} \quad \Omega^{\zeta} \tag{2.6c}$$

$$\sigma_{ij}^{\zeta}n_j^{\zeta}=\overline{t}_i^{\zeta} \quad \text{on} \quad \partial\Omega^{t\zeta} \tag{2.6d}$$

$$u_i^{\zeta}=\overline{u}_i^{\zeta} \quad \text{on} \quad \partial\Omega^{u\zeta} \tag{2.6e}$$

$$\partial\Omega^{t\zeta}\cup\partial\Omega^{u\zeta}=\partial\Omega^{\zeta} \quad \text{and} \quad \partial\Omega^{t\zeta}\cap\partial\Omega^{u\zeta}=0 \tag{2.6f}$$

where the summation convention over repeated subscripts is employed. Lowercase subscripts denote spatial dimensions, with commas denoting the spatial derivative and parentheses (i,j) denoting the symmetric part of the spatial derivative; $\partial\Omega^{u\zeta}$ and $\partial\Omega^{t\zeta}$ denote the essential (displacement) and natural (traction) boundary of the composite domain, respectively; $\boldsymbol{\sigma}^{\zeta}$ and $\boldsymbol{\varepsilon}^{\zeta}$ denote the stress and strain tensors, respectively; $\boldsymbol{b}^{\zeta}$ and $\boldsymbol{t}^{\zeta}$ denote the body force and traction vectors, respectively; and $\boldsymbol{L}^{\zeta}$ denotes the constitutive tensor satisfying conditions of symmetry

$$L_{ijkl}^{\zeta}=L_{jikl}^{\zeta}=L_{ijlk}^{\zeta}=L_{klij}^{\zeta} \tag{2.7}$$

and positivity

$$\exists c>0 \quad L_{ijkl}^{\zeta}\eta_{ij}\eta_{kl}\geq c\eta_{ij}\eta_{ij} \quad \forall\eta_{ij}=\eta_{ji} \tag{2.8}$$

In the two-scale mathematical homogenization theory to be subsequently developed, various fields are assumed to depend on two coordinates: $\boldsymbol{x}$ as the coarse-scale position vector in the coarse-scale domain Ω; and $\boldsymbol{y}$ as the fine-scale position vector in the unit cell domain Θ.

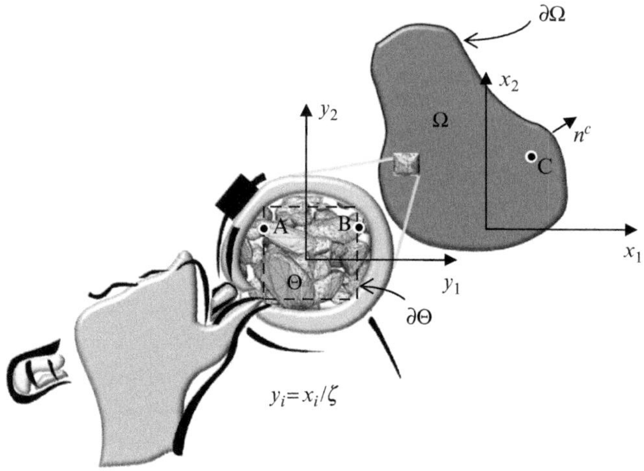

Figure 2.1 Definition of coarse and fine (unit cell) domains

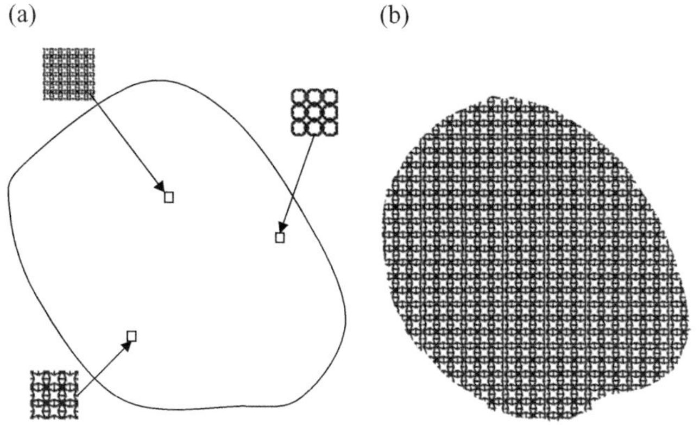

Figure 2.2 Definition of (a) locally periodic domains and (b) globally periodic domains

These two coordinates are related by $\boldsymbol{y} = \boldsymbol{x}/\zeta$ with $0 < \zeta \ll 1$. The unit cell domain Θ is typically chosen to be an open rectangular parallelepiped

$$\Theta =]0, l_1[\times]0, l_2[\times]0, l_3[\tag{2.9}$$

The fine-scale structure of the unit cell can be either periodic or random, but the solution is assumed to be locally periodic. In the neighboring points in the coarse-scale domain, *homologous* by periodicity (points A and B in Figure 2.1), the value of the response function is the same, but at points corresponding to different points in the coarse-scale domain (points A and C in Figure 2.1), the value of the function can be different by local periodicity assumption. The distinction between local periodicity and global periodicity is depicted in Figure 2.2.

Various response fields $f^{\zeta}(\boldsymbol{x})$ are assumed to be locally periodic in the unit cell domain Θ

$$f^{\zeta}(\boldsymbol{x})=f(\boldsymbol{x},\boldsymbol{y})=f(\boldsymbol{x},\boldsymbol{y}+\boldsymbol{kl}) \tag{2.10}$$

in which vector $\boldsymbol{l}=[l_1\ l_2\ l_3]^T$ denotes the basic period of the microstructure and $\boldsymbol{k}$ is a 3×3 diagonal matrix (in 3D) with integer components. Thus, various fields are defined in the product space $\Omega\times\Theta$.

The indirect coarse-scale spatial derivatives of the response function $f^{\zeta}(\boldsymbol{x})$ can be calculated by the chain rule as

$$f^{\zeta}{}_{,i}=\underbrace{\frac{\partial f(\boldsymbol{x},\boldsymbol{y})}{\partial x_i}}_{f_{,x_i}}+\frac{1}{\zeta}\underbrace{\frac{\partial f(\boldsymbol{x},\boldsymbol{y})}{\partial y_i}}_{f_{,y_i}} \tag{2.11}$$

where the comma followed by a subscript variable denotes a partial derivative with respect to the subscript variable ($f_{,x_i}=\partial f/\partial x_i$).

As a starting point, the displacement field $u_i^{\zeta}(\boldsymbol{x})=u_i(\boldsymbol{x},\boldsymbol{y})$ is approximated in terms of double-scale asymptotic expansions on $\Omega\times\Theta$ (often called a homogenization ansatz)

$$u_i^{\zeta}(\boldsymbol{x})=u_i^{(0)}(\boldsymbol{x},\boldsymbol{y})+\zeta u_i^{(1)}(\boldsymbol{x},\boldsymbol{y})+\zeta^2 u_i^{(2)}(\boldsymbol{x},\boldsymbol{y})+O\left(\zeta^3\right) \tag{2.12}$$

Strain expansion in the composite domain is obtained by inserting an ansatz (2.12) into (2.6c) with consideration of the indirect differentiation rule (2.11)

$$\varepsilon_{ij}^{\zeta}(\boldsymbol{x})=\frac{1}{\zeta}\varepsilon_{ij}^{(-1)}(\boldsymbol{x},\boldsymbol{y})+\varepsilon_{ij}^{(0)}(\boldsymbol{x},\boldsymbol{y})+\zeta\varepsilon_{ij}^{(1)}(\boldsymbol{x},\boldsymbol{y})+\zeta^2\varepsilon_{ij}^{(2)}(\boldsymbol{x},\boldsymbol{y})+O\left(\zeta^3\right) \tag{2.13}$$

where various order strain components are given as

$$\varepsilon_{ij}^{(-1)}=u_{(i,y_j)}^{(0)} \tag{2.14a}$$

$$\varepsilon_{ij}^{(s)}=u_{(i,x_j)}^{(s)}+u_{(i,y_j)}^{(s+1)} \quad s=0,1\ldots \tag{2.14b}$$

and the subscript pairs with parentheses denote the symmetric gradients defined as

$$u_{(i,y_j)}^{(k)}=\frac{1}{2}\left(\frac{\partial u_i^{(k)}}{\partial y_j}+\frac{\partial u_j^{(k)}}{\partial y_i}\right);\quad u_{(i,x_j)}^{(k)}=\frac{1}{2}\left(\frac{\partial u_i^{(k)}}{\partial x_j}+\frac{\partial u_j^{(k)}}{\partial x_i}\right) \tag{2.15}$$

Stresses and strains for different orders are related by the constitutive equation (2.6b)

$$\sigma_{ij}^{\zeta}(\boldsymbol{x})=\frac{1}{\zeta}\sigma_{ij}^{(-1)}(\boldsymbol{x},\boldsymbol{y})+\sigma_{ij}^{(0)}(\boldsymbol{x},\boldsymbol{y})+\zeta\sigma_{ij}^{(1)}(\boldsymbol{x},\boldsymbol{y})+\zeta^2\sigma_{ij}^{(2)}(\boldsymbol{x},\boldsymbol{y})+O\left(\zeta^3\right) \tag{2.16}$$

where

$$\sigma_{ij}^{(k)}(\boldsymbol{x},\boldsymbol{y})=L_{ijkl}(\boldsymbol{x},\boldsymbol{y})\,\varepsilon_{kl}^{(k)}(\boldsymbol{x},\boldsymbol{y}) \tag{2.17}$$

In general, a constitutive tensor can have a slow variation (from one layer in a composite to another) as well as a fast variation (within the unit cell). For simplicity, we will assume here that the constitutive tensor depends on a fine-scale coordinate only, that is, $L^{\zeta}_{ijkl}(\boldsymbol{x}) = L_{ijkl}(\boldsymbol{y})$.

Inserting the stress expansion (2.16) into equilibrium equation (2.6a) and making use of equation (2.11) yields the following equilibrium equations

$$\zeta^{-2}\sigma^{(-1)}_{ij,y_j} + \zeta^{-1}\left(\sigma^{(-1)}_{ij,x_j} + \sigma^{(0)}_{ij,y_j}\right) + \left(\sigma^{(0)}_{ij,x_j} + \sigma^{(1)}_{ij,y_j} + b^{\zeta}_i\right) + O(\zeta) = 0 \tag{2.18}$$

The leading order $O(\zeta^{-2})$ equilibrium is obtained by multiplying (2.18) by ζ^2 and then taking the limit of $\zeta \to 0^+$. Similarly, the $O(\zeta^{-1})$ equilibrium equation is obtained by multiplying the remaining equation (2.18) by ζ and then taking the limit of $\zeta \to 0^+$. This process is repeated to obtain various order equilibrium equations. The resulting leading order equilibrium equations are

$$O\left(\zeta^{-2}\right): \quad \sigma^{(-1)}_{ij,y_j} = 0 \tag{2.19a}$$

$$O\left(\zeta^{-1}\right): \quad \sigma^{(-1)}_{ij,x_j} + \sigma^{(0)}_{ij,y_j} = 0 \tag{2.19b}$$

$$O(1): \quad \sigma^{(0)}_{ij,x_j} + \sigma^{(1)}_{ij,y_j} + b^{\zeta}_i = 0 \tag{2.19c}$$

In the following, we will assume that the body force $\boldsymbol{b}^{\zeta}$ and boundary conditions $\bar{\boldsymbol{t}}^{\zeta}, \bar{\boldsymbol{u}}^{\zeta}$ are order one functions. Otherwise, these functions would have to be expanded in asymptotic expansion (see Problem 2.8). Consider the $O(\zeta^{-2})$ equilibrium equation (2.19a) in more detail. Premultiplying (2.19a) by $u^{(0)}_i$ and integrating over Θ yields

$$\int_{\Theta} u^{(0)}_i \sigma^{(-1)}_{ij,y_j} d\Theta = 0 \tag{2.20}$$

and subsequently integrating by parts gives

$$\int_{\partial\Theta} u^{(0)}_i \sigma^{(-1)}_{ij} n^{\Theta}_j d\gamma - \int_{\Theta} u^{(0)}_{i,y_j} \sigma^{(-1)}_{ij} d\Theta = 0 \tag{2.21}$$

where $\boldsymbol{n}^{\Theta}$ denotes the unit normal to the boundary $\partial\Theta$.

The boundary integral term in (2.21) vanishes due to local periodicity on $\partial\Theta$. This is because $u^{(0)}_i$ is equal on the opposite sides of the unit cell, while traction, $t^{(-1)}_i = \sigma^{(-1)}_{ij} n^{\Theta}_j$, is equal and opposite on the opposite sides of the unit cell. Further inserting (2.17) and (2.14a) into (2.21) yields

$$\int_{\Theta} u^{(0)}_{i,y_j} L_{ijkl} u^{(0)}_{k,y_l} d\Theta = 0 \tag{2.22}$$

From the positivity (2.8), the symmetry of constitutive tensor (2.7), and equation (2.22), we can infer that

$$u^{(0)}_{i,y_j} = 0 \tag{2.23}$$

and

$$u_i^{(0)} = u_i^{(0)}(\boldsymbol{x}) \tag{2.24a}$$

$$\sigma_{ij}^{(-1)} = 0 \tag{2.24b}$$

In other words, the first term in the asymptotic expansion of displacements (2.12) represents the coarse-scale displacement field.

We now proceed to the $O(\zeta^{-1})$ equilibrium equation (2.19b), which, when combined with (2.24b), can be written as

$$\sigma_{ij,y_j}^{(0)} = 0 \tag{2.25}$$

Inserting the constitutive (2.17) and kinematics (2.14) equations into (2.25) yields

$$\left(L_{ijkl}\left(u_{(k,x_l)}^{(0)} + u_{(k,y_l)}^{(1)}\right)\right)_{,y_j} = 0 \tag{2.26}$$

To solve (2.26) for $\boldsymbol{u}^{(1)}$ up to a constant, we introduce the following separation of variables

$$u_k^{(1)}(\boldsymbol{x},\boldsymbol{y}) = H_k^{mn}(\boldsymbol{y})\, u_{(m,x_n)}^{(0)}(\boldsymbol{x}) \tag{2.27}$$

where $H_k^{mn}(\boldsymbol{y})$ is termed as a first-order displacement influence function, or simply a displacement influence function. Note that the displacement influence function is C^0 continuous, locally periodic, and symmetric, that is, $H_k^{mn}(\boldsymbol{y}) = H_k^{nm}(\boldsymbol{y})$. Inserting the decomposition (2.27) into the $O(\zeta^{-1})$ equilibrium equation (2.26) and requiring the resulting equation to hold for arbitrary $u_{(m,x_n)}^{(0)}$ gives

$$\left(L_{ijkl}(\boldsymbol{y})\left(I_{klmn} + H_{(k,y_l)}^{mn}(\boldsymbol{y})\right)\right)_{,y_j} = 0 \tag{2.28}$$

where

$$I_{klmn} = \frac{1}{2}\left(\delta_{km}\delta_{ln} + \delta_{kn}\delta_{lm}\right) \tag{2.29}$$

and where δ_{km} is the Kronecker delta. Equation (2.29), together with the locally periodic boundary conditions, comprises a linear boundary value problem for $H_i^{mn}(\boldsymbol{y})$ up to a constant. To uniquely define $H_i^{mn}(\boldsymbol{y})$, two approaches are commonly used in practice:

$$\begin{aligned} &\text{Approach (a)}: \quad H_i^{mn}(\boldsymbol{y}) = 0 \quad \text{on} \quad \partial\Theta^{vert} \\ &\text{Approach (b)}: \quad \int_\Theta H_i^{mn}(\boldsymbol{y})\, d\Theta = 0 \end{aligned} \tag{2.30}$$

In Approach (a) in (2.30), $\partial\Theta^{vert}$ denotes the vertices of the unit cell. Approach (a) is simple to enforce, but Approach (b) has the advantage of identifying $\boldsymbol{u}^{(0)}(\boldsymbol{x})$ with an average (or coarse-scale) displacement denoted by $\boldsymbol{u}^c(\boldsymbol{x})$. This can be seen by integrating the asymptotic expansion of the unit cell

$$u_i^c(\boldsymbol{x}) = \frac{1}{|\Theta|}\int_\Theta u_i^\zeta(\boldsymbol{x},\boldsymbol{y})d\Theta \tag{2.31}$$

where $|\Theta|$ denotes the volume of the unit cell domain. Note that if $u_i^{(s)}$ for $s \geq 1$ is defined by a tensor product of functions in $\boldsymbol{y}$ and $\boldsymbol{x}$ [as in (2.27)], and if the integral of functions in $\boldsymbol{y}$ over the unit cell domain is defined to vanish [as in Approach (b) in (2.30)], then

$$u_i^{(0)}(\boldsymbol{x}) = u_i^c(\boldsymbol{x}) \tag{2.32}$$

The resulting leading order strain $\varepsilon_{ij}^{(0)}(\boldsymbol{x},\boldsymbol{y})$ is given as

$$\varepsilon_{kl}^{(0)}(\boldsymbol{x},\boldsymbol{y}) = E_{kl}^{mn}(\boldsymbol{y})\, u_{(m,x_n)}^{(0)}(\boldsymbol{x}) \equiv \varepsilon_{kl}^f(\boldsymbol{x},\boldsymbol{y}) \tag{2.33}$$

where $E_{kl}^{mn}(\boldsymbol{y})$ is the elastic strain influence function defined as

$$E_{kl}^{mn}(\boldsymbol{y}) = I_{klmn} + H_{(k,y_l)}^{mn}(\boldsymbol{y}) \tag{2.34}$$

The leading order strain will be also referred to as the fine-scale strain, denoted as $\varepsilon_{kl}^f(\boldsymbol{x},\boldsymbol{y})$. The overall (or coarse-scale) strain denoted by $\varepsilon_{mn}^c(\boldsymbol{x})$ is defined as an average of the leading order (or fine-scale) strain computed over the unit cell domain

$$\varepsilon_{mn}^c(\boldsymbol{x}) = \frac{1}{|\Theta|}\int_\Theta \varepsilon_{mn}^f(\boldsymbol{x},\boldsymbol{y})\,d\Theta \tag{2.35}$$

Since the integrals of $H_{(k,y_l)}^{mn}(\boldsymbol{y})$ over the unit cell domain vanish due to the local periodicity assumption, it follows that

$$\varepsilon_{mn}^c = u_{(m,x_n)}^{(0)} \tag{2.36}$$

and thus equation (2.33) can be written as

$$\varepsilon_{kl}^f(\boldsymbol{x},\boldsymbol{y}) = E_{kl}^{mn}(\boldsymbol{y})\varepsilon_{mn}^c(\boldsymbol{x}) \tag{2.37}$$

Likewise, the leading stress $\sigma_{ij}^{(0)}(\boldsymbol{x},\boldsymbol{y})$ is given by

$$\sigma_{ij}^{(0)}(\boldsymbol{x},\boldsymbol{y}) = \Sigma_{ij}^{mn}(\boldsymbol{y})\varepsilon_{mn}^c(\boldsymbol{x}) \equiv \sigma_{ij}^f(\boldsymbol{x},\boldsymbol{y}) \tag{2.38}$$

where $\Sigma_{ij}^{mn}(\boldsymbol{y})$ is the stress influence function defined by

$$\Sigma_{ij}^{mn}(\boldsymbol{y}) = L_{ijkl}(\boldsymbol{y})E_{kl}^{mn}(\boldsymbol{y}) \tag{2.39}$$

The leading order stress, which has a definition similar to that of fine-scale strain, will be referred to as the fine-scale stress, denoted by $\sigma_{kl}^f(\boldsymbol{x},\boldsymbol{y})$.

Consider the $O(1)$ equilibrium equation (2.19c) next. Integrating the $O(1)$ equilibrium equation over the unit cell domain and taking advantage of the local periodicity condition yields

$$\sigma^c_{ij,x_j} + b^c_i = 0 \tag{2.40}$$

where the coarse-scale σ^c_{ij} is defined as

$$\sigma^c_{ij}(\boldsymbol{x}) = \frac{1}{|\Theta|}\int_\Theta \sigma^f_{ij}(\boldsymbol{x},\boldsymbol{y})\,d\Theta \tag{2.41}$$

and the coarse-scale body force b^c_i is given by

$$b^c_i(\boldsymbol{x}) = \frac{1}{|\Theta|}\int_\Theta b_i(\boldsymbol{x},\boldsymbol{y})\,d\Theta \tag{2.42}$$

Inserting (2.17) and (2.37) into (2.41) yields

$$\sigma^c_{ij}(\boldsymbol{x}) = \frac{1}{|\Theta|}\int_\Theta L_{ijkl}(\boldsymbol{y})\,E^{mn}_{kl}(\boldsymbol{y})\,d\Theta\;\varepsilon^c_{mn}(\boldsymbol{x}) \tag{2.43}$$

from which we identify the overall or coarse-scale constitutive tensor L^c_{ijmn} as

$$L^c_{ijmn} = \frac{1}{|\Theta|}\int_\Theta L_{ijkl}(\boldsymbol{y})E^{mn}_{kl}(\boldsymbol{y})d\Theta = \frac{1}{|\Theta|}\int_\Theta \Sigma^{mn}_{ij}(\boldsymbol{y})d\Theta \tag{2.44}$$

Consequently, the coarse-scale constitutive relation can be expressed as

$$\sigma^c_{ij}(\boldsymbol{x}) = L^c_{ijmn}\varepsilon^c_{mn}(\boldsymbol{x}) \tag{2.45}$$

Note that L^c_{ijmn} is symmetric (has both major and minor symmetries), provided that the constitutive tensor in fine-scale $L_{ijkl}(\boldsymbol{y})$ is symmetric. This observation can be made due to the following consideration. Premultiplying (2.28) by $H^{st}_i(\boldsymbol{y})$, then integrating over the unit cell domain, followed by integration by parts and use of local periodicity, yields

$$\frac{1}{|\Theta|}\int_\Theta H^{st}_{(i,y_j)}(\boldsymbol{y})L_{ijkl}(\boldsymbol{y})E^{mn}_{kl}(\boldsymbol{y})d\Theta = 0 \tag{2.46}$$

Adding equation (2.46) and equation (2.44) yields the symmetric form of L^c_{stmn}

$$L^c_{stmn} = \frac{1}{|\Theta|}\int_\Theta E^{st}_{ij}(\boldsymbol{y})\,L_{ijkl}(\boldsymbol{y})\,E^{mn}_{kl}(\boldsymbol{y})\,d\Theta \tag{2.47}$$

Remark 2.1 For simplicity, we assumed a steady-state (static) equation in (2.6a). If the inertia is not negligible, (2.6a) has to be replaced by the equation of motion

$$\sigma_{ij,j}^{\zeta}+b_i^{\zeta}-\rho^{\zeta}\ddot{u}_i^{\zeta}=0 \tag{2.48}$$

where a superimposed dot denotes the time derivative. Assuming that the length of the wave propagating in the coarse-scale domain is much larger than the RVE size, the inertia term in (2.48) is $O(1)$ and therefore enters the coarse-scale equation of motion only, that is,

$$\sigma_{ij,x_j}^{c}+b_i^{c}-\rho^{c}\ddot{u}_i^{c}=0 \tag{2.49}$$

where

$$\rho^c(\boldsymbol{x})=\frac{1}{|\Theta|}\int_{\Theta}\rho(\boldsymbol{x},\boldsymbol{y})d\Theta \tag{2.50}$$

2.2.2 Two-Scale Formulation – Variational Form

In this section we present an alternative derivation of the two-scale problem (2.40) and (2.28), which is based on the weak form defined on the composite domain. This form is advantageous in the derivation of the higher order homogenization theories considered in section 2.5.

The weak form defined on the composite domain Ω^{ζ} is obtained by multiplying (2.6a) by the arbitrary test function $w_i^{\zeta}\in W_{\Omega}^{\zeta}$ and integrating by parts over the composite domain, which yields

$$\int_{\Omega^{\zeta}}w_{i,j}^{\zeta}\sigma_{ij}^{\zeta}d\Omega^{\zeta}=\int_{\Omega^{\zeta}}w_i^{\zeta}b_i^{\zeta}d\Omega^{\zeta}+\int_{\partial\Omega^{t\zeta}}w_i^{\zeta}\overline{t}_i^{\zeta}d\Gamma^{\zeta} \tag{2.51}$$

where the space of the test (or weight) function W_{Ω}^{ζ} is defined as

$$W_{\Omega}^{\zeta}=\left\{\boldsymbol{w}^{\zeta}\text{ defined in }\Omega^{\zeta},C^0\left(\Omega^{\zeta}\right),\ \boldsymbol{w}^{\zeta}=\boldsymbol{0}\ \text{ on }\partial\Omega^{u\zeta}\right\} \tag{2.52}$$

Inserting the stress asymptotic expansion (2.16) into (2.51) and using the definition of the spatial derivative of the two-scale weight function $w_i^{\zeta}(\boldsymbol{x})=w_i(\boldsymbol{x},\boldsymbol{y})$ in (2.11) yields

$$\begin{aligned}&\frac{1}{\zeta^2}\int_{\Omega^{\zeta}}\frac{\partial w_i}{\partial y_j}\sigma_{ij}^{(-1)}d\Omega^{\zeta}+\frac{1}{\zeta}\int_{\Omega^{\zeta}}\left(\frac{\partial w_i}{\partial y_j}\sigma_{ij}^{(0)}+\frac{\partial w_i}{\partial x_j}\sigma_{ij}^{(-1)}\right)d\Omega^{\zeta}\\&+\int_{\Omega^{\zeta}}\left(\frac{\partial w_i}{\partial y_j}\sigma_{ij}^{(1)}+\frac{\partial w_i}{\partial x_j}\sigma_{ij}^{(0)}\right)d\Omega^{\zeta}-\int_{\Omega^{\zeta}}w_i^{\zeta}b_i^{\zeta}d\Omega^{\zeta}-\int_{\partial\Omega^{t\zeta}}w_i^{\zeta}\overline{t}_i^{\zeta}d\Gamma^{\zeta}+O(\zeta)=0\end{aligned} \tag{2.53}$$

The leading order $O(\zeta^{-2})$ weak form is obtained by multiplying (2.53) by ζ^2 and then taking the limit of $\zeta\rightarrow0^+$. Similarly, the $O(\zeta^{-1})$ weak form is obtained by multiplying the remaining equation (2.53) by ζ and then taking the limit of $\zeta\to 0^+$. This process is repeated to obtain higher order equilibrium equations. The resulting leading order equilibrium equations are

$$O(\zeta^{-2}):\quad \int_{\Omega^\zeta} \frac{\partial w_i}{\partial y_j}\sigma_{ij}^{(-1)}\, d\Omega^\zeta = 0 \tag{2.54a}$$

$$O(\zeta^{-1}):\quad \int_{\Omega^\zeta} \left(\frac{\partial w_i}{\partial y_j}\sigma_{ij}^{(0)} + \frac{\partial w_i}{\partial x_j}\sigma_{ij}^{(-1)}\right) d\Omega^\zeta = 0 \tag{2.54b}$$

$$O(1):\quad \int_{\Omega^\zeta} \left(\frac{\partial w_i}{\partial y_j}\sigma_{ij}^{(1)} + \frac{\partial w_i}{\partial x_j}\sigma_{ij}^{(0)}\right) d\Omega^\zeta - \int_{\Omega^\zeta} w_i^\zeta b_i^\zeta\, d\Omega^\zeta - \int_{\partial\Omega^{t\zeta}} w_i^\zeta \bar{t}_i^\zeta\, d\Gamma^\zeta = 0 \tag{2.54c}$$

The integral of the two-scale function over the composite domain Ω^ζ is defined as

$$\lim_{\zeta\to 0^+} \int_{\Omega^\zeta} \Psi(\boldsymbol{x},\boldsymbol{y})\, d\Omega^\zeta = \lim_{\zeta\to 0^+} \int_{\Omega} \left(\frac{1}{|\Theta|}\int_{\Theta} \Psi(\boldsymbol{x},\boldsymbol{y})\, d\Theta\right) d\Omega \tag{2.55}$$

A similar two-scale integration scheme is employed over the boundary $\partial\Omega^\zeta$ of the composite domain

$$\lim_{\zeta\to 0^+} \int_{\partial\Omega^\zeta} \Psi(\boldsymbol{x},\boldsymbol{y})\, d\Gamma^\zeta = \lim_{\zeta\to 0^+} \int_{\partial\Omega} \left(\frac{1}{|\partial\omega|}\int_{\partial\omega} \Psi(\boldsymbol{x},\boldsymbol{y})\, ds\right) d\Gamma \tag{2.56}$$

where $\partial\omega$ is a characteristic length (or surface area) over which the averaging on the boundary $\partial\Omega^\zeta$ is carried out.

Since $\boldsymbol{w}^\zeta$ is arbitrary, we first choose $\boldsymbol{w}^\zeta = \boldsymbol{u}^{(0)}(\boldsymbol{x},\boldsymbol{y})$ in (2.54a) and then use the definition of $\sigma_{ij}^{(-1)}$ in (2.17) and (2.14a) in combination with the two-scale integration scheme defined in (2.55), which yields (2.22) and then (2.24).

The weak form of the unit cell problem (2.25) is obtained by choosing $\boldsymbol{w}^\zeta = \boldsymbol{w}^{(1)}(\boldsymbol{y}) \in W_\Theta$ and inserting it into (2.54b) where

$$W_\Theta = \left\{\boldsymbol{w}^{(1)}(\boldsymbol{y})\,\text{defined in}\,\Theta,\ C^0(\Theta),\ \text{periodic in}\,\Theta\right\} \tag{2.57}$$

Finally, the weak form coarse-scale problem (2.40) is obtained by choosing $\boldsymbol{w}^\zeta = \boldsymbol{w}^{(0)}(\boldsymbol{x}) \equiv \boldsymbol{w}^c(\boldsymbol{x}) \in W_\Omega$ in (2.54c) and utilizing the two-scale integration scheme (2.55) and (2.56) where

$$W_\Omega = \left\{\boldsymbol{w}^c(\boldsymbol{x})\ \text{defined in}\,\Omega,\ C^0(\Omega),\ \boldsymbol{w}^c = \boldsymbol{0}\ \text{on}\ \partial\Omega^u\right\} \tag{2.58}$$

The resulting coarse-scale weak form for $\boldsymbol{u}^c(\boldsymbol{x}) \in C^0(\Omega)$ is given by

$$\begin{aligned} &\int_{\Omega} \frac{\partial w_i^c}{\partial x_j}\sigma_{ij}^c\, d\Omega - \int_{\Omega} w_i^c b_i^c\, d\Omega - \int_{\partial\Omega^t} w_i^c \bar{t}_i^c\, d\Gamma = 0 \quad \forall \boldsymbol{w}^c \in W_\Omega \\ &u_i^c = \bar{u}_i^c \end{aligned} \tag{2.59}$$

where σ_{ij}^c and b_i^c are given in (2.41) and (2.42). The definitions of $\overline{u}_i^c$ and $\overline{t}_i^c$ follow from (2.56)

$$\overline{t}_i^c = \frac{1}{|\partial\omega|}\int_{\partial\omega} \overline{t}_i^\zeta \, ds$$
$$\overline{u}_i^c = \frac{1}{|\partial\omega|}\int_{\partial\omega} \overline{u}_i^\zeta \, ds \tag{2.60}$$

2.2.3 Hill–Mandel Macrohomogeneity Condition and Hill–Reuss–Voigt Bounds

The Hill–Mandel macrohomogeneity condition states that coarse-scale internal energy can be computed from fine-scale energy

$$\frac{1}{|\Theta|}\int_{\Theta} \sigma_{ij}^f \varepsilon_{ij}^f \, d\Theta = \sigma_{ij}^c \varepsilon_{ij}^c \tag{2.61}$$

Although the Hill–Mandel macrohomogeneity condition [6] was developed before the mathematical homogenization theory discussed so far, it can be derived from the mathematical homogenization theory as follows. Consider the fine-scale strain energy density defined as

$$\frac{1}{|\Theta|}\int_{\Theta} \sigma_{ij}^f \varepsilon_{ij}^f \, d\Theta = \frac{1}{|\Theta|}\int_{\Theta} \sigma_{ij}^f \left(I_{ijkl} + H_{(i,y_j)}^{kl} \right) d\Theta \varepsilon_{kl}^c$$
$$= \sigma_{ij}^c \varepsilon_{ij}^c + \frac{1}{|\Theta|}\int_{\Theta} \sigma_{ij}^f H_{(i,y_j)}^{kl} \, d\Theta \varepsilon_{kl}^c \tag{2.62}$$

where we make use of (2.33), (2.34), and (2.41). Integration by parts of the second term, and exploiting periodicity and the unit cell equilibrium $\partial \sigma_{ij}^f / \partial y_j = 0$, yields the Hill–Mandel macrohomogeneity condition (2.61).

The Hill–Reuss–Voigt bounds discussed next is a useful tool to bound the coarse-scale properties. It can be also used as an approximation of the coarse-scale material properties without solving a unit cell problem. Consider an additive decomposition of the leading order strain into the coarse-scale strain ε_{ij}^c and the fluctuation ε_{ij}^*

$$\varepsilon_{ij}^f = \varepsilon_{ij}^c + \varepsilon_{ij}^* \tag{2.63}$$

where

$$\varepsilon_{ij}^* = H_{(i,y_j)}^{kl} \varepsilon_{kl}^c \tag{2.64}$$

To obtain the bounds we first consider

$$\begin{aligned}
0 \le \frac{1}{|\Theta|}\int_{\Theta} \varepsilon_{ij}^{*} L_{ijkl}\varepsilon_{kl}^{*}\, d\Theta &= \frac{1}{|\Theta|}\int_{\Theta}\left(\varepsilon_{ij}^{f} L_{ijkl}\varepsilon_{kl}^{f} - 2\varepsilon_{ij}^{c} L_{ijkl}\varepsilon_{kl}^{f} + \varepsilon_{ij}^{c} L_{ijkl}\varepsilon_{kl}^{c}\right) d\Theta \\
&= \varepsilon_{ij}^{c} L_{ijkl}^{c}\varepsilon_{kl}^{c} - 2\varepsilon_{ij}^{c}\sigma_{ij}^{c} + \varepsilon_{ij}^{c}\left(\frac{1}{|\Theta|}\int_{\Theta} L_{ijkl}\, d\Theta\right)\varepsilon_{kl}^{c} \\
&= \varepsilon_{ij}^{c}\left(\frac{1}{|\Theta|}\int_{\Theta} L_{ijkl}\, d\Theta - L_{ijkl}^{c}\right)\varepsilon_{kl}^{c}
\end{aligned} \tag{2.65}$$

or

$$\alpha_{ij}\left(\frac{1}{|\Theta|}\int_{\Theta} L_{ijkl}\, d\Theta\right)\alpha_{kl} \ge \alpha_{ij} L_{ijkl}^{c}\alpha_{kl} \quad \forall \alpha_{ij} \tag{2.66}$$

where (2.17), (2.41), (2.61), (2.63), and the positive definiteness of L_{ijkl} were exercised.

For a complementary case, consider an additive decomposition of the leading order stress into coarse-scale stress σ_{ij}^{c} and fluctuation σ_{ij}^{*}, then

$$\begin{aligned}
0 \le \frac{1}{|\Theta|}\int_{\Theta} \sigma_{ij}^{*} M_{ijkl}\sigma_{kl}^{*}\, d\Theta &= \frac{1}{|\Theta|}\int_{\Theta}\left(\sigma_{ij}^{f} M_{ijkl}\sigma_{kl}^{f} - 2\sigma_{ij}^{c} M_{ijkl}\sigma_{kl}^{f} + \sigma_{ij}^{c} M_{ijkl}\sigma_{kl}^{c}\right) d\Theta \\
&= \sigma_{ij}^{c} M_{ijkl}^{c}\sigma_{kl}^{c} - 2\sigma_{ij}^{c}\varepsilon_{ij}^{c} + \sigma_{ij}^{c}\left(\frac{1}{|\Theta|}\int_{\Theta} M_{ijkl}\, d\Theta\right)\sigma_{kl}^{c} \\
&= \sigma_{ij}^{c}\left(\frac{1}{|\Theta|}\int_{\Theta} M_{ijkl}\, d\Theta - M_{ijkl}^{c}\right)\sigma_{kl}^{c}
\end{aligned} \tag{2.67}$$

where $M_{ijkl} = L_{ijkl}^{-1}$ and $M_{ijkl}^{c} = \left(L_{ijkl}^{c}\right)^{-1}$. From (2.67) it follows that

$$\alpha_{ij} L_{ijkl}^{c}\alpha_{kl} \ge \alpha_{ij}\left(\frac{1}{|\Theta|}\int_{\Theta} L_{ijkl}^{-1}\, d\Theta\right)^{-1}\alpha_{kl} \quad \forall \alpha_{ij} \tag{2.68}$$

In summary, the Hill–Reuss–Voigt condition bounds the eigenvalues of the coarse-scale constitutive tensor L_{ijkl}^{c} as follows

$$\alpha_{ij}\left(\frac{1}{|\Theta|}\int_{\Theta} M_{ijkl}\, d\Theta\right)^{-1}\alpha_{kl} \le \alpha_{ij} L_{ijkl}^{c}\alpha_{kl} \le \alpha_{ij}\left(\frac{1}{|\Theta|}\int_{\Theta} L_{ijkl}\, d\Theta\right)\alpha_{kl} \quad \forall \alpha_{ij} \tag{2.69}$$

In 1889 Voigt [15] assumed that the strain field in a polycrystalline material is uniform, that is, $\varepsilon_{ij}^{f} = \varepsilon_{ij}^{c}$. From the Hill–Mandel condition (2.61), it follows that

$$L_{ijkl}^{c} = \frac{1}{|\Theta|}\int_{\Theta} L_{ijkl}\, d\Theta$$

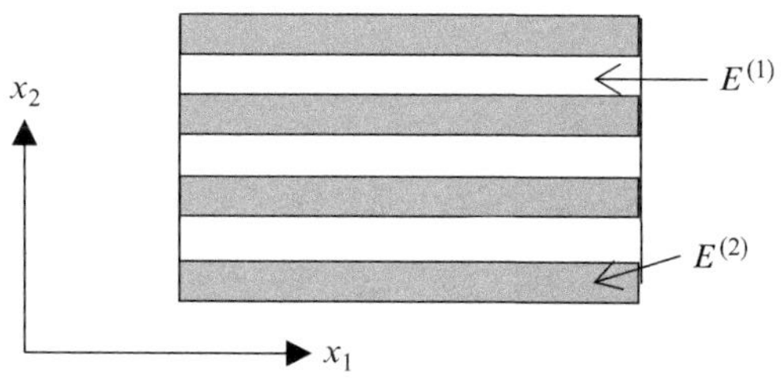

Figure 2.3 Layered elastic medium for interpretation of the Hill–Reuss–Voigt bounds assuming zero Poisson ratio

The dual assumption was made by Reuss [16], who in 1929 approximated the stress fields within the aggregate of polycrystalline material as uniform, that is, $\sigma_{ij}^f = \sigma_{ij}^c$. Again utilizing the Hill–Mandel condition (2.61) yields

$$L_{ijkl}^c = \left(\frac{1}{|\Theta|} \int_\Theta M_{ijkl} d\Theta \right)^{-1} \tag{2.70}$$

It is instructive to point out that if we consider a layered elastic medium consisting of two materials with zero Poisson ratio as shown in Figure 2.3, then the Hill–Reuss–Voigt bounds will provide an accurate coarse-scale Young's moduli E_1^c and E_2^c in the x_1 and x_2 directions, respectively.

If a load is applied in the x_1 direction, then the strain field is uniform and we have a Voigt bound

$$E_1^c = \frac{1}{|\Theta|} \int_\Theta E(\boldsymbol{x}) d\Theta = \phi^{(1)} E^{(1)} + \phi^{(2)} E^{(2)}$$

where $\phi^{(\alpha)}$ is a volume fraction of phases, such that $\phi^{(1)} + \phi^{(2)} = 1$. Likewise, if a load is applied in the x_2 direction, then the stress field is uniform and we have a Reuss bound

$$E_2^c = \left(\frac{1}{|\Theta|} \int_\Theta \frac{1}{E(\boldsymbol{x})} d\Theta \right)^{-1} = \left(\phi^{(1)}/E^{(1)} + \phi^{(2)}/E^{(2)} \right)^{-1}$$

Tighter bounds have been developed by Hashin and Shtrikman [40] based on variational principles using the concept of polarization or "filtering" of fine-scale fields. We refer to [41] for an elegant exposition of these ideas.

2.2.4 Numerical Implementation

We start by summarizing the two-scale governing equations derived in the previous section.

a. Coarse-scale problem

$$\begin{aligned}&\text{Find } u_i^c \text{ on } \Omega \text{ such that:}\\&\sigma_{ij,x_j}^c + b_i^c = 0 \quad \text{on} \quad \Omega\\&\sigma_{ij}^c = L_{ijmn}^c \varepsilon_{mn}^c \quad \text{on} \quad \Omega\\&u_i^c = \bar{u}_i^c \quad \text{on} \quad \partial\Omega^u\\&\sigma_{ij}^c n_j^c = \bar{t}_i^c \quad \text{on} \quad \partial\Omega^t\end{aligned} \tag{2.71}$$

b. Unit cell problem

$$\begin{aligned}&\text{Find } H_i^{mn}(\boldsymbol{y}) \text{ on } \Theta \text{ such that:}\\&\left[L_{ijkl}\left(H_{(k,y_l)}^{mn} + I_{klmn}\right)\right]_{,y_j} = 0 \quad \text{on} \quad \Theta\\&H_i^{mn}(\boldsymbol{y}) = H_i^{mn}(\boldsymbol{y}+\boldsymbol{l}) \quad \text{on} \quad \partial\Theta\\&H_i^{mn}(\boldsymbol{y}) = 0 \quad \text{on} \quad \partial\Theta^{vert}\end{aligned} \tag{2.72}$$

where

$$L_{ijmn}^c = \frac{1}{|\Theta|}\int_\Theta \Sigma_{ij}^{mn}(\boldsymbol{y})d\Theta \tag{2.73}$$

and $\boldsymbol{l}$ is a period.

In (2.72), for simplicity, we will consider the nonintegral normalization condition [Approach (a) in (2.30)]. Note that the stress influence functions $\Sigma_{ij}^{mn}(\boldsymbol{y})$ can be interpreted as a fine-scale stress in the unit cell stress induced by an overall unit strain ε_{mn}^c. The algorithmic details are discussed in section 2.2.4.1, section 2.2.4.2 and section 2.2.4.3.

Alternatively, we can construct a unit cell problem for the total displacement influence function U_i^{mn} defined as

$$\begin{aligned}&U_i^{mn} = H_i^{mn} + W_i^{mn}\\&W_i^{mn} = \frac{1}{2}\left(\delta_{im} y_n + \delta_{in} y_m\right)\end{aligned} \tag{2.74}$$

in which case the unit cell problem is given by

$$\begin{aligned}&\text{Find } U_i^{mn}(\boldsymbol{y}) \text{ on } \Theta \text{ such that:}\\&\left(L_{ijkl} U_{(k,y_l)}^{mn}\right)_{,y_j} = 0 \quad \text{on} \quad \Theta\\&U_i^{mn}(\boldsymbol{y}) - W_i^{mn}(\boldsymbol{y}) = U_i^{mn}(\boldsymbol{y}+\boldsymbol{l}) - W_i^{mn}(\boldsymbol{y}+\boldsymbol{l}) \quad \text{on} \quad \partial\Theta\\&U_i^{mn}(\boldsymbol{y}) = W_i^{mn}(\boldsymbol{y}) \quad \text{on} \quad \partial\Theta^{vert}\end{aligned} \tag{2.75}$$

2.2.4.1 Discrete Coarse-Scale Problem

We will consider a Galerkin approximation where the coarse-scale trial u_i^c and test w_i^c functions are discretized using the same $C^0(\Omega)$ continuous coarse-scale shape functions $N_{i\alpha}^c$

$$\begin{aligned} u_i^c &= N_{i\alpha}^c d_\alpha^c \\ w_i^c &= N_{i\alpha}^c c_\alpha^c \end{aligned} \tag{2.76}$$

where d_α^c and c_α^c denote nodal degrees of freedom of trial and test functions. Greek subscripts are reserved for degrees of freedom, and the summation convention over repeated indices is employed.

Writing the weak form of (2.71) and employing discretization (2.76) gives the discrete coarse-scale problem:

Given L_{ijkl}^c, b_i^c, $\overline{t}_i^c$, and $\overline{d}_\beta^c$, find d_β^c such that

$$\begin{aligned} &\underbrace{\int_\Omega B_{ij\alpha}^c L_{ijkl}^c B_{kl\beta}^c d\Omega}_{K_{\alpha\beta}^c} \cdot d_\beta^c = \underbrace{\int_\Omega N_{i\alpha}^c b_i^c d\Omega + \int_{\partial\Omega^t} N_{i\alpha}^c \overline{t}_i^c d\Gamma}_{f_\alpha^c} \\ &d_\beta^c = \overline{d}_\beta^c \quad \text{on} \quad \partial\Omega^u \end{aligned} \tag{2.77}$$

where $B_{ij\alpha}^c = N_{(i,x_j)\alpha}^c$ is the symmetric gradient of the coarse-scale shape functions.

2.2.4.2 Discrete Unit Cell Problem

Consider a Galerkin approximation of the unit cell trial H_i^{mn} and test w_i^f functions

$$\begin{aligned} H_i^{mn} &= N_{i\alpha}^f(\mathbf{y}) d_\alpha^{mn} \\ w_i^f &= N_{i\alpha}^f(\mathbf{y}) c_\alpha \end{aligned} \tag{2.78}$$

where d_α^{mn} and c_α denote nodal degrees of freedom of trial and test functions, respectively; $N_{i\alpha}^f(\mathbf{y})$ is a $C^0(\Theta)$ continuous fine-scale shape function; and $B_{ij\alpha}^f(\mathbf{y}) = N_{(i,y_j)\alpha}^f(\mathbf{y})$ is a corresponding symmetric gradient of the shape function.

Alternatively, we can directly discretize U_i^{mn} as

$$U_i^{mn} = N_{i\alpha}^f(\mathbf{y}) d_\alpha^{mn} \tag{2.79}$$

We now discuss the boundary conditions for the unit cell problem. Consider first a *periodic mesh* with nodes partitioned into four groups as shown in Figure 2.4.

Writing the weak form of the unit cell equations (2.72) and employing a Galerkin approximation (2.78) gives the discrete unit cell problem for H_i^{mn}, which states:

Given L_{ijkl}, find d_β^{mn} such that

$$\begin{aligned} &c_\alpha \left(K_{\alpha\beta}^f d_\beta^{mn} - f_\alpha^{mn} \right) = 0 \quad \forall c_\alpha - \text{periodic} \\ &\mathbf{d}_S^{mn} = \mathbf{d}_M^{mn}; \quad \mathbf{d}_V^{mn} = 0 \end{aligned} \tag{2.80}$$

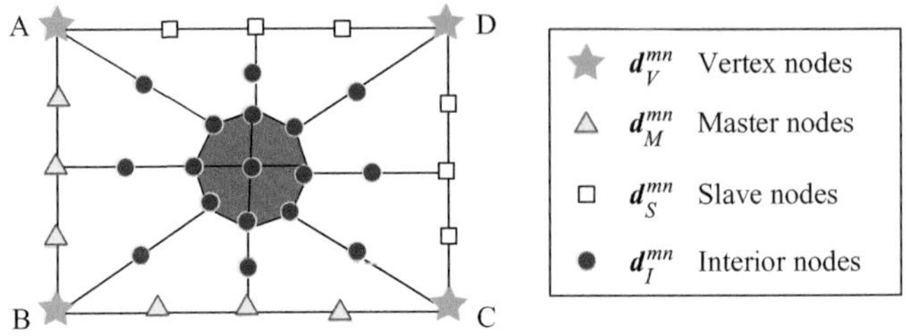

Figure 2.4 Partition of the unit cell nodes

where

$$K^f_{\alpha\beta} = \int_\Theta B^f_{ij\alpha} L_{ijkl} B^f_{kl\beta} d\Theta$$
$$f^{mn}_\alpha = -\int_\Theta B^f_{ij\alpha} L_{ijmn} d\Theta \tag{2.81}$$

Expressing c_α and d^{mn}_β in terms of independent (master) degrees-of-freedom $\tilde{c}_\delta$ and $\tilde{d}^{mn}_\gamma$

$$c_\alpha = T_{\alpha\delta}\tilde{c}_\delta \quad d^{mn}_\beta = T_{\beta\gamma}\tilde{d}^{mn}_\gamma \tag{2.82}$$

and inserting (2.82) into (2.80) yields the system of equations for independent degrees of freedom

$$\underbrace{T_{\alpha\delta} K^f_{\alpha\beta} T_{\beta\gamma}}_{\tilde{K}^f_{\delta\gamma}} \tilde{d}^{mn}_\gamma = \underbrace{T_{\alpha\delta} f^{mn}_\alpha}_{\tilde{f}^{mn}_\delta}$$
$$\tilde{K}^f_{\delta\gamma} \tilde{d}^{mn}_\gamma = \tilde{f}^{mn}_\delta \tag{2.83}$$

For periodic meshes, the system of equations (2.83) can be enforced *by assigning the same node number to corresponding master and slave nodes*. In this case, the master node becomes de facto an equilibrated internal node due to the contribution of the internal force from the elements surrounding the corresponding slave node as shown by the shaded area in Figure 2.5(a).

For nonperiodic meshes, periodic boundary conditions can be implemented by constructing surface-to-surface constraints. For example, nodes 1, 2, and 3 in Figure 2.5(b) can be used to express the displacement of node c using the element shape functions. Note that when master and slave surfaces have different mesh densities, the master surface should be chosen as the surface with the coarser mesh density. Constrained equations can be implemented in one of three ways: (i) by explicitly constructing the transformation matrix $T_{\alpha\delta}$ in (2.83); (ii) by using a penalty approach; or (iii) via Lagrange multipliers.

Remark 2.2 The question is whether the unit cell solution of stresses and strains depends on the unit cell size, that is, the value of ζ. If we were to transform the unit cell problem [equation (2.80) and equation (2.81)] into the physical domain, the unit cell displacements computed in the

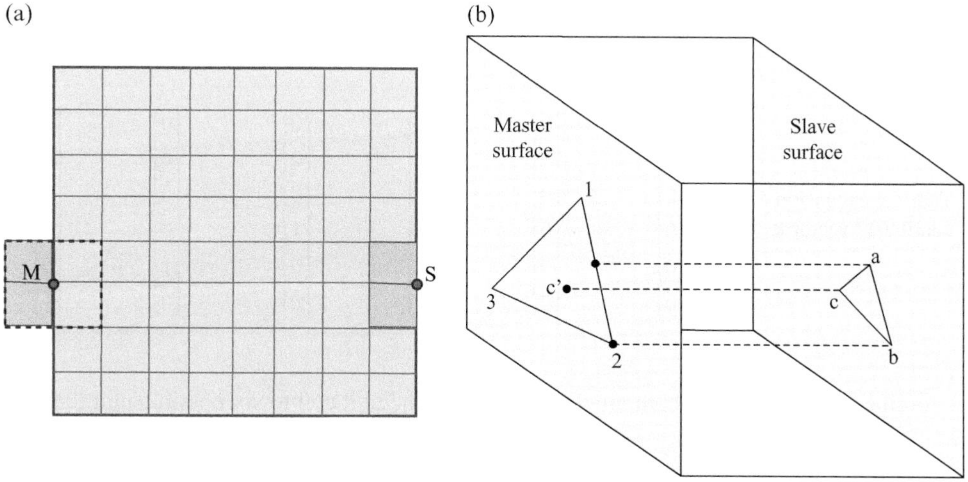

Figure 2.5 Master–slave relations: (a) a periodic mesh in a two-dimensional unit cell; and (b) a nonperiodic mesh in a three-dimensional unit cell

physical domain would be scaled by ζ in comparison with those computed in the unit cell coordinates. Moreover, the derivative with respect to the physical coordinate would scale by ζ^{-1} in comparison with the derivative with respect to the unit cell coordinates. Consequently, strains and stresses in the physical coordinate would be identical to those in the unit cell coordinate. The upshot of this observation is that: (i) we may construct and solve the unit cell problem in its physical domain; and (ii) the solution of the two-scale problem is independent of the unit cell size. The latter, however, points out the main limitation of the classical $O(1)$ homogenization theory.

2.2.4.3 Implementation in Commercial Software

In the following, we comment on the implementation of the two-scale analysis in a commercial package of choice. From the algorithmic point of view, the two-scale linear elasticity analysis consists of the following four steps:

1. Solve a unit cell problem with multiple right-hand sides (RHSs) $f_\alpha^{mn} = -\int_\Theta B_{ij\alpha}^f L_{ijmn} d\Theta$ and compute the stress influence functions Σ_{ij}^{mn}.
2. Evaluate the overall constitutive tensor components $L_{ijmn}^c = \frac{1}{|\Theta|}\int_\Theta \Sigma_{ij}^{mn}(\boldsymbol{y}) d\Theta$.
3. Solve the coarse-scale problem.
4. Postprocess the stresses in critical (or all) unit cells $\sigma_{ij}^f = \Sigma_{ij}^{mn}(\boldsymbol{y})\, \varepsilon_{mn}^c(\boldsymbol{x})$. The critical unit cells can identified by the largest norm of the coarse-scale strain.

We start with Step 1, the solution of a unit cell problem subjected to multiple RHS vectors (six in 3D due to the symmetry of indices mn). In the matrix implementation, L_{ijmn} is a 6×6 matrix where ij represents six rows and mn represents six columns due to minor symmetry (see Voigt notation in Chapter 1, section 1.5). For implementation in a commercial package,

Table 2.1 Thermal expansion coefficients (*mn*) corresponding to six loading cases in 3D

Indices (*mn*)	11	22	33	23	13	12
Thermal expansion coefficient vector $\boldsymbol{\kappa}$	$\begin{Bmatrix}1\\0\\0\\0\\0\\0\end{Bmatrix}$	$\begin{Bmatrix}0\\1\\0\\0\\0\\0\end{Bmatrix}$	$\begin{Bmatrix}0\\0\\1\\0\\0\\0\end{Bmatrix}$	$\begin{Bmatrix}0\\0\\0\\1\\0\\0\end{Bmatrix}$	$\begin{Bmatrix}0\\0\\0\\0\\1\\0\end{Bmatrix}$	$\begin{Bmatrix}0\\0\\0\\0\\0\\1\end{Bmatrix}$

the forcing term f_α^{mn} can be computed by subjecting the unit cell to six possible unit thermal strains ε_{mn}^{therm} defined as

$$\varepsilon_{mn}^{therm} = \kappa_{mn} \cdot \Delta T \tag{2.84}$$

where κ_{mn} and $\Delta T = -1$ are appropriately chosen thermal expansion coefficients and temperature changes. The thermal expansion coefficients for each loading case are chosen as indicated in Table 2.1.

We now turn to Step 2, an implementation of the overall constitutive tensor L_{ijmn}^c through the integral in equation (2.73). Recall that the stress influence functions Σ_{ij}^{mn} are fine-scale stresses σ_{ij}^f obtained by subjecting a unit cell to f_α^{mn}. In Step 1, we have already solved the unit cell problem subjected to the six loading cases. Thus, the stress influence functions Σ_{ij}^{mn} are stress outputs obtained for the aforementioned six loading cases. Equation (2.73) is integrated numerically over all the elements in the unit cell domain.

Finally, the coarse-scale analysis is carried out using the overall coefficients computed in Step 2. For the postprocessing (Step 4), the coarse-scale strains obtained in Step 3 are used in combination with the stress influence functions calculated in Step 1 to compute the fine-scale stresses in critical (or all) unit cells as

$$\sigma_{ij}^f = \Sigma_{ij}^{mn} \varepsilon_{mn}^c \tag{2.85}$$

2.2.4.4 Classical Boundary Conditions

One alternative to periodic boundary conditions is a uniform essential boundary condition by which the fine-scale displacement u_i^f is prescribed at the unit cell boundary $\partial\Theta$ in such a way as to coincide with the coarse-scale displacements obtained by integrating coarse-scale strain

$$u_i^f = \varepsilon_{ik}^c y_k \quad \text{on } \partial\Theta \tag{2.86}$$

Note that y_k is defined with respect to the unit cell centroid, and $\varepsilon_{ik}^c y_k$ is the portion of the coarse-scale displacement that does not include the rigid body motion of the unit cell centroid. Likewise, u_i^f is free of unit cell rigid body translation.

Multiplying (2.86) by n_j^{Θ}, then integrating over the boundary of the unit cell domain and applying Green's theorem yields

$$\varepsilon_{ij}^{c} = \frac{1}{|\Theta|}\int_{\Theta} \varepsilon_{ij}^{f}\, d\Theta \tag{2.87}$$

Prescribing essential boundary conditions on the unit cell boundary (2.86) is a special case of periodic boundary conditions by which the perturbation $u_i^{(1)}$ is assumed to vanish on the unit cell boundary.

The coarse-scale constitutive tensor L_{mnij}^{c} is computed as before

$$L_{mnij}^{c} = \frac{1}{|\Theta|}\int_{\Theta} \Sigma_{mn}^{ij}\, d\Theta \tag{2.88}$$

An alternative is to prescribe coarse-scale natural boundary conditions at the unit cell boundary

$$\sigma_{ij}^{f} n_j^{\Theta} = \sigma_{ij}^{c} n_j^{\Theta} \quad \text{on } \partial\Theta \tag{2.89}$$

Premultiplying (2.89) by y_k and integrating over the boundary of the unit cell domain yields

$$\int_{\Theta}\left(\sigma_{ij,y_j}^{f} y_k + \sigma_{ik}^{f}\right) d\Theta = \int_{\Theta}\left(\sigma_{ij,y_j}^{c} y_k + \sigma_{ik}^{c}\right) d\Theta \tag{2.90}$$

where we have taken advantage of integration by parts. Since $\sigma_{ij,y_j}^{f} = 0$ due to equilibrium in the unit cell and $\sigma_{ij,y_j}^{c} = 0$ since σ_{ij}^{c} is not a function of $\mathbf{y}$, then

$$\sigma_{ij}^{c} = \frac{1}{|\Theta|}\int_{\Theta} \sigma_{ij}^{f}\, d\Theta \tag{2.91}$$

Let $\tilde{E}_{mn}^{ij}$ be the strain field in a unit cell obtained by subjecting the unit cell to the natural boundary condition (2.89) with the value of prescribed traction corresponding to a coarse-scale stress equal to one. The resulting compliance matrix M_{mnij} is defined as

$$M_{mnij} = \frac{1}{|\Theta|}\int_{\Theta} \tilde{E}_{mn}^{ij}\, d\Theta \tag{2.92}$$

The coarse-scale properties obtained by the application of coarse-scale essential boundary conditions (2.86) on a unit cell boundary usually overestimate the real effective coarse-scale properties, while the coarse-scale natural boundary conditions (2.89) lead to their underestimation.

2.2.4.5 Plane Stress

For composite plate or shell elements, it is necessary to enforce a plane stress condition. This can be accomplished in one of the three ways. One possibility is to conduct a three-scale asymptotic homogenization approach along the lines proposed by Caillerie [42] and Kohn and Vogelius [43]. By using this approach, the microstructure is assumed to be periodic in the plane of the plate or shell, whose thickness is assumed to be infinitesimally small, that is, of the same order of magnitude as the unit cell in-plane dimensions. The second possibility is to assume that the unit coarse-scale strain component ε_{33}^{c} is unknown and to force the overall coarse-scale stress component σ_{33}^{c} to vanish (where subscript 3 denotes the direction normal to the plate or shell surface). Finally, the third and the simplest variant entails the nodes on the unit cell boundary normal to the surface of the plate to be free except for constraining a rigid body motion of the unit cell. In this latter scenario, the unit cell is subjected to five overall deformation modes (excluding ε_{33}^{c}).

2.2.4.6 Random Microstructure

In general, microstructure randomness can be accounted for using stochastic processes. As the emphasis of this book is on practical methods, attention is restricted to statistically homogeneous media. The notion of the unit cell representing periodic media is replaced by the so-called representative volume element (RVE), which is assumed to be macroscopically homogeneous. Consideration of random microstructure poses two main questions: (i) What is the minimum RVE size? and (ii) How can it be constructed?

One common approach [44] in determining RVE size is to study the convergence of the coarse-scale properties as a function of RVE size. Figure 2.6 shows that increasing the size of the RVE leads to better estimation of coarse-scale properties, and, when the unit cell is sufficiently large, results obtained by different boundary conditions converge to real effective

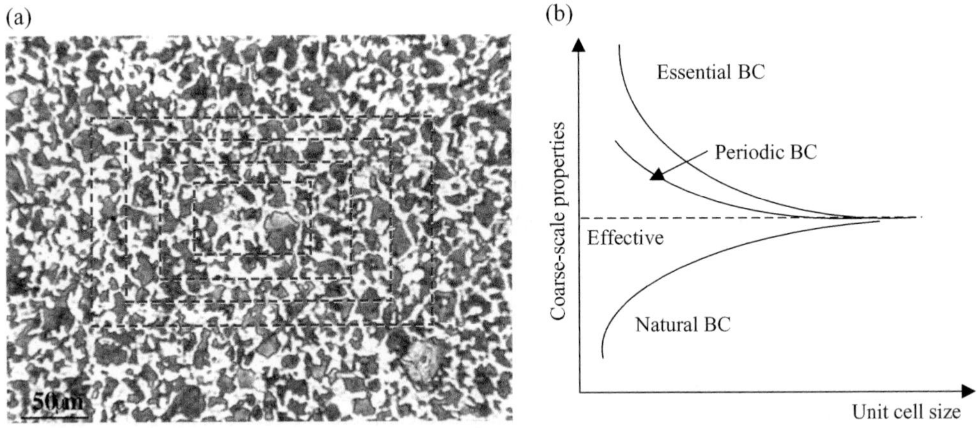

Figure 2.6 Microstructure of steel: (a) dashed lines depicting various unit cell sizes; and (b) convergence of coarse-scale properties with increasing unit cell size as obtained with different types of boundary conditions. BC, boundary condition

properties of the composite material. Note that the fastest convergence is most often obtained with periodic boundary conditions.

For a given unit cell size, the periodic boundary conditions provide a better estimation of the coarse-scale properties than the coarse-scale essential or natural boundary conditions.

The RVE can be generated in two principal ways: (i) direct use of scanned microstructural images; or (ii) reconstructing artificial microstructures. While the former procedure is relatively straightforward, the latter has the potential of creating a smaller RVE size, although the question of reproducibility of its main features still has to be addressed. Reconstruction (e.g., [45, 46]) is a process of generating the microstructure by employing the correlation functions characterizing its morphology.

2.2.4.7 Coarse-Scale Stress and Strain Obtained from Surface Measurements

It has been pointed out by Hill [47] that "Experimental determinations of mechanical behavior rest ultimately on measured loads or mean displacements over pairs of opposite faces of a representative cube. Macro-variables intended for constitutive laws should thus be capable of definition in terms of surface data alone, either directly or indirectly."

In view of Hill's statement, we can reformulate the definition of coarse-scale strain (2.35) and stress (2.41) in terms of unit cell boundary data.

Integration by parts of (2.35) and assuming perfect interfaces yields

$$\varepsilon_{ij}^{c} = \frac{1}{|\Theta|}\int_{\Theta} \varepsilon_{ij}^{f}\, d\Theta = \frac{1}{|\Theta|}\int_{\partial\Theta} \mathrm{sym}\left(u_i^f n_j^{\Theta}\right) d\gamma \tag{2.93}$$

Consider the integration of $\left(\sigma_{ij}^{f} y_k\right)_{,y_j}$ over the unit cell domain. Applying Green's theorem yields

$$\frac{1}{|\Theta|}\int_{\Theta} \left(\sigma_{ij}^{f} y_k\right)_{,y_j} d\Theta = \frac{1}{|\Theta|}\int_{\partial\Theta} \underbrace{\sigma_{ij}^{f} n_j^{\Theta}}_{t_i^f}\, y_k\, d\gamma \tag{2.94}$$

Further exploiting the unit cell equilibrium equation $\sigma_{ij,y_j}^{f} = 0$ allows coarse-scale stress to be expressed in terms of fine-scale stresses on the unit cell boundary

$$\sigma_{ik}^{c} = \frac{1}{|\Theta|}\int_{\Theta} \sigma_{ik}^{f}\, d\Theta = \frac{1}{|\Theta|}\int_{\partial\Theta} t_i^{f}\, y_k\, d\gamma \tag{2.95}$$

Note that point forces or high frequency inertia, which may have an $O(\zeta^{-1})$ contribution to the source equilibrium equation (2.6a) (or equation of motion), will contribute to the unit cell problem (2.25). In this case, the boundary integral term in (2.95), considered to be the "true" coarse-scale stress according to Hill [47], would no longer be equal to the average fine-scale stress.

2.2.4.8 Relation to Heterogeneous Multiscale Method

Homogenization of continua is closely related to the so-called heterogeneous multiscale method (HMM) developed by E *et al.* [48]. Here we will outline the application of HMM to elliptic partial differential equations (PDEs) with oscillatory coefficients (2.6). We will follow the presentation of the method outlined in [48]. The goal is to compute the coarse-scale stiffness matrix given by

$$K^c_{\alpha\beta}=\int_\Omega N^c_{(s\alpha,x_t)}L^c_{stmn}N^c_{(m\beta,x_n)}d\Omega \tag{2.96}$$

Denote the integrand of (2.96) as $g_{\alpha\beta}=N^c_{(s\alpha,x_t)}L^c_{stmn}N^c_{(m\beta,x_n)}$. The Gauss quadrature of equation (2.96) is then given by

$$K^c_{\alpha\beta}=\sum_{e=1}^{N^c_{el}}\int_\square Jg_{\alpha\beta}d\square=\sum_{e=1}^{N^c_{el}}\sum_{I=1}^{N^c_{\text{int}}}W_IJ_I(\hat{\boldsymbol{x}}_I)g_{\alpha\beta}(\hat{\boldsymbol{x}}_I) \tag{2.97}$$

where $\square$ is the parent element domain; J is the Jacobian; N^c_{el} and N^c_{int} are the number of coarse-scale elements and quadrature points, respectively; and $\hat{\boldsymbol{x}}_I$ and W_I are the coordinates of quadrature points and corresponding weights, respectively.

The HMM philosophy is based on solving the coarse-scale problem $K^c_{\alpha\beta}d^c_\beta=f^c_\alpha$, assuming that the coarse-scale model is completely known and valid everywhere, and then focusing on the question of how to find the missing constitutive relations (L^c_{stmn} in the present case) by solving the unit cell problem locally around each quadrature point $\hat{\boldsymbol{x}}_I$, as shown in Figure 2.7.

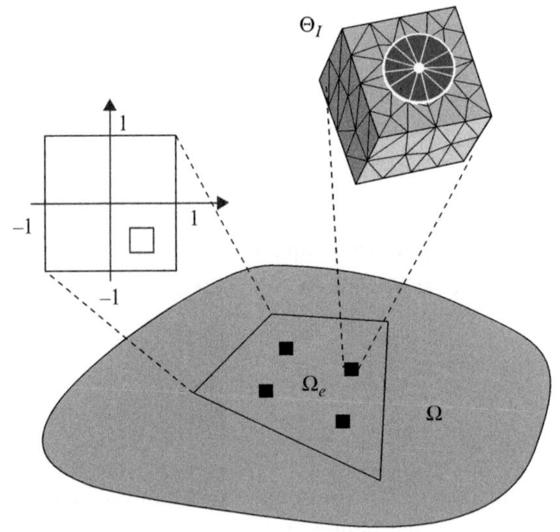

Figure 2.7 Two-scale integration scheme. Reproduced with permission from [35], © 2008 John Wiley & Sons

This is accomplished by expressing the coarse-scale stress in terms of coarse- and fine-scale fields and properties as follows

$$\begin{aligned}\sigma_{ij}^{c}(\hat{\boldsymbol{x}}_I) &= L_{ijkl}^{c}\underbrace{\left(\frac{1}{|\Theta|}\int_{\Theta}\varepsilon_{kl}^{f}(\hat{\boldsymbol{x}}_I,\boldsymbol{y})\,d\Theta\right)}_{\varepsilon_{kl}^{c}(\hat{\boldsymbol{x}}_I)} \\ &= \frac{1}{|\Theta|}\left(\int_{\Theta}L_{ijkl}(\hat{\boldsymbol{x}}_I,\boldsymbol{y})\,\varepsilon_{kl}^{f}(\hat{\boldsymbol{x}}_I,\boldsymbol{y})\,d\Theta\right)\end{aligned} \tag{2.98}$$

Equation (2.98) follows from the definition of coarse-scale strain (2.35), the definition of coarse-scale stress (2.41), and the coarse-scale constitutive equation (2.45).

The coarse-scale properties are found by solving (2.98) for six unit coarse-scale strains $\varepsilon_{kl}^{c}(\hat{\boldsymbol{x}}_I)=e_{kl}$, which yields

$$\begin{aligned}L_{ijkl}^{c} &= \frac{1}{|\Theta|}\left(\int_{\Theta}L_{ijkl}(\hat{\boldsymbol{x}}_I,\boldsymbol{y})\,\varepsilon_{kl}^{f}(\hat{\boldsymbol{x}}_I,\boldsymbol{y})\,d\Theta\right) \\ \varepsilon_{kl}^{c}(\hat{\boldsymbol{x}}_I) &= e_{kl}\end{aligned} \tag{2.99}$$

2.2.4.9 Consideration of Imperfect Interfaces

The formulation presented so far assumes perfect interfaces in the microstructure. If the interfaces are imperfect, the integration by parts has to account for these internal interfaces. Consider a unit cell consisting of two fine-scale phases occupying volumes Θ^1 and Θ^2 with corresponding boundaries $\partial\Theta^1$ and $\partial\Theta^2$, respectively. Note that $\partial\Theta^1 \cup \partial\Theta^2 = \partial\Theta \cup S$ where the internal interface is defined by $S=\partial\Theta^1 \cap \partial\Theta^2$. Let Θ^- be the domain of the unit cell problem that excludes the interfaces, such that $\Theta^- = \Theta^1 \cup \Theta^2$.

We now derive a modification to (2.87) that assumes imperfect interfaces. We follow the exposition given in [41]

$$\begin{aligned}\frac{1}{|\Theta|}\int_{\Theta^-}\varepsilon_{ij}^{f}\,d\Theta &= \frac{1}{|\Theta|}\left\{\int_{\Theta^1}u_{(i,y_j)}^{f}\,d\Theta+\int_{\Theta^2}u_{(i,y_j)}^{f}\,d\Theta\right\} \\ &= \frac{1}{|\Theta|}\left\{\int_{\partial\Theta^1}\mathrm{sym}\left(u_i^f n_j^1\right)d\gamma_1+\int_{\partial\Theta^2}\mathrm{sym}\left(u_i^f n_j^2\right)d\gamma_2\right\}\end{aligned} \tag{2.100}$$

where $\boldsymbol{n}^1$ and $\boldsymbol{n}^2$ denote the unit normals on the corresponding boundaries of the fine-scale phases. Inserting $u_i^f=\varepsilon_{ik}^{c}y_k$ on $\partial\Theta$ into (2.100) and noting that $\partial\Theta^1 \cup \partial\Theta^2 = \partial\Theta \cup S$ yields

$$\frac{1}{|\Theta|}\int_{\Theta^-}\varepsilon_{ij}^{f}\,d\Theta=\frac{1}{|\Theta|}\int_{\partial\Theta}\mathrm{sym}\left(\varepsilon_{ik}^{c}y_k n_j^{\Theta}\right)d\gamma+\frac{1}{|\Theta|}\int_{S}\mathrm{sym}\left(\left[\!\left[u_i^f\right]\!\right]\tilde{n}_j\right)dS \tag{2.101}$$

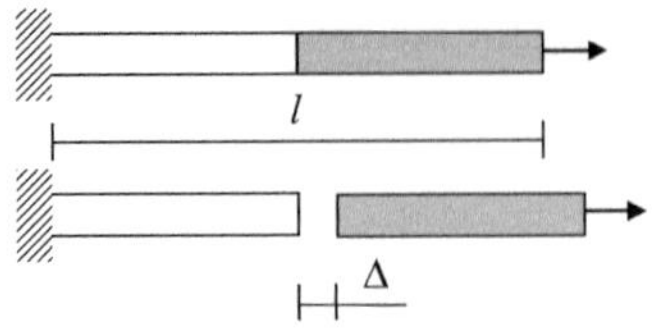

Figure 2.8 Imperfect interfaces – an illustrative example

where $\llbracket u_i^f \rrbracket$ denotes the displacement jumps at the interface S. $\boldsymbol{n}^\Theta$ and $\breve{\boldsymbol{n}}$ are the unit normals on $\partial\Theta$ and S, respectively. Applying Green's theorem to the first term on the right-hand side of (2.101) yields

$$\frac{1}{|\Theta|}\int_{\Theta^-} \varepsilon_{ij}^f d\Theta = \varepsilon_{ij}^c + \frac{1}{|\Theta|}\int_S \text{sym}\left(\llbracket u_i^f \rrbracket \breve{n}_j\right) dS \tag{2.102}$$

This result has been noted in a number of publications (see for instance [49]).

As an illustrative example, consider a 1D unit cell domain depicted in Figure 2.8. The unit cell is constrained on the left and is subjected to the coarse-scale axial strain $\varepsilon^c = \Delta/l$. The unit cell separates at the interface between the two fine-scale phases, introducing no strain in each of its phases, and thus the left-hand side term in (2.102) vanishes. The second term on the right-hand side of equation (2.102) is evaluated as follows: $|\Theta| = l$; the interface consists of a single point and thus the integral reduces to the integrand; $\llbracket u^f \rrbracket \breve{n} = u^L \breve{n}^L + u^R \breve{n}^R = 0 \cdot 1 + \Delta \cdot (-1) = -\Delta$; and consequently, equation (2.102) is satisfied.

Alternatively, instead of considering the strain in each fine-scale phase and the displacement jump at the interface, we can define the fine-scale strain over the entire unit cell domain Θ, which would account for the existence of displacement discontinuity at the interface. For the model problem considered in Figure 2.8, such fine-scale strain is given by

$$\varepsilon^f = \varepsilon^c + d(y - \breve{y})\llbracket u^f \rrbracket \breve{n} \tag{2.103}$$

where $d(y - \breve{y})$ is a Dirac delta function and $\breve{y}$ is the position of the interface.

2.2.5 Boundary Layers

The formulation presented so far assumes solution periodicity. Clearly, such periodicity is valid in the interior of the body, provided that the coarse-scale solution gradients are small. In the vicinity of free edges, the fine-scale structure is no longer periodic, and consequently, the solution is not periodic either. Furthermore, the boundary condition on the composite domain cannot be satisfied by a periodic unit cell solution.

One approach to account for nonperiodicity in the boundary layers is to introduce enrichment that would asymptotically decrease as it moves toward the interior of the body [50,51,52]. Alternatively, the upscaling methods based on enhanced kinematics [10,11,13,14] as outlined in section 2.3 can be employed.

Here we adopt the exposition of ideas based on [51,52]. At the boundary layer, an auxiliary coordinate system is introduced with the origin on the external surface, the axis z_3 aligned

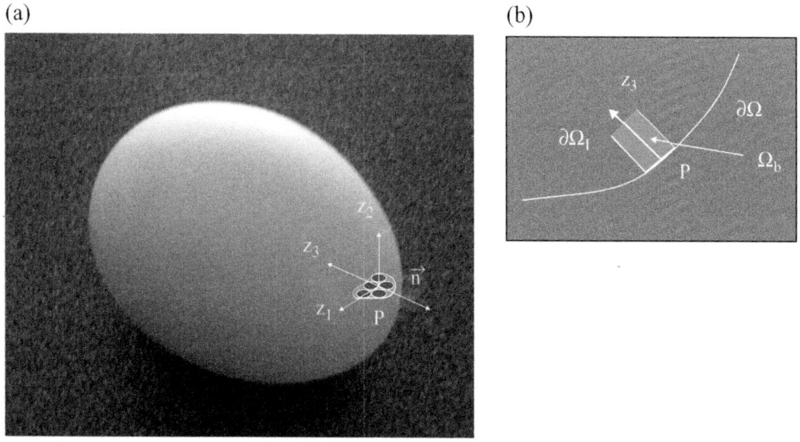

Figure 2.9 Boundary layer region at point P on the boundary: (a) coarse-scale domain; and (b) cross section around point P on the boundary

normal to this surface, and z_α ($\alpha=1,2$) oriented tangentially as shown in Figure 2.9. The boundary layer coordinate system is related to the coarse-scale coordinate system by

$$z_i = \frac{x_i}{\zeta} \tag{2.104}$$

The asymptotic expansion of the displacement field at the boundary layer is given by

$$u_i^\zeta(\boldsymbol{x}) = u_i^{(0)}(\boldsymbol{x},\boldsymbol{y}) + \zeta\left(u_i^{(1)}(\boldsymbol{x},\boldsymbol{y}) + v_i^{(1)}(\boldsymbol{x},\boldsymbol{z})\right) + O\left(\zeta^2\right) \tag{2.105}$$

where $v_i^{(1)}(\boldsymbol{x},\boldsymbol{z})$ is the leading order boundary layer enrichment term. It should vanish exponentially as we move away from the boundary in the physical domain or, in other words, as z_3 approaches infinity. In a tangential direction, the solution is assumed to be periodic in z_α for $\alpha = 1,2$.

The spatial derivative of function $f^\zeta(\boldsymbol{x})=f(\boldsymbol{x},\boldsymbol{y},\boldsymbol{z})$ is defined as

$$f_{,i}^\zeta = \frac{\partial f(\boldsymbol{x},\boldsymbol{y},\boldsymbol{z})}{\partial x_i} + \frac{1}{\zeta}\frac{\partial f(\boldsymbol{x},\boldsymbol{y},\boldsymbol{z})}{\partial y_i} + \frac{1}{\zeta}\frac{\partial f(\boldsymbol{x},\boldsymbol{y},\boldsymbol{z})}{\partial z_i} \tag{2.106}$$

The resulting strains are given by

$$\varepsilon_{ij}^\zeta = u_{(i,j)}^\zeta = \varepsilon_{ij}^{(0)} + v_{(i,z_j)}^{(1)} + O(\zeta) \tag{2.107}$$

Inserting (2.107) into constitutive equation (2.6b) gives

$$\sigma_{ij}^\zeta = \sigma_{ij}^{(0)} + \sigma_{ij}^{(0)b} + \zeta\left(\sigma_{ij}^{(1)} + \sigma_{ij}^{(1)b}\right) + O\left(\zeta^2\right) \tag{2.108}$$

where $\sigma_{ij}^{(0)b}$ is the leading order stress enrichment in the boundary layer given by

$$\sigma_{ij}^{(0)b} = L_{ijkl} v_{(k,z_l)}^{(1)} \tag{2.109}$$

The leading order equilibrium equation is obtained by inserting (2.108) into (2.6a), which yields

$$O\left(\zeta^{-1}\right):\quad \sigma_{ij,y_j}^{(0)} + \sigma_{ij,z_j}^{(0)b} = 0 \tag{2.110a}$$

$$O(1):\quad \frac{\partial}{\partial x_j}\left(\sigma_{ij}^{(0)} + \sigma_{ij}^{(0)b}\right) + \sigma_{ij,y_j}^{(1)} + \sigma_{ij,z_j}^{(1)b} + b_i^{\zeta} = 0 \tag{2.110b}$$

By choosing $H_i^{jk}(y)$ to be the solution of the unit cell problem [the $O(\zeta^{-1})$ equation] in the interior of the body in (2.28), we obtain

$$O\left(\zeta^{-1}\right):\quad \sigma_{ij,z_j}^{(0)b} = 0 \tag{2.111}$$

Equation (2.109) and equation (2.111) comprise the boundary value problem for the enrichment $v_i^{(1)}$ at the boundary layer with the following boundary conditions

$$v_i^{(1)}, \sigma_{ij}^{(0)b} - \text{periodic in } z_\alpha \quad \text{for } \alpha = 1,2 \tag{2.112a}$$

$$v_i^{(1)} = -u_i^{(1)} \quad \text{on } \partial\Omega^u \tag{2.112b}$$

$$\sigma_{ij}^{(0)b} n_j^c = \bar{t}_i^c - \sigma_{ij}^{(0)} n_j^c \quad \text{on } \partial\Omega^t \tag{2.112c}$$

$$v_i^{(1)} = 0 \quad z_3 \to \infty \tag{2.112d}$$

where (2.112b) follows from the assumption that the leading order displacement $u_i^{(0)}$ satisfies the essential boundary conditions. Equation (2.112d) can be satisfied by discretizing the boundary element region with infinite elements [52].

Integrating equation (2.110b) over the boundary layer domain Ω_b, which consists of the union of unit cell domains $\Omega_b = \bigcup_{\Theta_i \in \Omega_b} \Theta_i$, yields (2.40) where the coarse-scale stress is defined as

$$\sigma_{ij}^c = \frac{1}{|\Theta|} \int_\Theta \left(\sigma_{ij}^{(0)} + \sigma_{ij}^{(0)b}\right) d\Theta \tag{2.113}$$

and we have taken advantage of $\sigma_{ij}^{(1)}$ periodicity and

$$\int_{\Omega_b} \sigma_{ij,z_j}^{(1)b} d\Omega_b = 0 \tag{2.114}$$

The above follows from a periodicity of $\sigma_{ij}^{(1)b}$ on the lateral surfaces of Ω_b and vanishing $\sigma_{ij}^{(1)b} n_j^c$ on the external boundary (free surface) and at the interface with the interior ($z_3 \to \infty$).

2.2.6 Convergence Estimates

Convergence estimates for two-scale second-order elliptic PDEs with periodic coefficients can be found in [2,53,54], among others. Consider an asymptotic expansion of displacement and stresses (or strains)

$$\begin{aligned} \boldsymbol{u}^{\zeta} &= \boldsymbol{u}^{(0)} + \zeta \boldsymbol{u}^{(1)} + \boldsymbol{r}^{u} \\ \boldsymbol{\sigma}^{\zeta} &= \boldsymbol{\sigma}^{(0)} + \boldsymbol{r}^{\sigma} \end{aligned} \tag{2.115}$$

Then

$$\begin{aligned} \left\| \boldsymbol{r}^{u} \right\|_{H^1(\Omega)} &\le C\sqrt{\zeta} \\ \left\| \boldsymbol{r}^{\sigma} \right\|_{L_2(\Omega)} &\le C\sqrt{\zeta} \end{aligned} \tag{2.116}$$

for some constant C. These estimates, which are somewhat lower than expected, can be attributed primarily to the boundary layers.

An alternative approach [36] consists of estimating the first term neglected in the asymptotic expansion of the stress field $\zeta \boldsymbol{\sigma}^{(1)}$ and comparing its magnitude to the leading order stress $\boldsymbol{\sigma}^{(0)}$ in some norm.

To obtain $\zeta \boldsymbol{\sigma}^{(1)}$, the third term in the asymptotic expansion of displacements is decomposed as

$$u_i^{(2)}(\boldsymbol{x}, \boldsymbol{y}) = \eta_i^{jmn}(\boldsymbol{y}) \varepsilon_{mn,x_j}^{c}(\boldsymbol{x}) \tag{2.117}$$

where $\eta_i^{jmn}(\boldsymbol{y})$ is the second-order displacement influence function. Following the derivation outlined in section 2.2.1 yields the boundary value problem for $\eta_i^{jmn}(\boldsymbol{y})$

$$\begin{aligned} &\left[L_{iskl} \left(H_k^{mn} \delta_{jl} + \eta_{(k,y_l)}^{jmn} \right) \right]_{,y_s} + L_{ijkl} \left(H_{(k,y_l)}^{mn} + I_{klmn} \right) - L_{ijmn}^{c} = 0 \quad \text{on} \quad \Theta \\ &\eta_k^{jmn}(\boldsymbol{y}) \ \ \text{– periodic on} \quad \partial\Theta \\ &\eta_k^{jmn}(\boldsymbol{y}) = 0 \quad \text{on} \quad \partial\Theta^{vert} \quad (\text{or average of } \eta_k^{jmn} \text{ vanish}) \end{aligned} \tag{2.118}$$

In [36], a normalized truncation error e^{tr} is defined as

$$e^{tr} = \frac{\left\| \zeta \boldsymbol{\sigma}^{(1)} \right\|_{L_2(\Omega^e)}}{\left\| \boldsymbol{\sigma}^{(0)} \right\|_{L_2(\Omega^e)}} \tag{2.119}$$

Figure 2.10 depicts truncation error in a composite plate with a hole. It can be seen that larger errors are obtained in the vicinity of high coarse-scale stress gradients.

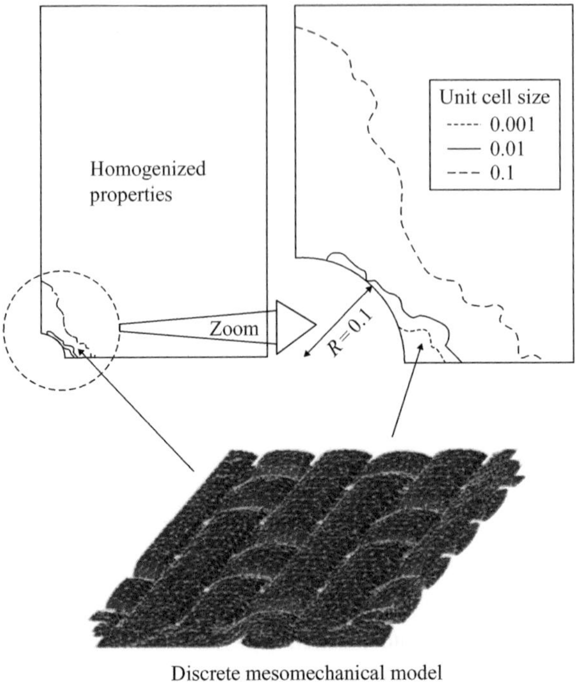

Figure 2.10 Distribution of the truncation error in a composite plate and the microstructure of the material

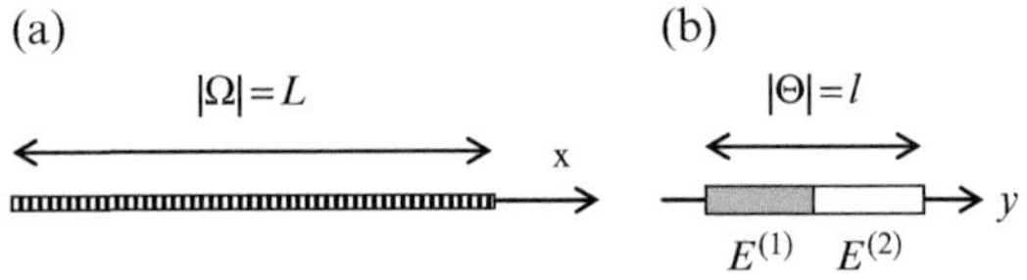

Figure 2.11 Model problem: (a) coarse-scale domain; and (b) a unit cell in the stretched coordinate system

2.2.6.1 One-Dimensional Model Problem

Consider a 1D composite domain of length $|\Omega|=L$ with a unit cell in the stretched coordinate system as shown in Figure 2.11.

Recall equation (2.26) for the unit cell problem in 1D, which can be written as

$$\frac{\partial}{\partial y}\left(L^{\zeta}\left(\frac{\partial u^{(0)}}{\partial x}+\frac{\partial u^{(1)}}{\partial y}\right)\right)=0 \tag{2.120}$$

Assuming $L^{\zeta}=E(y)$ and integrating (2.120) gives

$$E(y)\left(\frac{\partial u^{(0)}}{\partial x}+\frac{\partial u^{(1)}}{\partial y}\right)=c(x) \tag{2.121}$$

Further integrating (2.121) over the unit cell domain and accounting for periodicity of $u^{(1)}$ yields

$$c(x)=\left(\frac{1}{|\Theta|}\int_{\Theta}\frac{1}{E(y)}d\Theta\right)^{-1}\frac{\partial u^{(0)}}{\partial x} \tag{2.122}$$

Inserting (2.122) into (2.121) yields

$$\frac{\partial u^{(1)}}{\partial y}=\frac{\partial H(y)}{\partial y}\frac{\partial u^{(0)}}{\partial x} \tag{2.123}$$

where

$$\frac{\partial H(y)}{\partial y}=\frac{1}{E(y)}\left(\frac{1}{|\Theta|}\int_{\Theta}\frac{1}{E(y)}d\Theta\right)^{-1}-1 \tag{2.124}$$

The coarse-scale material properties (Young's modulus) follow from equation (2.44)

$$E^{c}=\frac{1}{|\Theta|}\int_{\Theta}\Sigma(y)d\Theta=\frac{1}{|\Theta|}\int_{\Theta}E(y)\left(1+\frac{\partial H}{\partial y}\right)d\Theta \tag{2.125}$$

Note that for 1D problems, the stress influence function Σ is constant and therefore (2.125) can be written as

$$E^{c}=\Sigma=E(y)\left(1+\frac{\partial H}{\partial y}\right) \tag{2.126}$$

Combining (2.124) and (2.126) gives

$$E^{c}=\left(\frac{1}{|\Theta|}\int_{\Theta}\frac{1}{E(y)}d\Theta\right)^{-1} \tag{2.127}$$

2.2.6.2 Three-Dimensional Problem

For verification, we consider a 3D fibrous unit cell. The phase properties of the microstructure are summarized in Table 2.2. The unit cell is discretized with 351 tetrahedral elements as shown in Figure 2.12.

The overall properties of the composite obtained in Step 2 (section 2.2.4.3) are depicted in Table 2.3. For comparison, the overall properties obtained using the Mori–Tanaka method [20] are also shown in Table 2.3.

For the coarse-scale analysis, we consider a cantilever beam subjected to a uniform distributed load along the top edge. For comparison, a reference solution is obtained using a single-scale finite element analysis on a fine mesh. Deformed meshes and von Mises stresses at a critical unit cell (left bottom corner) are shown in Figure 2.13. The stresses in a unit cell are obtained by postprocessing.

Table 2.2 Material properties for a fibrous unit cell

Materials	Young's modulus (GPa)	Poisson's ratio	Volume fraction
Titanium matrix	68.9	0.33	0.733
SiC fiber	379.2	0.21	0.267

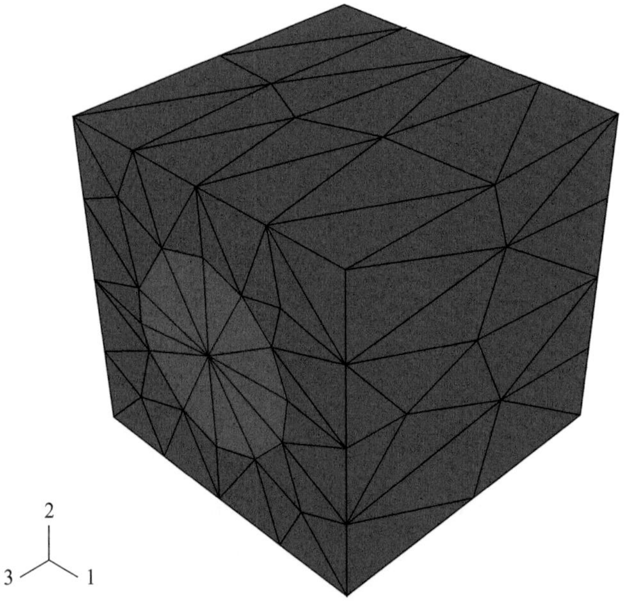

Figure 2.12 Finite element mesh of a fibrous unit cell. Reproduced with permission from [35], © 2008 John Wiley & Sons

Table 2.3 The coefficients for a homogenized stiffness matrix compared with the Mori–Tanaka method

11	22	33	23	13	12
140.3 (134.2)	57.3 (61.4)	57.7 (57.3)	0.0 (0.0)	0.0 (0.0)	0.1 (0.0)
	140.0 (134.2)	57.6 (57.3)	0.0 (0.0)	0.0 (0.0)	0.1 (0.0)
		185.6 (185.6)	0.0 (0.0)	0.0 (0.0)	0.0 (0.0)
			39.5 (38.2)	0.0 (0.0)	0.0 (0.0)
	SYM.			39.4 (38.2)	0.0
					36.5 (36.4)

2.2.6.3 Preliminary Material Design

Let $\boldsymbol{\beta}$ denote the parameters describing unit cell geometry. The goal of the preliminary material design is to construct a unit cell parametric geometry $\boldsymbol{\beta}$ such that the stresses in the unit cell are minimized, subject to manufacturing constraints on the unit cell geometry $g_i(\boldsymbol{\beta}) \leq 0$. We will refer to this process as a preliminary material design to distinguish it from a detailed material design where the inelastic behavior has to be accounted for.

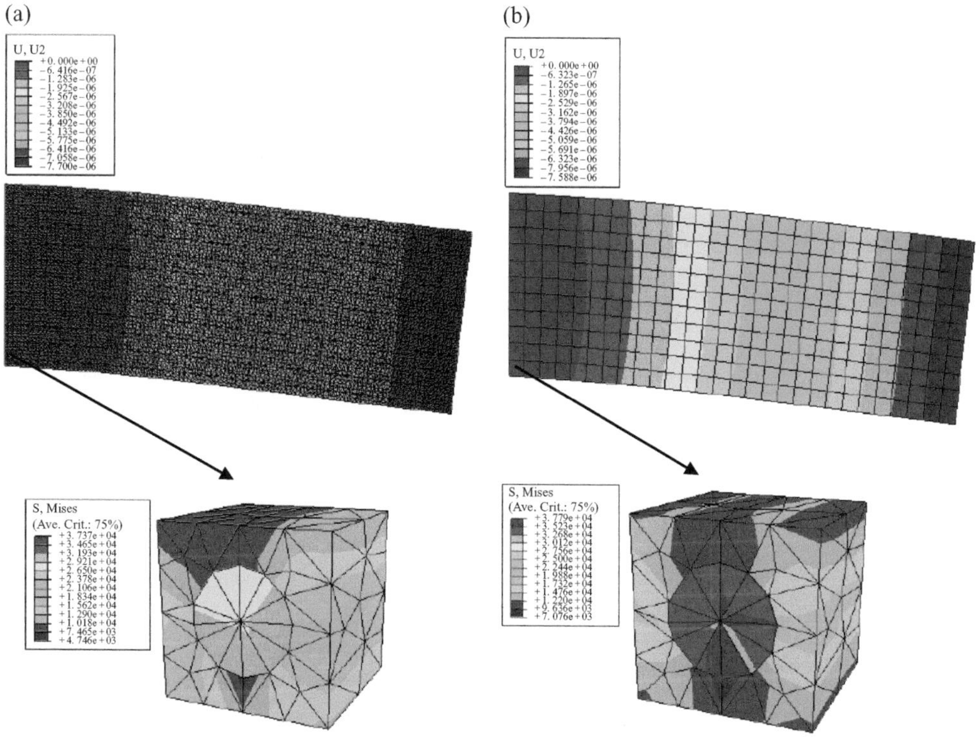

Figure 2.13 Comparison of the homogenization solution and the reference solution: (a) reference solution; and (b) upscaling solution. Reproduced with permission from [35], © 2008 John Wiley & Sons

Consider the fine-scale stress σ_{ij}^{f} expressed in terms of the coarse-scale strain ε_{kl}^{c}

$$\sigma_{ij}^{f} = \Sigma_{ij}^{mn}(\boldsymbol{y},\boldsymbol{\beta})\varepsilon_{mn}^{c} \tag{2.128}$$

where

$$\Sigma_{ij}^{mn}(\boldsymbol{y},\boldsymbol{\beta}) = L_{ijkl}(\boldsymbol{y},\boldsymbol{\beta})\left(I_{klmn} + H_{(k,y_l)}^{mn}(\boldsymbol{y},\boldsymbol{\beta})\right) \tag{2.129}$$

The goal is to find $\boldsymbol{\beta} \in \mathrm{M}$ such that the norm of maximum stress in the unit cell is minimized, subject to arbitrary unit coarse-scale strain

$$\min_{\boldsymbol{\beta}\in \mathrm{M}} \max_{y\in\Theta} \left\| \Sigma_{ij}^{mn}(\boldsymbol{y},\boldsymbol{\beta})\varepsilon_{mn}^{c} \right\|_2 \quad \forall \left\| \varepsilon_{mn}^{c} \right\|_2 = 1 \tag{2.130}$$

where M denotes the space of real valued functions that satisfy manufacturing constraints. Taking advantage of the inequality

$$\left\| \Sigma_{ij}^{mn}(\boldsymbol{y},\boldsymbol{\beta})\varepsilon_{mn}^{c} \right\|_2 \le \left\| \Sigma_{ij}^{mn}(\boldsymbol{y},\boldsymbol{\beta}) \right\|_2 \underbrace{\left\| \varepsilon_{mn}^{c} \right\|_2}_{1} = \left\| \Sigma_{ij}^{mn}(\boldsymbol{y},\boldsymbol{\beta}) \right\|_2 \tag{2.131}$$

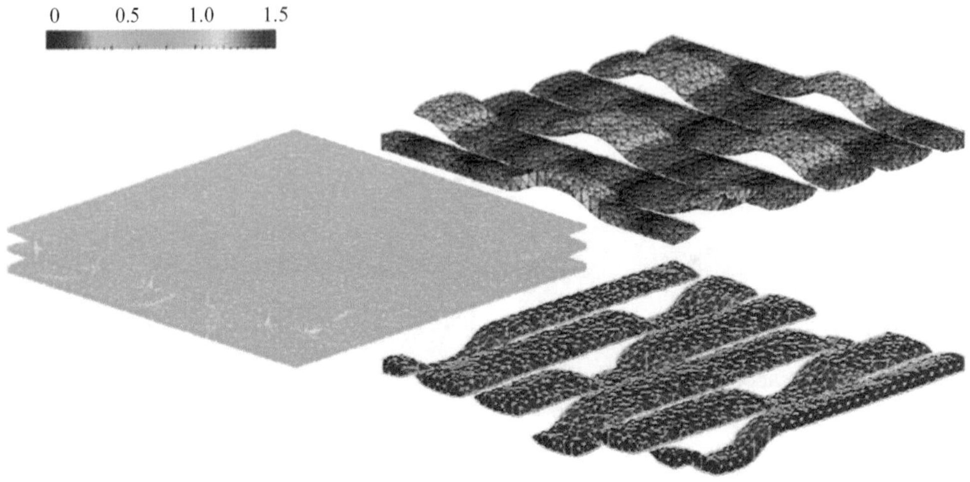

Figure 2.14 Axial stresses σ_{11}^{f} obtained in a unit cell subjected to coarse-scale stress $\sigma_{11}^{c} = 1$

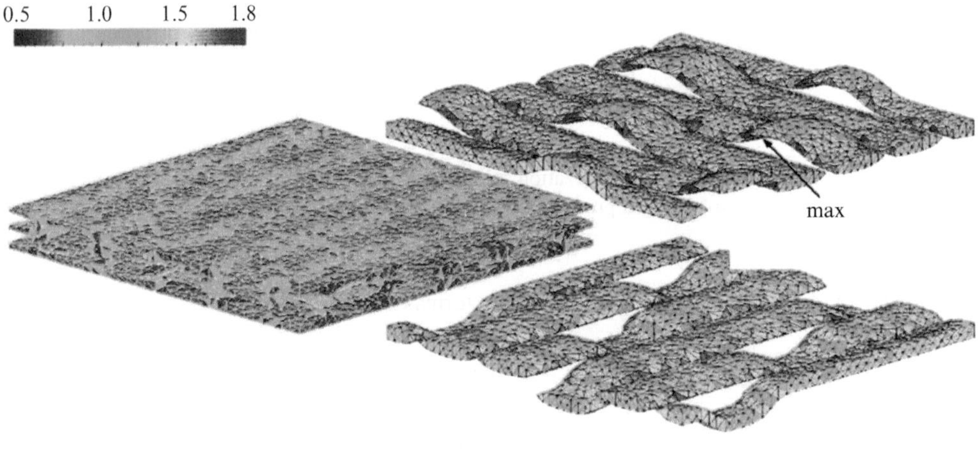

Figure 2.15 Spectral norm of the concentration tensor Σ_{ij}^{mn}

yields the following preliminary design problem. Find $\boldsymbol{\beta} \in \mathrm{M}$ such that

$$\min_{\boldsymbol{\beta} \in \mathrm{M}} \max_{y \in \Theta} \left\| \Sigma_{ij}^{mn}(\boldsymbol{y}, \boldsymbol{\beta}) \right\|_2 \tag{2.132}$$

Figure 2.14 depicts a woven composite subjected to coarse-scale stress $\sigma_{11}^{c} = 1$ and a spectral norm of Σ_{ij}^{mn}. The maximum axial stress σ_{11}^{f} developed in the unit cell is 1.57, which can be interpreted as a stress concentration factor. On the other hand, the maximum spectral norm of Σ_{ij}^{mn} is 1.77, which can be interpreted as a maximum stress norm obtained by subjecting the unit cell to arbitrary unit coarse-scale strain (Figure 2.15).

Remark 2.3 If the failure criterion in each phase is defined, then rather than minimizing the maximum stress, it is more appropriate to maximize the distance to failure.

2.3 Upscaling Based on Enhanced Kinematics

Coarse-scale equations can be upscaled from fine-scale equations either by constructing coarse-scale effective properties (or fields, in the case of nonlinear problems) or by enhancing the coarse-scale kinematics to reflect the oscillatory response of a heterogeneous medium. A key feature of enhanced kinematics-based upscaling methods is that construction of the enhanced basis functions is a local operation and therefore the process can be parallelized. Prior to describing various approaches, which differ in how the enhanced kinematics is formulated and incorporated into the coarse-scale equations and how the coarse-scale equations are integrated, we start by pinpointing the link between the homogenization-based and enhanced kinematics-based upscaling methods.

Consider the coarse-scale stiffness matrix in (2.77) derived using homogenization

$$K^c_{\alpha\beta} = \int_\Omega N^c_{(s\alpha,x_t)} L^c_{stmn} N^c_{(m\beta,x_n)} d\Omega \tag{2.133}$$

Substituting the expression for the coarse-scale material properties (2.47) gives

$$K^c_{\alpha\beta} = \int_\Omega \frac{1}{|\Theta|} \int_\Theta N^c_{(s\alpha,x_t)} E^{st}_{ij}(\mathbf{y}) L_{ijkl}(\mathbf{y}) E^{mn}_{kl}(\mathbf{y}) N^c_{(m\beta,x_n)} d\Theta d\Omega \tag{2.134}$$

Let $B^{\zeta}_{ij\alpha}$ be the enhanced strain-displacement matrix defined as

$$B^{\zeta}_{ij\alpha}(\mathbf{x},\mathbf{y}) = N^c_{(s\alpha,x_t)}(\mathbf{x}) E^{st}_{ij}(\mathbf{y}) \tag{2.135}$$

Then the stiffness matrix can be written as

$$K^c_{\alpha\beta} = \int_\Omega \frac{1}{|\Theta|} \int_\Theta B^{\zeta}_{ij\alpha}(\mathbf{x},\mathbf{y}) L_{ijkl}(\mathbf{y}) B^{\zeta}_{kl\beta}(\mathbf{x},\mathbf{y}) d\Theta d\Omega \tag{2.136}$$

Equation (2.135) and equation (2.136) suggest that rather than constructing effective material properties, we can construct enhanced kinematical relations instead.

One of the salient features of the $O(1)$ homogenization-based upscaling methods is that the fine-scale solution $u^{(1)}_i(\mathbf{x},\mathbf{y})$ is completely described by the coarse-scale solution $\varepsilon^c_{ij}(\mathbf{x})$ and the influence functions $H^{jk}_i(\mathbf{y})$, which can be precomputed at a material point. The microstructure in the unit cell can be periodic or random, but the solution has to be locally periodic. For these assumptions to be valid, a unit cell has to "experience" a constant variation of ε^c_{ij} over its domain Θ. This assumption breaks down in the high gradient regions, such as in the vicinity of free edges. Enhanced kinematics-based upscaling methods provide a general framework that does not require ε^c_{ij} to be constant over the unit cell domain.

2.3.1 Multiscale Finite Element Method

The main idea of the multiscale finite element method (MsFEM) [10] is to build the local small-scale information of the leading order differential operator into the finite element base functions. By using this approach, the enhanced coarse-scale basis functions $N^f_{i\alpha}$ are precomputed locally by solving the boundary value problem on a single element domain Ω_{el}

$$\begin{aligned} &\left(L_{ijkl} N^f_{(k\alpha,l)}\right)_{,j} = 0 \quad \text{on} \quad \Omega_{el} \\ &N^f_{k\alpha} = N^c_{k\alpha} \quad \text{on} \quad \partial\Omega_{el} \end{aligned} \tag{2.137}$$

The solution of (2.137) gives rise to the conformal finite element formulation. Alternatively, we can employ a nonconformal finite element formulation based on the oversampling idea [10]. By using this approach, the boundary value problem (2.137) is solved on a larger domain Ω^+_{el} such that $\Omega_{el} \subset \Omega^+_{el}$.

2.3.2 Variational Multiscale Method

The variational multiscale method (VMS) [11] gives a general framework for approximating the coarse-scale solution u^c_i and simultaneously reconstructing some fine-scale features of the solution. VMS, like the homogenization method, is based on a direct sum decomposition of the displacements $u^\zeta_i(\boldsymbol{x})$ into the coarse-scale u^c_i and fine-scale correction $u^{(1)}_i$ displacements. However, unlike the homogenization method, VMS assumes the two fields to be of the same order, and a single coordinate system is used to describe the two fields

$$u^\zeta_i(\boldsymbol{x}) = u^c_i(\boldsymbol{x}) + u^{(1)}_i(\boldsymbol{x}) \tag{2.138}$$

Furthermore, an additive decomposition is assumed for the weighting function

$$w^\zeta_i(\boldsymbol{x}) = w^c_i(\boldsymbol{x}) + w^{(1)}_i(\boldsymbol{x}) \tag{2.139}$$

and $u^c_i(\boldsymbol{x})$, $w^c_i(\boldsymbol{x})$ and $u^{(1)}_i(\boldsymbol{x})$, $w^{(1)}_i(\boldsymbol{x})$ can be discretized on the coarse and fine grids, respectively, as

$$\begin{aligned} &u^c_i = N^c_{i\alpha} d^c_\alpha; \quad w^c_i = N^c_{i\alpha} c^c_\alpha \\ &u^{(1)}_i = N^f_{i\alpha} d^{(1)}_\alpha; \quad w^{(1)}_i = N^f_{i\alpha} c^{(1)}_\alpha \end{aligned} \tag{2.140}$$

A similar decomposition of the displacements is employed in a number of other related methods [55,56,57]. Inserting (2.138), (2.139), and (2.140) into the weak form (2.51) and requiring arbitrariness of c^c_α and $c^{(1)}_\alpha$ yields the discrete coarse- and fine-scale problems

$$\underbrace{\int_\Omega N^c_{(i\alpha,j)} L^\zeta_{ijkl} N^c_{(k\beta,l)} d\Omega}_{K^{cc}_{\alpha\beta}} \cdot d^c_\beta + \underbrace{\int_\Omega N^c_{(i\alpha,j)} L^\zeta_{ijkl} N^f_{(k\beta,l)} d\Omega}_{K^{cf}_{\alpha\beta}} \cdot d^{(1)}_\beta = \underbrace{\int_\Omega N^c_{i\alpha} b^\zeta_i \, d\Omega + \int_{\partial\Omega^t} N^c_{i\alpha} \bar{t}^\zeta_i \, d\Gamma}_{f^c_\alpha} \tag{2.141a}$$

$$\underbrace{\int_\Omega N^f_{(i\alpha,j)} L^\zeta_{ijkl} N^c_{(k\beta,l)} d\Omega}_{K^{fc}_{\alpha\beta}} \cdot d^c_\beta + \underbrace{\int_\Omega N^f_{(i\alpha,j)} L^\zeta_{ijkl} N^f_{(k\beta,l)} d\Omega}_{K^{ff}_{\alpha\beta}} \cdot d^{(1)}_\beta = \underbrace{\int_\Omega N^f_{i\alpha} b^\zeta_i \, d\Omega + \int_{\partial\Omega^t} N^f_{i\alpha} \bar{t}^\zeta_i \, d\Gamma}_{f^f_\alpha} \tag{2.141b}$$

Equation (2.141) represents a coupled coarse-fine-scale problem having a computational complexity comparable with a direct numerical simulation of the fine-scale problem. To reduce the computational cost, the fine-scale test and trial functions are assumed to vanish on the element boundaries

$$w_i^{(1)} = u_i^{(1)} = 0 \quad \text{on} \quad \partial\Omega_{el} \tag{2.142}$$

Consequently, we can solve a local problem on each element domain and express fine-scale nodal displacements in terms of the coarse-scale nodal displacements of the element. This allows condensing out fine-scale degrees of freedom and deriving a coarse-scale element stiffness matrix.

Remark 2.4 Conformal MsFEM and VMS with homogeneous boundary conditions on the element boundary yield similar results, provided that the body forces are treated the same in both methods, that is, body forces have to be either included in (2.137) or excluded from (2.141b).

The discontinuous enrichment method (DEM) [12] is another approach aimed at relaxing the constraint equation (2.142). In this approach, the continuity of the enrichment across element interfaces is enforced weakly by Lagrange multipliers.

2.3.3 Multiscale Enrichment Based on Partition of Unity

To allow for variation in the coarse-scale solution over the unit cell domain, multiscale enrichment based on the partition of unity method (MEPU) [13,14,52] removes the dependency of the fine-scale functions on the coarse-scale solution. This is accomplished by replacing $\varepsilon_{kl}^c(\boldsymbol{x})$ and $H_i^{kl}(\boldsymbol{y})$ in equation (2.27) with the independent set of degrees-of-freedom a_A^{kl} and the influence functions defined over the local supports $H_i^{kl}(\boldsymbol{x})N_A(\boldsymbol{x})$, respectively, where uppercase Roman subscripts denote the node numbers. The resulting solution approximation states

$$u_i^\zeta = u_i^c + H_i^{kl}(\boldsymbol{x})N_A(\boldsymbol{x})a_A^{kl} \tag{2.143}$$

where the summation convention is employed for the repeated indices.

The discrete system of equations is obtained using a standard Galerkin method. MEPU exploits the special structure of the influence functions by utilizing a two-scale integration scheme. Element stiffness matrix $K_{\alpha\beta}^c = \int_{\Omega_{el}} g_{\alpha\beta} d\Omega$ is integrated as

$$K_{\alpha\beta}^c = \sum_{e=1}^{N_{el}^c} \int_\square J g_{\alpha\beta} d\square = \sum_{e=1}^{N_{el}^c} \sum_{I=1}^{N_{\text{int}}^c} W_I J_I(\hat{\boldsymbol{x}}_I)\, g_{\alpha\beta}(\hat{\boldsymbol{x}}_I) = \sum_{e=1}^{N_{el}^c} \sum_{I=1}^{N_{\text{int}}^c} W_I J_I(\hat{\boldsymbol{x}}_I) \frac{1}{|\Theta|} \int_{\Theta_I} g_{\alpha\beta}(\hat{\boldsymbol{x}}_I, \boldsymbol{y})\, d\Theta \tag{2.144}$$

The two-scale integration scheme, schematically depicted in Figure 2.7, centers a unit cell at each of the coarse-scale Gauss points. The value of the integrand at each Gauss point is replaced by its integral over the unit cell domain normalized by the volume of the unit cell. For coarse-scale elements encompassing numerous unit cells, there is a significant cost savings as compared with using a direct numerical integration over all the unit cells contained in the coarse-scale element.

Compared with the mathematical homogenization theory, there is an overhead associated with (i) the additional degrees-of-freedom a_A^{kl}, which cannot be condensed out on the element level, and (ii) the need for integration over all the coarse-scale Gauss points as opposed to the single integration over the unit cell domain required by the homogenization theory for linear problems. Therefore, MEPU should be primarily used in the boundary layer regions, whereas the $O(1)$ homogenization should be employed elsewhere. The issue of transitioning between the two regions is discussed in [13,14].

2.4 Homogenization of Nonlinear Heterogeneous Media

2.4.1 Asymptotic Expansion for Nonlinear Problems

We will consider the most general form of nonlinear formulation that includes both geometric (large deformation) and material nonlinearity. In deriving a two-scale problem, the governing equations at the fine scale of interest will be stated in terms of the deformation gradient and the first Piola–Kirchhoff stress. The choice of the deformation gradient (and its conjugate stress measure) is convenient in dealing with unit cell boundary conditions. In the discretization phase of the two-scale problem, the finite element equations will be reformulated in terms of Cauchy stress for problems where constitutive equations at the coarse scale are more conveniently defined in terms of Cauchy stress.

The strong form of the boundary value problem is expressed in the undeformed composite domain Ω_X^ζ

$$\frac{\partial P_{ij}^\zeta}{\partial X_j} + B_i^\zeta = 0 \quad \text{on} \quad \Omega_X^\zeta \tag{2.145a}$$

$$F_{ik}^\zeta = \delta_{ik} + \frac{\partial u_i^\zeta}{\partial X_k} \tag{2.145b}$$

$$P_{ij}^\zeta N_j^\zeta = \bar{T}_i^\zeta \quad \text{on} \quad \partial\Omega_X^{t\zeta} \tag{2.145c}$$

$$u_i^\zeta = \bar{u}_i \quad \text{on} \quad \partial\Omega_X^{u\zeta} \tag{2.145d}$$

$$\partial\Omega_X^{t\zeta} \cup \partial\Omega_X^{u\zeta} = \partial\Omega_X^\zeta \quad \text{and} \quad \partial\Omega_X^{t\zeta} \cap \partial\Omega_X^{u\zeta} = 0 \tag{2.145e}$$

where $\boldsymbol{F}^\zeta$ denotes the deformation gradient; $\boldsymbol{P}^\zeta$ denotes the first Piola–Kirchhoff stress tensor; $\boldsymbol{P}^\zeta$ depends on the deformation gradient $\boldsymbol{F}^\zeta$, its complete history, and the constitutive equation on the fine scale; $\boldsymbol{B}^\zeta$ and $\bar{\boldsymbol{T}}^\zeta$ denote the body force and traction vectors, respectively; and $\boldsymbol{N}^\zeta$ denotes the unit normal to the boundary $\partial\Omega_X^\zeta$. Lowercase subscripts i and j denote spatial dimensions, and subscripts X and x refer to the initial and deformed configurations, respectively. Summation convention over repeated subscripts is employed, except for the subscripts X and x.

The undeformed coordinates are denoted by $\boldsymbol{X}$ for the coarse-scale domain Ω_X and $\boldsymbol{Y}$ for the unit cell Θ_Y domain. These two coordinates are related by $\boldsymbol{Y} \equiv \boldsymbol{X} - \hat{\boldsymbol{X}}/\zeta$ with $0 < \zeta \ll 1$

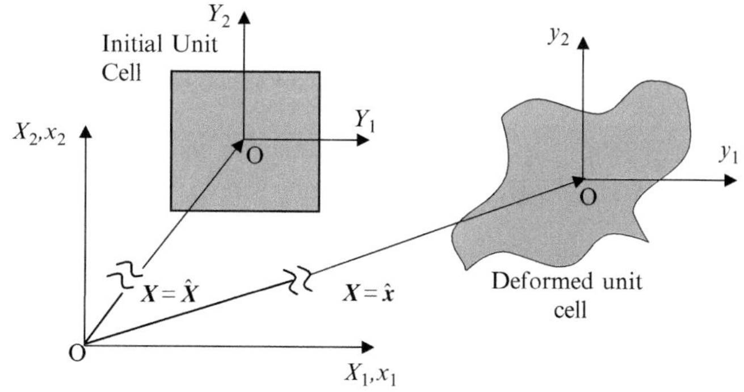

Figure 2.16 Coarse- and fine-scale coordinate systems

where $\hat{\boldsymbol{X}}$ denotes unit cell centroid coordinates. The initial composite domain Ω_X^ζ is defined as a product space, $\Omega_X \times \Theta_Y$. The coordinates in the deformed (or current) configuration are $\boldsymbol{x}$ and $\boldsymbol{y}$, which correspond to the deformed composite domain Ω_x^ζ and are defined by the product space of the coarse-scale domain Ω_x and the unit cell domain Θ_y. The dependence of various fields on the two-scale coordinates is denoted by $\boldsymbol{P}^\zeta(\boldsymbol{X})=\boldsymbol{P}(\boldsymbol{X},\boldsymbol{Y})$, $\boldsymbol{B}^\zeta(\boldsymbol{X})=\boldsymbol{B}(\boldsymbol{X},\boldsymbol{Y})$, $\overline{\boldsymbol{T}}^\zeta(\boldsymbol{X})=\overline{\boldsymbol{T}}(\boldsymbol{X},\boldsymbol{Y})$, and $\boldsymbol{u}^\zeta(\boldsymbol{X})=\boldsymbol{u}(\boldsymbol{X},\boldsymbol{Y})$.

Using an approach similar to that employed for linear problems, displacements are expanded as

$$u_i^\zeta(\boldsymbol{X}) = u_i^{(0)}(\boldsymbol{X}) + \zeta u_i^{(1)}(\boldsymbol{X},\boldsymbol{Y}) + \zeta^2 u_i^{(2)}(\boldsymbol{X},\boldsymbol{Y}) + O\left(\zeta^3\right) \tag{2.146}$$

where it is assumed that the size of the unit cell is infinitesimally small and therefore the leading-order displacement $u_i^{(0)}(\boldsymbol{X})$ is considered to be constant over the unit cell domain (see Remark 2.5). For large deformation problems, it is convenient to rewrite equation (2.146) as follows. Expanding $u_i^{(0)}(\boldsymbol{X})$ in a Taylor series around the unit cell centroid $\boldsymbol{X}=\hat{\boldsymbol{X}}$ as shown in Figure 2.16 gives

$$u_i^{(0)}(\boldsymbol{X}) = u_i^{(0)}(\hat{\boldsymbol{X}}) + \left.\frac{\partial u_i^{(0)}}{\partial X_j}\right|_{\hat{\boldsymbol{X}}} \left(X_j - \hat{X}_j\right) + \frac{1}{2}\left.\frac{\partial^2 u_i^{(0)}}{\partial X_j \partial X_k}\right|_{\hat{\boldsymbol{X}}} \left(X_j - \hat{X}_j\right)\left(X_k - \hat{X}_k\right) + \cdots \tag{2.147}$$

Since $X_j - \hat{X}_j = \zeta Y_j$, we can write

$$u_i^{(0)}(\boldsymbol{X}) = u_i^{(0)}\left(\hat{\boldsymbol{X}}\right) + \zeta \left.\frac{\partial u_i^{(0)}}{\partial X_j}\right|_{\hat{\boldsymbol{X}}} Y_j + \zeta^2 \frac{1}{2}\left.\frac{\partial^2 u_i^{(0)}}{\partial X_j \partial X_k}\right|_{\hat{\boldsymbol{X}}} Y_j Y_k + O\left(\zeta^3\right) \tag{2.148}$$

Similarly, higher order terms $u_i^{(n)}(\boldsymbol{X},\boldsymbol{Y})$, for $n \geq 1$ are expanded around $(\boldsymbol{X}=\hat{\boldsymbol{X}})$ as

$$u_i^{(n)}(\boldsymbol{X},\boldsymbol{Y}) = u_i^{(n)}(\hat{\boldsymbol{X}},\boldsymbol{Y}) + \zeta \left.\frac{\partial u_i^{(n)}}{\partial X_j}\right|_{\hat{\boldsymbol{X}}} Y_j + \zeta^2 \frac{1}{2}\left.\frac{\partial^2 u_i^{(n)}}{\partial X_j \partial X_k}\right|_{\hat{\boldsymbol{X}}} Y_j Y_k + O\left(\zeta^3\right) \tag{2.149}$$

Inserting (2.149) and (2.148) into (2.146) gives

$$u_i^{\zeta}(\boldsymbol{X}) = u_i(\hat{\boldsymbol{X}},\boldsymbol{Y}) = \hat{u}_i^{(0)}(\hat{\boldsymbol{X}}) + \zeta \hat{u}_i^{(1)}(\hat{\boldsymbol{X}},\boldsymbol{Y}) + \zeta^2 \hat{u}_i^{(2)}(\hat{\boldsymbol{X}},\boldsymbol{Y}) + O(\zeta^3) \tag{2.150}$$

where

$$\begin{aligned}
&\hat{u}_i^{(0)}(\hat{\boldsymbol{X}}) = u_i^{(0)}(\hat{\boldsymbol{X}}) \equiv u_i^c(\hat{\boldsymbol{X}}) \\
&\hat{u}_i^{(1)}(\hat{\boldsymbol{X}},\boldsymbol{Y}) = u_i^{(1)}(\hat{\boldsymbol{X}},\boldsymbol{Y}) + \left.\frac{\partial u_i^{(0)}}{\partial X_j}\right|_{\hat{\boldsymbol{X}}} Y_j \\
&\hat{u}_i^{(2)}(\hat{\boldsymbol{X}},\boldsymbol{Y}) = u_i^{(2)}(\hat{\boldsymbol{X}},\boldsymbol{Y}) + \left.\frac{\partial u_i^{(1)}}{\partial X_j}\right|_{\hat{\boldsymbol{X}}} Y_j + \frac{1}{2}\left.\frac{\partial^2 u_i^{(0)}}{\partial X_j \partial X_k}\right|_{\hat{\boldsymbol{X}}} Y_j Y_k
\end{aligned} \tag{2.151}$$

and $u_i^c(\hat{\boldsymbol{X}})$ denotes the rigid body motion of the unit cell centroid.

The spatial derivative of $f(\hat{\boldsymbol{X}},\boldsymbol{Y})$ is given by

$$\frac{\partial f^{\zeta}}{\partial X_i} = \frac{1}{\zeta}\frac{\partial f(\hat{\boldsymbol{X}},\boldsymbol{Y})}{\partial Y_i} \tag{2.152}$$

The displacement gradient components may be written as

$$\frac{\partial u_i^{\zeta}}{\partial X_k} = \frac{1}{\zeta}\frac{\partial u_i(\hat{\boldsymbol{X}},\boldsymbol{Y})}{\partial Y_k} = \frac{\partial \hat{u}_i^{(1)}(\hat{\boldsymbol{X}},\boldsymbol{Y})}{\partial Y_k} + \zeta \frac{\partial \hat{u}_i^{(2)}(\hat{\boldsymbol{X}},\boldsymbol{Y})}{\partial Y_k} + O\left(\zeta^2\right) \tag{2.153}$$

where

$$\begin{aligned}
&\frac{\partial \hat{u}_i^{(1)}(\hat{\boldsymbol{X}},\boldsymbol{Y})}{\partial Y_k} = \frac{\partial u_i^{(1)}(\hat{\boldsymbol{X}},\boldsymbol{Y})}{\partial Y_k} + \left.\frac{\partial u_i^{(0)}(\boldsymbol{X})}{\partial X_k}\right|_{\hat{\boldsymbol{X}}} \\
&\frac{\partial \hat{u}_i^{(2)}(\hat{\boldsymbol{X}},\boldsymbol{Y})}{\partial Y_k} = \frac{\partial u_i^{(2)}(\hat{\boldsymbol{X}},\boldsymbol{Y})}{\partial Y_k} + \left.\frac{\partial u_i^{(1)}(\boldsymbol{X},\boldsymbol{Y})}{\partial X_k}\right|_{\hat{\boldsymbol{X}}} + \left.\frac{\partial^2 u_i^{(1)}(\boldsymbol{X},\boldsymbol{Y})}{\partial X_j \partial Y_k}\right|_{\hat{\boldsymbol{X}}} Y_j + \left.\frac{\partial^2 u_i^{(0)}}{\partial X_j \partial X_k}\right|_{\hat{\boldsymbol{X}}} Y_j
\end{aligned} \tag{2.154}$$

The deformation gradient can be expressed as

$$F_{ik}^{\zeta} = \delta_{ik} + \frac{\partial u_i^{\zeta}}{\partial X_k} = F_{ik}^{(0)}(\hat{\boldsymbol{X}},\boldsymbol{Y}) + \zeta F_{ik}^{(1)}(\hat{\boldsymbol{X}},\boldsymbol{Y}) + O\left(\zeta^2\right) \tag{2.155}$$

where

$$\begin{aligned}
&F_{ik}^{(0)}(\hat{\boldsymbol{X}},\boldsymbol{Y}) = F_{ik}^c(\hat{\boldsymbol{X}}) + F_{ik}^*(\hat{\boldsymbol{X}},\boldsymbol{Y}) \equiv F_{ik}^f(\hat{\boldsymbol{X}},\boldsymbol{Y}) \\
&F_{ik}^c(\hat{\boldsymbol{X}}) = \delta_{ik} + \left.\frac{\partial u_i^c(\boldsymbol{X})}{\partial X_k}\right|_{\hat{\boldsymbol{X}}} \\
&F_{ik}^*(\hat{\boldsymbol{X}},\boldsymbol{Y}) = \frac{\partial u_i^{(1)}(\hat{\boldsymbol{X}},\boldsymbol{Y})}{\partial Y_k}
\end{aligned} \tag{2.156}$$

It can be seen that the coarse-scale deformation gradient F^c_{ik} is related to the average of the fine-scale deformation gradient $F^f_{ik}(\hat{\boldsymbol{X}},\boldsymbol{Y})$ over the unit cell domain

$$F^c_{ik}(\hat{\boldsymbol{X}})=\frac{1}{|\Theta_Y|}\int_{\Theta_Y} F^f_{ik}(\hat{\boldsymbol{X}},\boldsymbol{Y})\,d\Theta_Y \tag{2.157}$$

where, for the time being (see section 2.4.3 for other boundary conditions), we assume a $\boldsymbol{Y}$-periodicity of $u^{(1)}_i(\hat{\boldsymbol{X}},\boldsymbol{Y})$.

The first Piola–Kirchhoff stress $P_{ij}(\boldsymbol{F}^{\zeta})$ formulation is adopted due to its conjugacy with the deformation gradient. Expanding $P_{ij}(\boldsymbol{F}^{\zeta})$ around the fine-scale deformation gradient F^f_{ik} yields

$$P_{ij}(\boldsymbol{F}^{\zeta})=P_{ij}(\boldsymbol{F}^f)+\zeta\left.\frac{\partial P_{ij}}{\partial F^{\zeta}_{mn}}\right|_{\boldsymbol{F}^f}F^{(1)}_{mn}+O\left(\zeta^2\right)=P^{(0)}_{ij}(\boldsymbol{X},\boldsymbol{Y})+\zeta P^{(1)}_{ij}(\boldsymbol{X},\boldsymbol{Y})+O\left(\zeta^2\right) \tag{2.158}$$

Further expanding equation (2.158) in a Taylor series around the centroid $\boldsymbol{X}=\hat{\boldsymbol{X}}$ yields

$$\begin{aligned} P_{ij}(\boldsymbol{X},\boldsymbol{Y}) &= P^{(0)}_{ij}(\hat{\boldsymbol{X}},\boldsymbol{Y})+\left.\frac{\partial P^{(0)}_{ij}}{\partial X_k}\right|_{\hat{\boldsymbol{X}}}\left(X_k-\hat{X}_k\right)+\zeta P^{(1)}_{ij}(\hat{\boldsymbol{X}},\boldsymbol{Y})+\zeta\left.\frac{\partial P^{(1)}_{ij}}{\partial X_k}\right|_{\hat{\boldsymbol{X}}}(X_k-\hat{X}_k)\cdots \\ &= P^f_{ij}(\hat{\boldsymbol{X}},\boldsymbol{Y})+\zeta\left(\left.\frac{\partial P^{(0)}_{ij}}{\partial X_k}\right|_{\hat{\boldsymbol{X}}}Y_k+P^{(1)}_{ij}(\hat{\boldsymbol{X}},\boldsymbol{Y})\right)+O\left(\zeta^2\right) \end{aligned} \tag{2.159}$$

where $P^f_{ij}(\hat{\boldsymbol{X}},\boldsymbol{Y})\equiv P^{(0)}_{ij}(\hat{\boldsymbol{X}},\boldsymbol{Y})$ is the fine-scale first Piola–Kirchhoff stress. Inserting (2.159) into equilibrium equation (2.145)a yields the two-scale equilibrium equations

$$\frac{\partial P^f_{ij}(\hat{\boldsymbol{X}},\boldsymbol{Y})}{\partial Y_j}=0 \tag{2.160a}$$

$$\left.\frac{\partial P^f_{ij}}{\partial X_j}\right|_{\hat{\boldsymbol{X}}}+\frac{\partial P^{(1)}_{ij}(\hat{\boldsymbol{X}},\boldsymbol{Y})}{\partial Y_j}+B_i(\hat{\boldsymbol{X}},\boldsymbol{Y})=0 \tag{2.160b}$$

Integrating (2.160b) over the unit cell domain and using a $\boldsymbol{Y}$-periodicity of stress gives the coarse-scale equilibrium equation

$$\left.\frac{\partial P^c_{ij}}{\partial X_j}\right|_{\hat{\boldsymbol{X}}}+B^c_i(\hat{\boldsymbol{X}})=0 \tag{2.161}$$

where

$$P^c_{ij}(\hat{\boldsymbol{X}})=\frac{1}{|\Theta_Y|}\int_{\Theta_Y} P^f_{ij}(\hat{\boldsymbol{X}},\boldsymbol{Y})\,d\Theta_Y \tag{2.162a}$$

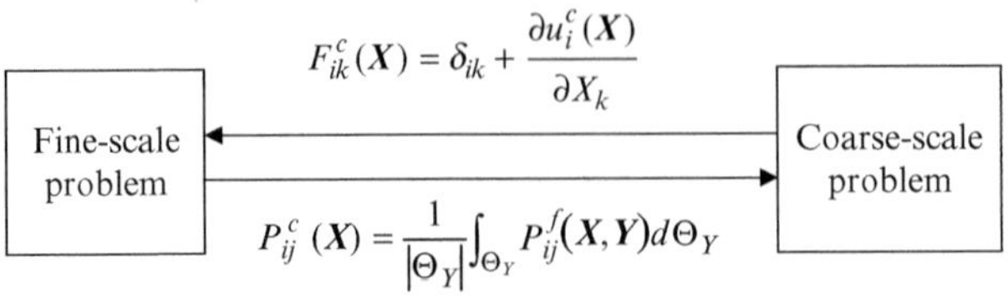

Figure 2.17 Information transfer between the coarse- and fine-scale problems

$$B^c_i(\hat{X}) = \frac{1}{|\Theta_Y|}\int_{\Theta_Y} B_i(\hat{X},Y)d\Theta_Y \tag{2.162b}$$

The two-scale problem, which consists of the unit cell equilibrium equation (2.160a) subjected to periodic (or other) boundary conditions and the coarse-scale equilibrium equation, is two-way coupled. The link between the two scales is schematically illustrated in Figure 2.17. The unit cell problem is driven by the coarse-scale deformation gradient $F^c_{ik}(\hat{X})$. Once the unit cell boundary value problem is solved, the coarse-scale stress $P^c_{ij}(\hat{X})$ is computed using (2.162a).

Remark 2.5 For nonlinear problems, we a priori assume that the leading order displacement is not a function of fine-scale coordinates. This has been shown to be a unique solution for linear elliptic problems. For a certain class of nonlinear problems, such as problems involving softening and localization as well as neutronic diffusion or radiative transport problems [58], we have to consider large oscillations in the leading term

$$u^\zeta_i(X) = u^{(0)}_i(X,Y) + \zeta u^{(1)}_i(X,Y) + O\left(\zeta^2\right) \tag{2.163}$$

2.4.2 *Formulation of the Coarse-Scale Problem*

We will consider a Galerkin approximation where the coarse-scale trial u^c_i and test w^c_i functions are discretized using the same approximation as for linear problems (2.76).

Remark 2.6 In deriving the coarse-scale weak form, we will make use of an argument of unit cell infinitesimality by which the unit cell centroid $\hat{X}$ can be positioned at an arbitrary point X. Therefore, the centroid $\hat{X}$ can be replaced by an arbitrary point X.

Writing the weak form of (2.161) and employing a Galerkin approximation (2.76) gives an incremental discrete coarse-scale problem, which states

$$\begin{aligned}
&\text{Given } {}_{n+1}B^c_i \text{ and } {}_{n+1}\bar{T}^c_i\text{, find } {}_{n+1}\Delta d^c_\alpha \text{ such that:}\\
&{}_{n+1}r^c_\alpha\left({}_{n+1}\Delta d^c\right) \equiv {}_{n+1}f^{\text{int}}_\alpha - {}_{n+1}f^{\text{ext}}_\alpha = 0\\
&{}_{n+1}d^c_\alpha = {}_{n+1}\bar{d}^c_\alpha \quad \text{on} \quad \partial\Omega^u_X\\
&n \leftarrow n+1\text{, Go to the next load increment}
\end{aligned} \tag{2.164}$$

where ${}_{n+1}r_{\alpha}^{c}$ and ${}_{n+1}\Delta d_{\alpha}^{c}$ are the coarse-scale residual and displacement increments in the load increment, respectively, and

$$
\begin{aligned}
{}_{n+1}f_{\alpha}^{\text{int}} &= \int_{\Omega_X} \frac{\partial N_{i\alpha}^{c}}{\partial X_j}\, {}_{n+1}P_{ij}^{c}\, d\Omega_X \\
{}_{n+1}f_{\alpha}^{\text{ext}} &= \int_{\Omega_X} N_{i\alpha}^{c}\, {}_{n+1}B_i^{c}\, d\Omega_X + \int_{\partial\Omega_X^t} N_{i\alpha}^{c}\, {}_{n+1}\overline{T}_i^{c}\, d\Gamma_X
\end{aligned}
\tag{2.165}
$$

where ${}_{n+1}P_{ij}^{c}$ and ${}_{n+1}B_i^{c}$ are given in (2.162). The coarse-scale traction ${}_{n+1}\overline{T}_i^{c}$ is defined similarly to (2.60)

$$
\overline{T}_i^{c}(\boldsymbol{X}) = \frac{1}{|\partial\omega_Y|}\int_{\partial\omega_Y} \overline{T}_i^{\zeta}(\boldsymbol{X},\boldsymbol{Y})\, ds_Y \tag{2.166}
$$

where $\partial\omega_Y$ is a characteristic surface over which prescribed traction is averaged and ${}_{n+1}f_{\alpha}^{\text{int}}$ and ${}_{n+1}f_{\alpha}^{\text{ext}}$ are the internal and external forces, respectively. The coarse-scale discrete problem (2.164) is solved using a Newton iteration at each increment. This requires updating the coarse-scale stress at each iteration $i+1$ as

$$
{}_{n+1}^{i+1}P_{ij}^{c}(\boldsymbol{X}) = \frac{1}{|\Theta_Y|}\int_{\Theta_Y} {}_{n+1}^{i+1}P_{ij}^{f}(\boldsymbol{X},\boldsymbol{Y})\, d\Theta_Y \tag{2.167}
$$

and computing the coarse-scale residual ${}_{n+1}^{i+1}r_{\alpha}^{c}$ at each iteration. The consistent linearization of the coarse-scale residual is described by

$$
K_{\alpha\beta}^{c} = \frac{\partial r_{\alpha}^{c}}{\partial d_{\beta}^{c}} = \frac{\partial f_{\alpha}^{\text{int}}}{\partial d_{\beta}^{c}} = \int_{\Omega_X} \frac{\partial N_{i\alpha}^{c}}{\partial X_j} \frac{\partial P_{ij}^{c}}{\partial F_{kl}^{c}} \frac{\partial N_{k\alpha}^{c}}{\partial X_l}\, d\Omega_X \tag{2.168}
$$

where $\partial P_{ij}^{c}/\partial F_{kl}^{c} = L_{ijkl}^{c}$ is a coarse-scale consistent tangent. Since there is no closed form relation between P_{ij}^{c} and F_{kl}^{c}, the coarse-scale consistent tangent is evaluated using finite differences. In this approach, ${}_{n+1}^{i+1}F_{kl}^{c}$ is perturbed by δF_{kl}^{c}, and a perturbation in fine-scale stress δP_{ij}^{f} is defined as

$$
\delta P_{ij}^{f} = P_{ij}^{f}\left({}_{n+1}^{i+1}\boldsymbol{F}^{c} + \delta\boldsymbol{F}^{c}\right) - P_{ij}^{f}\left({}_{n+1}^{i+1}\boldsymbol{F}^{c}\right) \tag{2.169}
$$

and is computed by solving nine unit cell problems. The coarse-scale consistent tangent is then calculated as

$$
L_{ijkl}^{c} = \frac{1}{\delta F_{kl}^{c}|\Theta_Y|}\int_{\Theta_Y} \delta P_{ij}^{f}\, d\Theta_Y \tag{2.170}
$$

If the fine-scale constitutive equations are expressed in terms of Cauchy stress, it is convenient to formulate the coarse-scale internal and external forces in the deformed configuration rather than in the reference configuration (here we omit the left superscripts) as

$$f_{\alpha}^{\text{int}} = \int_{\Omega_x} \frac{\partial N_{i\alpha}^c}{\partial x_j} \sigma_{ij}^c d\Omega_x$$
$$f_{\alpha}^{\text{ext}} = \int_{\Omega_x} N_{i\alpha}^c b_i^c d\Omega_x + \int_{\partial\Omega_x^t} N_{i\alpha}^c \bar{t}_i^c d\Gamma_x \tag{2.171}$$

where

$$b_i^c = B_i^c / J^c; \qquad \bar{t}_i^c d\Gamma_x = \bar{T}_i^c d\Gamma_X \tag{2.172}$$

$$J^f \sigma_{ji}^f = F_{jk}^f P_{ik}^f \tag{2.173a}$$

$$J^c \sigma_{ji}^c = F_{jk}^c P_{ik}^c \tag{2.173b}$$

and J^f and J^c are the determinants of F_{ij}^c and F_{jk}^f, respectively.

We now focus on deriving a closed form expression for the coarse-scale Cauchy stress σ_{ij}^c. Inserting (2.173a) into (2.162a) and recalling $d\Theta_y = J^f d\Theta_Y$, we have

$$P_{ik}^c = \frac{1}{|\Theta_Y|} \int_{\Theta_y} \left(F_{km}^f\right)^{-1} \sigma_{mi}^f d\Theta_y \tag{2.174}$$

Inserting (2.174) into (2.173b) and defining the volume of the *deformed coarse-scale configuration* as $|\Theta_y^c| = J^c |\Theta_Y|$, the overall Cauchy stress can be expressed as

$$\sigma_{ji}^c = \frac{1}{|\Theta_y^c|} \int_{\Theta_y} \Delta F_{jm}^{-1} \sigma_{mi}^f d\Theta_y \tag{2.175}$$

where $\Delta F_{jm}^{-1} = F_{jk}^c \left(F_{km}^f\right)^{-1}$ maps the unit cell deformed configuration Θ_y into the coarse-scale deformed configuration Θ_y^c as illustrated in Figure 2.18. Note that for periodic unit cell boundary conditions, $|\Theta_y^c| = |\Theta_y|$.

We now show that for periodic boundary conditions, equation (2.175) reduces to

$$\sigma_{ij}^c = \frac{1}{|\Theta_y|} \int_{\Theta_y} \sigma_{ij}^f d\Theta_y \tag{2.176}$$

The two deformed coordinates are related by $\boldsymbol{y} \equiv \boldsymbol{x}/\zeta$, which is similar to the way in which undeformed coordinates are related. The inverse of the deformation gradient $\left(F_{km}^{\zeta}\right)^{-1}$ is then given as

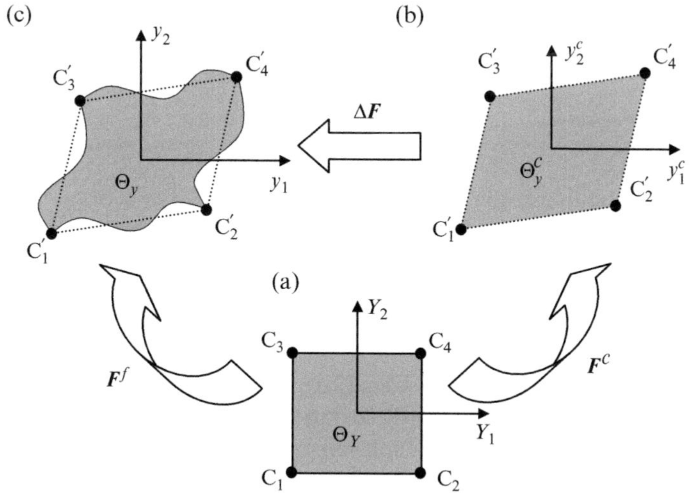

Figure 2.18 Unit cell configurations: (a) initial; (b) coarse-scale deformed (intermediate); and (c) fine-scale deformed (final)

$$\left(F_{ij}^{\zeta}\right)^{-1} = \delta_{ij} - \left(\frac{\partial}{\partial x_j} + \frac{1}{\zeta}\frac{\partial}{\partial y_j}\right) u_i^{\zeta}(\boldsymbol{x}) \tag{2.177}$$

Inserting the asymptotic expansion of displacements expressed in the deformed coordinates

$$u_i^{\zeta}(\boldsymbol{x}) = u_i^c(\boldsymbol{x}) + \zeta u_i^{(1)}(\boldsymbol{x},\boldsymbol{y}) + \zeta^2 u_i^{(2)}(\boldsymbol{x},\boldsymbol{y}) + O\left(\zeta^3\right) \tag{2.178}$$

into (2.177) and writing $\left(F_{ij}^{\zeta}\right)^{-1} = \left(F_{ij}^{f}\right)^{-1} + O(\zeta)$ yields

$$\left(F_{ij}^{f}\right)^{-1} = \left(F_{ij}^{c}\right)^{-1} - \frac{\partial u_i^{(1)}}{\partial y_j} \tag{2.179}$$

where $\left(F_{ij}^{c}\right)^{-1} = \delta_{ij} - \dfrac{\partial u_i^c}{\partial x_j}$. Inserting (2.179) into (2.175) yields

$$\begin{aligned}\sigma_{ji}^c &= \frac{1}{\left|\Theta_y\right|} F_{jk}^c \int_{\Theta_y} \left(\left(F_{km}^c\right)^{-1} - \frac{\partial u_k^{(1)}}{\partial y_m} \right) \sigma_{mi}^f d\Theta_y \\ &= \frac{1}{\left|\Theta_y\right|} \int_{\Theta_y} \sigma_{ji}^f d\Theta_y - \frac{1}{\left|\Theta_y\right|} F_{jk}^c \int_{\Theta_y} \frac{\partial u_k^{(1)}}{\partial y_m} \sigma_{mi}^f d\Theta_y\end{aligned} \tag{2.180}$$

Integrating by parts the second term in (2.180) and assuming solution periodicity and fine-scale equilibrium $\sigma_{mi,y_m}^f = 0$ yields (2.176).

Note that (2.176) is consistent with (2.162a) for periodic or essential unit cell boundary conditions only (see also [59]).

2.4.3 Formulation of the Unit Cell Problem

Consider an asymptotic expansion of displacements expressed in terms of the coarse-scale deformation gradient (2.150)

$$u_i^{\zeta} = u_i^c(\hat{\boldsymbol{X}}) + \zeta\left[\left(F_{ij}^c(\hat{\boldsymbol{X}}) - \delta_{ij}\right)Y_j + u_i^{(1)}(\hat{\boldsymbol{X}},\boldsymbol{Y})\right] + O\left(\zeta^2\right) \tag{2.181}$$

where the coarse-scale displacements $u_i^c(\hat{\boldsymbol{X}})$ in (2.181) represent a rigid body translation of the unit cell. Just as in the linear problem considered in section 2.2, it is necessary to constrain the rigid body translation of the unit cell. The unit cell problem will be solved for the leading $O(\zeta)$ order term, which is the translation-free fine-scale displacement u_i^f

$$u_i^f(\hat{\boldsymbol{X}},\boldsymbol{Y}) = \left(F_{ij}^c(\hat{\boldsymbol{X}}) - \delta_{ij}\right)Y_j + u_i^{(1)}(\hat{\boldsymbol{X}},\boldsymbol{Y}) \tag{2.182}$$

Proceeding to the discretization, there are two possibilities. We can discretize either (i) the perturbation $u_i^{(1)}$ or (ii) the rigid body free fine-scale solution u_i^f

$$u_i^{(1)}(\hat{\boldsymbol{X}},\boldsymbol{Y}) = N_{i\beta}^f(\boldsymbol{Y})d_\beta^{(1)}(\hat{\boldsymbol{X}})$$
$$u_i^f(\hat{\boldsymbol{X}},\boldsymbol{Y}) = N_{i\beta}^f(\boldsymbol{Y})d_\beta^f(\hat{\boldsymbol{X}})$$

where $N_{i\beta}^f(\boldsymbol{Y})$ denotes the fine-scale shape functions, which are typically defined in the parent element domain, and d_β^f and $d_\beta^{(1)}$ are the corresponding displacement degrees of freedom, respectively. The unit problem can be solved directly either for $d_\beta^f(\hat{\boldsymbol{X}})$ or for $d_\beta^{(1)}(\hat{\boldsymbol{X}})$. The corresponding two approaches are referred to hereafter as the *total* approach [60,61,62] and the *correction-based* [35] approach, respectively. The correction-based approach is more convenient for linear problems where the unit cell problem is stated in terms of fine-scale influence functions for periodic meshes. For nonlinear large deformation problems, the total discretization approach considered hereafter is more transparent, even though the imposition of boundary conditions is somewhat more cumbersome since it requires accounting for coarse-scale deformation.

We consider a Galerkin approximation of the test function

$$w_i^{(1)}(\hat{\boldsymbol{X}},\boldsymbol{Y}) = N_{i\eta}^f(\boldsymbol{Y})c_\eta^{(1)}(\hat{\boldsymbol{X}}) = N_{i\eta}^f(\boldsymbol{Y})T_{\eta\beta}\tilde{c}_\beta^{(1)}(\hat{\boldsymbol{X}}) \tag{2.183}$$

where $T_{\eta\beta}$ is a linear transformation operator that relates independent degrees-of-freedom $\tilde{c}_\beta^{(1)}(\hat{\boldsymbol{X}})$ and $c_\eta^{(1)}(\hat{\boldsymbol{X}})$ [see equation (2.82)].

Writing the weak form of (2.160a) and employing a Galerkin approximation for the test functions (2.183) gives the discrete unit cell problem at quadrature point $\hat{\boldsymbol{X}}$, which states

$$
\begin{aligned}
&\text{Given } {}^{i+1}_{n+1}F^c_{ij}(\hat{\boldsymbol{X}}) \text{ and prior converged solution } {}_n\boldsymbol{d}^f,\ {}_n\boldsymbol{P}^f \\
&\text{Find } {}^{i+1}_{n+1}\Delta\boldsymbol{d}^f \text{ such that:} \\
&{}^{i+1}_{n+1}r^f_\beta\left({}^{i+1}_{n+1}\Delta\boldsymbol{d}^f\right) = \int_{\Theta_Y} T_{\eta\beta}\frac{\partial N^f_{i\eta}}{\partial Y_j}\,{}^{i+1}_{n+1}P^f_{ij}\,d\Theta_Y = 0 \\
&\text{Subjected to unit cell boundary conditions}
\end{aligned}
\tag{2.184}
$$

The unit cell boundary conditions will be discussed next. The left superscript in (2.184) denotes the iteration count of the coarse-scale problem, and ${}_n\boldsymbol{d}^f$ and ${}_n\boldsymbol{P}^f$ denote the displacement and stress solutions that were previously converged (at load step n).

If the constitutive equations at the fine scale are expressed in terms of the Cauchy stress σ^f_{ij}, it is convenient to restate the unit cell problem as

$$
\begin{aligned}
&\text{Given } {}^{i+1}_{n+1}F^c_{ij}(\hat{\boldsymbol{X}}) \text{ and prior converged solution } {}_n\boldsymbol{d}^f,\ {}_n\boldsymbol{\sigma}^f \\
&\text{Find } {}^{i+1}_{n+1}\Delta\boldsymbol{d}^f \text{ such that:} \\
&{}^{i+1}_{n+1}r^f_\beta\left({}^{i+1}_{n+1}\Delta\boldsymbol{d}^f\right) = \int_{\Theta_y} T_{\eta\beta}\frac{\partial N^f_{i\eta}}{\partial y_j}\,{}^{i+1}_{n+1}\sigma^f_{ij}\,d\Theta_y = 0 \\
&\text{Subjected to periodic (or weakly) boundary conditions}
\end{aligned}
\tag{2.185}
$$

where we exploited (2.173a).

We now consider the unit cell boundary conditions. Figure 2.19(a) and (b) show the initial and the deformed unit cell configuration, respectively. The dotted line in Figure 2.19(b) depicts the deformed shape of the unit cell due to coarse-scale deformation only, whereas the solid line in Figure 2.19(b) depicts the final configuration. Figure 2.19(a) depicts the two nodes M and S on the opposite faces of the unit cell, with M and S being the master and slave nodes, respectively. The fine-scale displacements at the two nodes are given by

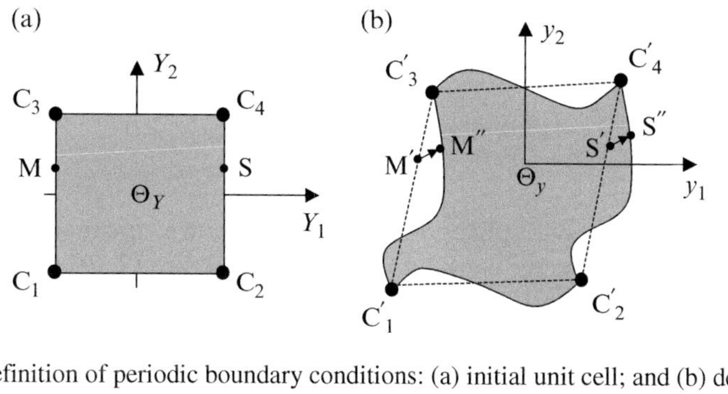

Figure 2.19 Definition of periodic boundary conditions: (a) initial unit cell; and (b) deformed unit cell

$$
\begin{aligned}
u_i^f\left(\hat{\boldsymbol{X}},\boldsymbol{Y}^{\mathrm{M}}\right) &= \left(F_{ij}^c(\hat{\boldsymbol{X}})-\delta_{ij}\right)Y_j^M + u_i^{(1)}\left(\hat{\boldsymbol{X}},\boldsymbol{Y}^{\mathrm{M}}\right) \\
u_i^f\left(\hat{\boldsymbol{X}},\boldsymbol{Y}^S\right) &= \left(F_{ij}^c(\hat{\boldsymbol{X}})-\delta_{ij}\right)Y_j^S + u_i^{(1)}\left(\hat{\boldsymbol{X}},\boldsymbol{Y}^S\right)
\end{aligned}
\tag{2.186}
$$

where $\boldsymbol{Y}^M$ and $\boldsymbol{Y}^S$ represent the coordinates of master and slave nodes on the unit cell boundary, respectively.

At the unit cell vertices $\partial\Theta_Y^{vert}$, the correction to the coarse-scale displacement field is assumed to vanish, that is, $u_i^{(1)}(\hat{\boldsymbol{X}},\boldsymbol{Y}^{vert})=0$. For periodic unit cells subjected to the uniform coarse-scale deformation gradient, $u_i^{(1)}(\hat{\boldsymbol{X}},\boldsymbol{Y})$ is assumed to be a periodic function, that is, $u_i^{(1)}(\hat{\boldsymbol{X}},\boldsymbol{Y}^M)=u_i^{(1)}(\hat{\boldsymbol{X}},\boldsymbol{Y}^S)$. Thus, subtracting the two equations in (2.186) yields the periodic boundary condition

$$
u_i^f(\hat{\boldsymbol{X}},\boldsymbol{Y}^{\mathrm{M}})-u_i^f(\hat{\boldsymbol{X}},\boldsymbol{Y}^S)=(F_{ij}^c(\hat{\boldsymbol{X}})-\delta_{ij})(Y_j^{\mathrm{M}}-Y_j^{\mathrm{S}})
\tag{2.187}
$$

We now consider alternative boundary conditions. In deriving the unit cell governing equation (2.160a), the following periodicity condition has been employed

$$
\int_{\Theta_Y}\frac{\partial u_i^{(1)}(\hat{\boldsymbol{X}},\boldsymbol{Y})}{\partial Y_j}d\Theta_Y=0
\tag{2.188}
$$

Applying Green's theorem and using (2.182), the above reduces to

$$
\int_{\partial\Theta_Y}\left(u_i^f(\hat{\boldsymbol{X}},\boldsymbol{Y})-\left(F_{ik}^c(\hat{\boldsymbol{X}})-\delta_{ik}\right)Y_k\right)N_j^\Theta d\gamma_Y=0
\tag{2.189}
$$

where $\boldsymbol{N}^\Theta$ is the unit normal to the unit cell boundary $\partial\Theta_Y$.

As an alternative to using equation (2.187) and equation (2.189), an essential boundary condition similar to (2.86) for small deformation problems can be considered

$$
u_i^f(\hat{\boldsymbol{X}},\boldsymbol{Y})-\left(F_{ik}^c(\hat{\boldsymbol{X}})-\delta_{ik}\right)Y_k=0
\tag{2.190}
$$

The essential boundary condition (2.190) can be enforced in the weak form as

$$
\int_{\partial\Theta_Y}\left(u_i^f(\hat{\boldsymbol{X}},\boldsymbol{Y})-\left(F_{ik}^c(\hat{\boldsymbol{X}})-\delta_{ik}\right)Y_k\right)\lambda_i d\gamma_Y=0
\tag{2.191}
$$

where λ_i is a Lagrange multiplier representing unknown tractions on $\partial\Theta_Y$. If we choose $\lambda_i=P_{ij}^cN_j^\Theta$ with P_{ij}^c being constant over Θ_Y and require (2.191) to be satisfied for arbitrary P_{ij}^c, this yields (2.189). We will refer to equation (2.189) as a natural boundary condition.

The essential (2.190) and natural (2.189) boundary conditions can be combined to define the so-called *mixed boundary condition* as follows

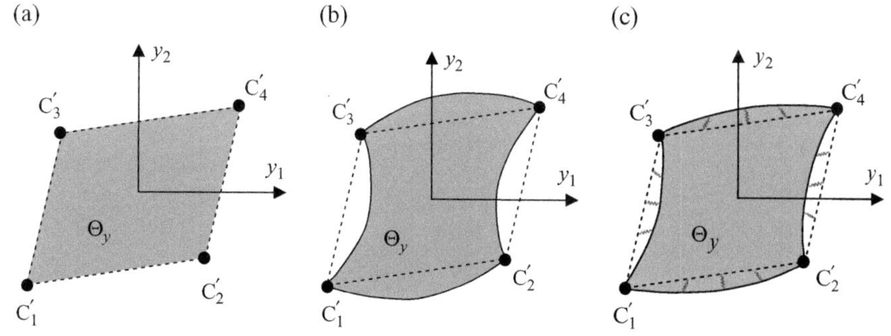

Figure 2.20 Definition of (a) essential, (b) natural, and (c) mixed boundary conditions

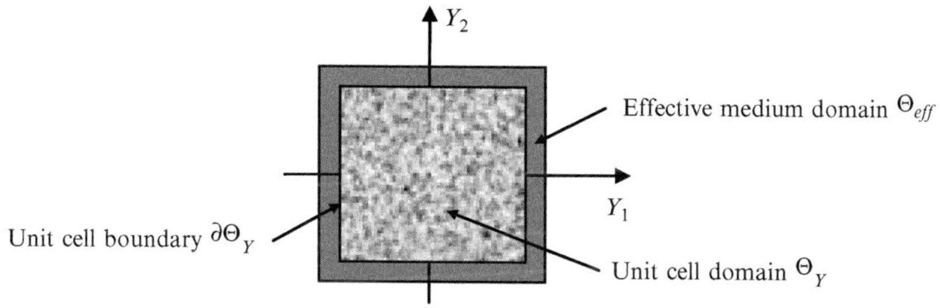

Figure 2.21 Unit cell and effective medium

$$\int_{\partial\Theta_Y}\left(u_i^f(\hat{\boldsymbol{X}},\boldsymbol{Y})-(F_{ik}^c(\hat{\boldsymbol{X}})-\delta_{ik})Y_k\right)N_j^{\Theta}d\gamma_Y=0 \tag{2.192a}$$

$$\left|\left(u_i^f(\hat{\boldsymbol{X}},\boldsymbol{Y})-(F_{ik}^c(\hat{\boldsymbol{X}})-\delta_{ik})Y_k\right)N_j^{\Theta}\right|\leq Tol \tag{2.192b}$$

Note that the mixed boundary condition (2.192) is more restrictive than the natural boundary condition (2.189) but is more compliant than the essential boundary condition (2.190), which is enforced up to a tolerance of *Tol*. The three types of unit cell boundary conditions are schematically illustrated in Figure 2.20. The mixed boundary condition can be implemented in several ways. Perhaps the simplest way is by defining double nodes on the boundary and placing linear or nonlinear springs between them as illustrated in Figure 2.20(c).

We will discuss the definition of the spring stiffness in section 2.4.4. Note that in the case of a very stiff spring, the mixed boundary condition (2.192) coincides with the essential boundary condition (2.190); for an infinitely compliant spring, it reduces to the natural boundary condition (2.189). Alternatively, condition (2.192b) can be enforced by placing a thin layer of effective medium around a unit cell to mimic its interaction with the surrounding medium as shown in Figure 2.21.

2.4.4 Example Problems

To study the accuracy of the nonlinear two-scale formulation, consider two test problems: a perforated plate and a plate with a center hole.

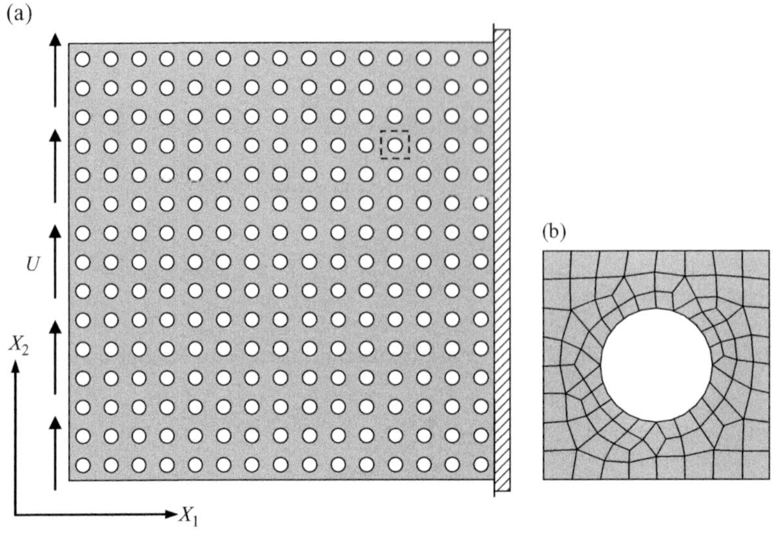

Figure 2.22 Geometry, boundary conditions and unit cell discretization: (a) macro geometry and boundary conditions; and (b) mesh of the unit cell. Reproduced with permission from [60], © 2008 John Wiley & Sons

2.4.4.1 Perforated Plate

The geometric configuration of the perforated plate is shown in Figure 2.22(a). The width W and the length L of the plate are both 60 mm. Circular holes with a radius of 1 mm are uniformly distributed in a rectangular arrangement. The right side of the plate is fixed. The top and bottom sides are free, and a displacement equal to 18 mm is applied at the left side. Plane strain condition is assumed. The material is assumed to be hyper-elastic. The strain energy is expressed in neo-Hookean form with an initial shear modulus of 160 MPa and a bulk modulus equal to 4000 MPa. The mass density of the material is $1.14\times10^3\,\mathrm{kg\,m^{-3}}$. For multiscale analysis, the macroscale mesh contains 25 four-node reduced integration quadrilateral elements; all elements are 12×12 mm. The size of the unit cell is 4×4 mm as shown in Figure 2.22(b). The reference solution is obtained by direct numerical simulation (DNS) of the plate with a sufficient number of elements.

Figure 2.23 compares the homogenization method with a reference solution in the subdomain denoted in Figure 2.22(a).

2.4.4.2 Plate with a Center Hole

Geometry and boundary conditions of the plate with a center hole are schematically shown in Figure 2.24. The width W and the length L of the plate are both 100 mm. A circular hole of radius $r = 1$ mm is placed at the center. Plane strain condition is assumed and the material properties are the same as in the first example. A displacement of 15 mm is applied at the left side. The right side of the plate is constrained in the horizontal direction. For the homogenization method, the size of the unit cell is 4×4 mm and the corresponding coarse mesh element is

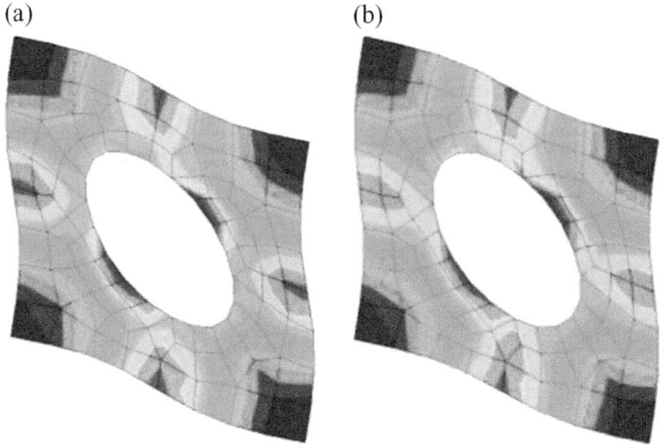

Figure 2.23 Unit cell deformation and von Mises stress obtained from (a) homogenization and (b) direct numerical simulation. Reproduced with permission from [60], © 2008 John Wiley & Sons

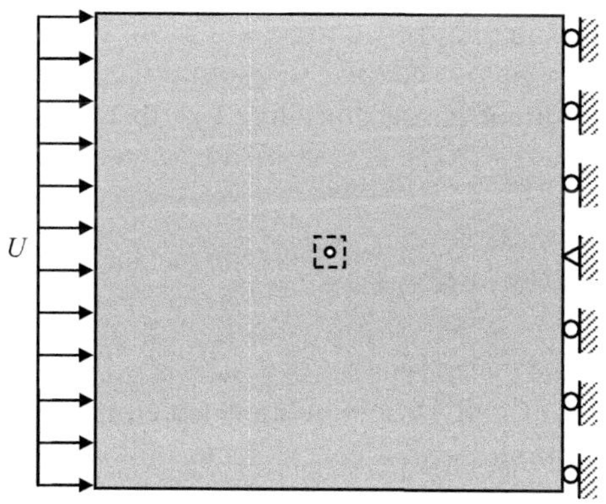

Figure 2.24 Plate with a hole in the center. Reproduced with permission from [60], © 2008 John Wiley & Sons

located at the center of the plate as shown by the dashed lines in Figure 2.24. The unit cell mesh is shown in Figure 2.22(b). Figure 2.25 depicts the deformed configuration and von Mises stress in the unit cell obtained by four different methods: (a) DNS; (b) homogenization with essential boundary conditions; (c) homogenization with natural boundary conditions; and (d) homogenization with mixed boundary conditions.

For the mixed boundary condition, the linear spring stiffness has been a priori determined as follows. At every node (one at a time) in the neighboring subdomain to the unit cell (without a hole), we prescribe a small displacement in the direction normal to the boundary and calculate the reaction force at that node. Assuming linear spring stiffness, the reaction divided by the prescribed displacement is equal to the spring stiffness at that node. Similarly, a nonlinear spring stiffness (not considered here) can be a priori defined.

It can be seen that homogenization with an essential boundary is surprisingly inaccurate because the unit cell is subjected to a uniform coarse-scale field on its boundary. The natural

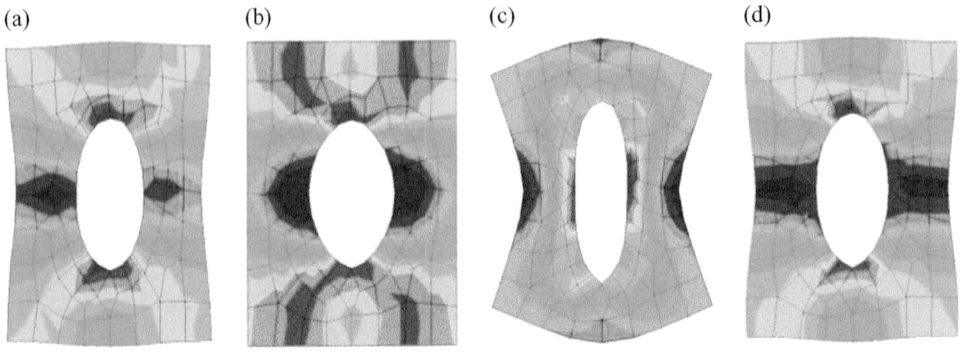

Figure 2.25 Comparison of the unit cell deformation and von Mises stress as obtained with (a) direct numerical simulation, (b) homogenization with essential boundary conditions, (c) homogenization with natural boundary conditions, and (d) homogenization with mixed boundary conditions. Reproduced with permission from [60], © 2008 John Wiley & Sons

boundary condition provides qualitatively correct kinematics but is too flexible. Finally, the mixed boundary condition seems to be able to capture the overall deformation correctly. The periodic boundary condition is not considered in this case since the deformation is clearly not periodic.

2.5 Higher Order Homogenization

2.5.1 *Introduction*

Higher order homogenization and closely related generalized continuum theories were developed to account for the existence of coarse-scale solution gradients within a unit cell. These generalized continuum theories can be classified into two main categories [63]: higher grade continua and higher order continua. Higher grade continuum is characterized by higher order spatial derivatives of the displacement field [64,65,66,67,68,69,70], whereas higher order continuum is endowed with additional degrees of freedom independent of the usual translational degrees of freedom, ranging from three rotational degrees of freedom in the Cosserat or polar continuum [71] to 12 degrees of freedom in the micromorphic continuum [72] and more in the so-called multiscale micromorphic continuum [73]. For background information, we refer to the review article by Green and Naghdi [74] and the monograph by Eringen [75].

The link between the generalized continua, on the one hand, and homogenization theories, on the other, was shown to exist by several investigators [76,77,78,79,80,81]. The unit cell in the first-order homogenization theories [23,27,28,30,32,41,82,83,84] experiences (or is subjected to) a constant macroscopic deformation gradient independent of its size. This is an anomaly that becomes unacceptable as the coarse-scale strain gradients become sufficiently large. Higher order theories [85,86], on the other hand, are equipped with a mechanism for subjecting the unit cell to "true" coarse-scale deformation. For example, in second-order theories, the coarse-scale deformation gradient is idealized to vary linearly over the unit cell domain.

2.5.2 Formulation

Consider the governing equations in (2.145). To account for the existence of high coarse-scale gradients, the coarse- and fine-scale coordinates $\boldsymbol{X}$ and $\boldsymbol{Y}$ will be related by $\boldsymbol{Y} \equiv \boldsymbol{X} - \hat{\boldsymbol{X}}/\zeta^{\alpha}$ with $0<\zeta \ll 1$ and $\alpha=1$ in the $O(1)$ theory.

Displacements are expanded similarly to (2.146), as

$$u_i^{\zeta}(\boldsymbol{X}) = u_i^{(0)}(\boldsymbol{X}) + \zeta^{\alpha} u_i^{(1)}(\boldsymbol{X},\boldsymbol{Y}) + \zeta^{2\alpha} u_i^{(2)}(\boldsymbol{X},\boldsymbol{Y}) + O(\zeta^{3\alpha}) \tag{2.193}$$

and the leading order term $u_i^{(0)}(\boldsymbol{X})$ is expanded in a Taylor series around the unit cell centroid $\boldsymbol{X}=\hat{\boldsymbol{X}}$

$$\begin{aligned} u_i^{(0)}(\boldsymbol{X}) = u_i^{(0)}(\hat{\boldsymbol{X}}) &+ \left.\frac{\partial u_i^0}{\partial X_j}\right|_{\hat{\boldsymbol{X}}} (X_j - \hat{X}_j) + \frac{1}{2}\left.\frac{\partial^2 u_i^{(0)}}{\partial X_j \partial X_k}\right|_{\hat{\boldsymbol{X}}} (X_j - \hat{X}_j)(X_k - \hat{X}_k) \\ &+ \frac{1}{6}\left.\frac{\partial^3 u_i^{(0)}}{\partial X_j \partial X_k \partial X_m}\right|_{\hat{\boldsymbol{X}}} (X_j - \hat{X}_j)(X_k - \hat{X}_k)(X_m - \hat{X}_m) + \cdots \end{aligned} \tag{2.194}$$

In the $O(1)$ theory considered in the previous section, it is assumed that the coarse-scale displacement $u_i^{(0)}(\boldsymbol{X})$ and its various order derivatives are $O(1)$ functions. Here we consider the existence of high coarse-scale gradients

$$\frac{\partial^{n+1} u_i^{(0)}}{\partial X_j \ldots \partial X_k} = O(\zeta^{-n\alpha}) \quad \text{for } n = 0,1,2\ldots \text{ and } 0<\alpha\leq 1 \tag{2.195}$$

such that

$$\begin{aligned} &\left.\frac{\partial u_i^{(0)}}{\partial X_j}\right|_{\hat{\boldsymbol{X}}} (X_j - \hat{X}_j) = O(\zeta^{\alpha}) \\ &\left.\frac{\partial^S u_i^{(0)}}{\partial X_j \partial X_k \ldots \partial X_t}\right|_{\hat{\boldsymbol{X}}} (X_j - \hat{X}_j)(X_k - \hat{X}_k)\ldots(X_t - \hat{X}_t) = O(\zeta^{\alpha}) \end{aligned} \tag{2.196}$$

From (2.195) and (2.196), it follows that $X_j - \hat{X}_j = \zeta^{\alpha} Y_j$ and thus $u_i^{(0)}(\boldsymbol{X})$ can be written as

$$u_i^{(0)}(\hat{\boldsymbol{X}},\boldsymbol{Y}) = u_i^c(\hat{\boldsymbol{X}}) + \zeta^{\alpha} \left(\begin{aligned} &\left.\frac{\partial u_i^{(0)}}{\partial X_j}\right|_{\hat{\boldsymbol{X}}} Y_j + \frac{1}{2}\zeta^{\alpha} \left.\frac{\partial^2 u_i^{(0)}}{\partial X_j \partial X_k}\right|_{\hat{\boldsymbol{X}}} Y_j Y_k \\ &+ \frac{1}{6}\zeta^{2\alpha} \left.\frac{\partial^3 u_i^{(0)}}{\partial X_j \partial X_k \partial X_m}\right|_{\hat{\boldsymbol{X}}} Y_j Y_k Y_m + \cdots \end{aligned} \right) \tag{2.197}$$

where $u_i^{(0)}(\hat{\boldsymbol{X}}) \equiv u_i^c(\hat{\boldsymbol{X}})$ represents the rigid body motion of the unit cell. Similarly, the higher order terms $u_i^{(n)}(\boldsymbol{X},\boldsymbol{Y})$, for $n \geq 1$ in (2.193) are expanded around $(\boldsymbol{X}=\hat{\boldsymbol{X}})$ as

$$u_i^{(n)}(\boldsymbol{X},\boldsymbol{Y}) = u_i^{(n)}(\hat{\boldsymbol{X}},\boldsymbol{Y}) + \zeta^{\alpha} \left.\frac{\partial u_i^{(n)}}{\partial X_j}\right|_{\hat{X}} Y_j + \zeta^{2\alpha} \frac{1}{2} \left.\frac{\partial^2 u_i^{(n)}}{\partial X_j \partial X_k}\right|_{\hat{X}} Y_j Y_k$$
$$+ \zeta^{3\alpha} \frac{1}{6} \left.\frac{\partial^3 u_i^{(n)}}{\partial X_j \partial X_k \partial X_m}\right|_{\hat{X}} Y_j Y_k Y_m \cdots \tag{2.198}$$

Inserting (2.197) and (2.198) into (2.193) gives

$$u_i^{\zeta}(\boldsymbol{X}) = u_i(\hat{\boldsymbol{X}},\boldsymbol{Y}) = u_i^c(\hat{\boldsymbol{X}}) + \zeta^{\alpha} \hat{u}_i^{(1)}(\hat{\boldsymbol{X}},\boldsymbol{Y}) + O\left(\zeta^{2\alpha}\right) \tag{2.199}$$

where

$$\hat{u}_i^{(1)}(\hat{\boldsymbol{X}},\boldsymbol{Y}) = u_i^{(1)}(\hat{\boldsymbol{X}},\boldsymbol{Y}) + \left.\frac{\partial u_i^{(0)}}{\partial X_j}\right|_{\hat{X}} Y_j + \frac{1}{2}\zeta^{\alpha} \left.\frac{\partial^2 u_i^{(0)}}{\partial X_j \partial X_k}\right|_{\hat{X}} Y_j Y_k$$
$$+ \frac{1}{6}\zeta^{2\alpha} \left.\frac{\partial^3 u_i^{(0)}}{\partial X_j \partial X_k \partial X_m}\right|_{\hat{X}} Y_j Y_k Y_m + \cdots \tag{2.200}$$

Recall that the spatial derivative is defined as

$$\frac{\partial f^{\zeta}(\boldsymbol{X})}{\partial X_i} = \frac{1}{\zeta^{\alpha}} \frac{\partial f(\hat{\boldsymbol{X}},\boldsymbol{Y})}{\partial Y_i} \tag{2.201}$$

The deformation gradient can be expressed as

$$F_{ik}^{\zeta} = \delta_{ik} + \frac{\partial u_i^{\zeta}}{\partial X_k} = \delta_{ik} + \frac{1}{\zeta^{\alpha}} \frac{\partial u_i(\hat{\boldsymbol{X}},\boldsymbol{Y})}{\partial Y_k} = F_{ik}^{(0)}(\hat{\boldsymbol{X}},\boldsymbol{Y}) + \zeta^{\alpha} F_{ik}^{(1)}(\hat{\boldsymbol{X}},\boldsymbol{Y}) + O\left(\zeta^{2\alpha}\right) \tag{2.202}$$

where

$$F_{ik}^{(0)}(\hat{\boldsymbol{X}},\boldsymbol{Y}) = F_{ik}^{c}(\hat{\boldsymbol{X}},\boldsymbol{Y}) + F_{ik}^{*}(\hat{\boldsymbol{X}},\boldsymbol{Y}) \equiv F_{ik}^{f}(\hat{\boldsymbol{X}},\boldsymbol{Y}) \tag{2.203}$$

$$F_{ik}^{c}(\hat{\boldsymbol{X}},\boldsymbol{Y}) = \delta_{ik} + \left.\frac{\partial u_i^{(0)}}{\partial X_k}\right|_{\hat{X}} + \zeta^{\alpha} \left.\frac{\partial^2 u_i^{(0)}}{\partial X_j \partial X_k}\right|_{\hat{X}} Y_j + \zeta^{2\alpha} \frac{1}{2} \left.\frac{\partial^3 u_i^{(0)}}{\partial X_j \partial X_m \partial X_k}\right|_{\hat{X}} Y_j Y_m \cdots \tag{2.204}$$

$$F_{ik}^{*}(\hat{\boldsymbol{X}},\boldsymbol{Y}) = \frac{\partial u_i^{(1)}(\hat{\boldsymbol{X}},\boldsymbol{Y})}{\partial Y_k} \tag{2.205}$$

$$F_{ik}^{(1)}(\hat{\boldsymbol{X}},\boldsymbol{Y}) = \frac{\partial u_i^{(2)}}{\partial Y_k} + \left.\frac{\partial u_i^{(1)}}{\partial X_k}\right|_{\hat{X}} \tag{2.206}$$

Following the derivation of $O(1)$ theory, the first Piola–Kirchhoff stress $P_{ij}(\boldsymbol{F}^{\zeta})$ formulation is adopted due to its conjugacy with the deformation gradient. Expanding $P_{ij}(\boldsymbol{F}^{\zeta})$ around the leading order deformation gradient $F_{ik}^{(0)}$ yields

$$P_{ij}\left(\boldsymbol{F}^{\zeta}\right)=P_{ij}\left(\boldsymbol{F}^{f}\right)+\zeta\left.\frac{\partial P_{ij}}{\partial F_{mn}^{\zeta}}\right|_{\boldsymbol{F}^{f}}F_{mn}^{(1)}+O\left(\zeta^{2}\right)=P_{ij}^{f}+\zeta P_{ij}^{(1)}+O\left(\zeta^{2}\right) \qquad (2.207)$$

To derive the fine- and coarse-scale weak forms, we adopt the variational approach outlined for linear problems in section 2.2.2. Using this approach, the weak form is first stated on the composite domain Ω_X^{ζ}. Premultiplying (2.145a) by arbitrary test function $w_i^{\zeta}(\boldsymbol{X})$ and integrating by parts over the composite domain yields

$$\int_{\Omega_X^{\zeta}} w_{i,j}^{\zeta}P_{ij}^{\zeta}\, d\Omega_X^{\zeta}=\int_{\Omega_X^{\zeta}} w_i^{\zeta}B_i^{\zeta}\, d\Omega_X^{\zeta}+\int_{\partial\Omega_X^{t\zeta}} w_i^{\zeta}\bar{T}_i^{\zeta}\, d\Gamma_X^{\zeta} \qquad (2.208)$$

where the space of test or weight function is defined in (2.52).

Following (2.197) and (2.199), the test function $w_i^{\zeta}(\boldsymbol{X})$ is decomposed as

$$w_i^{\zeta}(\boldsymbol{X})=w_i(\hat{\boldsymbol{X}},\boldsymbol{Y})=\hat{w}_i^{(0)}(\hat{\boldsymbol{X}})+\zeta^{\alpha}\hat{w}_i^{(1)}(\hat{\boldsymbol{X}},\boldsymbol{Y})+O\left(\zeta^{2\alpha}\right) \qquad (2.209)$$

where

$$\begin{aligned}
&\hat{w}_i^{(0)}(\hat{\boldsymbol{X}})=w_i^{c}(\hat{\boldsymbol{X}})\\
&\hat{w}_i^{(1)}(\hat{\boldsymbol{X}},\boldsymbol{Y})=w_i^{(1)}(\hat{\boldsymbol{X}},\boldsymbol{Y})+\left.\frac{\partial w_i^{(0)}}{\partial X_j}\right|_{\hat{\boldsymbol{X}}}Y_j+\frac{1}{2}\zeta^{\alpha}\left.\frac{\partial^2 w_i^{(0)}}{\partial X_j\partial X_k}\right|_{\hat{\boldsymbol{X}}}Y_jY_k\\
&\qquad+\frac{1}{6}\zeta^{2\alpha}\left.\frac{\partial^3 w_i^{(0)}}{\partial X_j\partial X_k\partial X_m}\right|_{\hat{\boldsymbol{X}}}Y_jY_kY_m+\cdots
\end{aligned} \qquad (2.210)$$

The coarse-scale test function $w_i^{(0)}$ gradients satisfy (2.195) and (2.196), and $w_{i,j}^{\zeta}$ is given by

$$w_{i,j}^{\zeta}=\frac{\partial w_i^{(1)}}{\partial Y_j}+\frac{\partial w_i^{(0)}}{\partial X_j}+\zeta^{\alpha}\frac{\partial^2 w_i^{(0)}}{\partial X_j\partial X_k}Y_k+\frac{1}{2}\zeta^{2\alpha}\frac{\partial^3 w_i^{(0)}}{\partial X_j\partial X_k\partial X_m}Y_kY_m+\ldots \qquad (2.211)$$

In the following derivation of the coarse-scale weak form, we will again make the argument of unit cell infinitesimality (see also Remark 2.6) by substituting $\hat{\boldsymbol{X}}\rightarrow\boldsymbol{X}$.

The leading order test $\boldsymbol{w}^{(0)}(\boldsymbol{X})\in W_{\Omega_X}$ is required to be $C^n(\Omega_X)$ continuous, defined as

$$W_{\Omega_X}=\left\{\boldsymbol{w}^{(0)}(\boldsymbol{X}) \text{ defined in } \Omega_X,\ \boldsymbol{w}^{(0)}(\boldsymbol{X})\in C^n(\Omega_X),\ \boldsymbol{w}^{(0)}=\boldsymbol{0} \text{ on } \partial\Omega_X^{u}\right\} \qquad (2.212)$$

and $\boldsymbol{w}^{(1)}(\boldsymbol{Y})\in W_{\Theta_Y}$ is required to be periodic C^0 continuous function on Θ_Y, defined as

$$W_{\Theta_Y}=\left\{\boldsymbol{w}^{(1)}(\boldsymbol{Y}) \text{ defined in } \Theta_Y,\ C^0(\Theta_Y),\ \text{periodic}\right\} \qquad (2.213)$$

Inserting the asymptotic expansions of stress (2.207) and the test functions (2.209), (2.210), and (2.211) into the weak form on the composite domain yields

$$\int_{\Omega_X^\zeta}\left(\begin{array}{l}\dfrac{\partial w_i^{(1)}}{\partial Y_j}+\dfrac{\partial w_i^{(0)}}{\partial X_j}+\zeta^\alpha\dfrac{\partial^2 w_i^{(0)}}{\partial X_j\partial X_k}Y_k\\ +\dfrac{1}{2}\zeta^{2\alpha}\dfrac{\partial^3 w_i^{(0)}}{\partial X_j\partial X_k\partial X_m}Y_kY_m+\cdots\end{array}\right)\left(P_{ij}^f+\zeta^\alpha P_{ij}^{(1)}\right)d\Omega_X^\zeta \qquad \forall \boldsymbol{w}^{(1)}\in W_{\Theta_Y},\boldsymbol{w}^{(0)}\in W_{\Omega_X}$$
$$=\int_{\Omega_X^\zeta}\left(w_i^{(0)}+\zeta^\alpha\hat{w}_i^{(1)}\right)B_i^\zeta d\Omega_X^\zeta+\int_{\partial\Omega_X^{t\zeta}}\left(w_i^{(0)}+\zeta^\alpha\hat{w}_i^{(1)}\right)\bar{T}_i^\zeta d\Gamma_X^\zeta+O\left(\zeta^{2\alpha}\right)$$

(2.214)

The integral of the two-scale function over the composite domain Ω_X^ζ is defined in the usual way as

$$\lim_{\zeta^\alpha\to0^+}\int_{\Omega_X^\zeta}\Psi(\boldsymbol{X},\boldsymbol{Y})d\Omega_X^\zeta=\lim_{\zeta^\alpha\to0^+}\int_{\Omega_X}\left(\frac{1}{|\Theta_Y|}\int_{\Theta_Y}\Psi(\boldsymbol{X},\boldsymbol{Y})d\Theta_Y\right)d\Omega_X \tag{2.215}$$

The two-scale integration scheme, which is similar to (2.56), will be employed for the integration of prescribed tractions and displacements over the boundary of the composite domain

$$\lim_{\zeta^\alpha\to0^+}\int_{\partial\Omega_X^\zeta}\Psi(\boldsymbol{X},\boldsymbol{Y})\,d\Gamma^\zeta=\lim_{\zeta^\alpha\to0^+}\int_{\partial\Omega_X}\left(\frac{1}{|\partial\omega|}\int_{\partial\omega}\Psi(\boldsymbol{X},\boldsymbol{Y})\,ds\right)d\Gamma_X \tag{2.216}$$

The leading order weak form of the unit cell problem is obtained by taking $w_i^{(0)}=0$ in the weak form (2.214) and utilizing (2.215), which yields

$$\int_{\Theta_Y}\frac{\partial w_i^{(1)}}{\partial Y_j}P_{ij}^f\;d\Theta_Y=0\quad\forall \boldsymbol{w}^{(1)}\in W_{\Theta_Y} \tag{2.217}$$

subjected to $F_{ik}^c(\boldsymbol{X},\boldsymbol{Y})$ in (2.204) and periodic boundary conditions on $u_i^{(1)}$. Note that in $O(1)$ homogenization, $F_{ik}^c(\boldsymbol{X},\boldsymbol{Y})$ is constant over the unit cell domain, whereas in higher order homogenization it contains higher order derivatives of $u_i^{(0)}$.

The leading order coarse-scale weak form is obtained by inserting $w_i^{(1)}=0$ into (2.214), which yields

$$\int_{\Omega_X}\frac{\partial w_i^{(0)}}{\partial X_k}P_{ik}^c d\Omega_X+\sum_{S=1}^{N-1}\frac{\zeta^{S\alpha}}{S!}\int_{\Omega_X}\frac{\partial^{S+1}w_i^{(0)}}{\partial X_k\partial X_j\ldots\partial X_t}Q_{ikj\ldots t}^{(S)}d\Omega_X=\int_{\partial\Omega_X^t}w_i^{(0)}\bar{T}_i^c d\Gamma_X$$
$$+\int_{\Omega_X}w_i^{(0)}B_i^c d\Omega_X+\sum_{S=1}^{N-1}\frac{\zeta^{S\alpha}}{S!}\left(\int_{\partial\Omega_X^t}\frac{\partial^S w_i^{(0)}}{\partial X_j\ldots\partial X_t}\bar{T}_{ij\ldots t}^{(S)}d\Gamma_X^t+\int_{\Omega_X}\frac{\partial^S w_i^{(0)}}{\partial X_j\ldots\partial X_t}B_{ij\ldots t}^{(S)}d\Omega_X\right)$$
$$\forall \boldsymbol{w}^{(0)}\in W_{\Omega_X}$$
$$u_i^c=\bar{u}_i^c\quad\text{on}\quad\partial\Omega_X^u \tag{2.218}$$

where

$$P_{ij}^c = \frac{1}{|\Theta_Y|}\int_{\Theta_Y} P_{ij}^f d\Theta_Y;\quad Q_{ikj}^{(1)} = \frac{1}{|\Theta_Y|}\int_{\Theta_Y} P_{ik}^f Y_j d\Theta_Y;\quad Q_{ikj\ldots t}^{(S)} = \frac{1}{|\Theta_Y|}\int_{\Theta_Y} P_{ik}^f Y_j \ldots Y_t d\Theta_Y \tag{2.219}$$

$$B_i^c = \frac{1}{|\Theta_Y|}\int_{\Theta_Y} B_i^\zeta d\Theta_Y;\quad \overline{T}_i^c = \frac{1}{|\partial\omega|}\int_{\partial\omega} \overline{T}_i^\zeta ds;\quad \overline{u}_i^c = \frac{1}{|\partial\omega|}\int_{\partial\omega} \overline{u}_i^\zeta ds \tag{2.220}$$

$$B_{ij\ldots t}^{(S)} = \frac{1}{|\Theta_Y|}\int_{\Theta_Y} B_i^\zeta \cdot Y_j \ldots Y_t d\Theta_Y;\quad \overline{T}_{ij\ldots t}^{(S)} = \frac{1}{|\partial\omega|}\int_{\partial\omega} \overline{T}_i^\zeta \cdot Y_j \ldots Y_t ds \tag{2.221}$$

B_i^c and $\overline{T}_i^c$ are the classical coarse-scale body forces and tractions; $B_{ij\ldots t}^{(S)}$ and $\overline{T}_{ij\ldots t}^{(S)}$ are the prescribed higher order moments and tractions.

Equation (2.218) defines the *higher grade continuum* boundary value problem [87,88]. Note that this formulation requires C^{N-1} continuity, while the *second grade continuum* requires C^1 continuity. The *higher order continuum* formulation can be constructed by defining the coarse-scale derivatives of test and trial function to be independent fields and then requiring the two to be equal in the weak form. While such a formulation alleviates computational difficulties arising from the C^{N-1} continuity, it introduces additional degrees of freedom and requires a mixed or multi-field formulation [87,89].

The strong form of the higher order/grade continua can be obtained by appropriate integration by parts of the weak form (2.218).

Remark 2.7 The last two terms in (2.218), which involve higher order moments and tractions, are of order $O(\zeta^\alpha)$ and therefore do not contribute to the leading order coarse-scale weak form. This is a consequence of the assumption that the rigid body motion of the unit cell $u_i^{(0)}(\hat{\boldsymbol{X}})$ in the asymptotic expansion (2.194) is of order $O(\zeta^{-\alpha})$ larger the remaining terms in the expansion. If all the terms in the asymptotic expansion were assumed to be of the same order, then the higher order moments and tractions would appear in the leading order coarse-scale weak form (2.218).

2.6 Multiple-Scale Homogenization

We now outline the formalism of the multiple-scale homogenization of the boundary problem described by equation (2.145). For simplicity, we consider the following linear elasticity problem

$$\sigma_{ij}^\zeta(\boldsymbol{x}) = L_{ijkl}^\zeta(\boldsymbol{x}) - \varepsilon_{kl}^\zeta(\boldsymbol{x}) \quad \boldsymbol{x}\in\Omega^\zeta \tag{2.222}$$

In this section, the left superscript in the brackets (I) denotes the scale, with $I=0$ denoting the coarsest scale, $n_{sc}-1$ denoting the finest scale, and n_{sc} denoting the total number of scales. Let ${}^{(I)}\boldsymbol{x}$ be the position vector at scale I related to the position vector at scale I-1 as

$${}^{(1)}\boldsymbol{x} \equiv {}^{(I-1)}\boldsymbol{x}/\zeta \quad \text{for } I=1,\ldots,n_{sc}-1 \tag{2.223}$$

Note that in general, various scaling parameters can be introduced for various scales. The volume of the unit cells at a scale I is denoted by ${}^{(I)}\Theta$.

Here we focus on a periodic theory by which any locally periodic function depends on position vectors at multiple scales

$$\phi^{\zeta}(\boldsymbol{x}) \equiv \phi\left({}^{(0)}\boldsymbol{x},\ldots,{}^{(n_{sc}-1)}\boldsymbol{x}\right) \tag{2.224}$$

The spatial derivative is given by

$$\phi^{\zeta}_{,i}(\boldsymbol{x}) = \sum_{I=0}^{n_{sc}-1} \zeta^{-I} \frac{\partial}{\partial\, {}^{(I)}x_i} \phi\left({}^{(0)}\boldsymbol{x},\ldots,{}^{(n_{sc}-1)}\boldsymbol{x}\right) \tag{2.225}$$

The asymptotic expansion of the displacement field can be expressed as

$$\begin{aligned} u_i\left({}^{(0)}\boldsymbol{x},\ldots,{}^{(n_{sc}-1)}\boldsymbol{x}\right) &= u_i^{(0)}\left({}^{(0)}\boldsymbol{x}\right) + \zeta u_i^{(1)}\left({}^{(0)}\boldsymbol{x},{}^{(1)}\boldsymbol{x}\right) + \zeta^2 u_i^{(2)}\left({}^{(0)}\boldsymbol{x},{}^{(1)}\boldsymbol{x},{}^{(2)}\boldsymbol{x}\right) \\ &+ \ldots + \zeta^{n_{sc}-1} u_i^{(n_{sc}-1)}\left({}^{(0)}\boldsymbol{x},\, {}^{(1)}\boldsymbol{x},\ldots,{}^{(n_{sc}-1)}\boldsymbol{x}\right) \end{aligned} \tag{2.226}$$

As in the two-scale theory, the right superscript in the brackets denotes the order of terms in the asymptotic expansion. Various terms in the asymptotic expansion (2.226) are associated with various scales, such that $u_i^{(I)} = {}^{(I)}u_i$. In (2.226) it is assumed that (i) the leading order term $u_i^{(0)}\left({}^{(0)}\boldsymbol{x}\right)$ is only a function of the coarsest scale coordinate, (ii) the leading order correction is a function of ${}^{(0)}\boldsymbol{x}, {}^{(1)}\boldsymbol{x}$, and (iii) the finer scale contributions depend on finer scale coordinates. This can be proved for linear problems (see Problem 2.7).

Based on the definition of the spatial derivative (2.225), the asymptotic expansion of the strain field is

$$\varepsilon^{\zeta}_{ij}(\boldsymbol{x}) = \varepsilon_{ij}\left({}^{(0)}\boldsymbol{x},\ldots,{}^{(n_{sc}-1)}\boldsymbol{x}\right) = \varepsilon^{f}_{ij}\left({}^{(0)}\boldsymbol{x},\ldots,{}^{(n_{sc}-1)}\boldsymbol{x}\right) + O(\zeta) \tag{2.227}$$

where the leading order term is given by

$$\varepsilon^{f}_{ij}\left({}^{(0)}\boldsymbol{x},\ldots,{}^{(n_{sc}-1)}\boldsymbol{x}\right) = \sum_{I=0}^{n_{sc}-1} u^{(I)}_{\left(i,{}^{(I)}x_j\right)}\left({}^{(0)}\boldsymbol{x},\ldots,{}^{(I)}\boldsymbol{x}\right) \tag{2.228}$$

The average strain at scale I-1 is obtained by averaging the strain at scale I

$${}^{(I-1)}\varepsilon_{ij}\left({}^{(0)}\boldsymbol{x},\ldots,{}^{(I-1)}\boldsymbol{x}\right) \equiv \frac{1}{\left|{}^{(I)}\Theta\right|} \int_{{}^{(I)}\Theta} {}^{(I)}\varepsilon_{ij}\left({}^{(0)}\boldsymbol{x},\ldots,{}^{(I)}\boldsymbol{x}\right) d\Theta \quad \text{for } I=1,\ldots,n_{sc}-1 \tag{2.229}$$

with the strain at the finest scale (n_{sc}-1) being the leading order (or fine-scale) strain

$${}^{(n_{sc}-1)}\varepsilon_{ij}\left({}^{(0)}\boldsymbol{x},\ldots,{}^{(n_{sc}-1)}\boldsymbol{x}\right) \equiv \varepsilon^{f}_{ij}\left({}^{(0)}\boldsymbol{x},\ldots,{}^{(n_{sc}-1)}\boldsymbol{x}\right) \tag{2.230}$$

With the consideration of periodicity at each scale, the relation between the average strains at two subsequent scales is given as

$$^{(I)}\varepsilon_{ij}\left({}^{(0)}\boldsymbol{x},\ldots,{}^{(1)}\boldsymbol{x}\right)={}^{(I-1)}\varepsilon_{ij}\left({}^{(0)}\boldsymbol{x},\ldots,{}^{(I-1)}\boldsymbol{x}\right)+u^{(I)}_{\left(i,{}^{(I)}x_j\right)}\left({}^{(0)}\boldsymbol{x},\ldots,{}^{(I)}\boldsymbol{x}\right)\ \text{for}\ I=1,\ldots,n_{sc}-1 \qquad (2.231)$$

The asymptotic expansion of the stress field is given by

$$\sigma^{\zeta}_{ij}(\boldsymbol{x})=\sigma_{ij}\left({}^{(0)}\boldsymbol{x},\ldots,{}^{(n_{sc}-1)}\boldsymbol{x}\right)=\sigma^{f}_{ij}\left({}^{(0)}\boldsymbol{x},\ldots,{}^{(n_{sc}-1)}\boldsymbol{x}\right)+O(\zeta) \qquad (2.232)$$

where the leading order (or fine-scale) stress is obtained from the constitutive equation (2.222), which yields

$$\sigma^{f}_{ij}\left({}^{(0)}\boldsymbol{x},\ldots,{}^{(n_{sc}-1)}\boldsymbol{x}\right)=L_{ijkl}\left({}^{(n_{sc}-1)}\boldsymbol{x}\right)\varepsilon^{f}_{kl}\left({}^{(0)}\boldsymbol{x},\ldots,{}^{(n_{sc}-1)}\boldsymbol{x}\right) \qquad (2.233)$$

The average stress at each scale is similar to the stress seen in (2.231) and is shown as

$$^{(I-1)}\sigma_{ij}\left({}^{(0)}\boldsymbol{x},\ldots,{}^{(I-1)}\boldsymbol{x}\right)\equiv\frac{1}{\left|{}^{(I)}\Theta\right|}\int_{{}^{(I)}\Theta}{}^{(I)}\sigma_{ij}\left({}^{(0)}\boldsymbol{x},\ldots,{}^{(I)}\boldsymbol{x}\right)d\Theta\quad \text{for}\ I=1,\ldots,n_{sc}-1 \qquad (2.234)$$

with the stress at the finest scale $(n_{sc}-1)$ of interest being the fine-scale stress

$$^{(n_{sc}-1)}\sigma_{ij}\left({}^{(0)}\boldsymbol{x},\ldots,{}^{(n_{sc}-1)}\boldsymbol{x}\right)\equiv\sigma^{f}_{ij}\left({}^{(0)}\boldsymbol{x},\ldots,{}^{(n_{sc}-1)}\boldsymbol{x}\right) \qquad (2.235)$$

The equilibrium equations at each scale follow from the equilibrium equation (2.145a) and spatial differentiation rule (2.225)

$$\begin{aligned}
&O\left(\zeta^{-I}\right):\quad {}^{(I)}\sigma_{ij,{}^{(I)}x_j}\left({}^{(0)}\boldsymbol{x},\ldots,{}^{(I)}\boldsymbol{x}\right)=0\quad \text{for}\ I=1,\ldots,n_{sc}-1\\
&O\left(\zeta^{0}\right):\quad {}^{(0)}\sigma_{ij,{}^{(0)}x_j}\left({}^{(0)}\boldsymbol{x}\right)+{}^{(0)}b_i=0
\end{aligned} \qquad (2.236)$$

where ${}^{(0)}b_i$ is a body force at the coarsest scale.

In the case of a three-scale formulation, which is a special case of the N-scale formulation, it is common to introduce the following notation: ${}^{(0)}\boldsymbol{x}=\boldsymbol{x}$, ${}^{(1)}\boldsymbol{x}=\boldsymbol{y}$, and ${}^{(2)}\boldsymbol{x}=\boldsymbol{z}$, as shown in Figure 2.26. The intermediate scale is often referred to as the mesoscale.

2.7 Going Beyond Upscaling – Homogenization-Based Multigrid

Since direct numerical simulation that resolves microstructural details often gives rise to a computationally intractable problem, the goal of upscaling (homogenization) discussed so far is to construct a coarse-scale problem that adequately reflects the fine-scale features. Upscaling presents a compromise between computational cost and accuracy demands. In the classical homogenization approaches, as well as in the HMM [7,48] and MEPU [13,14] upscaling methods, the local problem is a UC or RVE domain that is sufficiently large for

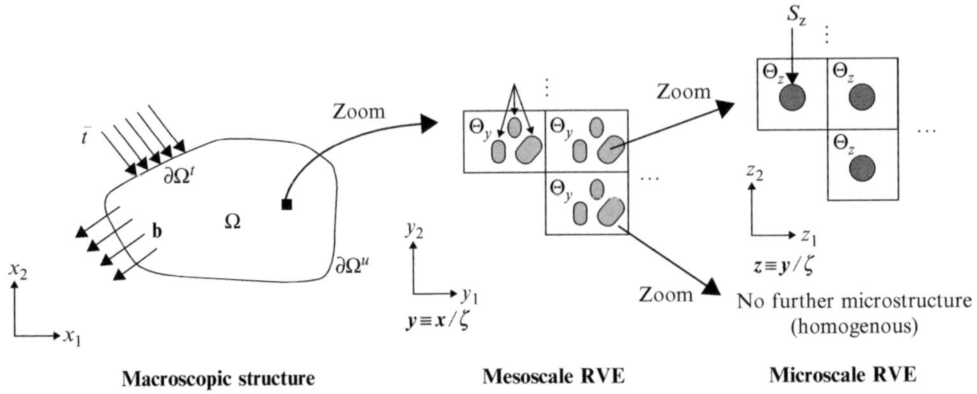

Figure 2.26 Three-scale coordinate system. Reproduced with permission from J. Fish, Q. Yu, Multiscale damage modelling for composite materials: theory and computational framework, International Journal for Numerical Methods in Engineering, Vol. 52, pp. 161–191 (2001), © 2001 John Wiley & Sons

representation of the heterogeneity scale, but small compared with the coarse-scale problem domain or wavelength of the traveling signal. In the VMS [11], conformal MsFEM [10], and DEM [12], the local problem is defined over the coarse-scale element domain, which is typically larger than the RVE domain. In nonconformal MsFEM [10], the local problem domain is larger than the coarse-scale element domain.

In many problems of practical interest, governing equations on the scale of material heterogeneity cannot be postulated due to a lack of knowledge of the precise material microstructure in the entire composite problem domain. Note that it is possible to scan a small device, but scanning an entire Boeing 787 is another matter. Thus, postulating a coarse-scale description either by upscaling or phenomenological modeling is the best that can be done. In the case of small devices that can be fully characterized on the scale of material heterogeneity, we will address the accuracy of upscaling methods.

Suppose that the coarse-scale problem is solved and the fine-scale solution is reconstructed by postprocessing [see equation (2.37) and equation (2.38)]. The question is how accurate the reconstructed fine-scale solution will be, that is, will it satisfy the fine-scale governing equations (2.6) on the composite domain? Clearly, if the exact fine-scale solution is such that the coarse-scale displacement gradients are constant over the RVE domain and if the RVE boundary tractions or displacements match those postulated a priori in the RVE problem, then both the postprocessed fine-scale solution and the coarse-scale solution are exact.

The primary objective of the so-called homogenization-based multigrid [90,91,92,93, 94,95,96] is to improve the quality of the upscaling solution by combining multigrid (see for instance [97,98]) and homogenization ideas. Qualitatively, the basic idea is as follows (note that all the computations are performed at the level of an algebraic system of equations):

1. Capture the coarse-scale system behavior by solving a discrete coarse-scale problem constructed by upscaling. This process is referred to as *coarse-grid correction*.
2. Postprocess the discrete fine-scale solution. This process is referred to as *prolongation*.

3. Subdivide the fine-scale problem domain into possibly overlapping subdomains. On each subdomain solve a local discrete problem by subjecting the boundary nodes of each local subdomain to the boundary displacement obtained from postprocessing in Step 2. This process is known as *relaxation*. This process may be repeated a few times by accounting for the interactions between local subdomains.
4. Insert the discrete fine-scale solution obtained after the relaxation in Step 3 into the fine-scale system of equations and compute the fine-scale residual. If the residual is small, the so-called two-grid cycle is considered to be converged. Otherwise, apply the fine-scale residual as an external load onto the coarse-scale problem and compute the coarse-scale residual. This step is called *restriction*.
5. Go to Step 1 with the residual obtained from Step 4.

The above procedure provides a physical motivation to the two-grid method, by which an oscillatory response is captured by solving a local problem (relaxation) and smooth behavior is obtained by solving a coarse-scale problem (coarse-grid correction). In subsequent sections, various mathematical aspects of the two-grid method are outlined in the context of a model problem.

2.7.1 Relaxation

In this section we outline the relaxation process, point out its limitations, and introduce a coarse-grid correction to capture both the fine- and coarse-scale behavior.

We start with the following model problem

$$\begin{aligned} &\frac{d}{dx}\left(E\frac{du}{dx}\right)+b=0 \quad 0<x<1 \\ &u(0)=u(1)=0 \end{aligned} \tag{2.237}$$

which corresponds to an elastic bar with a constant cross-sectional area $A=1$ and Young's modulus $E=1$, constrained at both ends. In the discrete system of equations in the matrix form, h is a finite element size, N is the number of unconstrained nodes, and nodes 0 and $N+1$ are constrained, which is shown as

$$\boldsymbol{Kd}=\boldsymbol{f} \tag{2.238}$$

where $\boldsymbol{K}$ is a tri-diagonal $N\times N$ stiffness matrix

$$\boldsymbol{K}=\frac{1}{h^2}\begin{bmatrix} 2 & -1 & & & \\ -1 & 2 & -1 & & \\ & & \ddots & & \\ & & & \ddots & \\ & & & -1 & 2 \end{bmatrix} \tag{2.239}$$

The linear system of equations (2.238) can be rewritten as

$$\boldsymbol{Dd}=\boldsymbol{Dd}+\boldsymbol{f}-\boldsymbol{Kd} \tag{2.240}$$

where $\boldsymbol{D}$ is a diagonal matrix of $\boldsymbol{K}$. The relaxation is an iterative process where the right-hand side of (2.240) is assumed to be known from the previous iteration, which gives

$$ {}^{i+1}\boldsymbol{d} = {}^{i}\boldsymbol{d} + \boldsymbol{D}^{-1}\left(\boldsymbol{f} - \boldsymbol{K}\,{}^{i}\boldsymbol{d}\right) \tag{2.241} $$

where the left lowercase superscript denotes the iteration count. Equation (2.241) is a local problem on a patch of elements with prescribed displacements and an external force defined by the residual

$$ {}^{i}\boldsymbol{r} = \boldsymbol{f} - \boldsymbol{K}\,{}^{i}\boldsymbol{d} \tag{2.242} $$

Inserting (2.242) into (2.241) and denoting $\boldsymbol{D}$ as the Jacobi preconditioner, $\boldsymbol{P}^J = \boldsymbol{D}$, yields

$$ {}^{i+1}\boldsymbol{d} = {}^{i}\boldsymbol{d} + \left(\boldsymbol{P}^J\right)^{-1}\,{}^{i}\boldsymbol{r} \tag{2.243} $$

Equation (2.241) can be also expressed as

$$ {}^{i+1}\boldsymbol{d} = \boldsymbol{R}^J\,{}^{i}\boldsymbol{d} + \boldsymbol{D}^{-1}\boldsymbol{f} \tag{2.244} $$

where $\boldsymbol{R}^J$ is the Jacobi iteration matrix defined as

$$ \boldsymbol{R}^J = \boldsymbol{I} - \left(\boldsymbol{P}^J\right)^{-1}\boldsymbol{K} \tag{2.245} $$

The fine-scale solution error ${}^{i+1}\boldsymbol{e}$ is defined as ${}^{i+1}\boldsymbol{e} = \boldsymbol{d} - {}^{i+1}\boldsymbol{d}$ and is given as

$$ {}^{i+1}\boldsymbol{e} = \boldsymbol{R}^J\,{}^{i}\boldsymbol{e} \tag{2.246} $$

An alternative relaxation scheme can be constructed as a weighted average between the Jacobi method and the previous solution

$$ \begin{aligned} {}^{i+1}\boldsymbol{d} &= (1-\omega)\,{}^{i}\boldsymbol{d} + \omega\left({}^{i}\boldsymbol{d} + \boldsymbol{D}^{-1}\,{}^{i}\boldsymbol{r}\right) \\ &= {}^{i}\boldsymbol{d} + \underbrace{\omega\boldsymbol{D}^{-1}}_{\left(\boldsymbol{P}^{\omega}\right)^{-1}}\,{}^{i}\boldsymbol{r} = {}^{i}\boldsymbol{d} + \left(\boldsymbol{P}^{\omega}\right)^{-1}\,{}^{i}\boldsymbol{r} \end{aligned} \tag{2.247} $$

This is known as a weighted Jacobi relaxation. Since relaxation provides an approximate solution of the fine-scale problem, the relaxation parameter $0 < \omega \leq 1$ gives an additional degree of freedom to improve its quality. We will comment on the choice of ω in the later discussion.

Remark 2.8 An improvement over the Jacobi (or weighted Jacobi) relaxation is to utilize a relaxed solution in one node (or collection of nodes) in solving the local problem for the next node (or collection of nodes) and so on. This is known as the Gauss–Seidel relaxation [99].

Let us now investigate the solution convergence of the weighted Jacobi relaxation. In a manner similar to the Jacobi method, the fine-scale error is given by

$$ {}^{i+1}\boldsymbol{e} = \boldsymbol{R}^{\omega}\,{}^{i}\boldsymbol{e} = (\boldsymbol{I} - (\boldsymbol{P}^{\omega})^{-1}\boldsymbol{K})\,{}^{i}\boldsymbol{e} \tag{2.248} $$

which can be bounded by

$$\left\| {}^{i+1}\boldsymbol{e} \right\|_2 \le \rho(\boldsymbol{R}^{\omega}) \left\| {}^{i}\boldsymbol{e} \right\|_2 \tag{2.249}$$

where $\rho(\boldsymbol{R}^{\omega})$ is the spectral norm of the iteration matrix, which quantifies the decay of error in a single relaxation. The spectral norm is defined by

$$\rho(\boldsymbol{R}^{\omega}) = \max_{1 \le k \le N} \left| \lambda_k(\boldsymbol{R}^{\omega}) \right| \tag{2.250}$$

where $\lambda_k(\boldsymbol{R}^{\omega})$ is the kth eigenvalue of the iteration matrix. Note that in general, the iteration matrix is not symmetric even for symmetric stiffness matrices.

Since the decay of error due to the sequence of k relaxations is governed by $(\rho(\boldsymbol{R}^{\omega}))^k$, the relaxation is considered convergent if

$$\rho(\boldsymbol{R}^{\omega}) < 1 \tag{2.251}$$

Consider now the stiffness matrix of the model problem (2.239). Since $\boldsymbol{D} = 2\boldsymbol{I}/h^2$, the iteration matrix $\boldsymbol{R}^{\omega}$ is symmetric and is given by

$$\boldsymbol{R}^{\omega} = \boldsymbol{I} - \frac{h^2 \omega}{2} \boldsymbol{K} \tag{2.252}$$

The eigenvalue problem, $(\boldsymbol{R}^{\omega} - \lambda(\boldsymbol{R}^{\omega})\boldsymbol{I})\boldsymbol{\phi} = \boldsymbol{0}$, can be written as

$$\left(\boldsymbol{K} - \frac{2}{\omega h^2}(1 - \lambda(\boldsymbol{R}^{\omega}))\boldsymbol{I} \right) \boldsymbol{\phi} = \boldsymbol{0} \tag{2.253}$$

which provides the relation between the eigenvalues of the iteration matrix $\boldsymbol{R}^{\omega}$ and the stiffness matrix $\boldsymbol{K}$

$$\lambda_k(\boldsymbol{R}^{\omega}) = 1 - \frac{\omega h^2}{2} \lambda_k(\boldsymbol{K}) \tag{2.254}$$

The eigenvalues of the stiffness matrix (2.239) are given by

$$\lambda_k(\boldsymbol{K}) = \frac{4}{h^2} \sin^2\left(\frac{k\pi}{2(N+1)} \right)$$

which yields the following iteration matrix eigenvalues

$$\lambda_k(\boldsymbol{R}^{\omega}) = 1 - 2\omega \sin^2\left(\frac{k\pi}{2(N+1)} \right) \tag{2.255}$$

Figure 2.27 depicts the eigenvalues of the iteration matrix as a function k and ω. In the case of a Jacobi relaxation (that is, $\omega = 1$), both smooth error modes (that is, small k values) and highly oscillatory modes converge very slowly, as evidenced by the value of the spectral radius $\lambda_k(\boldsymbol{R}^{\omega})$, which is close to one. The choice of $\omega = 2/3$, $\lambda_k(\boldsymbol{R}^{\omega}) \le 1/3$ for all values of $k \ge (N+1)/2$ means that

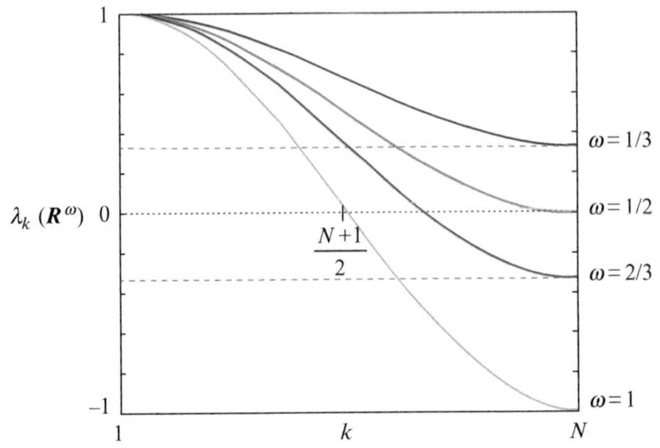

Figure 2.27 Eigenvalues of a weighted Jacobi iteration matrix as a function of k and ω

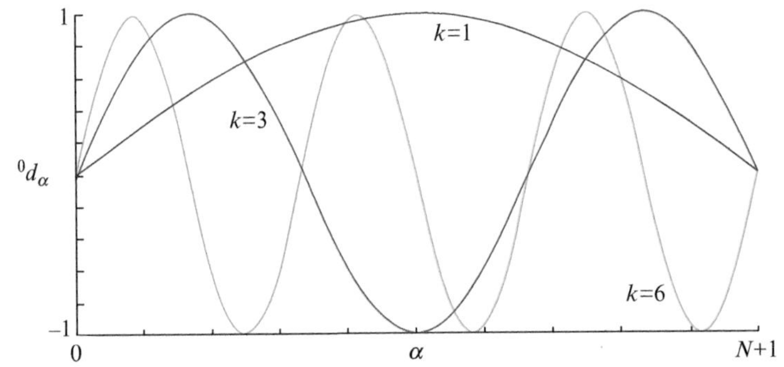

Figure 2.28 Initial errors

oscillatory modes of error will be reduced by a factor of one-third in each iteration [see (2.249)]. This is often referred to as a smoothing property of the relaxation methods, but it has little effect on the smooth modes of error that decay very slowly, independent of the values of ω.

The smoothing effect is further illustrated by considering a model problem (2.239) with zero body forces $\boldsymbol{Kd}=\boldsymbol{0}$ for which $\boldsymbol{d}=\boldsymbol{0}$ is the exact solution. We will consider initial guesses in the form suggested in [100]

$$^0d_\alpha = \sin\frac{\pi k\alpha}{N+1} \quad 0\leq\alpha\leq N+1, \quad 1\leq k\leq N \tag{2.256}$$

Figure 2.28 depicts three initial guesses, with $k=1$ and $k=6$ being the smoothest and the most oscillatory modes considered, respectively.

The solution error is given by ${}^i\boldsymbol{e}=\boldsymbol{d}-{}^i\boldsymbol{d}=-{}^i\boldsymbol{d}$. Figure 2.29 depicts $\left|{}^i\boldsymbol{e}\right|_\infty = \max_{0\leq\alpha\leq N+1}\left|{}^id_\alpha\right|$ as a function of the iteration count for the weighted Jacobi method, with $\omega=2/3$. The smoothing property of the weighted Jacobi relaxation is evidenced by the rapid decay of the oscillatory components of error.

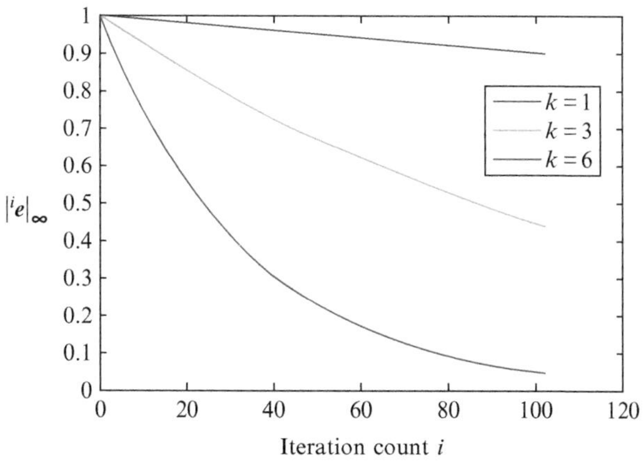

Figure 2.29 Convergence of a weighted Jacobi relaxation

The behavior of relaxation methods suggests the following two-grid solution strategy: capture the smooth components of error by solving an auxiliary problem on a coarse grid, which can resolve smooth components of error, while removing the remaining oscillatory components of error by relaxation. The question that remains to be answered is how to effectively define the auxiliary problem. In particular, what should the auxiliary problem be for heterogeneous media? Furthermore, what does it mean to have a smooth error for heterogeneous media? These issues are discussed in the subsequent sections.

2.7.2 Coarse-grid Correction

Let $N_B^c(\boldsymbol{x})$ be the coarse-grid shape function, with uppercase right subcripts denoting the node numbers. The discretized coarse-scale displacements are denoted as $u_i^c(\boldsymbol{x}) = N_B^c(\boldsymbol{x}) d_{iB}^c$, and d_{iB}^c denotes the nodal coarse-grid degrees of freedom. The coarse-grid displacements computed at the fine-grid nodes $\boldsymbol{x}_A$ are given by

$$u_i^c(\boldsymbol{x}_A) \equiv d_{iA} = N_B^c(\boldsymbol{x}_A) d_{iB}^c \tag{2.257}$$

The prolongation (or interpolation) operator Q_{AB} is defined as

$$Q_{AB} = N_B^c(\boldsymbol{x}_A) \tag{2.258}$$

such that $d_{iA} = Q_{AB} d_{iB}^c$. Without loss of generality, the prolongation operator can be expressed as

$$d_\alpha = Q_{\alpha\beta} d_\beta^c \quad \text{or} \quad \boldsymbol{d} = \boldsymbol{Q}\boldsymbol{d}^c \tag{2.259}$$

where, as before, Greek subscripts denote degrees of freedom rather than nodes, and matrices are denoted in bold.

Given the prolongation operator $\boldsymbol{Q}$ and the fine-scale solution matrix from the previous two-grid cycle I, ${}^I\boldsymbol{d}$, we seek a coarse-grid correction of error ${}^{I+1}\boldsymbol{e}^c$, which is a minimizer of the potential energy on the subspace of coarse-scale functions

$$\min_{^{I+1}e^C}\left(\frac{1}{2}\left({}^I d+Q\,{}^{I+1}e^c\right)^T K\left({}^I d+Q\,{}^{I+1}e^c\right)-\left({}^I d+Q\,{}^{I+1}e^c\right)^T f\right) \tag{2.260}$$

Minimization of the above with respect to the coarse-grid correction ${}^{I+1}e^c$ yields the discrete coarse-grid problem

$$\tilde{K}^c\,{}^{I+1}e^c={}^I r^c \tag{2.261}$$

where $\tilde{K}^c$ is a variational coarse-grid stiffness matrix given by

$$\tilde{K}^c=Q^T KQ \tag{2.262}$$

${}^I r^c$ is a coarse-grid residual matrix given by the so-called restriction

$${}^I r^c=Q^T\,{}^I r \tag{2.263}$$

and

$${}^I r=f-K\,{}^I d=Kd-K\,{}^I d=K\,{}^I e \tag{2.264}$$

Remark 2.9 It is important to distinguish between the variational coarse-grid stiffness matrix $\tilde{K}^c$ and the coarse-grid stiffness matrix K^c, computed directly from the coarse grid. For some model problems the two matrices may coincide, but in general they are different. In subsequent derivations we will denote the coarse-grid stiffness matrix as K^c, constructed either directly or variationally (2.262).

Let ${}^{I+1}\hat{d}$ be the fine-grid displacement matrix after the coarse-grid correction

$${}^{I+1}\hat{d}={}^I d+Q\,{}^{I+1}e^c={}^I d+Q(K^c)^{-1}\,{}^I r^c \tag{2.265}$$

Inserting (2.263) into (2.265) and recalling that ${}^I d=d-{}^I e$ yields

$$d-{}^{I+1}\hat{e}=d-{}^I e+Q(K^c)^{-1}Q^T\,{}^I r \tag{2.266}$$

where ${}^{I+1}\hat{e}=d-{}^{I+1}\hat{d}$ is the fine-grid error after the coarse-grid correction. Inserting (2.264) into (2.266) gives

$${}^{I+1}\hat{e}=T\,{}^I e \tag{2.267}$$

where T is a coarse-grid iteration matrix given by

$$T=I-Q(K^c)^{-1}Q^T K \tag{2.268}$$

From (2.268) we can identify the inverse of the coarse-grid preconditioner $(P^c)^{-1}$, which is a prolongation of the coarse-grid stiffness matrix inverse

$$(P^c)^{-1}=Q(K^c)^{-1}Q^T \tag{2.269}$$

The error following a single two-grid cycle consisting of a coarse-grid correction and post-relaxation (relaxation coming after the coarse-grid correction) ${}^{I+1}e$ is given by

$$^{I+1}\boldsymbol{e} = \boldsymbol{R}\boldsymbol{T}\ {}^{I}\boldsymbol{e} \tag{2.270}$$

The coarse-grid iteration matrix possesses the so-called $\boldsymbol{K}$-orthogonality property defined as

$$(\boldsymbol{I}-\boldsymbol{T})^T\boldsymbol{K}\boldsymbol{T} = \boldsymbol{0} \tag{2.271}$$

The above follows from

$$\begin{aligned}(\boldsymbol{I}-\boldsymbol{T})^T\boldsymbol{K}\boldsymbol{T} &= \left(\boldsymbol{Q}(\boldsymbol{K}^c)^{-1}\boldsymbol{Q}^T\boldsymbol{K}\right)^T\boldsymbol{K}\left(\boldsymbol{I}-\boldsymbol{Q}(\boldsymbol{K}^c)^{-1}\boldsymbol{Q}^T\boldsymbol{K}\right)\\ &= \boldsymbol{K}^T\boldsymbol{Q}\left((\boldsymbol{K}^c)^{-1}\right)^T\boldsymbol{Q}^T\boldsymbol{K}\left(\boldsymbol{I}-\boldsymbol{Q}(\boldsymbol{K}^c)^{-1}\boldsymbol{Q}^T\boldsymbol{K}\right)\\ &= \boldsymbol{K}^T\boldsymbol{Q}\left((\boldsymbol{K}^c)^{-1}\right)^T\boldsymbol{Q}^T\boldsymbol{K} - \boldsymbol{K}^T\boldsymbol{Q}\left((\boldsymbol{K}^c)^{-1}\right)^T\underbrace{\boldsymbol{Q}^T\boldsymbol{K}\boldsymbol{Q}}_{\boldsymbol{K}^c}(\boldsymbol{K}^c)^{-1}\boldsymbol{Q}^T\boldsymbol{K} = \boldsymbol{0}\end{aligned} \tag{2.272}$$

The upshot of the $\boldsymbol{K}$-orthogonality property is that the coarse-grid correction reduces the overall error in the energy norm, which follows from the following simple analysis. The error prior to coarse-grid correction ${}^{I}\boldsymbol{e}$ can be decomposed as

$${}^{I}\boldsymbol{e} = (\boldsymbol{I}-\boldsymbol{T})\ {}^{I}\boldsymbol{e} + \boldsymbol{T}\ {}^{I}\boldsymbol{e} \tag{2.273}$$

Define the square of the energy norm of error $\left\|{}^{I}\boldsymbol{e}\right\|_{\boldsymbol{K}}^2$ as

$$\begin{aligned}\left\|{}^{I}\boldsymbol{e}\right\|_{\boldsymbol{K}}^2 &= \left({}^{I}\boldsymbol{e}\right)^T\boldsymbol{K}\ {}^{I}\boldsymbol{e} = \left\|(\boldsymbol{I}-\boldsymbol{T})\,{}^{I}\boldsymbol{e}\right\|_{\boldsymbol{K}}^2 + \left\|\boldsymbol{T}\ {}^{I}\boldsymbol{e}\right\|_{\boldsymbol{K}}^2 + \left({}^{I}\boldsymbol{e}\right)^T(\boldsymbol{I}-\boldsymbol{T})^T\boldsymbol{K}\boldsymbol{T}\ {}^{I}\boldsymbol{e}\\ &\quad + \left({}^{I}\boldsymbol{e}\right)^T\boldsymbol{T}^T\boldsymbol{K}(\boldsymbol{I}-\boldsymbol{T})\ {}^{I}\boldsymbol{e}\end{aligned} \tag{2.274}$$

Exploiting the $\boldsymbol{K}$-orthogonality condition (2.272) and (2.267) yields

$$\left\|{}^{I+1}\hat{\boldsymbol{e}}\right\| = \left\|\boldsymbol{T}\ {}^{I}\boldsymbol{e}\right\|_{\boldsymbol{K}}^2 < \left\|{}^{I}\boldsymbol{e}\right\|_{\boldsymbol{K}}^2 \tag{2.275}$$

2.7.3 Two-grid Convergence for a Model Problem in a Periodic Heterogeneous Medium

We will consider a 1D model problem (2.237) with oscillatory periodic piecewise constant coefficients $E^{(1)}, E^{(2)}$ and a 0.5 volume fraction [90,91]. The model problem of length l is discretized with 2 $(m-1)$ elements, each element having constant coefficients. The eigenvalues can be computed in a closed form [90,91]

$$\lambda_k = \frac{\dfrac{4E^c}{h}\sin^2\left(\dfrac{\pi k}{2m}\right)}{1+\sqrt{1-q\sin^2\left(\dfrac{\pi k}{2m}\right)}};\quad \lambda_{2m-k} = \frac{\dfrac{4E^c}{h}\sin^2\left(\dfrac{\pi k}{2m}\right)}{1-\sqrt{1-q\sin^2\left(\dfrac{\pi k}{2m}\right)}};\quad \lambda_m = \frac{4E^c}{hq} \tag{2.276}$$

where $1 \le k \le m-1$, h is the unit cell size, E^c is the coarse-scale Young's modulus, and q is the ratio between the geometric and arithmetic averages of the coefficients given as

$$E^c = \frac{2E^{(1)}E^{(2)}}{E^{(1)}+E^{(2)}}; \quad q = \left(\frac{\sqrt{E^{(1)}E^{(2)}}}{(E^{(1)}+E^{(2)})/2} \right)^2 \tag{2.277}$$

Note that in many applications of interest $E^{(1)} \gg E^{(2)}$ or $E^{(2)} \gg E^{(1)}$, which yields $0 < q \ll 1$. Consequently, the eigenvalues are clustered at the two ends of the spectrum, with one-half being $O(1)$ and the other half being $O(1/q)$. More importantly, the $O(1)$ eigenvalues are identical to those obtained from the coarse-scale problem with homogenized coefficients, that is,

$$\lambda_k = \lambda_k^c \quad 1 \leq k \leq m-1 \tag{2.278}$$

Furthermore, the corresponding eigenvectors are related by

$$\boldsymbol{\phi}_k = \boldsymbol{Q}\boldsymbol{\phi}_k^c \quad 1 \leq k \leq m-1 \tag{2.279}$$

where $\boldsymbol{Q}$ is the prolongation operator obtained from the discretization of the displacement influence functions (see section 2.7.4).

This character of the spectrum suggests a two-grid strategy where a relaxation scheme is engineered to capture the higher frequency response of the fine-scale model as represented by a linear combination of the $O(1/q)$ eigenmodes. The auxiliary coarse grid is then engineered to effectively capture the remaining lower frequency response of the fine-scale problem. For a periodic heterogeneous medium, such an auxiliary coarse model coincides with the boundary value problem, with homogenized coefficients as evidenced by the identical eigenvalues.

Consider a smooth error component expressed in terms of a linear combination of smooth eigenmodes on the fine scale. We now show that, provided (2.278) and (2.279) hold, then

$$\boldsymbol{T}\boldsymbol{\phi}_k = \boldsymbol{0} \quad 1 \leq k \leq m-1 \tag{2.280}$$

We start by expressing the inverse of the coarse-scale (or coarse-grid) stiffness matrix in terms of its eigenvectors and eigenvalues

$$(\boldsymbol{K}^c)^{-1} = \sum_{i=1}^{m-1} \frac{1}{\lambda_i^c} \boldsymbol{\phi}_i^c (\boldsymbol{\phi}_i^c)^T \tag{2.281}$$

The inverse of the coarse-grid preconditioner is given by

$$(\boldsymbol{P}^c)^{-1} = \boldsymbol{Q} \left(\sum_{i=1}^{m-1} \frac{1}{\lambda_i^c} \boldsymbol{\phi}_i^c (\boldsymbol{\phi}_i^c)^T \right) \boldsymbol{Q}^T = \sum_{i=1}^{m-1} \frac{1}{\lambda_i^c} (\boldsymbol{Q}\boldsymbol{\phi}_i^c)(\boldsymbol{Q}\boldsymbol{\phi}_i^c)^T = \sum_{i=1}^{m-1} \frac{1}{\lambda_i} \boldsymbol{\phi}_i (\boldsymbol{\phi}_i)^T \tag{2.282}$$

Inserting (2.282) into (2.268) and (2.280) yields

$$\boldsymbol{T}\boldsymbol{\phi}_k = \boldsymbol{\phi}_k - \sum_{i=1}^{m-1} \frac{1}{\lambda_i} \boldsymbol{\phi}_i \left(\boldsymbol{\phi}_i\right)^T \underbrace{\boldsymbol{K}\boldsymbol{\phi}_k}_{\lambda_k \boldsymbol{\phi}_k} \tag{2.283}$$

Note that $(\boldsymbol{\phi}_i)^T \boldsymbol{\phi}_k = \delta_{ik}$ where δ_{ik} is a Kronecker delta and, therefore, (2.280) is satisfied. Similarly, it can be shown that

$$\boldsymbol{T}\boldsymbol{\phi}_{2m-k} = \boldsymbol{\phi}_{2m-k} \quad 1 \le k \le m-1 \tag{2.284}$$

which implies that the coarse grid defined by (2.278) and (2.279) does not affect the oscillatory modes of error.

We now consider a single post-relaxation using the weighted Jacobi relaxation method. It can be shown [90] that the eigenvalues corresponding to the oscillatory modes of error of the weighted Jacobi iteration matrix are given by

$$\max_{1\le k\le m-1} \left|\lambda_{2m-k}\left(\boldsymbol{R}^{\omega}\right)\right| = \frac{q}{4-q} \tag{2.285a}$$

$$\omega = \frac{4}{4-q} \tag{2.285b}$$

The spectral radius of the two-grid method $\rho(\boldsymbol{R}^{\omega}\boldsymbol{T})$, with a coarse grid constructed to satisfy equation (2.278) and equation (2.279) and a single post-relaxation based on the weighted Jacobi method with a relaxation parameter selected based on (2.285b), is given by [90]

$$\rho(\boldsymbol{R}^{\omega}\boldsymbol{T}) = \frac{q}{4-q} \tag{2.286}$$

For example, if either $E^{(1)}/E^{(2)}$ or $E^{(2)}/E^{(1)}$ is 100, then the two-scale process converges in three iterations up to a tolerance of 10^{-5}. Note that for homogeneous materials, $q=1$; this yields an optimal value of $\omega=2/3$ and the classical two-grid spectral radius of $\rho(\boldsymbol{R}^{\omega}\boldsymbol{T})=1/3$.

The estimate for the model problem (2.286) suggests that the aforementioned two-grid scheme falls into the category of multiscale methods that offer a sublinear rate of convergence.

2.7.4 Upscaling-Based Prolongation and Restriction Operators

Consider a two-term asymptotic expansion of displacements with a local coordinate system $\boldsymbol{x}-\hat{\boldsymbol{x}}=\boldsymbol{y}$ (here $\hat{\boldsymbol{x}}$ is the unit cell centroid), that is, $\zeta=1$ and the displacement influence functions H_i^{jk} extend over the entire composite problem domain by periodicity

$$u_i^{\zeta} \equiv u_i = u_i^c + H_i^{jk}\varepsilon_{jk}^c \tag{2.287}$$

Inserting the discretized coarse-scale displacements and the displacement influence function into (2.287) yields

$$u_i = N_A^c d_{iA}^c + N_B d_{iB}^{jk} N_{A,k}^c d_{jA}^c = \left(N_A^c \delta_{ij} + N_B d_{iB}^{jk} N_{A,k}^c\right) d_{jA}^c \tag{2.288}$$

We assume that the coarse-scale shape functions N_A^c can be expressed by a linear combination of fine-scale shape functions N_B

$$N_A^c = N_B \bar{Q}_{BA} \tag{2.289}$$

Calculating u_i at the fine-grid nodes and using (2.289) yields

$$\begin{aligned} u_i(\boldsymbol{x}_C) \equiv d_{iC} &= \left(N^c_A(\boldsymbol{x}_C)\delta_{ij} + N_B(\boldsymbol{x}_{\underline{C}})d^{jk}_{iB}N^c_{A,k}(\boldsymbol{x}_{\underline{C}})\right)d^c_{jA} \\ &= \underbrace{N_B(\boldsymbol{x}_{\underline{C}})}_{\delta_{BC}}\left(\bar{Q}_{BA}\delta_{ij} + d^{jk}_{iB}N^c_{A,k}(\boldsymbol{x}_{\underline{C}})\right)d^c_{jA} = \left(\bar{Q}_{CA}\delta_{ij} + d^{jk}_{i\underline{C}}N^c_{A,k}(\boldsymbol{x}_{\underline{C}})\right)d^c_{jA} \end{aligned} \tag{2.290}$$

or

$$\begin{aligned} d_{iC} &= Q_{iCjA}d^c_{jA}; \quad \boldsymbol{d} = \boldsymbol{Q}\boldsymbol{d}^c \\ Q_{iAjC} &= \bar{Q}_{CA}\delta_{ij} + d^{jk}_{i\underline{C}}N^c_{A,k}\left(\boldsymbol{x}_{\underline{C}}\right) \end{aligned} \tag{2.291}$$

The resulting two-scale prolongation $\boldsymbol{Q}$ operator consists of the usual smooth prolongation [first term in (2.291)] and the fine-scale correction [second term in (2.291)]. Recall that the restriction operator, which constructs the right-hand side for coarse-scale problem (2.263), is given by $\boldsymbol{Q}^T$.

Remark 2.10 The second term in (2.291) is discontinuous along the coarse-scale element boundaries, which will give rise to discontinuity in the fine-scale solution after prolongation. This can be circumvented either by averaging the prolongated solution [91] or by employing conformal upscaling methods based on enhanced kinematics.

To this end, we will show that the coarse-grid stiffness matrix with homogenized material properties is an approximation of the variational restriction of the fine-scale stiffness matrix using the two-scale prolongation operator (2.291).

Consider the coarse-scale stiffness matrix $\boldsymbol{K}^c$ computed using a two-scale integration scheme

$$\begin{aligned} K^c_{iAjB} &= \int_\Omega N^c_{A,k}L^c_{ikjl}N^c_{B,l}d\Omega \\ &= \int_\Omega \frac{1}{|\Theta|}\int_\Theta N^c_{A,k}\left(I_{stik} + H^{ik}_{(s,t)}\right)L_{stmn}\left(I_{mnjl} + H^{jl}_{(m,n)}\right)N^c_{B,l}d\Theta d\Omega \end{aligned} \tag{2.292}$$

Inserting discretization $H^{ik}_{(s,t)} = N_{C,t}d^{ik}_{sC}$ and (2.289) into (2.292) yields

$$\begin{aligned} K^c_{iAjB} &= \int_\Omega \frac{1}{|\Theta|}\int_\Theta \left(N^c_{A,t}\delta_{is} + N^c_{A,k}N_{C,t}d^{ik}_{sC}\right)L_{stmn}\left(N^c_{B,j}\delta_{mn} + N^c_{B,l}N_{D,n}d^{jl}_{mD}\right)d\Theta d\Omega \\ &= \int_\Omega \frac{1}{|\Theta|}\int_\Theta \left(N_{C,t}\bar{Q}_{CA}\delta_{is} + N^c_{A,k}N_{C,t}d^{ik}_{sC}\right)L_{stmn}\left(N_{D,n}\bar{Q}_{DB}\delta_{jm} + N^c_{B,l}N_{D,n}d^{jl}_{mD}\right)d\Theta d\Omega \end{aligned} \tag{2.293}$$

We further assume that derivatives of the coarse-scale shape functions can be approximated by their average $\bar{N}^c_{A,k}$ over the coarse-scale element domain Ω^c_e, which yields

$$K^c_{iAjB} = \int_{\Omega} \frac{1}{|\Theta|} \int_{\Theta} \left(\bar{Q}_{CA}\delta_{is} + N^c_{A,k} d^{ik}_{sC} \right) N_{C,t} L_{stmn} N_{D,n} \left(\bar{Q}_{DB}\delta_{jm} + N^c_{B,l} d^{jl}_{mD} \right) d\Theta d\Omega$$

$$\approx \sum_e \underbrace{\left(\bar{Q}_{CA}\delta_{is} + \bar{N}^c_{A,k} d^{ik}_{sC} \right)}_{Q^e_{iAsC}} \underbrace{\left(\int_{\Omega^c_e} \frac{1}{|\Theta|} \int_{\Theta} N_{C,t} L_{stmn} N_{D,n} d\Theta d\Omega^c_e \right)}_{K^{ef}_{sCmD}} \underbrace{\left(\bar{Q}_{DB}\delta_{jm} + \bar{N}^c_{B,l} d^{jl}_{mD} \right)}_{Q^e_{jBmD}}$$

$$\approx Q^e_{iAsC} K^{ef}_{sCmD} Q^e_{jBmD} \tag{2.294}$$

where K^{ef}_{sCmD} is the fine-scale stiffness matrix computed over the coarse-scale element domain and Q^e_{iAsC} is its prolongation operator.

2.7.5 Homogenization-based Multigrid and Multigrid Acceleration

It is convenient to recast a two-grid cycle into a symmetric form, where the cycle starts by n pre-relaxations, followed by a coarse-grid correction and finally by n post-relaxations, in which case the two-grid iteration matrix $\boldsymbol{R}^{MG}$ is given by

$$\boldsymbol{R}^{MG} = (\boldsymbol{R})^n \boldsymbol{T} (\boldsymbol{R})^n \tag{2.295}$$

For large scale problems and/or problems involving multiple scales, the two-grid method can be extended to the so-called multigrid method. Consider, for instance, the three scales denoted as macro, meso, and micro in Figute 2.26. A single three-scale V cycle consists of:

- Relaxation at the microscale
- Restriction of the microscale residual to the mesoscale (microscale operators (the restriction and prolongation matrices) are obtained by discretizing the microscale influence function)
- Relaxation at the mesoscale
- Restriction of the mesoscale residual to the macroscale (mesoscale operators are obtained by discretizing the mesoscale influence function)
- Direct solution at the macroscale
- Prolongation of the macroscale correction to the mesoscale using the mesoscale prolongation operator
- Relaxation at the mesoscale
- Prolongation of the mesoscale correction to the microscale using the microscale prolongation operator
- Relaxation at the microscale

The multigrid cycle can be accelerated using the conjugate gradient method (CGM) or other acceleration schemes [101]. Let ${}^I\boldsymbol{d}$ be the solution obtained at the end of previous cycle I, and let ${}^{I+1}\breve{\boldsymbol{d}}$ be the new solution obtained by the end of cycle $I+1$. The conjugate gradient method seeks an improved solution ${}^{I+1}\boldsymbol{d}$ along the search direction defined as

$${}^{I+1}\breve{\boldsymbol{s}} = {}^{I+1}\breve{\boldsymbol{d}} - {}^I\boldsymbol{d} \tag{2.296}$$

and the search direction from the previous cycle ${}^{I}\boldsymbol{s} = {}^{I}\boldsymbol{d} - {}^{I-1}\boldsymbol{d}$. The CGM algorithm is as follows. Find scalars α, β such that

$$^{I+1}\boldsymbol{d} = {}^{I}\boldsymbol{d} + \alpha \ {}^{I+1}\boldsymbol{s} \tag{2.297a}$$

$$^{I+1}\boldsymbol{s} = {}^{I+1}\breve{\boldsymbol{s}} + \beta \ {}^{I}\boldsymbol{s} \tag{2.297b}$$

$$^{I+1}\boldsymbol{s}^{T}\boldsymbol{K} \ {}^{I}\boldsymbol{s} = 0 \tag{2.297c}$$

Multiplying (2.297b) by ${}^{I}\boldsymbol{s}^{T} \ \boldsymbol{K}$ yields

$$\beta = -\frac{{}^{I}\boldsymbol{s}^{T} \ \boldsymbol{K} \ {}^{I+1}\breve{\boldsymbol{s}}}{{}^{I}\boldsymbol{s}^{T} \ \boldsymbol{K} \ {}^{I}\boldsymbol{s}} \tag{2.298}$$

Once the search direction is calculated, the value of search direction α can be obtained by minimizing potential energy (2.260) with respect to α.

2.7.6 Nonlinear Multigrid

There are two basic multilevel approaches for solving a nonlinear discrete system of equations arising from PDEs: (i) the Newton-multigrid; and (ii) the full approximation storage (FAS) multigrid. In the Newton-multigrid, the nonlinear problem is solved using Newton's method, where a standard linear multigrid is applied to solve a linearized system of equations. In this section we will focus on the latter, the FAS approach, which directly utilizes multigrid principles to solve a nonlinear system of equations.

The FAS scheme was introduced in the seminal paper by Brandt [102] in 1977. The algorithm has been developed to solve nonlinear discrete problems of the form

$$\boldsymbol{f}^{\text{int}}(\boldsymbol{d}) = \boldsymbol{f}^{\text{ext}} \tag{2.299}$$

We begin by denoting ${}^{I}\boldsymbol{d}$ as the approximation to the exact solution $\boldsymbol{d}$ and by denoting ${}^{I}\boldsymbol{e}$ as the error

$${}^{I}\boldsymbol{e} = \boldsymbol{d} - {}^{I}\boldsymbol{d} \tag{2.300}$$

In the linear case $\boldsymbol{f}^{\text{int}}(\boldsymbol{d}) = \boldsymbol{K}\boldsymbol{d}$, the residual ${}^{I}\boldsymbol{r} = \boldsymbol{f}^{\text{ext}} - \boldsymbol{K} \ {}^{I}\boldsymbol{d}$ satisfies the residual equation

$$\boldsymbol{K} \ {}^{I}\boldsymbol{e} = {}^{I}\boldsymbol{r} \tag{2.301}$$

In the nonlinear case, given the approximation ${}^{I}\boldsymbol{d}$, the residual is

$${}^{I}\boldsymbol{r} = \boldsymbol{f}^{\text{ext}} - \boldsymbol{f}^{\text{int}}\left({}^{I}\boldsymbol{d}\right) \tag{2.302}$$

Subtracting (2.299) from (2.302) gives

$$\boldsymbol{f}^{\text{int}}(\boldsymbol{d}) - \boldsymbol{f}^{\text{int}}\left({}^{I}\boldsymbol{d}\right) = {}^{I}\boldsymbol{r} \tag{2.303}$$

Clearly, $\boldsymbol{f}^{\text{int}}(\boldsymbol{d}-{}^{I}\boldsymbol{d})\neq{}^{I}\boldsymbol{r}$ and the coarse-grid correction cannot be written in the form of (2.301). In FAS, the multigrid is directly applied to (2.299), with (2.303) used as the coarse-grid correction. The process begins by applying a relaxation technique to (2.299) to obtain the approximate solution $\boldsymbol{d}$. The nonlinear relaxation consists of solving a local nonlinear discrete system of equations rather than a linear system [103].

Based on (2.303), we define the coarse-grid equation as

$$\boldsymbol{f}_C^{\text{int}}\left({}^{I}\boldsymbol{d}^C+{}^{I}\boldsymbol{e}^C\right)-\boldsymbol{f}_C^{\text{int}}\left({}^{I}\boldsymbol{d}^C\right)={}^{I}\boldsymbol{r}^C \tag{2.304}$$

where $\boldsymbol{f}_C^{\text{int}}$ is the internal force on the coarse grid and ${}^{I}\boldsymbol{r}^C$ is the restriction of the fine-grid residual onto the coarse problem

$${}^{I}\boldsymbol{r}^C=\boldsymbol{Q}^T\ {}^{I}\boldsymbol{r}=\boldsymbol{Q}^T\left(\boldsymbol{f}^{\text{ext}}-\boldsymbol{f}^{\text{int}}\left({}^{I}\boldsymbol{d}\right)\right) \tag{2.305}$$

where $\boldsymbol{Q}^T$ is the restriction operator. We also define an injection operator, $\boldsymbol{A}$, which is a local averaging of fine-scale displacements. The coarse-scale displacements ${}^{I}\boldsymbol{d}^C$ are computed by locally averaging fine-scale displacements

$${}^{I}\boldsymbol{d}^C=\boldsymbol{A}\ {}^{I}\boldsymbol{d} \tag{2.306}$$

Note that in general, $\boldsymbol{Q}^T$ and $\boldsymbol{A}$ are different. Inserting (2.306) into (2.304) yields

$$\boldsymbol{f}_C^{\text{int}}\left(\boldsymbol{A}\ {}^{I}\boldsymbol{d}+{}^{I}\boldsymbol{e}^C\right)=\boldsymbol{f}_C^{\text{int}}\left(\boldsymbol{A}\ {}^{I}\boldsymbol{d}\right)+\boldsymbol{Q}^T\left(\boldsymbol{f}^{\text{ext}}-\boldsymbol{f}^{\text{int}}\left({}^{I}\boldsymbol{d}^f\right)\right) \tag{2.307}$$

Defining

$${}^{I}\boldsymbol{u}^C=\boldsymbol{A}\ {}^{I}\boldsymbol{d}+{}^{I}\boldsymbol{e}^C \tag{2.308}$$

$$\boldsymbol{f}_C^{\text{ext}}=\boldsymbol{f}_C^{\text{int}}\left(\boldsymbol{A}\ {}^{I}\boldsymbol{d}\right)+\boldsymbol{Q}^T\left(\boldsymbol{f}^{\text{ext}}-\boldsymbol{f}^{\text{int}}\left({}^{I}\boldsymbol{d}\right)\right) \tag{2.309}$$

we obtain the following nonlinear coarse-grid problem

$$\boldsymbol{f}_C^{\text{int}}\left({}^{I}\boldsymbol{u}^C\right)=\boldsymbol{f}_C^{\text{ext}}\left({}^{I}\boldsymbol{d}\right) \tag{2.310}$$

which is in the form of the original fine-scale equation. Alternatively, instead of evaluating the internal force $\boldsymbol{f}_C^{\text{int}}$ on the coarse mesh directly, it can be computed using a more expensive Galerkin projection

$$\boldsymbol{f}_C^{\text{int}}(\boldsymbol{v}_C)=\boldsymbol{Q}^T\boldsymbol{f}^{\text{int}}(\boldsymbol{Q}\boldsymbol{v}_C) \tag{2.311}$$

where $\boldsymbol{Q}$ is the prolongation operator and $\boldsymbol{v}_C$ represents an arbitrary displacement vector on the coarse grid.

Once the nonlinear problem for ${}^{I}\boldsymbol{u}^C$ in (2.310) has been solved, the coarse-grid error is computed from (2.308)

$$ {}^{I}\boldsymbol{e}^{c} = {}^{I}\boldsymbol{u}^{c} - \boldsymbol{A}\,{}^{I}\boldsymbol{d}^{f} \tag{2.312} $$

and interpolated back to the fine grid

$$ {}^{I+1}\boldsymbol{d} = {}^{I}\boldsymbol{d} + \boldsymbol{Q}\,{}^{I}\boldsymbol{e}^{c} = {}^{I}\boldsymbol{d} + \mathbf{Q}\left({}^{I}\boldsymbol{u}^{c} - \boldsymbol{A}\,{}^{I}\boldsymbol{d}\right) \tag{2.313} $$

This completes a single cycle of the two-level FAS method.

2.7.7 Multigrid for Indefinite Systems

For indefinite linearized systems, such as Navier–Stokes equations, saddle-point and least squares problems with constraints, and problems involving an indefinite constitutive tensor arising from damage/localization in solids or shocks in fluids, multilevel or single level iterative methods may not converge. For weakly indefinite systems, remedies exist in the form of multilevel preconditioners with a very fine coarse grid [104] in combination with the generalized minimal residual (GMRES) or quasi-minimal residual (QMR) acceleration schemes [105]. For highly indefinite systems, the global-basis filter [106,107,108,109] provides a general framework to filter out the indefinite portion of the system behavior. This method identifies the eigenvalues of the iteration matrix that fall outside the region of convergence and constructs an auxiliary coarse model from the corresponding eigenvectors.

Let $\boldsymbol{R}$ be the iteration matrix of a single or multilevel method. Identify all modes of $\boldsymbol{R}$ that are outside the region of convergence (the unit circle) by solving for the largest eigenvalues $|\lambda_i(\boldsymbol{R})| > 1$ of the iteration matrix

$$ \boldsymbol{R}\phi_i = \lambda_i \phi_i \quad i = 1,\ldots,N \tag{2.314} $$

for which either the Lanczos method or the Arnoldi method [110] is used to approximate the largest eigenvectors. The iteration matrix of a single cycle without an external accelerator, as shown in Figure 2.30, can be written as

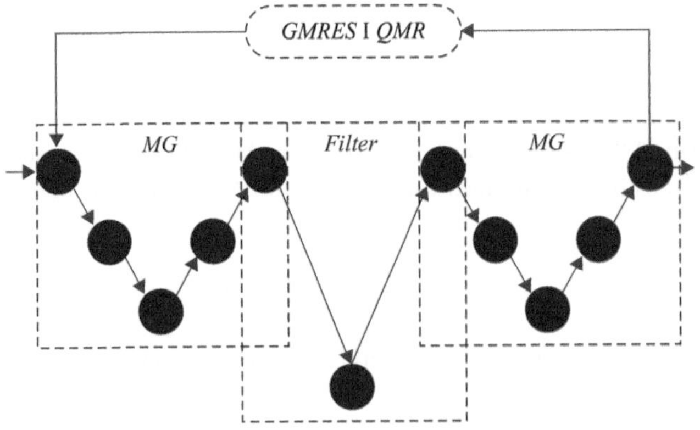

Figure 2.30 Accelerated multigrid for an indefinite system

$$^{I+1}\boldsymbol{e}=\boldsymbol{R}^{\upsilon}\boldsymbol{F}\boldsymbol{R}^{\upsilon\ I}\boldsymbol{e} \tag{2.315}$$

where υ is the number of multigrid cycles, and $\boldsymbol{F}$ is a filter defined as

$$\boldsymbol{F}=\boldsymbol{I}-\boldsymbol{Q}_F\left(\boldsymbol{Q}_F^*\boldsymbol{K}\boldsymbol{Q}_F\right)^{-1}\boldsymbol{Q}_F^*\boldsymbol{K} \tag{2.316}$$

and where the prolongation operator $\boldsymbol{Q}_F$ is spanned by eigenvectors corresponding to the m eigenvalues of $\boldsymbol{R}$ that are greater than one

$$\boldsymbol{Q}_F=\text{span}\{\boldsymbol{\phi}_i\}_{i=1}^{m}=\begin{bmatrix} | & & | \\ \boldsymbol{\phi}_1 & \cdots & \boldsymbol{\phi}_m \\ | & & | \end{bmatrix} \tag{2.317}$$

For indefinite nonsymmetric systems, the iteration matrix $\boldsymbol{R}$ may have complex eigenvectors. For this reason, adjoint (star) notation is used.

Problems

Problem 2.1: Consider the following initial value problem for $u(t)$

$$\begin{aligned} &\ddot{u}^{\zeta}+2\zeta\dot{u}^{\zeta}+u\zeta=0 \\ &u^{\zeta}(0)=0 \\ &\dot{u}^{\zeta}(0)=1 \end{aligned}$$

where ζ is a small positive parameter.

a. Solve the problem by expanding $u^{\zeta}(t)$ in asymptotic expansion

$$u^{\zeta}(t)=u^{(0)}(t)+\zeta u^{(1)}(t)+\cdots$$

b. Is the solution valid for all t?

Problem 2.2: Consider a temperature field in composite domain Ω^{ζ} with boundary $\partial\Omega^{\zeta}$. Temperature T is at a point $\boldsymbol{x}$ in time $t\in(0;\infty)$ and is assumed to be a continuous function of space and time $T=T(\boldsymbol{x},t)$. Assume that (i) there is no mechanical deformation accompanying thermal process, (ii) material parameters are independent of the temperature, and (iii) there are no phase changes or latent heat effects. Consider the following governing equations of transient heat transfer:

a. flow equilibrium conditions: $\rho^{\zeta}c^{\zeta}\dot{T}^{\zeta}+q_{i,i}^{\zeta}-g^{\zeta}=0$
b. Fourier's law: $q_i^{\zeta}=-k_{ij}^{\zeta}T_{,j}^{\zeta}$
c. temperature (essential) boundary conditions: $T^{\zeta}=\overline{T}$ on $\partial\Omega^{T\zeta}$

d. heat flux (natural) boundary conditions: $q_i^{\zeta} n_j^{\zeta} = -\bar{q}$ on $\partial\Omega^{t\zeta}$ where $\partial\Omega^{q\zeta} \cup \partial\Omega^{T\zeta} = \partial\Omega^{\zeta}$ and $\partial\Omega^{q\zeta} \cap \partial\Omega^{T\zeta} = 0$.
e. initial condition: $T_0^{\zeta} = T(\boldsymbol{x},0)$; $t=0$

where $c^{\zeta}(\boldsymbol{x}/\zeta)$ denotes the specific heat capacity of the system, $\rho^{\zeta}(\boldsymbol{x}/\zeta)$ denotes material density, $k_{ij}^{\zeta}(\boldsymbol{x}/\zeta)$ represents the oscillatory periodic heat conductivity tensor, and g^{ζ} is the heat generated per unit volume.

Develop coarse-scale equations for the heat conduction problem.

Problem 2.3: Consider an infinitesimally thin laminate plate in bending. Assume that the thickness of the plate is given by 2ζ and consider an asymptotic expansion of the displacements in the form of

$$u_i^{\zeta}(\boldsymbol{x}) = u_i^{(0)}(x_1, x_2, y_3) + \zeta u_i^{(1)}(x_1, x_2, y_3) + \cdots$$

where $x_3/y_3 = \zeta$. Show that

$$u_i = u_i^{(0)} - \frac{\partial u_3^{(0)}}{\partial x_i} x_3 \qquad i = 1,2$$

$$u_3 = u_3^{(0)} - \frac{L_{33ij}}{L_{3333}} \frac{\partial u_i^{(0)}}{\partial x_j} x_3$$

Problem 2.4: Consider a composite plate in bending that can be approximated by the classical Love–Kirchhoff assumption of the 3D displacement field of a plate

$$u_i^{\zeta}(\boldsymbol{x}, z) = u_i^{\zeta}(\boldsymbol{x}) - z\frac{\partial w^{\zeta}(\boldsymbol{x})}{\partial x_i}$$

$$u_z^{\zeta}(\boldsymbol{x}, z) = w^{\zeta}(\boldsymbol{x})$$

where $\boldsymbol{x} = [x_1, x_2]$ denotes the in-plane coordinates of the flat plate while the z direction coincides with the thickness direction of the laminate. Assume linear elasticity with infinitesimal strains. For the constitutive equations, assume plane stress conditions within each layer. Assume the following truncated asymptotic expansions

$$u_i^{\zeta}(\boldsymbol{x}) = u_i^{(0)}(\boldsymbol{x}, \boldsymbol{y}) + \zeta u_i^{(1)}(\boldsymbol{x}, \boldsymbol{y})$$
$$w^{\zeta}(\boldsymbol{x}) = w^{(0)}(\boldsymbol{x}, \boldsymbol{y}) + \zeta w^{(1)}(\boldsymbol{x}, \boldsymbol{y}) + \zeta^2 w^{(1)}(\boldsymbol{x}, \boldsymbol{y})$$

where $\boldsymbol{y} = \boldsymbol{x}/\zeta$. Derive the governing equations for the laminate plate.

Problem 2.5: Consider a porous medium with a periodic microstructure saturated with a viscous fluid. We denote the domain of this overall porous medium by Ω^{ζ} in the 3D space that can be divided into the fluid phase B^{ζ} and a solid phase. Assume that the fluid flowing in this porous medium is a Newtonian fluid with the governing equations for the steady-state Stokes flow given by

Equilibrium: $\sigma^{\zeta}_{ij,j} + \rho b_i = 0$ on B^{ζ}

Constitutive law: $\sigma^{\zeta}_{ij} = -p^{\zeta}\delta_{ij} + \mu^{\zeta}\dfrac{\partial v^{\zeta}_i}{\partial x_j}$ on B^{ζ}

Continuity: $v^{\zeta}_{i,i} = 0$ on B^{ζ}

No-slip condition: $v^{\zeta}_i = 0$ on S^{ζ}

where p^{ζ} is pressure, μ^{ζ} is fluid viscosity, v^{ζ}_i is fluid velocity, ρ is fluid density, ρb_i is the body force vector per unit volume, and S^{ζ} indicates the interface between the solid and fluid phases. Also, assume that the fluid viscosity is of order ζ^2, that is, $\mu^{\zeta} = \zeta^2\mu$, so that the effect of viscosity is negligible for the overall flow field. Consider the following asymptotic expansions for the velocity and pressure fields

$$v^{\zeta}_i(\boldsymbol{x}) = v^{(0)}_i(\boldsymbol{x},\boldsymbol{y}) + \zeta v^{(1)}_i(\boldsymbol{x},\boldsymbol{y}) + \zeta^2 v^{(2)}_i(\boldsymbol{x},\boldsymbol{y}) + \cdots$$

$$p^{\zeta}(\boldsymbol{x}) = p^{(0)}(\boldsymbol{x},\boldsymbol{y}) + \zeta p^{(1)}(\boldsymbol{x},\boldsymbol{y}) + \zeta^2 p^{(2)}_i(\boldsymbol{x},\boldsymbol{y}) + \cdots$$

a. Show that the leading order governing equations are given by

$$\frac{\partial p^{(0)}}{\partial x_i} - \mu\frac{\partial^2 v^{(0)}_i}{\partial y_p \partial y_p} = \rho b_i \quad \text{on} \quad \Omega\times\Theta$$

$$\frac{\partial v^{(0)}_i}{\partial y_i} = 0 \quad \text{on} \quad \Omega\times\Theta$$

b. Assume the solution of the form $v^{(0)}_i = -\left(\rho b_i - \dfrac{\partial p^{(0)}(\boldsymbol{y})}{\partial x_i}\right)H^j_i(\boldsymbol{y})$

show that the governing equations for influence function $H^j_i(\boldsymbol{y})$ are

$$\mu\frac{\partial^2 H^j_i}{\partial y_p \partial y_p} = \delta_{ij} \quad \text{on} \quad \Theta_B$$

where Θ_B is the fluid part of the domain restricted to a unit cell domain Θ.

c. Show that the generalized Darcy's law is given by

$$V_i \equiv \frac{1}{|\Theta|}\int_{\Theta} v^{(0)}_i d\Theta = K_{ij}\frac{\partial P}{\partial x_j}$$

where permeability K_{ij} and generalized pressure P are defined as

$$K_{ij} = \frac{1}{|\Theta|}\int_{\Theta} H_i^j d\Theta$$

$$P = p^0 - \rho b_i x_i$$

d. Show that the above generalized Darcy's law together with coarse-scale conservation of mass

$$\frac{\partial V_i}{\partial x_i} = 0 \quad \text{on} \quad \Omega$$

and boundary conditions

$$p^0 = \overline{p} \quad \text{on} \quad \partial\Omega^p$$

$$V_i n_i = \overline{q} \quad \text{on} \quad \partial\Omega^q$$

comprise the coarse-scale problem where $\overline{p}$ and $\overline{q}$ are the prescribed pressure and flux defined on the boundaries $\partial\Omega^p$ and $\partial\Omega^q$, respectively.

Problem 2.6: Show that coarse-grid iteration matrix $\boldsymbol{T}$ is a projection operator, that is,

$$\boldsymbol{T} = \boldsymbol{TT}$$

Problem 2.7: For multiple-scale linear homogenization, prove that the leading order term $u_i^{(0)}\left({}^{(0)}\boldsymbol{x}\right)$ in the asymptotic expansion (2.226) is a function of the coarsest scale coordinate only. Also, prove that the leading order correction $u_i^{(1)}\left({}^{(0)}\boldsymbol{x},{}^{(1)}\boldsymbol{x}\right)$ is a function of ${}^{(0)}\boldsymbol{x},{}^{(1)}\boldsymbol{x}$, and so on.

Problem 2.8: Assume that the body forces $\boldsymbol{b}^{\zeta}$ and tractions $\overline{\boldsymbol{t}}^{\zeta}$ are highly oscillatory functions

$$\boldsymbol{b}^{\zeta} = \zeta^{-1}\boldsymbol{b}^{(-1)}(\boldsymbol{x},\boldsymbol{y}) + \boldsymbol{b}^{(0)}(\boldsymbol{x},\boldsymbol{y}) + \cdots$$

$$\overline{\boldsymbol{t}}^{\zeta} = \zeta^{-1}\overline{\boldsymbol{t}}^{(-1)}(\boldsymbol{x},\boldsymbol{y}) + \overline{\boldsymbol{t}}^{(0)}(\boldsymbol{x},\boldsymbol{y}) + \cdots$$

where $\overline{\boldsymbol{t}}^{(-1)}$ denotes the pore pressure in the unit cell as shown in Figure 2.31. Derive the unit cell and coarse-scale equations. Is there any constraint on the choice of $\boldsymbol{b}^{(-1)}$ and $\overline{\boldsymbol{t}}^{(-1)}$?

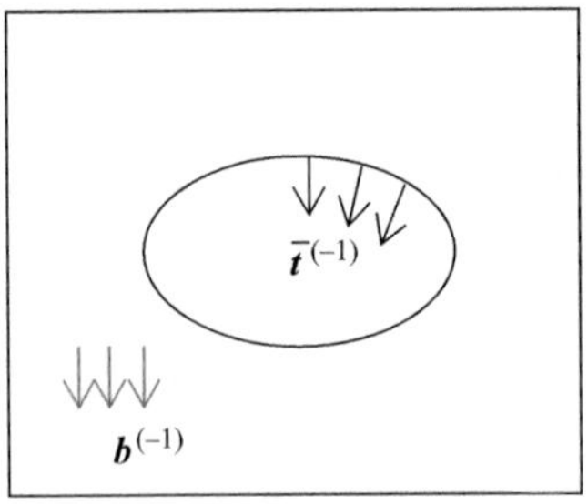

Figure 2.31 Unit cell with pore pressure

References

[1] Babuska, I. Upscaling and application. Mathematical and computational problems. In Numerical Solutions of Partial Differential Equations – III, SYNSPADE, ed. B Hubbard, Academic Press, 1976.

[2] Benssousan, A., Lions, J.L. and Papanicoulau, G. Asymptotic Analysis for Periodic Structures. North-Holland, 1978.

[3] Sanchez-Palencia, E. Non-homogeneous Media and Vibration Theory. Lecture Notes in Physics, Vol. 127. Springer-Verlag, 1980.

[4] Suquet, P. Elements of Upscaling for Inelastic Solid Mechanics. Lecture Notes in Physics, Vol. 272. Springer-Verlag, 1987, pp. 193–278.

[5] Bakhvalov, N.S. and Panassenko, G.P. Upscaling: Averaging Processes in Periodic Media. Kluwer Academic Publishers, 1989.

[6] Hill, R. Elastic properties of reinforced solids: some theoretical principles. Journal of the Mechanics and Physics of Solids 1963, 11, 357–372.

[7] E, W. and Engquist, B. The heterogeneous multi-scale methods. Communications in Mathematical Sciences 2003, 1, 87–133

[8] Babuska, I. Homogenization and its applications, mathematical and computational problems. In Numerical Solutions of Partial Differential Equations – III, SYNSPADE, ed. B. Hubbard. Academic Press, 1976, pp. 89–116.

[9] Babuska, I., Caloz, G. and Osborn, J. Special finite element methods for a class of second order elliptic problems with rough coefficients. SIAM Journal on Numerical Analysis 1994, 31, 945–981.

[10] Hou, T. and Wu, X. A multiscale finite element method for elliptic problems in composite materials and porous media. Journal of Computational Physics 1997, 134, 169–189.

[11] Hughes, T.J.R. Multiscale phenomena: Greens functions, the Dirichlet to Neumann formulation, subgrid scale models, bubbles and the origin of stabilized methods. Computer Methods in Applied Mechanics and Engineering 1995, 127, 387–401.

[12] Farhat, C., Harari, I. and Franca, L.P. The discontinuous enrichment method, Computer Methods in Applied Mechanics and Engineering 2001, 190, 6455–6479.

[13] Fish, J. and Yuan, Z. Multiscale enrichment based on the partition of unity. International Journal for Numerical Methods in Engineering 2005, 62(10), 1341–1359.

[14] Fish, J. and Yuan, Z. Multiscale enrichment based on partition of unity for nonperiodic fields and nonlinear problems. Computational Mechanics 2007, 40, 249–259.

[15] Voigt, W. Über die Beziehung zwischen den beiden Elastizitätskonstanten isotroper, Körper. Wiedemanns Annalen der Physik und Chemie 1889, 38, 573–587.

[16] Reuss, A. Berechnung der Fliessgrenze von Mischkristallen auf Grund der Plastizitätsbedingung für Einkristalle. Zeitschrift für Angewandte Mathematik und Mechanik 1929, 9, 49–58.

[17] Geers, M.G.D., Kouznetsova, V. and Brekelmans, W.A.M.Gradient-enhanced computational upscaling for the micro-macro scale transition. Journal de Physique IV 2001, 11, 145–152.

[18] Eshelby, J.D. The determination of the field of an ellipsoidal inclusion and related problems. Proceedings of the Royal Society London, Series A 1957, 241, 376–396.

[19] Hashin, Z. The elastic moduli of heterogeneous materials. Journal of Applied Mechancs 1962, 29, 143–150.

[20] Mori, T. and Tanaka, K. Average stress in the matrix and average elastic energy of materials with misfitting inclusions. Acta Metallurgica 1973, 21, 571–574.

[21] Hill, R. A self-consistent mechanics of composite materials. Journal of the Mechanics and Physics of Solids 1965, 13, 357–372.

[22] Christensen, R.M. and Lo, K.H. Solution for effective shear properties in three phase sphere and cylinder models. Journal of the Mechanics and Physics of Solids 1979, 27, 315–330.

[23] Feyel, F. and Chaboche, J.L. FE^2 multiscale approach for modelling the elastoviscoplastic behaviour of long fiber SiC/Ti composite materials. Computer Methods in Applied Mechanics and Engineering 2000, 183, 309–330.

[24] Suquet, P.M. Local and global aspects in the mathematical theory of plasticity. In Plasticity Today: Modelling, Methods and Applications, eds A. Sawczuk and G. Bianchi. Elsevier, 1985, pp. 279–310.

[25] Guedes, J.M. and Kikuchi, N. Preprocessing and postprocessing for materials based on the upscaling method with adaptive finite element methods. Computer Methods in Applied Mechanics and Engineering 1990, 83, 143–198.

[26] Terada, K. and Kikuchi, N. Nonlinear upscaling method for practical applications. In Computational Methods in Micromechanics, eds S. Ghosh and M. Ostoja-Starzewski. ASME, 1995, Vol. AMD-212/MD-62, pp. 1–16.

[27] Terada, K. and Kikuchi, N. A class of general algorithms for multi-scale analysis of heterogeneous media. Computer Methods in Applied Mechanics and Engineering 2001, 190, 5427–5464.
[28] Matsui, K., Terada, K. and Yuge, K.Two-scale finite element analysis of heterogeneous solids with periodic microstructures, Computers and Structures 2004, 82(7–8), 593–606.
[29] Ghosh, S., Lee, K. and Moorthy, S. Multiple scale analysis of heterogeneous elastic structures using upscaling theory and Voronoi cell finite element method. International Journal of Solids and Structures 1995, 32, 27–62.
[30] Ghosh, S., Lee, K. and Moorthy, S. Two scale analysis of heterogeneous elasticplastic materials with asymptotic upscaling and Voronoi cell finite element model. Computer Methods in Applied Mechanics and Engineering 1996, 132, 63–116.
[31] Miehe, C. and Koch, A. Computational micro-to-macro transition of discretized microstructures undergoing small strain. Archive of Applied Mechanics 2002, 72, 300–317.
[32] Smit, R.J.M., Brekelmans, W.A.M. and Meijer, H.E.H. Prediction of the mechanical behavior of nonlinear heterogeneous systems by multilevel finite element modeling. Computer Methods in Applied Mechanicsand Engineering 1998, 155, 181–192.
[33] Kouznetsova, V., Brekelmans, W.-A. and Baaijens, F.P.-T. An approach to micro-macro modeling of heterogeneous materials. Computational Mechanics. 2001, 27, 37–48.
[34] Michel, J.-C., Moulinec, H. and Suquet, P. Effective properties of composite materials with periodic microstructure: a computational approach. Computer Methods in Applied Mechanics and Engineering 1999, 172, 109–143.
[35] Yuan, Z. and Fish, J. Towards realization of computational homogenization in practice. International Journal for Numerical Methods in Engineering 2008, 73(3), pp. 361–380.
[36] Fish, J., Nayak, P. and Holmes, M.H. Microscale reduction error indicators and estimators for a periodic heterogeneous medium. Computational Mechanics: The International Journal 1994, 14, 1–16.
[37] Ostoja-Starzewski, M., Boccara, S.D. and Jasiuk, I. Couple-stress moduli and characteristic length of a two-phase composite. Mechanics Research Communications 1999, 26(4), 387–396.
[38] van der Sluis, O., Vosbeek, P.H.J., Schreurs, P.J.G. and Meijer, H.E.H. Upscaling of heterogeneous polymers. International Journal of Solids and Structures 1999, 36, 3193–3214.
[39] Kouznetsova, V., Geers, M.G.D. and Brekelmans, W.A.M. Multi-scale constitutive modelling of heterogeneous materials with a gradient-enhanced computational homogenization scheme. International Journal for Numerical Methods in Engineering 2002, 54, 1235–1260.
[40] Hashin, Z. and Shtrikman, S. On some variational principles in anisotropic and nonhomogeneous elasticity. Journal of the Mechanics and Physics of Solids 1962, 10, 335–342.
[41] Zohdi, T.J. and Wriggers, P. An Introduction to Computational Micromechanics. Springer, 2008.
[42] Caillerie, D. Thin elastic and periodic plates. Mathematical Methods in the Applied Sciences, 1984, 6(3), 159–191.
[43] Kohn, R.V. and Vogelius, M. A new model for thin plates with rapidly varying thickness. International Journal of Solids and Structures 1984, 20(4), 333–350.
[44] Drugan, W.J. and Willis, J.R. A micromechanics-based nonlocal constitutive equation and estimates of representative volume element size for elastic composites. Journal of the Mechanics and Physics of Solids 1996, 44, 497–524.
[45] Yeong., C.L.Y. and Torquato, S. Reconstructing random media. Physical Review E 1998, 57, 495–506.
[46] Lee, H., Brandyberry, M., Tudor, A. and Matous, K.Three-dimensional reconstruction of statistically optimal unit cells of polydisperse particulate composites from microtomography, Physical Review E 2009, 80, 061301.
[47] Hill, R. On constitutive macro-variables for heterogeneous solids at finite strain. Proceedings of the Royal Society of London, Series A 1972, 326, 131–147.
[48] E, W., Engquist, B., Li, X., Ren, W. and Vanden-Eijnden, E. Heterogeneous multiscale methods: a review. Communications in Computational Physics 2007, 2(3), 367–450.
[49] Malvern, L. Introduction to the Mechanics of a Continuous Medium. Prentice Hall, 1968.
[50] Ladeveze, P., ed., Local Effects in the Analysis of Structures. Elsevier, 1985.
[51] Dumontet, H. Homogeneisation et effets de bords dans les materiaux composites. Doctoral thesis, l'Université Pierre et Marie Curie, Paris, 1990.
[52] Lefik, M. and Schrefler, B. FE modelling of a boundary layer corrector for composites using the upscaling theory. Engineering Computations 1996, 13(6), 31–42.
[53] Jikov, V.V., Kozlov, S.M. and Oleinik, O.A. Upscaling of Differential Operators and Integral Functionals. Springer Verlag, 1994.
[54] Allaire, G. Upscaling and two-scale convergence. SIAM Journal on Mathematical Analysis 1992, 23(6), 1482–1518.

[55] Fish, J. The s-version of the finite element method. Computers and Structures 1992, 43(3), 539–547.
[56] Fish, J. and Markolefas, S. The s-version of the finite element method for multilayer laminates. International Journal for Numerical Methods in Engineering 1992, 33(5), 1081–1105.
[57] Fish, J. Hierarchical modeling of discontinuous fields. Communications in Applied Numerical Methods 1992, 8, 443–453.
[58] Allaire, G. and Bal, G. Upscaling of the criticality spectral equation in neutron transport. Mathematical Modelling and Numerical Analysis 1999, 33, 721–746.
[59] de Souza Neto, E.A. and Feijóo, R.A. On the equivalence between spatial and material volume averaging of stress in large strain multi-scale solid constitutive models, Mechanics of Materials 2008, 40, 803–811.
[60] Fish, J. and Fan, R. Mathematical homogenization of nonperiodic heterogeneous media subjected to large deformation transient loading. International Journal for Numerical Methods in Engineering 2008, 76(7), 1044–1064.
[61] Kouznetsova, V., Brekelmans, W.-A. and Baaijens, F.P.-T. An approach to micro-macro modeling of heterogeneous materials. Computational Mechanics 2001, 27, 37–48.
[62] Geers, M., Kouznetsova, V. and Brekelmans, W.-A.Multiscale first-order and second-order computational upscaling of microstructures towards continua. International Journal for Multiscale Computational Engineering 2003, 1, 371–386.
[63] Forest, S. and Sievert, R. Nonlinear microstrain theories. International Journal of Solids and Structures 2006, 43, 7224–7245.
[64] Aero, E.L. and Kuvshinskii, E.V. 'Fundamental equations of the theory of elastic materials with rotationally interacting particles.' Fizika Tverdogo Tela (St. Peterburg) 1960, 2, 1399–1409.
[65] Grioli, G. Elasticità asimmetrica.' Annali di Matematica Pura ed Applicata 1960, 50, 389–417.
[66] Truesdell, C. and Toupin, R.A. Classical Field Theories of Mechanics. Handbuch der Physik. Springer, 1960, Vol. III, p. 1.
[67] Mindlin, R. Micro-structure in linear elasticity. Archive for Rational Mechanics and Analysis 1964, 16, 51–78.
[68] Germain, P. The method of virtual power in continuum mechanics. Part 2: Microstructure. SIAM Journal on Applied Mathematics 1973, 25, 556–575.
[69] Maugin, G. Nonlocal theories or gradient-type theories: A matter of convenience? Archives of Mechanics 1979, 31, 15–26.
[70] Aifantis, E. The physics of plastic deformation. International Journal of Plasticity 1987, 3, 211–248.
[71] Cosserat, E. and Cosserat, F. Sur la mecanique generale. Comptes Rendus de l'Académie des Sciences Paris 1907, 145, 1139.
[72] Eringen, A. and Suhubi, E. Nonlinear theory of simple microelastic solids. International Journal of Engineering Science 1964, 2, 189–203 (389–404).
[73] Vernerey, F.J., Liu, W.K. and Moran, B. Multi-scale micromorphic theory for hierarchical materials. Journal of the Mechanics and Physics of Solids 2007, 55(12), 2603–265.
[74] Green, A.E. and Naghdi, P.M. A unified procedure for construction of theories of deformable media. II: Generalized continua. Proceedings of the Royal Society London, Series A 1995, 448(1934), 357–377.
[75] Eringen, A. Microcontinuum Field Theories: Vol. I Foundations and Solids; Vol. II Fluent Media. Springer, 1999 (Vol. I); 2001 (Vol. II).
[76] Diener, G., Hurrich, A. and Weissbarth, J. Bounds on the non-local effective elastic properties of composites. Journal of the Mechanics and Physics of Solids 1984, 32, 21–39.
[77] Triantafyllidis, N. and Bardenhagen, S. The influence of scale size on the stability of periodic solids and the role of associated higher order gradient continuum models. Journal of the Mechanics and Physics of Solids 1996, 44, 1891–1928.
[78] Forest, S. Upscaling methods and the mechanics of generalized continua. In International Seminar on Geometry, Continuum and Microstructure, ed. G. Maugin. Travaux en Cours No. 60. Hermann, 1999, pp. 35–48.
[79] Chen, W. and Fish, J. A dispersive model for wave propagation in periodic heterogeneous media based on homogenization with multiple spatial and temporal scales. Journal of Applied Mechanics 2001, 68(2), 153–161.
[80] Fish, J. and Chen, W. Uniformly valid multiple spatial-temporal scale modeling for wave propagation in heterogeneous media. Mechanics of Composite Materials and Structures 2001, 8, 81–99.
[81] Fish, J., Chen, W. and Nagai, G. Nonlocal dispersive model for wave propagation in heterogeneous media. Part 1: One-dimensional case. International Journal for Numerical Methods in Engineering 2002, 54, 331–346.
[82] Michel, J.-C., Moulinec, H. and Suquet, P. Effective properties of composite materials with periodic microstructure: a computational approach. Computer Methods in Applied Mechanics and Engineering 1999, 172, 109–143.

[83] McVeigh, C., Vernerey, F., Liu, W.K. and Brinson, L.C. Multiresolution analysis for material design. Computer Methods in Applied Mechanics and. Engineering 2006, 195, 5053–5076.

[84] Miehe, C. Strain-driven upscaling of inelastic microstructures and composites based on an incremental variational formulation. International Journal for Numerical Methods in Engineering 2002, 55, 1285–1322.

[85] Ostoja-Starzewski, M., Boccara, S.D. and Jasiuk, I. Couple-stress moduli and characteristic length of a two-phase composite. Mechanics Research Communications 1999, 26(4), 387–396.

[86] van der Sluis, O., Vosbeek, P.H.J., Schreurs, P.J.G. and Meijer, H.E.H. Upscaling of heterogeneous polymers. International Journal of Solids and Structures 1999, 36, 3193–3214.

[87] Kouznetsova, V.G., Geers, M.G.D. and Brekelmans, W.A.M. Multi-scale second-order computational upscaling of multi-phase materials: a nested finite element solution strategy. Computer Methods in Applied Mechanics and Engineering 2004, 193, 5525–5550.

[88] Fleck, N.A. and Hutchinson, J.W. Strain gradient plasticity. Advances in Applied Mechanics 1997, 33, 295–361.

[89] Shu, J.Y., King, W.E. and Fleck, N.A. Finite elements for materials with strain gradient effects. International Journal of Numerical Methods in Engineering 1999, 44, 373–391.

[90] Fish, J. and Belsky, V. Multigrid method for periodic heterogeneous media. Part 1: Convergence studies for one dimensional case. Computer Methods in Applied Mechanics and Engineering 1995, 126, 1–16.

[91] Fish, J. and Belsky, V. Multigrid method for periodic heterogeneous media. Part 2: Multiscale modeling and quality control in multidimensional case. Computer Methods in Applied Mechanics and Engineering 1995, 126, 17–38.

[92] Bayreuther, C.G. and Miehe, C. Coupling of homogenization techniques with multigrid solvers for unstructured meshes. In Lecture Notes in Applied and Computational Mechanics, eds W.L. Wendland and M. Effendiev. Springer, 2003, pp. 67–72.

[93] Miehe, C. and Bayreuther, G. On multiscale FE analyses of heterogeneous structures: from homogenization to multigrid solvers. International Journal for Numerical Methods in Engineering 2007, 71(10), 1135–1180.

[94] Neuss, N., Jager, W. and Wittum, G. Homogenization and multigrid. Computing 2001, 66, 1–26.

[95] Knapek, S. Matrix-dependent multigrid-homogenization for diffusion problems. SIAM Journal on Scientific Computing 1999, 20, 515–533.

[96] Moulton, J.D., Dendy, J.E. and Hyman, J.M. The black box multigrid numerical homogenization algorithm. Journal of Computational Physics 1998, 142, 80–108.

[97] Brandt, A. Multi-level adaptive solutions to boundary-value problems. Mathematics of Computations 1977, 31, 333–390.

[98] Hackbusch, W. Multigrid Methods and Applications. Springer, 1985.

[99] Hageman, L. and Young, D. Applied Iterative Methods. Academic Press, 1981.

[100] Briggs, W.L., Henson, V.E.and McCormick, S.F. A Multigrid Tutorial. Society for Industrial and Applied Mathematics, 2000.

[101] Saad, Y. Iterative Methods for Sparse Linear Systems, 2nd edition. SIAM, 2000.

[102] Brandt, A. Multi-level adaptive solutions to boundary-value problems. Mathematics of Computation 1977, 31(138), 333–390.

[103] Ortega, J.M. and Rheinboldt, W.C. Iterative Solution of Nonlinear Equations in Several Variables. Academic Press, 1970.

[104] Bramble, J.H., Kwak, D.Y. and Pasciak, J.E. Uniform convergence V-cycle iterations for indefinite nonsymmetric problems. SIAM Journal of Numerical Analysis 1994, 31(6), 1746–1763.

[105] Giddings, T.E. and Fish, J. An algebraic two-level preconditioner for asymmetric, positive-definite systems. International Journal for Numerical Methods in Engineering 1999, 45, 1433–1456.

[106] Fish, J. and Qu, Y. Global basis two-level method for indefinite systems. Part 1: Convergence studies. International Journal for Numerical Methods in Engineering 2000, 49, 439–460.

[107] Qu, Y. and Fish, J.Global basis two-level method for indefinite systems. Part 2: Computational issues. International Journal for Numerical Methods in Engineering 2000, 49, 461–478.

[108] Waisman, H., Fish, J., Tuminaro, R.S. and Shadid, J. The Generalized Global-Basis (GGB) method. International Journal for Numerical Methods in Engineering 2004, 61(8), 1243–1269.

[109] Waisman, H., Fish, J., Tuminaro, R.S. and Shadid, J. Accelerated Generalized Global-Basis (GGB) method for nonlinear problems. Journal of Computational Physics 2005, 210(1), 274–291.

[110] Sorensen, D.C. Implicitly restarted Arnoldi/Lanczos method for large scale eigenvalue calculations. In Parallel Numerical Algorithms, eds D.E. Keyes, A. Sameh and V. Venkatakrishnan. Kluwer, 1996, pp. 119–166.

3

Upscaling/Downscaling of Atomistic/Continuum Media

3.1 Introduction

Constructing thermo-mechanical continuum equations based on the motion of atoms has been a subject of significant interest in the physics, materials science, and mechanics communities. However, numerous challenges and major obstacles must be overcome before a link between atomistic and continuum media can be fully established. We will start by outlining some of the major challenges, then briefly review the state of the art in the field and enumerate a subset of issues addressed in this chapter.

The first difficulty is conceptual in nature, since the physics describing continuum media and atomistic media is different. While a continuum description of mechanical deformation can be explicitly derived from atomistics (and this, to a certain extent, has been successfully demonstrated), the thermal process can only be accounted for phenomenologically in the form of a heat transfer equation.

The second difficulty is associated with the formulation of the basic atomistic model required for developing phenomenological heat transfer equations. The mechanism by which heat is transferred depends on the material system. Gases, for example, transfer heat by direct collisions between molecules. Nonmetallic solids, such as ceramics, transfer heat by lattice vibrations so that there is no net motion of the media as the energy propagates through. Such heat transfer is often described in terms of *phonons* (the quanta of lattice vibrations). Metals, on the other hand, have free electrons that are not bound to any particular atom. As the electrons move, they undergo a series of collisions in which the faster electrons (on the hot side of the solid) give off some of their energy to the slower electrons. Conduction through electron collision is more effective than through lattice vibration; this is why metals generally are better heat conductors than ceramic materials, which do not have many free electrons. This implies that for metals, the base mathematical model that describes the motion of atoms using Newton's laws does not

Practical Multiscaling, First Edition. Jacob Fish.

contain sufficient information for development of a complete phenomenological model of heat transfer. Quantum mechanical considerations are unavoidable in this case.

The third obstacle is purely computational. Deterministic atomistic level computations, which numerically solve Newton's equations of motion, can model systems up to the order of 4×10^9 atoms for timescales of the order of nanoseconds [1], but this is still orders of magnitude below continuum length and timescales, which are of the order of millimeters and milliseconds, respectively. Continuum-level simulations can operate in the latter regimes, but they do so at the expense of explicit atomistic resolution. This difficulty can be partially circumvented by introducing an intermediate so-called coarse-grained model (or mesoscale model). Well-known examples of such coarse-grained models include dislocation dynamics and coarse-grained molecular dynamics.

There have been numerous attempts to reconcile the fine-scale descriptions and the continuum thermo-mechanical descriptions. One of the most fundamental approaches is based on statistical mechanics, which converts atomistic data to macroscopic observables such as pressure, energy, and heat capacities. In a somewhat related effort, Zhou [2] developed an equivalent deterministic thermo-mechanical continuum theory based on decomposing atomistic velocity into a structural deformation and thermal oscillation parts. Also noteworthy is an elegant exposition of the linkage between homogenization of continua and atomistic media given by Costanzo *et al.* [3] for mechanical fields. A similar starting point has been employed by Li and E [4] within the framework of the heterogeneous multiscale method (HMM) [5]. This method consists of solving the thermo-mechanical equations of continuum numerically and then finding the missing constitutive data (mechanical and thermal) by performing atomistic simulations subjected to local boundary conditions extracted from the continuum. An extension of the quasi-continuum method to a finite temperature regime has been proposed by Dupuy *et al.* [6] by incorporating the potential of mean force (PMF) originally introduced by Kirkwood in 1935 [7]. Several other noteworthy approaches originally developed for zero-temperature applications have been extended to finite temperatures. These include the coupled atomistics and discrete dislocation (CADD) method [8], the bridging scale method [9], and the bridging domain method [10].

This chapter is based on the generalized mathematical homogenization (GMH) theory [11,12,13,14,15,16], which is a generalization of the spatial homogenization approaches considered in Chapter 2, as applied to discrete media. Molecular statics homogenization procedures were devised in [17,18,19]. We will focus on linking molecular dynamics (MD), which describes the motion of atoms or coarse-grained discrete media, with thermo-mechanical continuum equations. Since the base model from which continuum equations are derived is molecular dynamics, only heat transfer due to lattice vibration (phonons) will be accounted for. The GMH approach resembles HMM [5], with the main difference being that the coarse-scale problem is derived directly from atomistics without making any a priori assumption about its mathematical structure. Numerical experiments will be conducted to verify GMH against the reference molecular dynamics solution.

3.2 Governing Equations

3.2.1 Molecular Dynamics Equation of Motion

We will consider a lattice of locally identical unit cells consisting of n atoms. The initial position of atom I in a reference configuration is denoted by $\boldsymbol{X}^I$. Lagrangian description is adopted. The displacement of atom I with respect to the reference position is designated by $\boldsymbol{u}^I$. Upon deformation of the lattice, the new position of atom I as denoted by $\boldsymbol{x}^I$ is given by

$$\begin{aligned}\boldsymbol{x}^I &= \boldsymbol{X}^I + \boldsymbol{u}^I \\ \boldsymbol{u}^I &= \boldsymbol{u}^I(\boldsymbol{X}^I, t)\end{aligned} \tag{3.1}$$

The vector separating atoms I and J in the reference configuration is given by

$$\boldsymbol{X}^{IJ} = \boldsymbol{X}^J - \boldsymbol{X}^I \tag{3.2}$$

The corresponding vector separating the two atoms in the deformed configuration is

$$\boldsymbol{x}^{IJ} = \boldsymbol{x}^J - \boldsymbol{x}^I = \boldsymbol{X}^{IJ} + \boldsymbol{u}^J(\boldsymbol{X}^J, t) - \boldsymbol{u}^I(\boldsymbol{X}^J, t) \tag{3.3}$$

In this chapter, Roman uppercase superscripts I and J are reserved for labels applied to atoms and will not be subject to summation convention. Spatial directions, for which summation convention over repeated indices is applied, will be denoted by lowercase subscripts.

For simplicity, we will focus our attention on pairwise potentials. For pairwise potentials, the interaction between atoms I and J is depicted by the interatomic potential $\Phi^{IJ}(\boldsymbol{x}^{IJ})$. The interatomic force $\boldsymbol{f}^{IJ}$ applied to atom I by atom J is derived as the gradient of the total potential energy of the system

$$\boldsymbol{f}^{IJ} = \frac{\partial \Phi^{IJ}(\boldsymbol{x}^{IJ})}{\partial x^{IJ}} \frac{\boldsymbol{x}^{IJ}}{x^{IJ}} \tag{3.4}$$

where $x^{IJ} = |\boldsymbol{x}^{IJ}|$ is the distance between atoms I and J and where $\boldsymbol{x}^{IJ}/x^{IJ}$ is the unit vector in the direction of $\boldsymbol{x}^{IJ}$. In view of how the force on a particle is calculated, and borrowing the language of the continua, we could say that the particle ensemble is typically taken to be a hyperelastic material, that is, a material whose internal response is completely characterized by a stored energy function.

The equation of motion for atom I can be written as

$$m^I \frac{d^2 \boldsymbol{u}^I}{dt^2} = \sum_{J \neq I} \boldsymbol{f}^{IJ}(\boldsymbol{x}^{IJ}) \tag{3.5}$$

where m^I is the mass of atom I; $d\boldsymbol{u}^I/dt$ represents the material time derivative of $\boldsymbol{u}^I$; and the superscript J denotes the neighboring atoms that interact with atom I, such that $|\boldsymbol{x}^J - \boldsymbol{x}^I| < r_c$, with r_c being the cutoff radius. For simplicity, external forces will not be considered.

Due to a locally periodic lattice structure, the mass of atom m^I and the interatomic force $\boldsymbol{f}^{IJ}$ are assumed to be locally periodic. Attention is restricted to cases where the wavelength of the traveling signal λ is much larger than the size of the unit cell l, that is, $\zeta = l/\lambda \ll 1$.

3.2.2 Multiple Spatial and Temporal Scales and Rescaling of the MD Equations

Due to the rapidly varying interatomic potentials, two distinct material coordinates are employed to describe the heterogeneity at the atomistic level: (i) the coarse scale denoted by $\boldsymbol{X}$, at which scale the atomistic features are invisible; and (ii) the atomistic scale or fine scale, denoted by $\boldsymbol{Y}$. The two scales are related by

$$\boldsymbol{Y} = \boldsymbol{X}/\zeta \qquad 0 < \zeta \ll 1 \tag{3.6}$$

The corresponding spatial scales are denoted by $\boldsymbol{x}$ and $\boldsymbol{y}$, respectively, and are related by $\boldsymbol{y} = \boldsymbol{x}/\zeta$.

In addition to the continuum (slow) timescale, we introduce a fast timescale τ in order to model lattice vibration (phonons) at a finite temperature. The fast timescale is related to the usual timescale by

$$\tau = t/\zeta \qquad 0 < \zeta \ll 1 \tag{3.7}$$

The resulting displacement field and its derivatives are functions of $\boldsymbol{X}, \boldsymbol{Y}, t$ and τ. Thus, the first two material time derivatives of the displacement field are given by

$$\begin{aligned} \frac{d\boldsymbol{u}(\boldsymbol{X},\boldsymbol{Y}^I,t,\tau)}{dt} &= \left(\zeta^{-1}\frac{\partial}{\partial\tau}+\frac{\partial}{\partial t}\right)\boldsymbol{u}^I \\ \frac{d^2\boldsymbol{u}(\boldsymbol{X},\boldsymbol{Y}^I,t,\tau)}{dt^2} &= \left(\zeta^{-2}\frac{\partial^2}{\partial\tau^2}+\zeta^{-1}2\frac{\partial^2}{\partial t\partial\tau}+\frac{\partial^2}{\partial t^2}\right)\boldsymbol{u}^I \end{aligned} \tag{3.8}$$

A multiple-scale asymptotic expansion is employed to approximate the displacement field as

$$\boldsymbol{u}(\boldsymbol{X},\boldsymbol{Y},t,\tau) = \boldsymbol{u}^{(0)}(\boldsymbol{X},t) + \zeta\boldsymbol{u}^{(1)}(\boldsymbol{X},\boldsymbol{Y},t,\tau) + \cdots \tag{3.9}$$

where the leading order displacement $\boldsymbol{u}^{(0)}$ is termed a coarse-scale displacement (see Chapter 2) and $\boldsymbol{u}^c \equiv \boldsymbol{u}^{(0)}$ is assumed to be independent of the fine-scale coordinates $\boldsymbol{Y}$ and τ. In the present context, $\boldsymbol{u}^c$ denotes continuum displacements. For the $O(1)$ homogenization pursued in this chapter, only the first two terms in the asymptotic expansion are taken into account; the remaining higher order terms are neglected. Note that all the displacement terms in the asymptotic expansion (3.9) are $O(1)$ quantities. Higher order theory can be developed by assuming the existence of higher order coarse-scale gradients within an atomistic unit cell along the lines outlined in Chapter 2, section 2.5.

Inserting the asymptotic expansion (3.9) into (3.8) yields

$$\begin{aligned} \frac{d\boldsymbol{u}(\boldsymbol{X},\boldsymbol{Y}^I,t,\tau)}{dt} &= \frac{\partial\boldsymbol{u}^c(\boldsymbol{X},t)}{\partial t}+\frac{\partial\boldsymbol{u}^{(1)}(\boldsymbol{X},\boldsymbol{Y}^I,t,\tau)}{\partial\tau}+O(\zeta) \\ \frac{d^2\boldsymbol{u}(\boldsymbol{X},\boldsymbol{Y}^I,t,\tau)}{dt^2} &= \zeta^{-1}\frac{\partial^2\boldsymbol{u}^{(1)}(\boldsymbol{X},\boldsymbol{Y}^I,t,\tau)}{\partial\tau^2}+\frac{\partial^2\boldsymbol{u}^c(\boldsymbol{X},t)}{\partial t^2}+2\frac{\partial^2\boldsymbol{u}^{(1)}(\boldsymbol{X},\boldsymbol{Y}^I,t,\tau)}{\partial t\partial\tau}+O(\zeta) \end{aligned} \tag{3.10}$$

The resulting displacement, velocity, and acceleration are functions of $\boldsymbol{X}, \boldsymbol{Y}, t$, and τ. It can be seen that the leading order velocity is of order $O(1)$, whereas the leading order acceleration is of order $O(\zeta^{-1})$ and arises from the oscillatory portion of the displacement field $\boldsymbol{u}^{(1)I} \equiv \boldsymbol{u}^{(1)}(\boldsymbol{X}, \boldsymbol{Y}^I, t, \tau)$ rather than from the coarse-scale displacement $\boldsymbol{u}^c(\boldsymbol{X}, t)$.

We further define the temporal averaging operator as

$$\langle \chi(\tau) \rangle = \frac{1}{\tau_0} \int_0^{\tau_0} \chi(\tau) d\tau \tag{3.11}$$

with τ_0 as a period. We assume that various fields $\chi(\tau)$ are τ-periodic, i.e., periodic in the fast timescale. Applying (3.11) to (3.10) and taking advantage of τ-periodicity yields the coarse-scale velocity $\boldsymbol{v}^c(\boldsymbol{X}, t)$ and acceleration $\boldsymbol{a}^c(\boldsymbol{X}, t)$

$$\begin{aligned} \boldsymbol{v}^c(\boldsymbol{X}, t) &= \frac{\partial \boldsymbol{u}^c(\boldsymbol{X}, t)}{\partial t} \equiv \dot{\boldsymbol{u}}^c(\boldsymbol{X}, t) \\ \boldsymbol{a}^c(\boldsymbol{X}, t) &= \frac{\partial^2 \boldsymbol{u}^c(\boldsymbol{X}, t)}{\partial t^2} \equiv \ddot{\boldsymbol{u}}^c(\boldsymbol{X}, t) \end{aligned} \tag{3.12}$$

Prior to carrying out the multiple-scale asymptotic analysis, it is necessary to *rescale* the molecular dynamics equation (3.5). Assuming that the nondimensional material density is $\rho_0 \sim O(1)$ and the characteristic size of the unit cell is $l \sim O(\zeta)$, then the volume of the unit cell is $|\Theta| \sim O(\zeta^3)$ and the mass is $m^I \sim \rho_0 |\Theta| \sim O(\zeta^3)$. And since $\ddot{\boldsymbol{u}}^I \sim O(\zeta^{-1})$, it follows from equation (3.5) that

$$\boldsymbol{f}^{IJ} = \boldsymbol{f}^{IJ}(\boldsymbol{x}, \boldsymbol{Y}^{IJ}, t, \tau) \sim O(\zeta^2) \tag{3.13}$$

where the coarse- and fine-scale coordinates are related by equation (3.6); thus, we have

$$\boldsymbol{X}^{IJ} = \zeta \boldsymbol{Y}^{IJ} = \zeta(\boldsymbol{Y}^J - \boldsymbol{Y}^I) \tag{3.14}$$

and $\boldsymbol{Y}^{IJ} \sim O(1)$.

To identify the terms of equal order, we introduce the following normalized quantities

$$\begin{aligned} \bar{m}^I &= m^I / \zeta^3 \sim O(1) \\ \bar{\boldsymbol{f}}^{IJ} &= \boldsymbol{f}^{IJ} / \zeta^2 \sim O(1) \end{aligned} \tag{3.15}$$

Further assuming the mass to be a function of the fast spatial coordinate only, $\bar{m}^I = \bar{m}^I(\boldsymbol{Y}^I)$, the Lagrangian description of the rescaled MD equation of motion is given by

$$\bar{m}^I \frac{d^2 \boldsymbol{u}^I}{dt^2} = \frac{1}{\zeta} \sum_{J \neq I} \bar{\boldsymbol{f}}^{IJ} \tag{3.16}$$

Note that based on (3.10), acceleration is of order $O(\zeta^{-1})$.

3.3 Generalized Mathematical Homogenization

3.3.1 Multiple-Scale Asymptotic Analysis

We will assume that (i) the coarse-scale coordinate $\boldsymbol{X}$ takes a continuous series of values, (ii) the displacement $\boldsymbol{u}^I$ is continuous and differentiable in $\boldsymbol{X}$, and (iii) the fine-scale coordinate $\boldsymbol{Y}$ is discrete. We denote the displacement of atom I by $\boldsymbol{u}(\boldsymbol{X},\boldsymbol{Y}^I,t,\tau)$. The displacements of the neighboring atoms $\boldsymbol{u}(\boldsymbol{X},\boldsymbol{Y}^J,t,\tau)$ can be expanded using a Taylor series around point $\boldsymbol{X}$ (which denotes the position of atom I) as

$$
\begin{aligned}
u_i^J &= u_i(\boldsymbol{X},\boldsymbol{Y}^J,t,\tau)+\frac{\partial u_i(\boldsymbol{X},\boldsymbol{Y}^J,t,\tau)}{\partial X_j}\cdot X_j^{IJ} \\
&+\frac{1}{2}\frac{\partial^2 u_i(\boldsymbol{X},\boldsymbol{Y}^J,t,\tau)}{\partial X_j \partial X_k}\cdot X_j^{IJ}X_k^{IJ}+\cdots
\end{aligned}
\tag{3.17}
$$

Applying the unit cell infinitesimality assumption made in Chapter 2, we assume infinitesimality of the distance between neighboring atoms. Thus, the position of atom I is arbitrary and, therefore, $\boldsymbol{X}^I$ is replaced by $\boldsymbol{X}$. From (3.17) we have

$$
\begin{aligned}
u_i^J-u_i^I &= u_i(\boldsymbol{X},\boldsymbol{Y}^J,t,\tau)-u_i(\boldsymbol{X},\boldsymbol{Y}^I,t,\tau)+\frac{\partial u_i(\boldsymbol{X},\boldsymbol{Y}^J,t,\tau)}{\partial X_j}\cdot X_j^{IJ} \\
&+\frac{1}{2}\frac{\partial^2 u_i(\boldsymbol{X},\boldsymbol{Y}^J,t,\tau)}{\partial X_j \partial X_k}\cdot X_j^{IJ}X_k^{IJ}+\cdots
\end{aligned}
\tag{3.18}
$$

Inserting (3.14) into (3.18) yields

$$
\begin{aligned}
u_i^J-u_i^I &= u_i(\boldsymbol{X},\boldsymbol{Y}^J,t,\tau)-u_i(\boldsymbol{X},\boldsymbol{Y}^I,t,\tau)+\zeta\frac{\partial u_i(\boldsymbol{X},\boldsymbol{Y}^J,t,\tau)}{\partial X_j}\cdot Y_j^{IJ} \\
&+\frac{1}{2}\zeta^2\frac{\partial^2 u_i(\boldsymbol{X},\boldsymbol{Y}^J,t,\tau)}{\partial X_j \partial X_k}\cdot Y_j^{IJ}Y_k^{IJ}+\cdots
\end{aligned}
\tag{3.19}
$$

Further inserting the asymptotic expansion (3.9) into (3.19) yields

$$
\begin{aligned}
u_i^J-u_i^I &= \zeta\left(u_i^{(1)}(\boldsymbol{X},\boldsymbol{Y}^J,t,\tau)-u_i^{(1)}(\boldsymbol{X},\boldsymbol{Y}^I,t,\tau)+\frac{\partial u_i^c(\boldsymbol{X},t)}{\partial X_j}\cdot Y_j^{IJ}\right) \\
&+\zeta^2\left(\frac{\partial u_i^{(1)}(\boldsymbol{X},\boldsymbol{Y}^J,t,\tau)}{\partial X_j}\cdot Y_j^{IJ}+\frac{1}{2}\frac{\partial^2 u_i^c(\boldsymbol{X},t)}{\partial X_j \partial X_k}\cdot Y_j^{IJ}Y_k^{IJ}\right)+\cdots
\end{aligned}
\tag{3.20}
$$

Inserting (3.20) into (3.3) yields

$$\boldsymbol{x}^{IJ} = \boldsymbol{X}^{IJ} + \boldsymbol{u}^{J} - \boldsymbol{u}^{I} = \zeta \boldsymbol{\phi}^{IJ} + \zeta^2 \boldsymbol{\psi}^{IJ} + \cdots \tag{3.21}$$

and

$$\boldsymbol{y}^{IJ} = \boldsymbol{x}^{IJ}/\zeta = \boldsymbol{\phi}^{IJ} + \zeta \boldsymbol{\psi}^{IJ} + \cdots \tag{3.22}$$

where

$$\begin{aligned}
\phi_i^{IJ} &= F_{ij}^c(\boldsymbol{X},t) Y_j^{IJ} + u_i^{(1)}(\boldsymbol{X},\boldsymbol{Y}^J,t,\tau) - u_i^{(1)}(\boldsymbol{X},\boldsymbol{Y}^I,t,\tau) \\
\psi_i^{IJ} &= \frac{\partial u_i^{(1)}(\boldsymbol{X},\boldsymbol{Y}^J,t,\tau)}{\partial X_j} Y_j^{IJ} + \frac{1}{2} \frac{\partial^2 u_i^c(\boldsymbol{X},t)}{\partial X_j \, \partial X_k} Y_j^{IJ} Y_k^{IJ}
\end{aligned} \tag{3.23}$$

with $F_{ij}^c(\boldsymbol{X},t)$ being the coarse-scale deformation gradient

$$F_{ij}^c(\boldsymbol{X},t) = \delta_{ij} + \frac{\partial u_i^c(\boldsymbol{X},t)}{\partial X_j} \tag{3.24}$$

Since $\boldsymbol{\phi}^{IJ} \sim O(1)$ and $\boldsymbol{\psi}^{IJ} \sim O(1)$, due to the fact that $u_i^{(1)}$, u_i^c and their derivatives are $O(1)$, we have

$$\frac{\left\| \zeta^2 \boldsymbol{\psi}^{IJ} \right\|}{\left\| \zeta \boldsymbol{\phi}^{IJ} \right\|} \sim O(\zeta) \tag{3.25}$$

where $\|\cdot\|$ denotes a vector norm. Following (3.25), the interatomic force field can be expanded around the leading order term

$$\begin{aligned}
\bar{f}_i^{IJ} &= \bar{f}_i^{IJ}(\zeta \boldsymbol{y}^{IJ}) = \bar{f}_i^{IJ}(\zeta \boldsymbol{\phi}^{IJ} + \zeta^2 \boldsymbol{\psi}^{IJ} + \cdots) \\
&= \bar{f}_i^{IJ}(\zeta \boldsymbol{\phi}^{IJ}) + \left. \frac{\partial \bar{f}_i^{IJ}}{\partial y_k^{IJ}} \frac{\partial y_k^{IJ}}{\partial x_j^{IJ}} \right|_{\boldsymbol{y}^{IJ} = \boldsymbol{\phi}^{IJ}} \zeta^2 \psi_j^{IJ} + O(\zeta^2) \\
&= \hat{f}_i^{IJ} + \zeta \left. \frac{\partial \bar{f}_i^{IJ}}{\partial y_j^{IJ}} \right|_{\boldsymbol{\phi}^{IJ}} \psi_j^{IJ} + O(\zeta^2)
\end{aligned} \tag{3.26}$$

where

$$\hat{f}_i^{IJ} = \bar{f}_i^{IJ}(\zeta \boldsymbol{\phi}^{IJ}) \tag{3.27}$$

Inserting (3.26) and (3.10) into the rescaled molecular dynamics equations of motion (3.16) yields

$$\bar{m}^I\left(\zeta^{-1}\frac{\partial^2 u_i^{(1)I}}{\partial\tau^2}+\frac{\partial^2 u_i^c}{\partial t^2}+2\frac{\partial^2 u_i^{(1)I}}{\partial t\,\partial\tau}+\zeta\frac{\partial^2 u_i^{(1)I}}{\partial t^2}+O(\zeta^2)\right)$$
$$=\sum_{J\neq I}\left(\zeta^{-1}\hat{\bar{f}}_i^{IJ}+\left.\frac{\partial\hat{\bar{f}}_i^{IJ}}{\partial y_j^{IJ}}\right|_{\phi^{IJ}}\psi_j^{IJ}+O(\zeta)\right) \tag{3.28}$$

Collecting terms of equal power of ζ gives different orders of equations of motion starting at $O(\zeta^{-1})$

$$O(\zeta^{-1}):\quad \bar{m}^I\frac{\partial^2 u_i^{(1)I}}{\partial\tau^2}=\sum_{J\neq I}\hat{\bar{f}}_i^{IJ} \tag{3.29}$$

$$O(1):\quad \bar{m}^I\left(\frac{\partial^2 u_i^c}{\partial t^2}+2\frac{\partial^2 u_i^{(1)I}}{\partial t\partial\tau}\right)=\sum_{J\neq I}\left.\frac{\partial\bar{f}_i^{IJ}}{\partial y_j^{IJ}}\right|_{\phi^{IJ}}\psi_j^{IJ} \tag{3.30}$$

3.3.2 The Dynamic Atomistic Unit Cell Problem

Consider the $O(\zeta^{-1})$ equation of motion (3.29) first. Inserting the normalized mass and interatomic force (3.15) into (3.29) yields the dynamic atomistic unit cell problem

$$m^I\frac{\partial^2\hat{\boldsymbol{u}}^{(1)}(\boldsymbol{X},\boldsymbol{Y}^I,t,\tau)}{\partial\tau^2}=\zeta^2\sum_{I\neq J}\hat{\boldsymbol{f}}^{IJ}\quad\forall I \tag{3.31}$$

where

$$\hat{\boldsymbol{u}}^{(1)}(\boldsymbol{X},\boldsymbol{Y}^I,t,\tau)=\zeta\boldsymbol{u}^{(1)}(\boldsymbol{X},\boldsymbol{Y}^I,t,\tau);\quad \hat{\boldsymbol{f}}^{IJ}=\boldsymbol{f}^{IJ}(\zeta\boldsymbol{\phi}^{IJ}) \tag{3.32}$$

In the above, $\hat{\boldsymbol{u}}^{(1)}$ can be interpreted as a correction to the classical Cauchy–Born rule [20]. Equation (3.31) reflects the fact that when a coarse-scale deformation gradient is imposed on an atomistic unit cell, the deformation field becomes nonuniform, that is, an internal relaxation occurs and the corresponding inhomogeneous atomic displacements are determined by the dynamic equilibrium condition of each atom in the unit cell. Note that the atomistic unit cell problem (3.31), which describes lattice vibration (phonons), depends on the fast time coordinate.

In accordance with Chapter 2, section 2.4, the fine-scale displacement of atoms u_i^f, which is defined to be free of rigid body displacements $u_i^c(\boldsymbol{X},t)$ and velocities $v_i^c(\boldsymbol{X},t)$, is given by

$$u_i^f(\boldsymbol{X},\boldsymbol{Y},t,\tau)=(F_{ij}^c(\boldsymbol{X},t)-\delta_{ij})Y_j+u_i^{(1)}(\boldsymbol{X},\boldsymbol{Y},t,\tau) \tag{3.33}$$

Note that the internal force in (3.31) depends on u_i^f, whereas the acceleration is computed for the deviation from the average $\hat{\boldsymbol{u}}^{(1)}$.

3.3.3 The Coarse-Scale Equations of Motion

We now proceed by considering the $O(1)$ equation of motion (3.30). Inserting the normalized mass and interatomic force (3.15) into (3.30) yields

$$m^I\left(\frac{\partial^2 u_i^c(\boldsymbol{X},t)}{\partial t^2}+2\frac{\partial^2 u_i^{(1)}(\boldsymbol{X},\boldsymbol{Y}^I,t,\tau)}{\partial t\partial\tau}\right)=\zeta\sum_{J\neq I}\left.\frac{\partial f_i^{IJ}}{\partial y_j^{IJ}}\right|_{\phi^{IJ}}\psi_j^{IJ} \tag{3.34}$$

Summing up equation (3.34) for all (n) atoms in the unit cell, and dividing the resulting equation by the volume of the unit cell $|\Theta|$, yields

$$\frac{1}{|\Theta|}\sum_{I=1}^{n}\left[m^I\left(\frac{\partial^2 u_i^c(\boldsymbol{X},t)}{\partial t^2}+2\frac{\partial^2 u_i^{(1)}(\boldsymbol{X},\boldsymbol{Y}^I,t,\tau)}{\partial t\partial\tau}\right)\right]=\frac{\zeta}{|\Theta|}\sum_{I=1}^{n}\sum_{J\neq I}\left.\frac{\partial f_i^{IJ}}{\partial y_j^{IJ}}\right|_{\phi^{IJ}}\psi_j^{IJ} \tag{3.35}$$

Applying the temporal averaging operator (3.11) to (3.35), we have

$$\frac{1}{|\Theta|}\sum_{I=1}^{n}m^I\ddot{u}_i^c(X,t)=\frac{\zeta}{|\Theta|}\left\langle\sum_{I=1}^{n}\sum_{J\neq I}\left.\frac{\partial f_i^{IJ}}{\partial y_j^{IJ}}\right|_{\phi^{IJ}}\psi_j^{IJ}\right\rangle \tag{3.36}$$

where we accounted for the fact that the second term vanishes due to τ-periodicity. Further defining mass density ρ_0 as

$$\rho_0=\frac{1}{|\Theta|}\sum_{I=1}^{n}m^I \tag{3.37}$$

yields

$$\rho_0\,\ddot{u}_i^c(\boldsymbol{X},t)=\frac{\zeta}{|\Theta|}\left\langle\sum_{I=1}^{n}\sum_{J\neq I}\left.\frac{\partial f_i^{IJ}}{\partial y_j^{IJ}}\right|_{\phi^{IJ}}\psi_j^{IJ}\right\rangle \tag{3.38}$$

Exploiting both the chain rule, (3.22), and (3.23) yields

$$\begin{aligned}f_{i,X_j}^{IJ}&=\frac{\partial f_i^{IJ}}{\partial y_k^{IJ}}\frac{\partial y_k^{IJ}}{\partial X_j}=\frac{\partial f_i^{IJ}}{\partial y_k^{IJ}}\frac{\partial\phi_k^{IJ}}{\partial X_j}+O(\zeta)\\&=\frac{\partial f_i^{IJ}}{\partial y_k^{IJ}}\left(\frac{\partial^2 u_k^c(\boldsymbol{X},t)}{\partial X_m\partial X_j}Y_m^{IJ}+\frac{\partial u_k^{(1)}(\boldsymbol{X},\boldsymbol{Y}^J,t,\tau)}{\partial X_j}-\frac{\partial u_k^{(1)}(\boldsymbol{X},\boldsymbol{Y}^I,t,\tau)}{\partial X_j}\right)+O(\zeta)\end{aligned} \tag{3.39}$$

From (3.23) and the $O(1)$ term of (3.39), we have

$$\begin{aligned}
\left.\frac{\partial f_i^{IJ}}{\partial y_j^{IJ}}\right|_{\phi^{IJ}} \psi_j^{IJ} &= \left.\frac{\partial f_i^{IJ}}{\partial y_j^{IJ}}\right|_{\phi^{IJ}} \left(\frac{\partial u_j^{(1)}(\mathbf{X},\mathbf{Y}^J,t,\tau)}{\partial X_m} + \frac{1}{2}\frac{\partial^2 u_j^c(\mathbf{X},t)}{\partial X_m \partial X_k} Y_k^{IJ}\right) Y_m^{IJ} \\
&= \left.\frac{\partial f_i^{IJ}}{\partial y_j^{IJ}}\right|_{\phi^{IJ}} \frac{1}{2}\left(\frac{\partial u_j^{(1)}(\mathbf{X},\mathbf{Y}^J,t,\tau)}{\partial X_m} - \frac{\partial u_j^{(1)}(\mathbf{X},\mathbf{Y}^I,t,\tau)}{\partial X_m} + \frac{\partial^2 u_j^c(\mathbf{X},t)}{\partial X_m \partial X_k} Y_k^{IJ}\right) Y_m^{IJ} \\
&\quad + \left.\frac{\partial f_i^{IJ}}{\partial y_j^{IJ}}\right|_{\phi^{IJ}} \frac{1}{2}\left(\frac{\partial u_j^{(1)}(\mathbf{X},\mathbf{Y}^I,t,\tau)}{\partial X_m} + \frac{\partial u_j^{(1)}(\mathbf{X},\mathbf{Y}^J,t,\tau)}{\partial X_m}\right) Y_m^{IJ} \\
&= \frac{1}{2}\frac{\partial}{\partial X_j}\left(f_i^{IJ} Y_j^{IJ}\right) + \frac{1}{2}\left.\frac{\partial f_i^{IJ}}{\partial y_j^{IJ}}\right|_{\phi^{IJ}} \frac{\partial}{\partial X_m}\left(u_j^{(1)}(\mathbf{X},\mathbf{Y}^I,t,\tau) + u_j^{(1)}(\mathbf{X},\mathbf{Y}^J,t,\tau)\right) Y_m^{IJ}
\end{aligned} \tag{3.40}$$

where we assume that Y_j^{IJ} is independent of X_j.

Inserting (3.40) into (3.38) yields

$$\begin{aligned}
\rho_0 \ddot{u}_i^c(\mathbf{X},t) &= \frac{1}{2|\Theta|}\left\langle \sum_{I=1}^{n}\sum_{J\neq I} \frac{\partial}{\partial X_j}\left(f_i^{IJ} X_j^{IJ}\right)\right\rangle \\
&\quad + \frac{\zeta}{2|\Theta|}\left\langle \sum_{I=1}^{n}\sum_{J\neq I} \left.\frac{\partial f_i^{IJ}}{\partial y_j^{IJ}}\right|_{\phi^{IJ}} \frac{\partial}{\partial X_m}\left(u_j^{(1)}(\mathbf{X},\mathbf{Y}^I,t,\tau) + u_j^{(1)}(\mathbf{X},\mathbf{X}^J,t,\tau)\right) Y_m^{IJ}\right\rangle
\end{aligned} \tag{3.41}$$

In the remainder of this section, we show that the second term on the right-hand side of (3.41) vanishes for a spatially periodic atomistic lattice. We start by recalling

$$\mathbf{x}^{JI} = \mathbf{x}^I - \mathbf{x}^J = \mathbf{X}^I - \mathbf{X}^J + \mathbf{u}^I - \mathbf{u}^J = -\mathbf{x}^{IJ} = \zeta\boldsymbol{\phi}^{JI} + \zeta^2\boldsymbol{\psi}^{JI}(\mathbf{X},\mathbf{Y},t,\tau) + \cdots \tag{3.42}$$

where

$$\phi_i^{JI} = F_{ij}^c(\mathbf{X},t) Y_j^{JI} + u_i^{(1)}(\mathbf{X},\mathbf{Y}^I,t,\tau) - u_i^{(1)}(\mathbf{X},\mathbf{Y}^J,t,\tau) = -\phi_i^{IJ} \tag{3.43}$$

$$\psi_i^{JI} = \frac{\partial u_i^{(1)}(\mathbf{X},\mathbf{Y}^J,t,\tau)}{\partial X_j} Y_j^{JI} - \frac{1}{2}\frac{\partial^2 u_i^c(\mathbf{X},t)}{\partial X_j \partial X_k} Y_j^{JI} Y_k^{JI} = -\psi_i^{IJ} \tag{3.44}$$

Newton's third law requires

$$\boldsymbol{f}^{IJ} = -\boldsymbol{f}^{JI} \tag{3.45}$$

From (3.43) and (3.45), we have the following relation

$$\frac{\partial f_i^{IJ}}{\partial y_j^{IJ}} = -\frac{\partial f_i^{JI}}{\partial y_j^{IJ}} = -\frac{\partial f_i^{JI}}{\partial\left(-y_j^{JI}\right)} = \frac{\partial f_i^{JI}}{\partial y_j^{JI}} \tag{3.46}$$

Let the interacting neighbor atoms of atom I be denoted as $N_1, N_2, \ldots, N_p, \ldots, N_K$, where K is the number of the interacting atoms such that $\left|\boldsymbol{x}^{N_p} - \boldsymbol{x}^I\right| < r_c$, $p = 1, 2, \ldots, K$ and where r_c is the cutoff radius. The second term on the right-hand side of (3.41) can then be written as

$$\begin{aligned}
&\sum_{I=1}^{n}\sum_{J\neq I} \left.\frac{\partial f_i^{IJ}}{\partial y_j^{IJ}}\right|_{\boldsymbol{\phi}^{IJ}} \frac{\partial}{\partial X_m}\left(u_j^{(1)}(\boldsymbol{X},\boldsymbol{Y}^I,t,\tau) + u_j^{(1)}(\boldsymbol{X},\boldsymbol{Y}^J,t,\tau)\right) Y_m^{IJ} \\
&= \sum_{I=1}^{n} \left(
\begin{aligned}
&\left.\frac{\partial f_i^{IN_1}}{\partial y_j^{IN_1}}\right|_{\boldsymbol{\phi}^{IN_1}} \frac{\partial}{\partial X_m}\left(u_j^{(1)}(\boldsymbol{X},\boldsymbol{Y}^I,t,\tau) + u_j^{(1)}(\boldsymbol{X},\boldsymbol{Y}^{N_1},t,\tau)\right)\left(Y_m^{N_1} - Y_m^I\right) \\
&+ \left.\frac{\partial f_i^{IN_2}}{\partial y_j^{IN_2}}\right|_{\boldsymbol{\phi}^{IN_2}} \frac{\partial}{\partial X_m}\left(u_j^{(1)}(\boldsymbol{X},\boldsymbol{Y}^I,t,\tau) + u_j^{(1)}(\boldsymbol{X},\boldsymbol{Y}^{N_2},t,\tau)\right)\left(Y_m^{N_2} - Y_m^I\right) \\
&+ \ldots \left.\frac{\partial f_i^{IN_p}}{\partial y_j^{IN_p}}\right|_{\boldsymbol{\phi}^{IN_2}} \frac{\partial}{\partial X_m}\left(u_j^{(1)}(\boldsymbol{X},\boldsymbol{Y}^I,t,\tau) + u_j^{(1)}(\boldsymbol{X},\boldsymbol{Y}^{N_p},t,\tau)\right)\left(Y_m^{N_p} - Y_m^I\right) \\
&+ \ldots \left.\frac{\partial f_i^{IN_K}}{\partial y_j^{IN_K}}\right|_{\boldsymbol{\phi}^{IN_2}} \frac{\partial}{\partial X_m}\left(u_j^{(1)}(\boldsymbol{X},\boldsymbol{Y}^I,t,\tau) + u_j^{(1)}(\boldsymbol{X},\boldsymbol{Y}^{N_K},t,\tau)\right)\left(Y_m^{N_K} - Y_m^I\right)
\end{aligned}
\right)
\end{aligned} \tag{3.47}$$

The summation in (3.47) is carried out over all atoms in the unit cell. First, we consider a case in which both the atom I and any of its interacting neighboring atoms N_p $(p = 1, 2, \ldots, K)$ are inside the unit cell. For each interacting atom pair (I, N_p), there are two terms in the summation (3.47), given by

$$\begin{aligned}
&\left.\frac{\partial f_i^{IN_p}}{\partial y_j^{IN_p}}\right|_{\boldsymbol{\phi}^{IN_p}} \frac{\partial}{\partial X_m}\left(u_j^{(1)}(\boldsymbol{X},\boldsymbol{Y}^I,t,\tau) + u_j^{(1)}(\boldsymbol{X},\boldsymbol{Y}^{N_p},t,\tau)\right)\left(Y_m^{N_p} - Y_m^I\right) \\
&+ \left.\frac{\partial f_i^{N_pI}}{\partial y_j^{N_pI}}\right|_{\boldsymbol{\phi}^{IN_p}} \frac{\partial}{\partial X_m}\left(u_j^{(1)}(\boldsymbol{X},\boldsymbol{Y}^{N_p},t,\tau) + u_j^{(1)}(\boldsymbol{X},\boldsymbol{Y}^I,t,\tau)\right)\left(Y_m^I - Y_m^{N_p}\right) = 0 \qquad \left(p = 1, 2, \ldots, K\right)
\end{aligned} \tag{3.48}$$

The above identity follows from (3.46). If any portion of the interacting atom N_p lies outside the unit cell, by periodicity, the displacement and force vector of atom N_p takes the same value as the corresponding atom in the unit cell, and thus the summation (3.48) holds. For a discussion of nonperiodic systems, we refer to [2,21].

In view of (3.47) and (3.48), we have

$$\sum_{I=1}^{n}\sum_{J\neq I}\left.\frac{\partial f_i^{IJ}}{\partial y_j^{IJ}}\right|_{\phi^{IJ}}\frac{\partial}{\partial X_m}\left(u_j^{(1)}(\boldsymbol{X},\boldsymbol{Y}^I,t,\tau)+u_j^{(1)}(\boldsymbol{X},\boldsymbol{Y}^J,t,\tau)\right)Y_m^{IJ}=0 \tag{3.49}$$

Inserting (3.49) into (3.41) yields the coarse-scale equation of motion

$$\rho_0\,\ddot{u}_i^c(\boldsymbol{X},t)-\frac{\partial\left\langle P_{ij}(\boldsymbol{X},t,\tau)\right\rangle}{\partial X_j}=0 \tag{3.50a}$$

$$P_{ij}(\boldsymbol{X},t,\tau)=\frac{1}{2|\Theta|}\sum_{I=1}^{n}\sum_{J\neq I}f_i^{IJ}(\zeta\phi^{IJ})X_j^{IJ} \tag{3.50b}$$

where $\boldsymbol{P}(\boldsymbol{X},t,\tau)$ is the first Piola–Kirchhoff stress tensor. This is similar to the virial stress except for the dynamic term, which is absent. The absence of the dynamic term in the virial formula has been pointed out by several investigators, including Srolovitz *et al.* [22], Horstemeyer and Baskes [23], and Alber *et al.* [24], among others. Equation (3.50a) represents the Lagrangian description of the conservation of linear momentum.

3.3.4 Continuum Description of Equation of Motion

Following the formalism introduced by Irving and Kirkwood [25] (see also [4,26]), the equation of motion of atoms (3.5) can be described using the following continuum description

$$\rho_0(\boldsymbol{X})=\sum_{I=1}^{n}m^I\delta(\boldsymbol{X}-\boldsymbol{X}^I) \tag{3.51a}$$

$$v_i(\boldsymbol{X},t)=\sum_{I=1}^{n}v_i^I(t)\delta(\boldsymbol{X}-\boldsymbol{X}^I) \tag{3.51b}$$

$$P_{ij}(\boldsymbol{X},t)=-\frac{1}{2|\Theta|}\sum_{I=1}^{n}\sum_{J\neq I}f_i^{IJ}(t)X_j^{IJ}(t)B^{IJ}(\boldsymbol{X}) \tag{3.51c}$$

where $\delta(\cdot)$ is the delta function and $B^{IJ}(\boldsymbol{X})$ represents a weighted fraction of the bond length segment between atoms I and J that lies within the unit cell

$$B^{IJ}(\boldsymbol{X})=\int_0^1\delta(\lambda\boldsymbol{X}^{IJ}+\boldsymbol{X}^J-\boldsymbol{X})d\lambda \tag{3.52}$$

It can be shown that

$$
\begin{aligned}
&-X_i^{IJ}\frac{\partial B^{IJ}}{\partial X_i}=\delta(\boldsymbol{X}-\boldsymbol{X}^I)-\delta(\boldsymbol{X}-\boldsymbol{X}^J)\\
&\frac{1}{|\Theta|}\int_\Theta B^{IJ}d\Theta=-1 \quad \text{for periodic BC}
\end{aligned}
\tag{3.53}
$$

Inserting (3.53) into (3.51) yields the equation of motion in the reference coordinate system. Furthermore, averaging $P_{ij}(\boldsymbol{X},t)$ in (3.51c) over the unit cell domain yields (3.50b).

It is instructive to point out that if the equation of motion is stated in Eulerian coordinates, then the resulting stress measure will contain the momentum flux, not just the mechanical force between different material points appearing in the Cauchy stress expression.

3.3.5 The Thermal Equation

According to the equipartition of energy theorem, temperature is directly related to the kinetic energy of the atomistic system

$$
\mathcal{K}=\sum_{I=1}^{n}\frac{\left|\boldsymbol{p}^{(1)I}\right|^2}{2m^I}=\frac{k_BT}{2}nN_d \tag{3.54}
$$

where k_B is the Boltzmann constant and $T(\boldsymbol{X},t,\tau)$ is an instantaneous temperature of the ensemble (see section 3.5 for the definition of various ensembles). According to the theorem of the equipartition of energy, each degree of freedom contributes $k_BT/2$. For n atoms, each with N_d degrees of freedom, the kinetic energy is $nN_dk_BT/2$. $\boldsymbol{p}^{(1)I}$ is a deviation from the momentum of atom I

$$
p_i^{(1)I}=\frac{\partial u_i^{(1)I}}{\partial \tau}m^I \tag{3.55}
$$

Remark 3.1 In (3.54), it is assumed that only the high-frequency (phonons) part of the atomic velocity contributes to temperature. Note that the effect of free electrons on heat conduction in metals and alloys is not accounted for. In order to account for the contributions of free electrons (metals, alloys, and semiconductors), modifications to the MD framework are required [27].

The square of the momentum of an atom associated with $u_i^{(1)I}$ is

$$
\left|p^{(1)I}\right|^2=p_i^{(1)I}p_i^{(1)I}=(m^I)^2\frac{\partial u_i^{(1)I}}{\partial \tau}\frac{\partial u_i^{(1)I}}{\partial \tau} \tag{3.56}
$$

Inserting (3.56) into (3.54), we have

$$\sum_{I=1}^{n} m^I \frac{\partial u_i^{(1)I}}{\partial \tau} \frac{\partial u_i^{(1)I}}{\partial \tau} = nk_B N_d T \tag{3.57}$$

Differentiating (3.57) with respect to slow time gives

$$\sum_{i=1}^{n} m^I \frac{\partial u_i^{(1)I}}{\partial \tau} \frac{\partial^2 u_i^{(1)I}}{\partial t \partial \tau} = \frac{nk_B N_d}{2} \frac{\partial T}{\partial t} \tag{3.58}$$

Multiplying both sides of (3.34) by $\partial u_i^{(1)I}/\partial \tau$ and summing up over all atoms in the unit cell yields

$$\sum_{I=1}^{n} m^I \left(\frac{\partial u_i^{(1)I}}{\partial \tau} \frac{\partial^2 u_i^c}{\partial t^2} + 2 \frac{\partial u_i^{(1)I}}{\partial \tau} \frac{\partial^2 u_i^{(1)I}}{\partial t \partial \tau} \right) = \zeta \sum_{I=1}^{n} \sum_{J \neq I} \frac{\partial u_i^{(1)I}}{\partial \tau} \left. \frac{\partial f_i^{IJ}}{\partial y_j^{IJ}} \right|_{\phi^{IJ}} \psi_j^{IJ} \tag{3.59}$$

Applying the temporal averaging operator to the above equation and accounting for the fact that the first term on the left-hand side vanishes due to τ-periodicity gives

$$\left\langle \sum_{I=1}^{n} m^I \frac{\partial u_i^{(1)I}}{\partial \tau} \frac{\partial^2 u_i^{(1)I}}{\partial t \partial \tau} \right\rangle = \left\langle \frac{\zeta}{2} \sum_{i=1}^{n} \frac{\partial u_i^{(1)I}}{\partial \tau} \sum_{J \neq I} \left. \frac{\partial f_i^{IJ}}{\partial y_j^{IJ}} \right|_{\phi^{IJ}} \psi_j^{IJ} \right\rangle \tag{3.60}$$

Inserting (3.60) into (3.58) and dividing the resulting equation by the volume of the unit cell $|\Theta|$ yields

$$\frac{nk_B N_d}{2|\Theta|} \frac{\partial \langle T \rangle}{\partial t} = \frac{\zeta}{2|\Theta|} \left\langle \sum_{I=1}^{n} \sum_{J \neq I} \frac{\partial u_i^{(1)I}}{\partial \tau} \left. \frac{\partial f_i^{IJ}}{\partial y_j^{IJ}} \right|_{\phi^{IJ}} \psi_j^{IJ} \right\rangle \tag{3.61}$$

From (3.40) follows

$$\begin{aligned} &\sum_{I=1}^{n} \sum_{J \neq I} \frac{\partial u_i^{(1)I}}{\partial \tau} \left. \frac{\partial f_i^{IJ}}{\partial y_j^{IJ}} \right|_{\phi^{IJ}} \psi_j^{IJ} \\ &= \frac{1}{2} \sum_{I=1}^{n} \sum_{J \neq I} \frac{\partial u_i^{(1)I}}{\partial \tau} \left(\frac{\partial}{\partial X_j} \left(f_i^{IJ} Y_j^{IJ} \right) + \left. \frac{\partial f_i^{IJ}}{\partial y_j^{IJ}} \right|_{\phi^{IJ}} \frac{\partial \left(u_j^{(1)I} + u_j^{(1)J} \right)}{\partial X_m} Y_m^{IJ} \right) \end{aligned} \tag{3.62}$$

Dividing (3.62) by the volume of the atomistic unit cell and inserting the result into (3.61) yields

$$\frac{nk_B N_d}{2|\Theta|}\frac{\partial\langle T\rangle}{\partial t}=\frac{\zeta}{4|\Theta|}\left\langle\sum_{I=1}^{n}\sum_{J\neq I}\frac{\partial u_i^{(1)I}}{\partial\tau}\left(\frac{\partial}{\partial X_j}\left(f_i^{IJ}Y_j^{IJ}\right)+\left.\frac{\partial f_i^{IJ}}{\partial y_j^{IJ}}\right|_{\phi^{IJ}}\frac{\partial\left(u_j^{(1)I}+u_j^{(1)J}\right)}{\partial X_m}Y_m^{IJ}\right)\right\rangle \tag{3.63}$$

Based on the chain rule of differentiation, we have

$$\frac{\partial f_i^{IJ}}{\partial\tau}=\left.\frac{\partial f_i^{IJ}}{\partial y_j^{IJ}}\right|_{\phi^{IJ}}\frac{\partial\phi_j^{IJ}}{\partial\tau}=\left.\frac{\partial f_i^{IJ}}{\partial y_j^{IJ}}\right|_{\phi^{IJ}}\left(\frac{\partial u_j^{(1)J}}{\partial\tau}-\frac{\partial u_j^{(1)I}}{\partial\tau}\right) \tag{3.64}$$

and

$$\frac{\partial u_i^{(1)I}}{\partial\tau}\frac{\partial}{\partial X_j}\left(f_i^{IJ}Y_j^{IJ}\right)=\frac{\partial}{\partial X_j}\left(\frac{\partial u_i^{(1)I}}{\partial\tau}\left(f_i^{IJ}Y_j^{IJ}\right)\right)-\left(f_i^{IJ}Y_j^{IJ}\right)\frac{\partial}{\partial X_j}\left(\frac{\partial u_i^{(1)I}}{\partial\tau}\right) \tag{3.65}$$

Consider the following derivative with respect to fast time

$$\left(f_i^{IJ}Y_j^{IJ}\right)\frac{\partial}{\partial X_j}\left(\frac{\partial u_i^{(1)I}}{\partial\tau}\right)=\frac{\partial}{\partial\tau}\left(\left(f_i^{IJ}Y_j^{IJ}\right)\frac{\partial u_i^{(1)I}}{\partial X_j}\right)-\left(\frac{\partial f_i^{IJ}}{\partial\tau}Y_j^{IJ}\right)\frac{\partial u_i^{(1)I}}{\partial X_j} \tag{3.66}$$

Inserting (3.66) into (3.65) yields

$$\frac{\partial u_i^{(1)I}}{\partial\tau}\frac{\partial}{\partial X_j}\left(f_i^{IJ}Y_j^{IJ}\right)=\frac{\partial}{\partial X_j}\left(\frac{\partial u_i^{(1)I}}{\partial\tau}\left(f_i^{IJ}Y_j^{IJ}\right)\right)-\frac{\partial}{\partial\tau}\left(\left(f_i^{IJ}Y_j^{IJ}\right)\frac{\partial u_i^{(1)I}}{\partial X_j}\right)+\left(\frac{\partial f_i^{IJ}}{\partial\tau}Y_j^{IJ}\right)\frac{\partial u_i^{(1)I}}{\partial X_j} \tag{3.67}$$

From (3.67) follows

$$\begin{aligned}&\frac{\partial u_i^{(1)I}}{\partial\tau}\left(\frac{\partial}{\partial X_j}\left(f_i^{IJ}Y_j^{IJ}\right)+\left.\frac{\partial f_i^{IJ}}{\partial y_j^{IJ}}\right|_{\phi^{IJ}}\frac{\partial\left(u_j^{(1)I}+u_j^{(1)J}\right)}{\partial X_m}Y_m^{IJ}\right)\\&=\frac{\partial}{\partial X_j}\left(\frac{\partial u_i^{(1)I}}{\partial\tau}\left(f_i^{IJ}Y_j^{IJ}\right)\right)-\frac{\partial}{\partial\tau}\left(\left(f_i^{IJ}Y_j^{IJ}\right)\frac{\partial u_i^{(1)I}}{\partial X_j}\right)+\left(\frac{\partial f_i^{IJ}}{\partial\tau}Y_j^{IJ}\right)\frac{\partial u_i^{(1)I}}{\partial X_j}\\&\quad+\frac{\partial u_i^{(1)I}}{\partial\tau}\left.\frac{\partial f_i^{IJ}}{\partial y_j^{IJ}}\right|_{\phi^{IJ}}\frac{\partial\left(u_j^{(1)I}+u_j^{(1)J}\right)}{\partial X_m}Y_m^{IJ}\end{aligned} \tag{3.68}$$

Using (3.64), the last two terms in (3.68) can be written as

$$\begin{aligned}
&\frac{\partial f_i^{IJ}}{\partial \tau}\frac{\partial u_i^{(1)I}}{\partial X_j}Y_j^{IJ}+\frac{\partial u_i^{(1)I}}{\partial \tau}\left.\frac{\partial f_i^{IJ}}{\partial y_j^{IJ}}\right|_{\phi^{IJ}}\frac{\partial\left(u_j^{(1)I}+u_j^{(1)J}\right)}{\partial X_m}Y_m^{IJ}\\
&=\left.\frac{\partial f_i^{IJ}}{\partial y_j^{IJ}}\right|_{\phi^{IJ}}\left(\frac{\partial u_j^{(1)J}}{\partial \tau}-\frac{\partial u_j^{(1)I}}{\partial \tau}\right)\frac{\partial u_i^{(1)I}}{\partial X_m}Y_m^{IJ}+\frac{\partial u_i^{(1)I}}{\partial \tau}\left.\frac{\partial f_i^{IJ}}{\partial y_j^{IJ}}\right|_{\phi^{IJ}}\frac{\partial\left(u_j^{(1)I}+u_j^{(1)J}\right)}{\partial X_m}Y_m^{IJ}\\
&=\left.\frac{\partial f_i^{IJ}}{\partial y_j^{IJ}}\right|_{\phi^{IJ}}\frac{\partial u_j^{(1)J}}{\partial \tau}\frac{\partial u_i^{(1)I}}{\partial X_m}Y_m^{IJ}+\left.\frac{\partial f_i^{IJ}}{\partial y_j^{IJ}}\right|_{\phi^{IJ}}\frac{\partial u_i^{(1)I}}{\partial \tau}\frac{\partial u_j^{(1)J}}{\partial X_m}Y_m^{IJ}
\end{aligned}\tag{3.69}$$

Further inserting (3.69) into (3.68) yields

$$\begin{aligned}
&\frac{\partial u_i^{(1)I}}{\partial \tau}\left(\frac{\partial}{\partial X_j}\left(f_i^{IJ}Y_j^{IJ}\right)+\left.\frac{\partial f_i^{IJ}}{\partial y_j^{IJ}}\right|_{\phi^{IJ}}\frac{\partial\left(u_j^{(1)I}+u_j^{(1)J}\right)}{\partial X_m}Y_m^{IJ}\right)\\
&=\frac{\partial}{\partial X_j}\left(\frac{\partial u_i^{(1)I}}{\partial \tau}\left(f_i^{IJ}Y_j^{IJ}\right)\right)-\frac{\partial}{\partial \tau}\left(\left(f_i^{IJ}Y_j^{IJ}\right)\frac{\partial u_i^{(1)I}}{\partial X_j}\right)\\
&\quad+\left.\frac{\partial f_i^{IJ}}{\partial y_j^{IJ}}\right|_{\phi^{IJ}}\frac{\partial u_j^{(1)J}}{\partial \tau}\frac{\partial u_i^{(1)I}}{\partial X_m}Y_m^{IJ}+\left.\frac{\partial f_i^{IJ}}{\partial y_j^{IJ}}\right|_{\phi^{IJ}}\frac{\partial u_i^{(1)I}}{\partial \tau}\frac{\partial u_j^{(1)J}}{\partial X_m}Y_m^{IJ}
\end{aligned}\tag{3.70}$$

From (3.70) follows

$$\begin{aligned}
&\left\langle\sum_{I=1}^{n}\sum_{J\neq I}\frac{\partial u_i^{(1)I}}{\partial \tau}\left(\frac{\partial}{\partial X_j}\left(f_i^{IJ}Y_j^{IJ}\right)+\left.\frac{\partial f_i^{IJ}}{\partial y_j^{IJ}}\right|_{\phi^{IJ}}\frac{\partial\left(u_j^{(1)I}+u_j^{(1)J}\right)}{\partial X_m}Y_m^{IJ}\right)\right\rangle\\
&=\frac{\partial}{\partial X_j}\left\langle\sum_{I=1}^{n}\sum_{J\neq I}\frac{\partial u_i^{(1)I}}{\partial \tau}\left(f_i^{IJ}Y_j^{IJ}\right)\right\rangle\\
&\quad+\left\langle\sum_{I=1}^{n}\sum_{J\neq I}\left(\left.\frac{\partial f_i^{IJ}}{\partial y_j^{IJ}}\right|_{\phi^{IJ}}\frac{\partial u_j^{(1)J}}{\partial \tau}\frac{\partial u_i^{(1)I}}{\partial X_m}Y_m^{IJ}+\left.\frac{\partial f_i^{IJ}}{\partial y_j^{IJ}}\right|_{\phi^{IJ}}\frac{\partial u_i^{(1)I}}{\partial \tau}\frac{\partial u_j^{(1)J}}{\partial X_m}Y_m^{IJ}\right)\right\rangle
\end{aligned}\tag{3.71}$$

where we have made use of the fact that the temporal average of the second term on the right-hand side of (3.70) vanishes due to periodicity in the fast timescale.

We proceed to prove that the second term in the right-hand side of (3.71) vanishes. As in the previous section, we assume that for atom I, its interacting neighbor atoms are

$N_1, N_2, \ldots, N_p, \ldots, N_K$, where K is the number of the interacting atoms such that $\left|\boldsymbol{x}^{N_p} - \boldsymbol{x}^I\right| < r_c$, $p = 1, 2, \ldots, K$. The following summation over the atomistic unit cell can be expanded as

$$\begin{aligned}
&\sum_{I=1}^{n}\sum_{J\neq I}\left(\left.\frac{\partial f_i^{IJ}}{\partial y_j^{IJ}}\right|_{\phi^{IJ}} \frac{\partial u_j^{(1)J}}{\partial \tau}\frac{\partial u_i^{(1)I}}{\partial X_m} Y_m^{IJ} + \left.\frac{\partial f_i^{IJ}}{\partial y_j^{IJ}}\right|_{\phi^{IJ}} \frac{\partial u_i^{(1)I}}{\partial \tau}\frac{\partial u_j^{(1)J}}{\partial X_m} Y_m^{IJ}\right)\\
&= \sum_{I=1}^{n}\left(\left.\frac{\partial f_i^{IN_1}}{\partial y_j^{IN_1}}\right|_{\phi^{IN_1}} \frac{\partial u_j^{(1)N_1}}{\partial \tau}\frac{\partial u_i^{(1)I}}{\partial X_m}\left(Y_m^{N_1} - Y_m^I\right) + \left.\frac{\partial f_i^{IN_1}}{\partial y_j^{IN_1}}\right|_{\phi^{IN_1}} \frac{\partial u_i^{(1)I}}{\partial \tau}\frac{\partial u_j^{(1)N_1}}{\partial X_m}\left(Y_m^{N_1} - Y_m^I\right)\right)\\
&+ \sum_{I=1}^{n}\left(\left.\frac{\partial f_i^{IN_2}}{\partial y_j^{IN_2}}\right|_{\phi^{IN_2}} \frac{\partial u_j^{(1)N_2}}{\partial \tau}\frac{\partial u_i^{(1)I}}{\partial X_m}\left(Y_m^{N_2} - Y_m^I\right) + \left.\frac{\partial f_i^{IN_2}}{\partial y_j^{IN_2}}\right|_{\phi^{IN_2}} \frac{\partial u_i^{(1)I}}{\partial \tau}\frac{\partial u_j^{(1)N_2}}{\partial X_m}\left(Y_m^{N_2} - Y_m^I\right)\right) \qquad (3.72)\\
&\ldots + \sum_{I=1}^{n}\left(\left.\frac{\partial f_i^{IN_p}}{\partial y_j^{IN_p}}\right|_{\phi^{IN_p}} \frac{\partial u_j^{(1)N_p}}{\partial \tau}\frac{\partial u_i^{(1)I}}{\partial X_m}\left(Y_m^{N_p} - Y_m^I\right) + \left.\frac{\partial f_i^{IN_p}}{\partial y_j^{IN_p}}\right|_{\phi^{IN_p}} \frac{\partial u_i^{(1)I}}{\partial \tau}\frac{\partial u_j^{(1)N_p}}{\partial X_m}\left(Y_m^{N_p} - Y_m^I\right)\right)\\
&\ldots + \sum_{I=1}^{n}\left(\left.\frac{\partial f_i^{IN_K}}{\partial y_j^{IN_K}}\right|_{\phi^{IN_K}} \frac{\partial u_j^{(1)N_K}}{\partial \tau}\frac{\partial u_i^{(1)I}}{\partial X_m}\left(Y_m^{N_K} - Y_m^I\right) + \left.\frac{\partial f_i^{IN_K}}{\partial y_j^{IN_K}}\right|_{\phi^{IN_K}} \frac{\partial u_i^{(1)I}}{\partial \tau}\frac{\partial u_j^{(1)N_K}}{\partial X_m}\left(Y_m^{N_K} - Y_m^I\right)\right)
\end{aligned}$$

The summation in (3.72) is carried out over all atoms in the unit cell. First, we consider a case in which both atom I and any of its interacting neighboring atoms N_p $(p = 1, 2, \ldots, K)$ are in the unit cell. For each interacting atom pair (I, N_p), there are two terms in the summation (3.72) given by

$$\begin{aligned}
&\left.\frac{\partial f_i^{IN_p}}{\partial y_j^{IN_p}}\right|_{\phi^{IN_p}} \frac{\partial u_j^{(1)N_p}}{\partial \tau}\frac{\partial u_i^{(1)I}}{\partial X_m}\left(Y_m^{N_p} - Y_m^I\right) + \left.\frac{\partial f_i^{IN_p}}{\partial y_j^{IN_p}}\right|_{\phi^{IN_p}} \frac{\partial u_i^{(1)I}}{\partial \tau}\frac{\partial u_j^{(1)N_p}}{\partial X_m}\left(Y_m^{N_p} - Y_m^I\right)\\
&+ \left.\frac{\partial f_i^{N_pI}}{\partial y_j^{N_pI}}\right|_{\phi^{N_pI}} \frac{\partial u_j^{(1)I}}{\partial \tau}\frac{\partial u_i^{(1)N_p}}{\partial X_m}\left(Y_m^I - Y_m^{N_p}\right) + \left.\frac{\partial f_i^{N_pI}}{\partial y_j^{N_pI}}\right|_{\phi^{N_pI}} \frac{\partial u_i^{(1)N_p}}{\partial \tau}\frac{\partial u_j^{(1)I}}{\partial X_m}\left(Y_m^I - Y_m^{N_p}\right) = 0
\end{aligned} \qquad (3.73)$$

for every $(p = 1, 2, \ldots, K)$ where we have exploited (3.46).

If any portion of the interacting neighbor atom N_p lies outside the unit cell, by periodicity, the displacement and force vector of atom N_p takes the same value as the corresponding atom in the unit cell, and thus the summation in (3.73) also vanishes.

In view of (3.72) and (3.73), we have

$$\left\langle \sum_{I=1}^{n}\sum_{J\neq I}\left(\left.\frac{\partial f_i^{IJ}}{\partial y_j^{IJ}}\right|_{\phi^{IJ}} \frac{\partial u_j^{(1)J}}{\partial \tau}\frac{\partial u_i^{(1)I}}{\partial X_m} Y_m^{IJ} + \left.\frac{\partial f_i^{IJ}}{\partial y_j^{IJ}}\right|_{\phi^{IJ}} \frac{\partial u_i^{(1)I}}{\partial \tau}\frac{\partial u_j^{(1)J}}{\partial X_m} Y_m^{IJ}\right)\right\rangle = 0 \qquad (3.74)$$

Inserting (3.74) into (3.71) and then inserting the resulting equation into (3.63) yields

$$C\frac{\partial\langle T\rangle}{\partial t}-\frac{\partial\langle q_j\rangle}{\partial X_j}=0 \tag{3.75}$$

where

$$C=nk_BN_d/|\Theta| \tag{3.76a}$$

$$q_j=\frac{1}{2|\Theta|}\sum_{I=1}^{n}\sum_{J\neq I}\frac{\partial u_i^{(1)I}}{\partial\tau}\left(f_i^{IJ}X_j^{IJ}\right) \tag{3.76b}$$

and $\boldsymbol{q}(\boldsymbol{X},t,\tau)$ is identified with the instantaneous heat flux vector.

3.3.6 Extension to Multi-Body Potentials

In the previous sections, we only considered pairwise potentials, which are usually inadequate for the modeling of solids. The methodology developed, however, is generic and is not limited to pairwise potentials. In the following, we will discuss generalization to the many-body potentials. More details can be found in [14].

The energy associated with atom I can be written as a sum of energies arising from the pairwise Φ^{IJ} and many-body V terms

$$E^I=\frac{1}{2}\sum_{J\neq I}\Phi^{IJ}(\boldsymbol{x}^{IJ})+\sum_{(I,N_1,N_2,\ldots,N_K)}V\underbrace{(\boldsymbol{x}^{IN_1},\boldsymbol{x}^{IN_2},\ldots,\boldsymbol{x}^{IN_K},\boldsymbol{x}^{N_1N_2},\ldots)}_{K(K+1)/2} \tag{3.77}$$

The many-body potentials term V is a function of $K(K+1)/2$ variables. Let the interacting neighboring atoms of atom I be denoted by $N_1, N_2, \ldots, N_p, \ldots, N_K$, where K is the number of the interacting atoms. The interatomic force acting on atom I by its neighbors is evaluated as

$$\boldsymbol{f}^I=-\frac{\partial E^I}{\partial\boldsymbol{x}^I}$$

The equation of motion for atom I can be written as

$$m^I\ddot{\boldsymbol{u}}^I=\boldsymbol{f}^I=\sum_{J\neq I}\boldsymbol{f}^{IJ}(\boldsymbol{x}^{IJ})+\sum_{(I,N_1,N_2,\ldots,N_K)}\boldsymbol{S}(\boldsymbol{x}^{IN_1},\boldsymbol{x}^{IN_2},\ldots,\boldsymbol{x}^{IN_K},\boldsymbol{x}^{N_1N_2},\ldots) \tag{3.78}$$

where $\boldsymbol{f}^{IJ}$ is a force exerted on atom I by atom J resulting from the pairwise potential Φ^{IJ} and $\boldsymbol{S}$ is the corresponding force resulting from the many-body potential V.

The homogenization process outlined in the previous sections can be repeated (see [14] for details).

3.4 Finite Element Implementation and Numerical Verification

3.4.1 Weak Forms and Semidiscretization of Coarse-Scale Equations

The initial and boundary conditions for the coarse-scale equations of motion (3.50) are given as

$$\boldsymbol{u}^c(\boldsymbol{X},0)=\boldsymbol{p}(\boldsymbol{X});\quad \frac{\partial u^c}{\partial t}(\boldsymbol{X},0)=\boldsymbol{g}(\boldsymbol{X}) \tag{3.79}$$

$$\boldsymbol{u}^c(\boldsymbol{X},t)=\bar{\boldsymbol{u}}(\boldsymbol{X},t)\quad \text{on}\quad \partial\Omega^u;\quad \langle \boldsymbol{P}\rangle\cdot\boldsymbol{n}=\bar{\boldsymbol{t}}\quad \text{on}\quad \partial\Omega^t \tag{3.80}$$

where $\bar{\boldsymbol{u}}(\boldsymbol{X},t)$ and $\bar{\boldsymbol{t}}$ are vectors of prescribed displacement and traction, respectively, $\partial\Omega^u\cap\partial\Omega^t=0$, $\partial\Omega^u\cup\partial\Omega^t=\partial\Omega$ is the boundary of the coarse-scale domain under consideration, and $\boldsymbol{n}$ is the outward norm of the boundary.

The initial and boundary conditions for the coarse-scale thermal equations (3.75) are given as

$$\langle T\rangle(\boldsymbol{X},0)=\langle T_0\rangle(\boldsymbol{X}) \tag{3.81}$$

$$\langle T\rangle(\boldsymbol{X},t)=\bar{T}(\boldsymbol{X},t)\quad \text{on}\quad \partial\Omega^T;\quad \langle \boldsymbol{q}\rangle\cdot\boldsymbol{n}=\bar{q}(\boldsymbol{X},t)\quad \text{on}\quad \partial\Omega^q \tag{3.82}$$

where $\bar{T}(\boldsymbol{X},t)$ and $\bar{q}(\boldsymbol{X},t)$ are the prescribed temperature and boundary heat flux, respectively, and $\partial\Omega^T\cap\partial\Omega^q=0$, $\partial\Omega^T\cup\partial\Omega^q=\partial\Omega$.

The weak form of the thermo-mechanical equation is stated as follows

For $t\in(0,T^{in})$, find $\boldsymbol{u}^c(\boldsymbol{X},t)\in\mathcal{U}_\Omega^d$, $\langle T\rangle(\boldsymbol{X},t)\in\mathcal{U}_\Omega^\theta$, such that for all $\boldsymbol{w}^d(\boldsymbol{X})\in W_\Omega^d$, $w^\theta(\boldsymbol{X})\in W_\Omega^\theta$ the following holds:

$$\int_\Omega \rho_0 w_i^d(\boldsymbol{X})\frac{\partial^2 u_i^c(\boldsymbol{X},t)}{\partial t^2}d\Omega-\int_\Omega w_i^d(\boldsymbol{X})\frac{\partial\langle P_{ij}\rangle}{\partial X_j}d\Omega=0 \tag{3.83}$$

$$\int_\Omega C w^\theta(\boldsymbol{X})\frac{\partial\langle T\rangle}{\partial t}d\Omega-\int_\Omega w^\theta(\boldsymbol{X})\frac{\partial\langle q_i\rangle}{\partial X_i}d\Omega=0 \tag{3.84}$$

subjected to the initial conditions (3.79) and (3.81), where trial spaces are defined as

$$\mathcal{U}_\Omega^d=\left\{\boldsymbol{u}^c(\boldsymbol{X},t)\middle|\boldsymbol{u}^c(\boldsymbol{X},t)\in C^0(\Omega),\boldsymbol{u}^c(\boldsymbol{X},t)=\bar{\boldsymbol{u}}(\boldsymbol{X},t)\quad \text{on}\quad \partial\Omega^u\right\} \tag{3.85}$$

$$\mathcal{U}_\Omega^\theta=\left\{\langle T\rangle(\boldsymbol{X},t)\middle|\langle T\rangle(\boldsymbol{X},t)\in C^0(\Omega),\langle T\rangle(\boldsymbol{X},t)=\bar{T}(\boldsymbol{X},t)\quad \text{on}\quad \partial\Omega^T\right\} \tag{3.86}$$

and the weight function spaces are given by

$$W_\Omega^d=\left\{\boldsymbol{w}^d(\boldsymbol{X})\middle|\boldsymbol{w}^d(\boldsymbol{X})\in C^0(\Omega),\boldsymbol{w}^d(\boldsymbol{X})=\boldsymbol{0}\quad \text{on}\quad \partial\Omega^u\right\} \tag{3.87}$$

$$W_\Omega^\theta = \left\{ w^\theta(\boldsymbol{X}) \middle| w^\theta(\boldsymbol{X}) \in H^1(\Omega), w^\theta(\boldsymbol{X}) = 0 \quad \text{on} \quad \partial\Omega^T \right\} \tag{3.88}$$

Integration by parts of the second term in (3.83) and (3.84) and making use of the boundary conditions (3.80) and (3.82) yields

$$\int_\Omega \rho_0 w_i^d(\boldsymbol{X}) \ddot{u}_i^c(\boldsymbol{X},t) d\Omega + \int_\Omega \frac{\partial w_i^d(\boldsymbol{X})}{\partial X_j} \langle P_{ij} \rangle d\Omega = \int_{\partial\Omega^t} w_i^d(\boldsymbol{X}) \bar{t}_i d\Gamma \tag{3.89}$$

$$\int_\Omega C w^\theta(\boldsymbol{X}) \frac{\partial \langle T \rangle}{\partial t} d\Omega + \int_\Omega \frac{\partial w^\theta(\boldsymbol{X})}{\partial X_i} \langle q_i \rangle d\Omega = \int_{\partial\Omega^q} w^\theta(\boldsymbol{X}) \, \bar{q}(\boldsymbol{X},t) d\Gamma \tag{3.90}$$

After introducing the Galerkin finite element discretization in space $u_i^c = N_{i\alpha}^d d_\alpha$, $\langle T \rangle = N_\beta^\theta \theta_\beta$, we have the following semidiscrete thermo-mechanical equations

$$M_{\alpha\beta} \ddot{d}_\beta(t) + f_\alpha^{\text{int}}(\boldsymbol{d}(t)) = f_\alpha^{\text{ext}}(t) \tag{3.91}$$

$$C_{\alpha\beta} \dot{\theta}_\beta(t) + Q_\alpha(\boldsymbol{\theta}(t)) = S_\alpha(t) \tag{3.92}$$

with the initial conditions

$$d_\alpha(0) = d_{\alpha 0}; \qquad \dot{d}_\alpha(0) = v_{\alpha 0}; \qquad \theta_\alpha(0) = \theta_{\alpha 0} \tag{3.93}$$

where

$$M_{\alpha\beta} = \int_\Omega \rho_0 N_{i\alpha}^d N_{i\beta}^d d\Omega; \qquad C_{\alpha\beta} = \int_\Omega C N_\alpha^\theta N_\beta^\theta d\Omega \tag{3.94}$$

are the mass and capacity matrices, respectively,

$$f_\alpha^{\text{int}}(\boldsymbol{d}(t)) = \int_\Omega N_{i\alpha,j}^d \langle P_{ij} \rangle d\Omega; \qquad Q_\alpha(\boldsymbol{\theta}(t)) = \int_\Omega N_{\alpha,i}^\theta \langle q_i \rangle d\Omega \tag{3.95}$$

are the nodal internal force and heat flux vectors, respectively, and

$$f_\alpha^{\text{ext}}(t) = \int_{\partial\Omega^t} N_{i\alpha}^d \bar{t}_i d\Gamma; \qquad S_\alpha(t) = \int_{\partial\Omega^q} N_\alpha^\theta \bar{q} d\Gamma \tag{3.96}$$

are the corresponding discrete source (force) vectors. Since nodal internal force $f_\alpha^{\text{int}}(\boldsymbol{d}(t))$ and heat flux $Q_\alpha(\boldsymbol{\theta}(t))$ vectors depend on the displacement and temperature fields, the semidiscrete equation of motion and the thermal equations are two-way coupled.

The discretized coarse-scale problem is stated as

$$\begin{aligned}
&\text{Given } {}_{n+1}\bar{t}_i,\ {}_{n+1}\bar{q},\ \text{find } {}_{n+1}\Delta d_\alpha,\ {}_{n+1}\Delta\theta_\alpha \ \text{ such that:}\\
&{}^{i+1}_{n+1}r^d_\alpha\left({}^{i+1}_{n+1}\boldsymbol{d}\right) \equiv M_{\alpha\beta}\,{}^{i+1}_{n+1}\ddot{d}_\beta + {}^{i+1}_{n+1}f^{\text{int}}_\alpha - {}^{i+1}_{n+1}f^{\text{ext}}_\alpha = 0\\
&{}^{i+1}_{n+1}r^\theta_\alpha\left({}^{i+1}_{n+1}\boldsymbol{\theta}\right) \equiv C_{\alpha\beta}\,{}^{i+1}_{n+1}\dot{\theta}_\beta + {}^{i+1}_{n+1}Q_\alpha - {}^{i+1}_{n+1}S_\alpha = 0\\
&{}^{i+1}_{n+1}d_\alpha = {}_{n+1}\bar{d}_\alpha \quad \text{on } \partial\Omega^u \qquad {}^{i+1}_{n+1}\theta_\alpha = {}_{n+1}\bar{\theta}_\alpha \quad \text{on } \partial\Omega^\theta\\
&n \leftarrow n+1, \quad \text{Go to the next load increment}
\end{aligned} \tag{3.97}$$

where ${}^{i+1}_{n+1}r^d_\alpha$ and ${}^{i+1}_{n+1}r^\theta_\alpha$ are the residuals associated with the mechanical and thermal problems, respectively, and ${}^{i+1}_{n+1}(\cdot)$ denotes the $(i+1)$th iteration during the load increment $(n+1)$. No summation convention is employed for left indices. The coarse-scale variables d_β and θ_β are integrated using the Newmark-β and backward Euler methods, respectively [28].

The flowchart in Figure 3.1 describes the overall framework integrating MD and thermo-mechanical finite element codes. The coarse-scale (continuum) and fine-scale (atomistic) analyses can be linked through the finite element user-defined mechanical and thermal functions. At each quadrature point, load increment $n+1$, and iteration $i+1$, finite element code invokes mechanical and thermal problem subroutines; these subroutines execute the atomistic unit cell problem subjected to the coarse-scale fields, including deformation gradient ${}^{i+1}_{n+1}\boldsymbol{F}^c$, temperature ${}^{i+1}_{n+1}\langle T\rangle$, and temperature gradient ${}^{i+1}_{n+1}\nabla_{\boldsymbol{X}}\langle T\rangle$. At each fast time step, the fine-scale (MD) solver computes instantaneous stresses (3.50b) and fluxes (3.76b), as well as computing the instantaneous tangent modulus ${}^{i+1}_{n+1}L_{ijkl}$ and conductivity ${}^{i+1}_{n+1}K_{ij}$. This is followed by fast time averaging of these fields

$$
{}^{i+1}_{n+1}\begin{bmatrix} \langle P_{ij} \rangle \\ \langle L_{klpq} \rangle \\ \langle q_i \rangle \\ \langle K_{jm} \rangle \end{bmatrix} = \frac{1}{\tau_0}\int_0^{\tau_0} {}^{i+1}_{n+1}\begin{bmatrix} P_{ij} \\ L_{klpq} \\ q_i \\ K_{jm} \end{bmatrix}\left({}^{i+1}_{n+1}\boldsymbol{F}^c,\ {}^{i+1}_{n+1}\langle T\rangle,\ {}^{i+1}_{n+1}\nabla_{\boldsymbol{X}}\langle T\rangle, \tau\right) d\tau \tag{3.98}
$$

3.4.2 The Fine-Scale (Atomistic) Problem

The coarse-scale quantities ${}^{i+1}_{n+1}\boldsymbol{F}^c$, ${}^{i+1}_{n+1}\langle T\rangle$, ${}^{i+1}_{n+1}\nabla_{\boldsymbol{X}}\langle T\rangle$ obtained from the solution of the coarse-scale problem are passed to the atomistic unit cell problem. The unit cell problem (3.31) is a MD-like problem, except that it is integrated in fast time τ for the perturbation $\hat{\boldsymbol{u}}^{(1)}$, rather than for the full atomistic motion. The unit cell problem (3.31) is subjected to the coarse-scale fields and appropriate mechanical and thermal boundary conditions and is evolved using an explicit MD-like integrator over a characteristic timescale τ_0, which has to be sufficiently long to obtain the statistically meaningful overall stress ${}^{i+1}_{n+1}\langle P_{ij}\rangle$ and heat flux ${}^{i+1}_{n+1}\langle q_i\rangle$ required for the advancement of the coarse-scale problem.

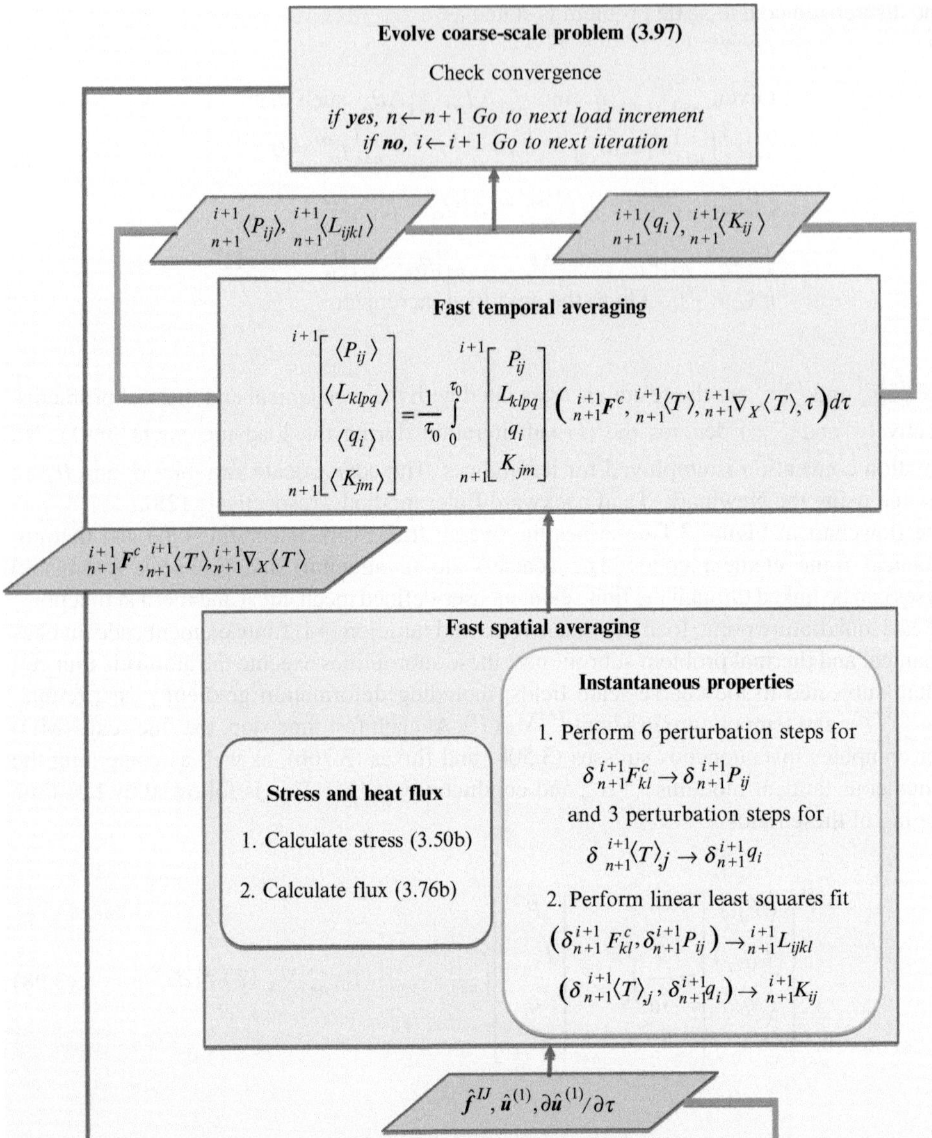

Figure 3.1 Program architecture for the two-scale continuum-atomistic simulation

One of the salient features of the GMH approach is that the two timescales (t,τ) are independent variables. Therefore τ_0 (which is generally much smaller than the slow time increment Δt) and the time evolution of the coarse-scale problem are completely independent. In general, the characteristic timescale τ_0 depends on the unit cell size [21,29], that is, larger cells can use the smaller characteristic timescale τ_0.

In the remainder of this section, we will focus on the following computational issues: (i) boundary and initial conditions for the mechanical problems; (ii) thermal boundary conditions;, and (iii) evaluation of the instantaneous mechanical and thermal properties needed for the implicit integration of the coarse-scale problem.

3.4.2.1 Displacement Boundary Conditions and Initial Conditions

Given the coarse-scale deformation gradient ${}_{n+1}^{i+1}\boldsymbol{F}^c$, we first determine the deformed configuration of the unit cell, which serves as the initial configuration for the MD-like simulation. In the *deformed* configuration, the position of the atoms in the unit cell vertices is prescribed to maintain the specified coarse-scale deformation gradient ${}_{n+1}^{i+1}\boldsymbol{F}^c$. The velocities $\partial u_i^{(1)}/\partial \tau$ of the atoms are initialized by a Boltzmann distribution corresponding to temperature ${}_{n+1}^{i+1}\langle T\rangle$ so that the average velocity is zero. At each time step, the value of the perturbation $\hat{u}_i^{(1)}$ is updated for all the atoms, and the *total separations* between the atoms are calculated for the interatomic force evaluation.

3.4.2.2 Temperature Boundary Conditions

To calculate the flux ${}_{n+1}^{i+1}q_i$, the unit cell has to be subjected to the temperature gradient. This is accomplished by considering neighboring unit cells as shown in Figure 3.2. Given the unit cell temperature ${}_{n+1}^{i+1}\langle T\rangle$ and the temperature gradient ${}_{n+1}^{i+1}\nabla_X\langle T\rangle$, the temperature of each unit cell is uniquely determined. A similar boundary region approach has been used in [21] for subjecting thermal boundary conditions to a local atomistic region in the context of a concurrent multiscale approach.

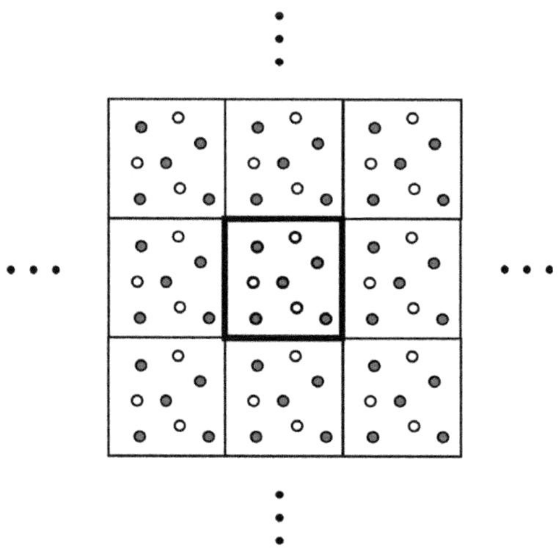

Figure 3.2 Temperature boundary conditions: a unit cell and its border region in the undeformed configuration

3.4.2.3 Instantaneous Mechanical and Thermal Properties

To calculate the instantaneous tangent modulus and the instantaneous conductivity, a linear perturbation approach is employed. Using this approach, six perturbations of the coarse-scale deformation gradient $\delta\,{}^{i+1}_{n+1}F^c_{kl}$ are applied to the unit cell, and the instantaneous tangent modulus ${}^{i+1}_{n+1}L_{ijkl}$ is evaluated by calculating the resulting perturbation of stress $\delta\,{}^{i+1}_{n+1}P_{ij}$. The instantaneous tangent modulus is computed as ${}^{i+1}_{n+1}L_{ijkl} = \delta\,{}^{i+1}_{n+1}P_{ij} / \delta\,{}^{i+1}_{n+1}F^c_{kl}$. Since different values of the deformation gradient perturbation $\delta\,{}^{i+1}_{n+1}F^c_{kl}$ may lead to different instantaneous tangent moduli, it is necessary to consider m perturbations for which $6\times m$ $\delta\,{}^{i+1}_{n+1}F^c_{kl}$ are constructed, and $\delta\,{}^{i+1}_{n+1}P_{ij}$ is evaluated $6\times m$ times. The instantaneous tangent modulus ${}^{i+1}_{n+1}L_{ijkl}$ is then obtained by a linear least square fit for each of its components.

The procedure for calculating instantaneous conductivity ${}^{i+1}_{n+1}K_{ij}$ is similar to ${}^{i+1}_{n+1}L_{ijkl}$, except that there are three components of the uniform temperature gradient $\delta\,{}^{i+1}_{n+1}\nabla_X\langle T\rangle$ in three dimensions. For each $\delta\,{}^{i+1}_{n+1}\nabla_X\langle T\rangle$, we calculate the perturbation of flux $\delta\,{}^{i+1}_{n+1}q_i$. The resulting instantaneous conductivity is determined either by ${}^{i+1}_{n+1}K_{ij} = \delta\,{}^{i+1}_{n+1}q_i / \delta\,{}^{i+1}_{n+1}\langle T\rangle_{,j}$ or by the linear least square fit described above.

3.5 Statistical Ensemble

A key concept in statistical mechanics is the *ensemble*. An ensemble is a collection of *microstates* of a system of atoms, all of which have one or more macroscopic properties (or thermodynamic states) in common. Therein, the thermodynamic state of a system is defined by a small set of macroscopic properties, in which the total energy is T, the total volume is V, and the total number of particles is N. Other thermodynamic properties can be derived from these quantities. A microstate of a system of atoms is defined by its atomic positions $\boldsymbol{x}$ and momenta $\boldsymbol{p} = m\dot{\boldsymbol{x}}$, which can also be considered as coordinates in the multidimensional phase space. For a system with N atoms, the phase space will have $6N$ dimensions. A single 6D point in this phase space is denoted by $\boldsymbol{\Gamma} = [\boldsymbol{x}, \boldsymbol{p}]^T$. There are different ensemble types with different characteristics, including the *microcanonical* ensemble, the *canonical* ensemble, and the *grand-canonical* ensemble. In the microcanonical (NVE) ensemble, the number of atoms N, the volume V, and the total energy E are conserved, such that the system is basically treated in isolation and exchanges nothing with its environment. In the canonical (NVT) ensemble, the temperature is controlled by a heat reservoir. The thermodynamic state of the ensemble is characterized by a fixed number of atoms N, a fixed volume V, and a constant temperature (instantaneous) T. Because of its interaction with the heat reservoir, the total energy is variable. A closely related ensemble has a fixed pressure P rather than a fixed volume V, and hence is termed an *isobaric-isothermal* (NPT) ensemble.

Finally, in the *grand-canonical* (μVT) ensemble, there is an exchange of both energy and atoms with the surrounding reservoir, where μ describes the fixed chemical potential.

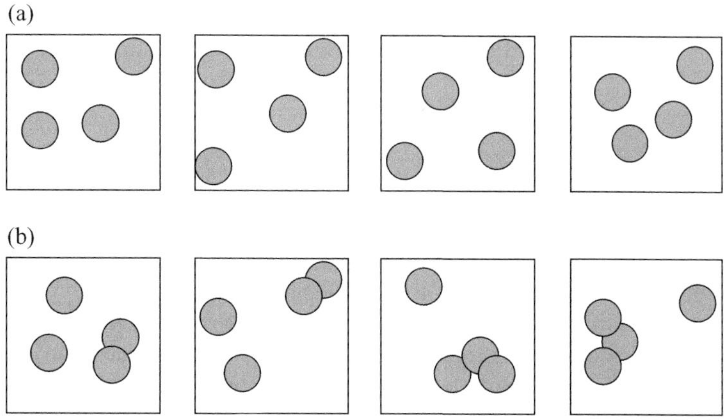

Figure 3.3 Microcanonical and canonical ensembles: (a) a microcanonical ensemble (in which each microstate has the same energy); and (b) a canonical ensemble [in which the energy is not preserved, as evidenced by the atoms being very close, and each microstate in the canonical ensemble has the same (instantaneous) temperature]

Macroscopic (homogenized) properties are defined in terms of ensemble averages. Let $\pi(\Gamma_i)$ be the probability of microstate i. A certain macroscopic property $\mathcal{P}$ can then be calculated as a weighted average

$$\langle \mathcal{P} \rangle_{\text{ensemble}} = \int_{\Gamma} \pi(\boldsymbol{\Gamma})\, \mathcal{P}(\boldsymbol{\Gamma}) d\Gamma \approx \sum_{i=1}^{M} \pi(\Gamma_i) \mathcal{P}(\Gamma_i) \tag{3.99}$$

where M is the number of microstates. The probability density function depends on the ensemble type. In the microcanonical ensemble, all microstates have equal probability, that is, $\pi(\Gamma_i) = 1/M$. In the canonical ensemble

$$\pi(\Gamma_i) = \frac{1}{Q} \exp\left(-\frac{E(\Gamma_i)}{k_B T} \right); \qquad Q = \sum_{i=1}^{M} \exp\left(-\frac{E(\Gamma_i)}{k_B T} \right) \tag{3.100}$$

where $E(\Gamma_i)$ denotes the Hamiltonian (or energy of the microstate). In general, the so-called partition function Q is extremely difficult to compute because all possible microstates of the system must be generated. Given that we must consider all microstates, regardless of their energy, unphysical microstates are excluded due to their low probability. Microstates with overlapping atoms have extremely high energy; the probability density function is practically zero in such instances, and thus $\pi(\Gamma_i)$ is negligible. As the temperature increases, higher-energy microstates have a proportionately larger influence on the ensemble averages.

In MD simulations, the points in the ensemble are computed sequentially in time. Thus, in order to compute an ensemble average, the MD simulation must pass through all possible states that correspond to the particular ensemble. In practice, we use a finite M number of microstates to obtain an ensemble averages. In the context of MD, the system is simulated over M time steps where $\mathcal{P}(\tau)$ denotes the instantaneous value of $\mathcal{P}$. The *ergodic postulate* relates the ensemble average to a time average, so it can be cast as a time average

$$\langle \mathcal{P} \rangle_{\text{time}} = \lim_{\tau_0 \to \infty} \frac{1}{\tau_0} \int_0^{\tau_0} \mathcal{P}(\Gamma(\tau))\, d\tau \approx \frac{1}{M} \sum_{\tau_i=1}^{M} \mathcal{P}(\Gamma(\tau_i)) \tag{3.101}$$

This time average depends on the initial values at $\tau = 0$. However, if the dynamics is ergodic (that is, it can reach all elements of the corresponding ensemble), then within the limit of infinite time, the initial conditions become irrelevant. Since there is a one-to-one relationship between the initial conditions of a system and its state at some other time, averaging over a large number of initial conditions is equivalent to averaging over the time-evolved states of the system. Thus, the ergodic hypothesis assumes that ensemble averages are equal to the time averages of the macroscopic property

$$\langle \mathcal{P} \rangle_{\text{time}} = \langle \mathcal{P} \rangle_{\text{ensemble}} \tag{3.102}$$

3.6 Verification

We will first consider a model problem with an atomistic chain consisting of 1201 atoms (schematically depicted in Figure 3.4). The atoms are initially equally spaced with spacing a, and each atom interacts with its nearest neighbors. The atomistic chain is assumed to possess a periodic microstructure with a unit cell of length l composed of three atoms with masses m_1, m_2, and m_1. The chain is subjected to two initial bell-shaped temperature distributions, one at each end, with amplitude T_{max} and width $2\delta = 1/4L$, where L is the total length of the atomic chain. The temperatures at the two ends of the atomistic chain are constant in time. The interatomic potentials take the form of the Lennard-Jones potential. In the unit cell, the interatomic potential between the first atom and the second atom is Φ_1, and the interatomic potential between the second atom and the third atom is Φ_2. The two potentials are given by

$$\Phi_1(r) = 4\varepsilon_1 \left[\left(\frac{\sigma}{r} \right)^{12} - \left(\frac{\sigma}{r} \right)^{6} \right]; \quad \Phi_2(r) = 4\varepsilon_2 \left[\left(\frac{\sigma}{r} \right)^{12} - \left(\frac{\sigma}{r} \right)^{6} \right] \tag{3.103}$$

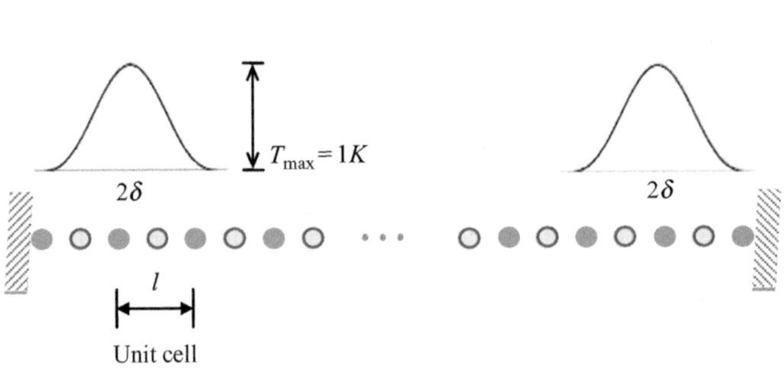

Figure 3.4 An atomic chain and a unit cell

where ε_1 and ε_2 are characteristic energy scales of the interaction, r is the distance between the particles, and σ is the characteristic length scale of the interaction. We assume that the initial configuration of the atomistic chain is in equilibrium, without external forces, so that $a=2^{1/6}\sigma$.

The interatomic forces are evaluated as

$$f_{01}=\frac{d\Phi_1}{dr}=\frac{24\varepsilon_1}{\sigma}\left[\left(\frac{\sigma}{r}\right)^7-2\left(\frac{\sigma}{r}\right)^{13}\right];\quad f_{12}=\frac{d\Phi_2}{dr}=\frac{24\varepsilon_2}{\sigma}\left[\left(\frac{\sigma}{r}\right)^7-2\left(\frac{\sigma}{r}\right)^{13}\right] \tag{3.104}$$

For the three-atom unit cell under consideration, the instantaneous first Piola–Kirchhoff stress (3.50b) and the heat flux (3.76b) are

$$P(X,t,\tau)=\frac{1}{2}\left[f_{01}(\zeta\phi_{01})\right]+\left[f_{12}(\zeta\phi_{12})\right];\quad q(X,t,\tau)=\frac{m_1-m_2}{2m_1}u^{(1)}(Y_1)P(X,t,\tau) \tag{3.105}$$

where

$$\zeta\phi_{01}=(1+u^c)a+\frac{m_1+m_2}{m_1}\hat{u}^1(Y_1);\quad \zeta\phi_{12}=(1+u^c)a-\frac{m_1+m_2}{m_1}\hat{u}^{(1)}(Y_1) \tag{3.106}$$

The linear mass density is given by (3.37)

$$\rho_0=\frac{1}{l}\sum_{i=1}^{2}m_i(Y)=(m_1+m_2)/l \tag{3.107}$$

The material parameters are $m_2/m_1=5$ and $\varepsilon_2/\varepsilon_1=2$. The amplitude of the initial temperature is set to be $T_{max}=1K$. The coarse-scale problem is solved using an explicit solver.

The CPU time of homogenization in comparison with that of MD at different atomistic chain sizes is plotted in Figure 3.5. For all cases considered, the continuum mesh is sufficiently fine to resolve the macroscopic solution gradients. It can be observed that as chain size increases, the ratio of the CPU time between MD and GMH simulations is increased without compromising on solution accuracy. For an atomistic chain of 120,001 atoms, the speed-up factor was approximately 1500.

For the second example, we consider a 3D model of a beam with the $100\times10\times10$ body-centered cubic lattice structure. The total number of atoms occupying the volume of the beam is 22,221. All atoms are assumed to have equal mass m. The beam is clamped at the two ends and is subjected to an initial bell-shaped temperature distribution with amplitude $T_{max}=40K$ and width $2\delta=L$, where L is the total length of the atomic chain. The interatomic potentials take the form of the Lennard-Jones potential.

The continuum is discretized with 25 hexahedral elements, whereas the atomistic unit is composed of 35 atoms. We consider hexahedral elements with an eight-point Gauss quadrature and a one-point Gauss-quadrature with stabilization [30]. In comparison with MD, the CPU speed-up factor is approximately 190 in the case of the eight-point quadrature, and over 1350 in the case of the one-point quadrature with stabilization. As in the first example, the coarse-scale problem is solved using an explicit solver. Figure 3.6 shows that the temperature field at $x=3/5L$, which was obtained by GMH, is in good agreement with the temperature field obtained by MD.

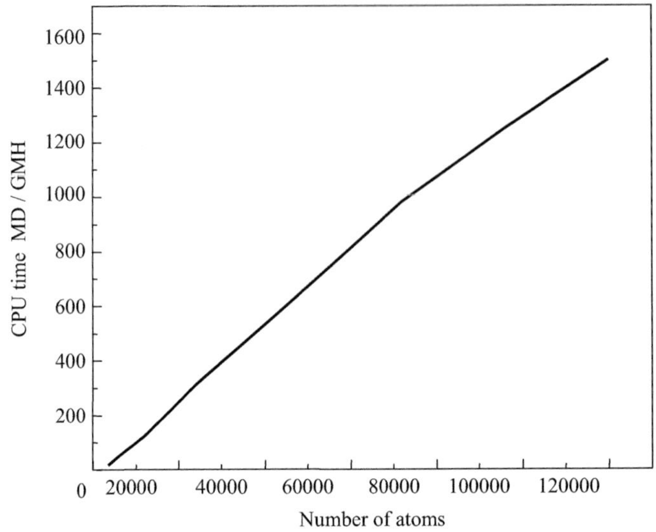

Figure 3.5 The speed-up of GMH over MD in terms of CPU time for atomistic chains of different lengths

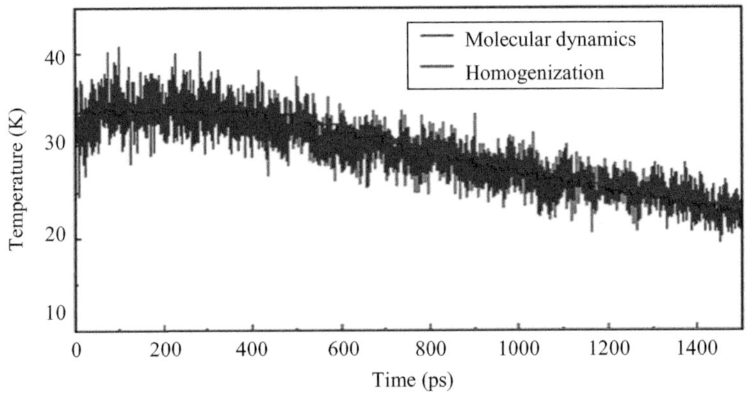

Figure 3.6 Temperatures at an observation window. Reproduced from [14], © 2008 Elsevier

For the third example, we consider a model problem of a silicon beam (depicted in Figure 3.7).

Silicon has a basic diamond structure. Each silicon crystal cube has 18 atoms, and each side of the cube is 0.543 nm. The size of the beam is 2000 × 2000 × 20000 length units, where each unit length σ is 2.059 Å. Thus, the dimensions of the beam are 1.08 μm × 1.08 μm × 10.8 μm. If not specified, all units are nondimensional. Other than the length units, the remaining basic nondimensional units are (i) energy (e = 2.315 eV) and (ii) mass ($m = 4.6654\times10^{-26}$ kg per atom, which is the mass of each silicon atom). All other units follow from these basic units. For example, the unit of time is $\sigma\sqrt{m/e} = 74.310\,\text{fs}$.

The beam is clamped at the left ($x = 0$) and pressure $p(t) = -0.00002\sin(6.28318\times10^{-5}\pi t)$ is applied on the top surface. The initial temperature is 20 K, and the surface heat flux $q(t) = 10^{-5}$

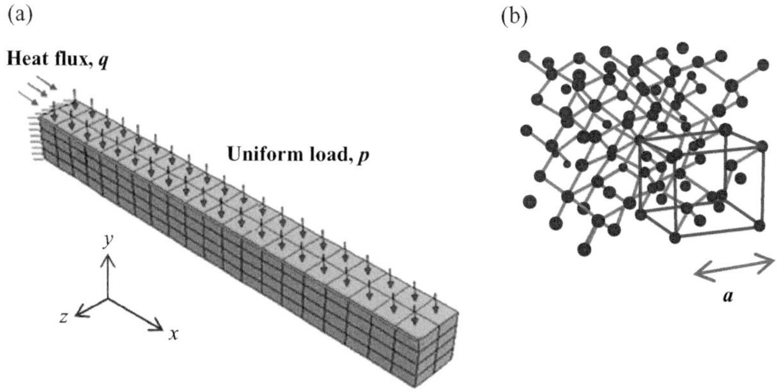

Figure 3.7 Three-dimensional silicon beam (a) and a unit cell (b), where $a = 0.543$ nm is the size of a single silicon crystal cube

is applied at $x = 0$. An eight-node coupled temperature–displacement brick element is used. The element has eight quadrature points. At each quadrature point, a unit cell problem involving a $2\times2\times2$ single silicon crystal cube is considered. Each unit cell consists of 95 atoms instead of 144(=18×8), since some atoms are common to neighboring crystal cubes. The bonded interactions for the silicon crystals are modeled using the Stillinger-Weber potential [31]. The stable time step of 3.7 fs is used for the integration of the fine-scale equations of motion with the Nosé-Poincaré thermostat [32].

To study the influence of mesh refinement and τ_0, we consider (i) 100, 200, 500, 800, and 1000 fast time steps over which the results are averaged and (ii) finite element meshes ranging from 40 to 160 elements. The results are illustrated in Figure 3.8.

In the final example, we study the failure properties of a silicon nanowire. The nanowire depicted in Figure 3.9 is 2 × 2 × 20, that is, 1.08 nm × 1.08 nm × 10.8 nm. One end of the nanowire is held fixed, while the other is pulled uniformly with an applied displacement of $u(t)=0.014t$. Temperature is kept at 20 K. For the unit cell, the atoms on the two surfaces normal to the axial direction have been constrained in the deformed configuration; the atoms on the lateral surfaces are free, which allows for relaxation at these free surfaces. As in the previous example, we have used the $2\times2\times2$ quadrature scheme to avoid consideration of the stabilization required for a one-point quadrature. The eight-point quadrature scheme has the effect of a variable macroscopic deformation gradient over a unit cell domain.

The axial stress–strain response is shown in Figure 3.10. The axial strain is defined as $\varepsilon_{11}= (l-l_0)/l_0$, where l is the current wire length and l_0 is the initial length.

To validate the results obtained by GMH, a full MD model is employed. Prior to extension, the nanowire is fully relaxed at 20 K. The relaxed model is considered as an initial undeformed state. The nanowire is subdivided into 10 segments; the segments are sequentially numbered starting from the fixed end. Each curve in Figure 3.10 corresponds to the stress in different segments. The difference between the curves is due to statistical scatter. The nanowire is pulled at a uniform rate of 2.67×10^9 s^{-1}. The strain increment size is chosen as 0.002. For the MD simulation, the number of time steps for the relaxation of each strain increment is taken as 500, with an explicit time integration step equal to 1.5 fs. Note that the number of micro time steps in GMH is five times fewer than in MD simulation, whereas the

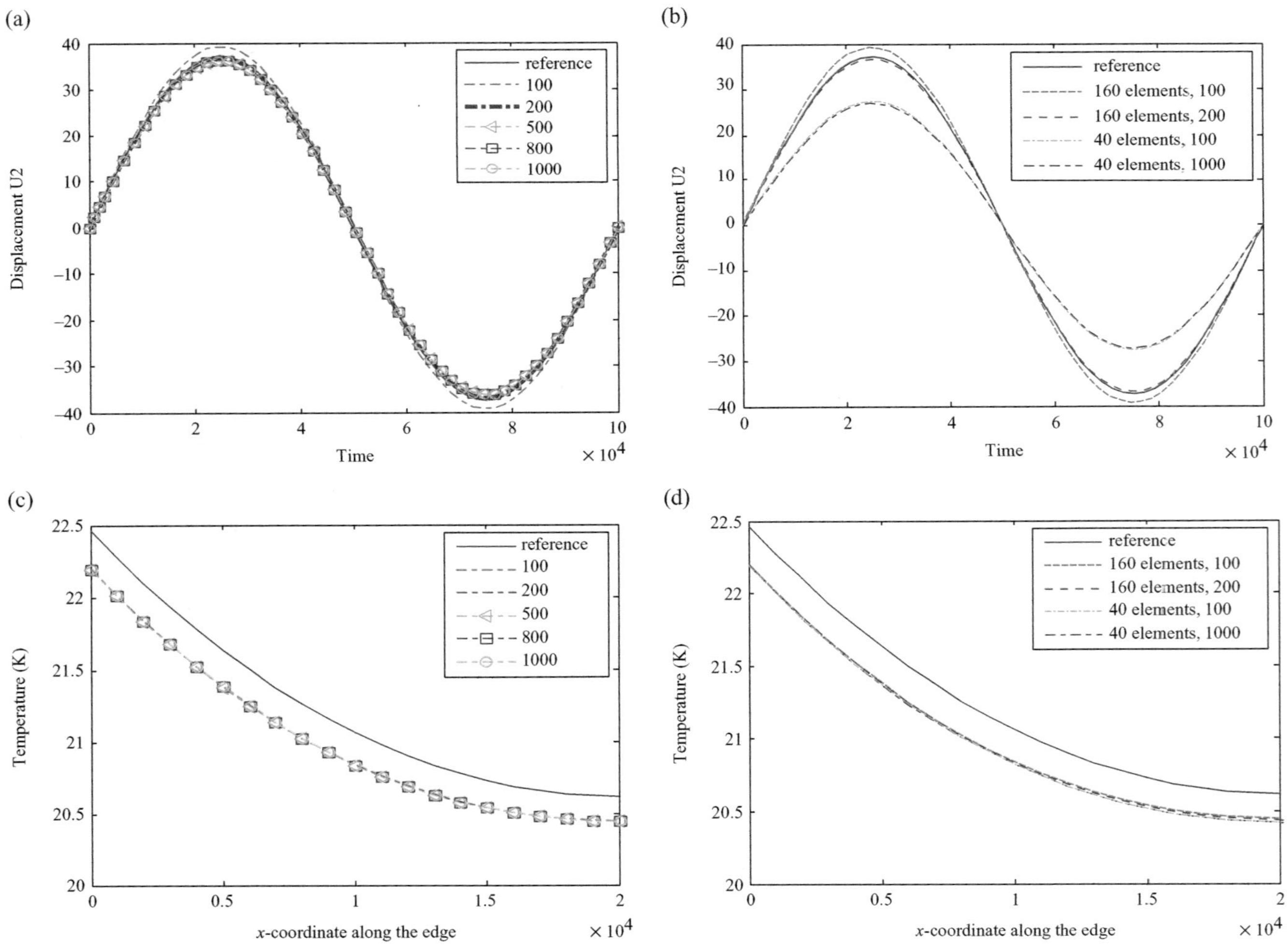

Figure 3.8 Silicon beam: (a) the influence of τ_0 on displacements (80 elements); (b) the influence of mesh refinement on displacements; (c) the influence of τ_0 on temperature (80 elements); and (d) the influence of mesh refinement on temperature

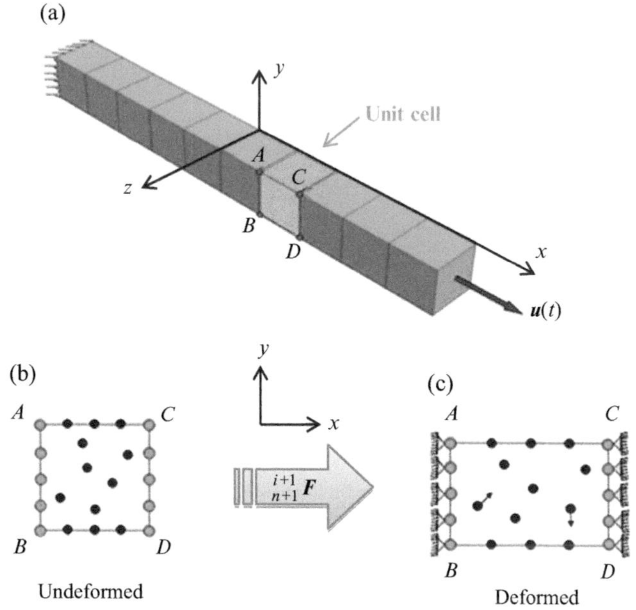

Figure 3.9 Silicon nanowire (a), and the unit cell, undeformed (b) and deformed (c)

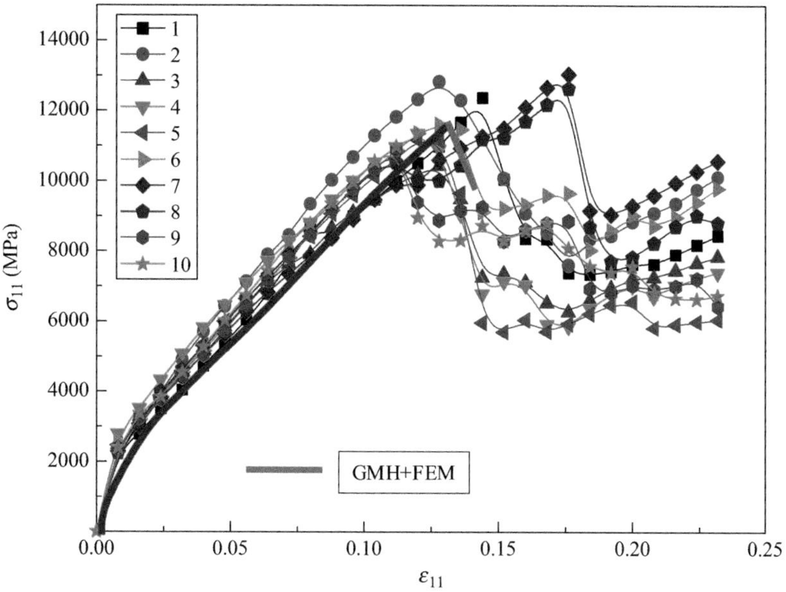

Figure 3.10 The stress–strain curve of silicon nanowire under extension loading. Pulling rate of 2.67×10^9 s^{-1}. Comparison of MD and GMH results

number of macro steps in GMH and MD is identical. After reaching the peak value of approximately 11,000 MPa, the stress decreases dramatically, indicating collapse. The collapse of the nanowire occurs at approximately $\varepsilon_{11} = 0.13$, which is in good agreement with the published results in [33].

3.7 Going Beyond Upscaling

For some nanoscale systems, it is necessary to carry out detailed molecular (atomistic) level simulations without upscaling to the continuum level. Furthermore, atomistic unit cell problems are often computationally taxing due to the large number of atoms and the long time windows over which they have to be simulated.

Current MD algorithms severely restrict modeling efforts to relatively small systems and/or short time intervals. The algorithmic challenges facing MD simulations stem from the difficulty of designing methods that are insensitive, in terms of the integration time step, to rapid fluctuations in chemical bonds and short-range forces. Most widely used algorithms in MD are explicit methods, such as Verlet [34,35] and Gear's predictor-corrector methods [36]. The severe limitation in the ability of the explicit methods to propagate numerical trajectories stems from a wide range of timescales spanning many orders of magnitude. In polymer chains, for instance, bond-stretching vibrations are the fastest atomic motions in a molecule, typically on the order of femtoseconds, whereas the relaxation of polymers in the form of segmental motions or terminal relaxations of chains spans the timescales in the range between 10^{-2}s and 10^{4}s [37]. The maximum time step is limited by the smallest oscillation period that can be found in the simulated system. This time step is necessary to maintain the stability of explicit numerical integration schemes [38].

Considerable efforts have been devoted to increasing the critical time step. Many insightful observations on the numerical properties of the algorithms have been reported, and numerous improvements have been proposed. One common approach is to constrain bond lengths using either the SHAKE or RATTLE algorithms [39]. In this approach, bonds are treated as being constrained to have a fixed length. Another approach to increasing the time step in MD simulations is to use the multiple-time-step (MTS) methods. These methods are based upon integration schemes that allow for time steps of differing lengths according to how rapidly a given type of interaction is evolving in time [40,41]. The temporal gap can be partially circumvented by parallel algorithms. However, due to the inherently sequential nature of time integration schemes, parallelism has been exploited in the spatial domain only, while computations in the time domain remain primarily sequential. One promising research avenue is to develop integration algorithms that are parallelizable in a time domain. A different (or complementary) approach is to exploit implicit integration. Implicit methods require the solution of coupled nonlinear equations at every time step, which is both expensive and difficult to implement. Nevertheless, they reduce stability restrictions and enable a much larger time step, typically governed by accuracy considerations [42,43]. In general, accuracy is determined by how close the solution of atomic trajectories obtained by the numerical integration is to the true trajectories. We refer to this as the *accuracy in local quantities*. Molecular systems, however, are highly nonlinear and exhibit sensitive dependence on perturbations. Moreover, the initial conditions are usually assigned randomly. Thus, the accurate approximation of local trajectories on meaningful time intervals is neither obtainable nor desired. This suggests analyzing the accuracy with respect to global quantities of interest, such as temperature, energy, and diffusion.

In this section, we present a multigrid-like approach based on multilevel principles for solving large molecular systems characterized by diverse spatial and temporal scales. This method is implicit in both space and time domains and typically uses 10 to 50 times larger time steps than explicit methods selected based on the accuracy of global fields of interest. The space–time character of this method provides an improved stability without directly solving the space–time problem. The multiscale nature of the system is exploited in two ways: (i) it substantially reduces the cost of multilevel iterations; and (ii) it is parallelizable in the time domain by exploiting the waveform relaxation (WR) scheme in the smoothing process.

The multilevel formulation presented here, termed as heterogeneous space–time full approximation scheme (HFAS) [44], is based on a variant of the nonlinear multigrid theory, an approach known as the full approximation scheme (FAS) [45]. FAS allows for consideration of different force fields at various scales and results in added flexibility and superior computational performance. The so-called multilevel waveform relaxation (MWR) for parabolic partial differential equations (PDEs) was developed in [46,47].

3.7.1 Spatial Multilevel Method Versus Space–Time Multilevel Method

We will start with a qualitative comparison between the classical spatial multilevel method and the space–time multilevel method presented in this section. For the space–time problem, we consider a 1D atomistic chain (which corresponds to the 2D problem in the space–time domain), whereas for the spatial structured geometric multilevel method, we consider a 2D spatial domain (as shown in Figure 3.11).

For the standard (spatial) multilevel method in 2D, the coarse-grid correction is provided by the auxiliary coarse grid [blue circles and blue dashed lines in Figure 3.11(a)]. The smoothing process (red hatched lines) is then equivalent to finding the local equilibrium positions of the remaining fine-scale degrees of freedom (white circles). On the other hand, in the space–time multilevel method, the coarse-grid problem corresponds to the solution on a spatial slab [blue dashed lines in Figure 3.15(b)], whereas smoothing corresponds to the solution of the spatial block on a temporal grid carried out window by window to reduce storage (red hatched lines) using the waveform relaxation outlined in section 3.7.2. Note that in the standard multilevel method, line smoothing corresponds to solving an algebraic system of equations corresponding to the degrees of freedom positioned along the line, while in the space–time variant, line smoothing is a time integration process.

One of the main challenges in devising an efficient multilevel approach is constructing the coarse-scale problem. For continuum systems governed by PDEs, it is possible to construct an auxiliary coarse grid based on PDEs and constitutive equations of the source problem. Thus, the coarse-grid problem can be directly formulated on the auxiliary coarse model. Employing an analogous approach to molecular equations with force fields of representative atoms computed from the same interatomic potentials will result in unrealistic solutions primarily because the physics at various scales is different. In polymers, for instance, the most common coarse-grained model is based on an equation of motion of the Langevin type (also known as Brownian dynamics), which adds additional terms to the usual MD equations. In section 3.7.3 we will describe a variant of the FAS to space–time systems.

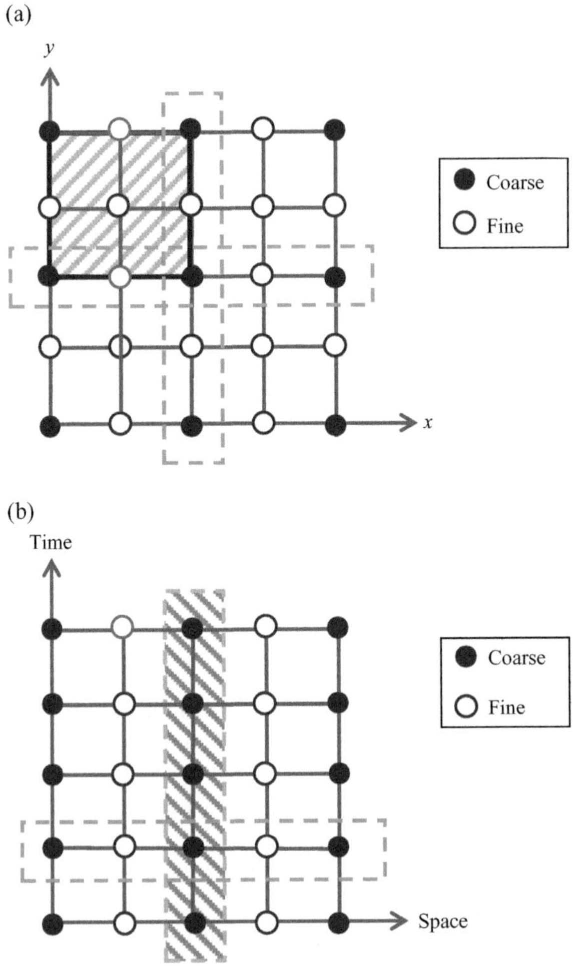

Figure 3.11 Standard multilevel (a) in 2D versus space–time multilevel (b)

In the remaining sections of this chapter, we will consider the following matrix notation to describe a dynamic system

$$\begin{aligned} \boldsymbol{M}^f \ddot{\boldsymbol{d}}^f &= \boldsymbol{f}_{\text{int}}^f(\boldsymbol{d}^f) + \boldsymbol{f}_{\text{ext}}^f \\ \boldsymbol{d}^f(0) &= \boldsymbol{d}_0^f \\ \dot{\boldsymbol{d}}^f(0) &= \boldsymbol{v}_0^f \end{aligned} \tag{3.108}$$

where $\boldsymbol{d}$ denotes the vector of atom positions, $\boldsymbol{M}^f$ denotes the mass matrix, $\boldsymbol{f}_{\text{ext}}^f$ denotes the external force vector (usually $\boldsymbol{f}_{\text{ext}}^f = \boldsymbol{0}$), and $\boldsymbol{f}_{\text{int}}^f$ denotes the internal force field defined as the gradient of the potential energy.

3.7.2 The WR Scheme

In the WR algorithm, the space–time domain is partitioned in space into smaller subsystems, thereafter referred to as *spatial blocks* or simply *blocks*. Each block is then integrated over the temporal grid. The spatial blocks can be integrated independently on different processors. The temporal grid is divided into intervals called *windows*. Windows are used to accelerate convergence and to reduce storage. At every time step, an algebraic system of equations is solved on a spatial grid. (See Figure 3.12 for definitions.)

Information transfer between different windows takes place once the integration over the corresponding windows is completed. The main advantages of the WR method are that it permits simultaneous integration of spatial blocks in each window and it allows unstructured integration. For instance, spatial blocks involving stiff connections will require smaller time steps than those having more compliant connections. The WR method has been used primarily in electrical network simulations and parabolic initial value problems. Limited studies have been conducted for hyperbolic [48] and second-order [49] systems.

In the remainder of this section, we will give some technical details of the WR method, focusing on nonlinear systems. For highly nonlinear systems, such as those arising from MD simulations, two versions of the WR method are common. The first is a direct extension of linear WR formulations, the so-called waveform relaxation Newton (WRN) method [27]. In the MD context, the basic idea is to solve the following nonlinear scalar equations

$$\begin{aligned} m_I^f \, {}^{v+1}\ddot{\boldsymbol{d}}_I^f &= \boldsymbol{f}_{\text{int}\,I}^f\left({}^{v}\boldsymbol{d}_1^f, \ldots, {}^{v}\boldsymbol{d}_{I-1}^f, {}^{v+1}\boldsymbol{d}_I^f, {}^{v}\boldsymbol{d}_{I+1}^f, \ldots, {}^{v}\boldsymbol{d}_N^f\right) \\ {}^{v+1}\boldsymbol{d}_I^f(0) &= \boldsymbol{d}_{0I}^f \\ {}^{v+1}\dot{\boldsymbol{d}}_I^f(0) &= v_{0I}^f \end{aligned} \tag{3.109}$$

for every atom I in the system. The left superscripts v and $v+1$ denote the iteration count within the time window $t \in [t_0, t_n]$. Each atom (or spatial block consisting of several atoms) in

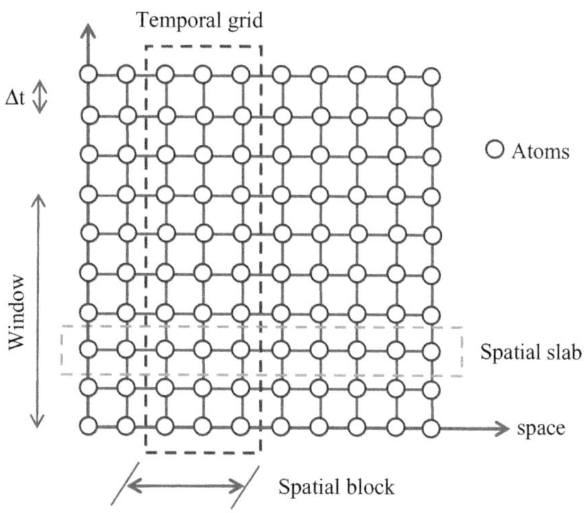

Figure 3.12 Slabs, windows, and blocks in the waveform relaxation method

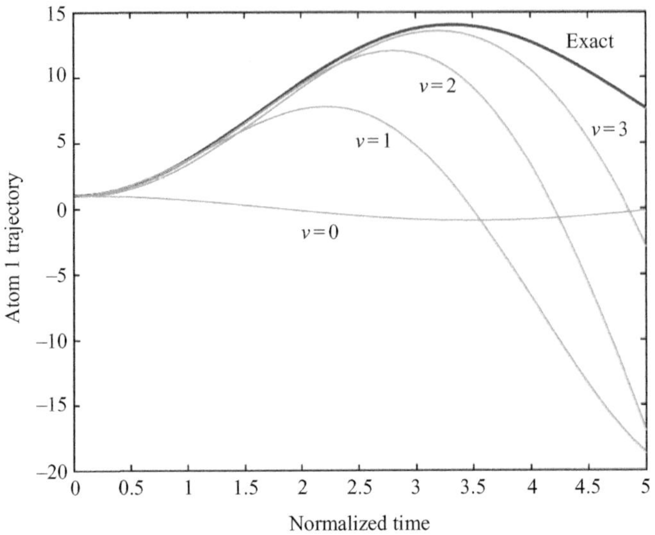

Figure 3.13 Space–time convergence of WR methods

iteration $\nu+1$ is integrated over a time window based on its previous position in time (in the same iteration $\nu+1$) and the information about its neighbor positions. As opposed to classical integrators, neighboring positions are taken from the iteration ν. This resembles the Jacobi smoothing method for steady-state problems. One variant is synonymous with the Gauss–Seidel splitting method and is based on updating internal forces using the information already available from the iteration $\nu+1$. Another variant is known as the waveform Newton [49], where the internal force in (3.109) is approximated as

$$\boldsymbol{f}_{\text{int}}^{f} = \boldsymbol{f}_{\text{int}}^{f}({}^{\nu}\boldsymbol{d}^{f}) + \boldsymbol{D}^{f}({}^{\nu}\boldsymbol{d}^{f})({}^{\nu}\boldsymbol{d}^{f} - {}^{\nu+1}\boldsymbol{d}^{f}) \tag{3.110}$$

where $\boldsymbol{D}^{f}({}^{\nu}\boldsymbol{d}^{f})$ is the diagonal of the stiffness matrix.

The WR iteration is terminated when the maximum residual in a time window is smaller than a specified tolerance.

An illustration of the convergence of WR methods with the trajectory of one atom is shown in Figure 3.13.

It can be seen that the method converges over the entire trajectory, as opposed to standard integration schemes (explicit and implicit) that march on the atomistic trajectory. The major drawback of the WR method is its slow convergence in the case of a strong coupling between spatial blocks and sizable windows [50]. In the space–time multilevel method, the WR plays the role of a smoother aimed at capturing the high frequency response of atomistic vibrations.

3.7.3 Space–Time FAS

In this section, we will develop the space–time FAS approach for MD simulations. The space–time FAS approach differs from the original FAS approach described in Chapter 2 in two respects. First, the method is used to solve a space–time problem. Secondly, different operators are employed at different levels (scales). This is in contrast to the classical FAS where the same operators are utilized at different levels.

We now focus on the formulation and algorithmic details of the space–time FAS method. The first step is pre-smoothing in the space–time domain. This is accomplished using the waveform Newton method described in the previous section. We define the space–time problem (which is similar to steady-state nonlinear problems) over a certain time window as follows. Let ${}^I\ddot{\boldsymbol{d}}^f(t), {}^I\boldsymbol{d}^f(t)$ be the solution from the previous cycle. The corresponding fine-scale residual over a space–time window is given by

$$ {}^I\boldsymbol{r}^f\left({}^I\ddot{\boldsymbol{d}}^f, {}^I\boldsymbol{d}^f\right) = \boldsymbol{M}^f\,{}^I\ddot{\boldsymbol{d}}^f(t) - \boldsymbol{f}^f_{\text{int}}\left({}^I\boldsymbol{d}^f(t)\right) \tag{3.111} $$

The coarse-scale solution is obtained by restriction of the residual and averaging of the fine-scale displacements

$$ \begin{aligned} {}^I\boldsymbol{r}^c(t) &= \boldsymbol{Q}^T\;{}^I\boldsymbol{r}^f(t) = \boldsymbol{Q}^T\left(\boldsymbol{M}^f\,{}^I\ddot{\boldsymbol{d}}^f(t) - \boldsymbol{f}^f_{\text{int}}\left({}^I\boldsymbol{d}^f(t)\right)\right) \\ {}^I\boldsymbol{d}^c(t) &= \boldsymbol{A}\;\;{}^I\boldsymbol{d}^f(t) \end{aligned} \tag{3.112} $$

where $\boldsymbol{Q}^T$ is a restriction operator and $\boldsymbol{A}$ is injection operator defined in section 2.7.6. The space–time FAS scheme is then defined as

$$ \begin{aligned} \boldsymbol{M}^c\,{}^I\ddot{\boldsymbol{u}}^c(t) - \boldsymbol{f}^c_{\text{int}}\left({}^I\boldsymbol{u}^c(t)\right) &= {}^I\boldsymbol{r}^c_{ext}\left({}^I\boldsymbol{d}^f\right) \\ {}^I\boldsymbol{u}^c(0) &= \boldsymbol{A}\;\;{}^I\boldsymbol{u}^f(0) \\ {}^I\dot{\boldsymbol{u}}^c(0) &= \boldsymbol{A}\;\;{}^I\boldsymbol{v}^f(0) \end{aligned} \tag{3.113} $$

where $\boldsymbol{M}^c$ is the mass matrix of the coarse problem, the initial conditions $\boldsymbol{d}_0$ and $\boldsymbol{v}_0$ are simply the restriction of the fine-scale initial conditions, and ${}^I\boldsymbol{u}^c$ and $\boldsymbol{r}^c_{ext}\left({}^I\boldsymbol{d}^f\right)$ are defined as in Chapter 2, which yields

$$ \begin{aligned} {}^I\boldsymbol{u}^c(t) &= \boldsymbol{A}\;\;{}^I\boldsymbol{d}^f(t) + {}^I\boldsymbol{e}^c(t) \\ {}^I\boldsymbol{r}^c_{ext}\left({}^I\boldsymbol{d}^f\right) &= \boldsymbol{M}^c\left(\boldsymbol{A}\,{}^I\ddot{\boldsymbol{d}}^f(t)\right) - \boldsymbol{f}^c_{\text{int}}\left(\boldsymbol{A}\;\;{}^I\boldsymbol{d}^f(t)\right) - \boldsymbol{Q}^T\left(\boldsymbol{M}^f\,{}^I\ddot{\boldsymbol{d}}^f(t) - \boldsymbol{f}^f\left({}^I\boldsymbol{d}^f(t)\right)\right) \end{aligned} \tag{3.114} $$

The coarse-scale operator $\boldsymbol{f}^c_{\text{int}}$ can take various forms. For instance, it can be represented by the coarse-grained molecular dynamics (CGMD) model [51].

A simplified variant can be obtained by simplifying the force field calculations rather than by spatial coarsening. In the absence of coarsening, $\boldsymbol{Q}=\boldsymbol{A}=\boldsymbol{I}$ and $\boldsymbol{M}^c=\boldsymbol{M}^f$. (See [45] for details.)

Problems

Problem 3.1: Consider the Lennard-Jones potential

$$ E(r) = 4\varepsilon\left[\left(\frac{\sigma}{r}\right)^{12} - \left(\frac{\sigma}{r}\right)^{6}\right] $$

where ε and σ are model parameters and $r = \sqrt{x_k^{IJ} x_k^{IJ}}$ is the distance between two atoms.

(a) Show that the distance $R = \sqrt{X_k^{IJ} X_k^{IJ}}$ for which the interatomic forces vanish is given by $R = \sqrt[6]{2}\ \sigma$. Hint: R is obtained by setting $dE/dr=0$.
(b) Show that a quadratic approximation of the Lennard-Jones potential is given by

$$E(r) = -\varepsilon + \frac{1}{2}k(r-R)^2$$

where $k = 36\sqrt[3]{4}\ \varepsilon/\sigma^2$ can be interpreted as a spring stiffness.

Problem 3.2: Consider a fine-scale position vector to be a continuous function of scalar λ such that [16]

$$Y_k(\lambda) = (1-\lambda)Y_k^I + \lambda Y_k^J$$

Taking the derivative with respect to λ yields

$$dY_k = Y_k^{IJ} d\lambda \quad \text{or} \quad \frac{d\lambda}{dY_k} = \frac{1}{Y_k^{IJ}}$$

Assuming that $\boldsymbol{x}^{IJ}(\boldsymbol{X}, \boldsymbol{Y}(\lambda))$ is a continuous function of $\boldsymbol{X}$ and λ, show that

$$\frac{\partial x_i^{IJ}}{\partial X_k} = F_{ik}^c(\boldsymbol{X},t) + \frac{1}{Y_k^{IJ}}\left(u_i^{(1)}(\boldsymbol{X},\boldsymbol{Y}^J,t,\tau) - u_i^{(1)}(\boldsymbol{X},\boldsymbol{Y}^I,t,\tau)\right) + O(\zeta)$$

Problem 3.3: Consider a 1D atomistic chain as shown in Figure 3.14.

(a) Let x^I be the position vector of atom I in the deformed coordinates. Write an equation of motion for atom I considering near neighbor interaction only.
(b) Expand the displacements of atoms x^{I+1} and x^{I+1} using a Taylor series expansion

$$u(x \pm l) = u(x) \pm \frac{1}{2}l^2\frac{\partial^2 u}{\partial x^2} \pm \cdots$$

and write continuum equations by utilizing an equation of motion.

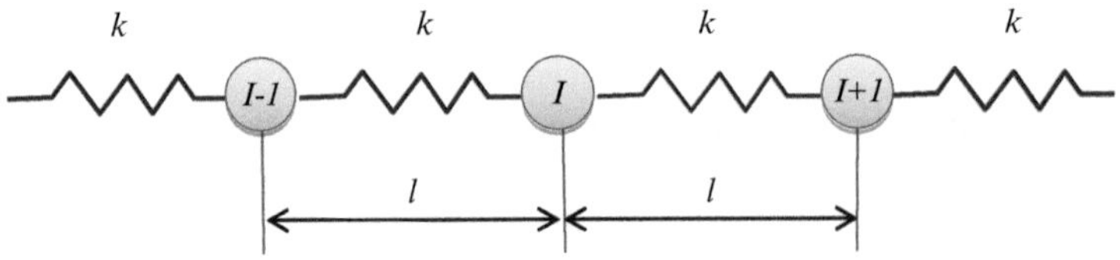

Figure 3.14 Atomistic chain

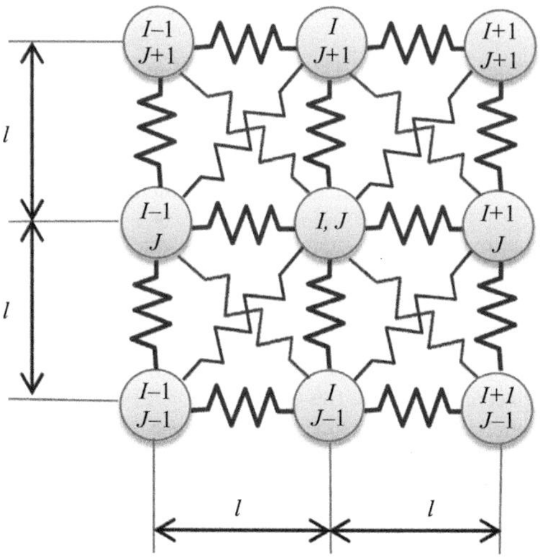

Figure 3.15 Two-dimensional square lattice

Problem 3.4: Consider the rectangular lattice structure shown in Figure 3.15. Assume the quadratic approximation of the Lennard-Jones potential considered in Problem 3.1(b).

(a) Find the effective spring stiffnesses.
(b) Repeat Problem 3.3 for the spring stiffnesses found in Problem 3.4(a).

Problem 3.5: Consider a unit cell of three atoms as depicted in Figure 3.14. Consider quadratic interatomic potentials giving rise to a linear stiffness equal to k_1 between atoms I and $I+1$ and k_2 between atoms I and $I-1$. Assume that the unit cell has been subjected to a macroscopic deformation gradient equal to 1.01.

(a) Calculate the atomistic displacements, the interatomic forces, and the first Piola–Kirchhoff stress using the definition in equation (3.50b).
(b) Calculate the energy stored in the unit cell $W(F^c)$ subjected to the coarse-scale deformation gradient. Show that the first Piola–Kirchhoff stress can be computed by $\partial W/\partial F^c$.

Problem 3.6: Repeat Problem 3.5 for the 2D square lattice unit cell shown in Figure 3.15.

References

[1] LAMMPS Molecular Dynamics Simulator. http://www.cs.sandia.gov/~sjplimp/ lammps.html.
[2] Zhou, M. Thermomechanical continuum interpretation of atomistic deformation. International Journal for Multiscale Computational Engineering 2005, (2), 177–197.
[3] Costanzo, F., Gray, G.L. and Andia, P.C. On the notion of average mechanical properties in MD simulation via homogenization. Modelling andSimulation in Materials Science and Engineering 2004, 12, S333–S345.
[4] Li, X. and E, W. Multiscale modeling of dynamics of solids at finite temperature. Journal of the Mechanics and Physics of Solids 2005, 53, 1650–1685.

[5] E, W. and Engquist, B. The heterogeneous multi-scale methods. Communications in Mathematical Sciences 2002, 1, 87–132.
[6] Dupuy, L.M., Tadmor, E.B., Miller, R.E. and Phillips, R. Finite temperature quasicontinuum: molecular dynamics. Physical Review Letters 2005, 95, 060202-1–060202-4.
[7] Kirkwood, J.G. Statistical mechanics of fluid mixtures. Journal of Chemical Physics 1935, 3, 300–313.
[8] Shiari, B., Miller, R.E. and Klug, D.D. Finite temperature multiscale computational modeling of materials at the nanoscale. In Proceedings of the 2005 International Conference on MEMS, NANO, and Smart Systems, Banff, Canada, July, 2005.
[9] Park, H.S., Karpov, E.G. and Liu, W.K.A temperature equation for coupled atomistic/continuum simulations. Computer Methods in Applied Mechanics and Engineering 2004, 193, 1713–1732.
[10] Xiao, S.P. and Belytschko, T. A bridging domain method for coupling continua with molecular dynamics. Computer Methods in Applied Mechanics and Engineering 2004, 193, 1645–1669.
[11] Chen, W. and Fish, J. A mathematical homogenization perspective of virial stress. International Journal for Numerical Methods in Engineering 2006, 67, 189–207.
[12] Chen, W. and Fish, J. A generalized space-time mathematical homogenization theory for bridging atomistic and continuum scales. International Journal for Numerical Methods in Engineering 2006, 67, 253–271.
[13] Fish, J., Chen, W. and Li, R. Generalized mathematical homogenization of atomistic media at finite temperatures in three dimensions. Computer Methods in Applied Mechanics and Engineering 2007, 196, 908–922.
[14] Li, A., Li, R. and Fish, J. Generalized mathematical homogenization: from theory to practice. Computer Methods in Applied Mechanics and Engineering 2008, 197, 245–269.
[15] Ghysels, P., Samaey, G., Van Liedekerke, P., Tijskens, B., Ramon, H. and Roose, D. Multiscale modeling of viscoelastic plant tissue. International Journal for Multiscale Computational Engineering 2010, 8, 379–396.
[16] Chockalingam, K. and Wellford, L.C. Multi-scale homogenization procedure for continuum–atomistic, thermo-mechanical problems, Computer Methods in Applied Mechanics and Engineering 2011, 200, 356–371.
[17] Chung, P.W. and Namburu, R.R. On a formulation for a multiscale atomistic-continuum homogenization method. International Journal of Solids and Structures 2003, 40(10), 2563–2588.
[18] Sunyk, R. and Steinmann. P. On higher gradients in continuum-atomistic modeling. International Journal of Solids and Structures 2003, 40, 6877–6896.
[19] Dettmar, J. Static and dynamic homogenization analyses of discrete granular and atomistic structures on different time and length scales. Dissertation, University of Stuttgart, 2006.
[20] Ericksen, J.L. On the Cauchy-Born rule. Mathematics and Mechanics of Solids 2008, 13, 199–220.
[21] Xiantao Li, and Weinan E, Multiscale modeling of the dynamics of solids at finite temperature. Journal of the Mechanics and Physics of Solids 2005, 53, 1650–1685.
[22] Srolovitz, D., Maeda, K., Vitek, V. and Egami, T. Structural defects in amorphous solids: statistical analysis of a computer model. Philosophical Magazine A 1981, 44, 847–866.
[23] Horstemeyer, M.F. and Baskes, M.I. Atomic finite deformation simulations: a discussion on length scale effects in relation to mechanical stress. Journal of Engineeringg Materials and Technology 1999, 121, 114–119.
[24] Alber, I., Bassani, J.L., Khantha, M., Vitek, V. and Wang, G.J. Grain boundaries as heterogeneous systems: atomic and continuum elastic properties. Philosophical Transactions of the Royal Society London 1992, A339, 555–586.
[25] Irving, J.H. and Kirkwood, J.G. The statistical mechanical theory of transport processes IV. Journal of Chemical Physics 1950, 18, 817–829.
[26] Zimmerman, J.A., Klein, P.A. and Webb III, E.B. Coupling and communicating between atomistic and continuum simulation methodologies. In Multiscaling in Molecular and Continuum Mechanics: Interaction of Time and Size from Macro to Nano, ed. G.C. Sih. Springer, 2007, pp. 439–455.
[27] Berman, R. Thermal Conduction in Solids. Oxford Studies in Physics. ClarendonPress/Oxford University Press, 1976.
[28] Belytschko, T., Liu, W.K. and Moran, B. Nonlinear Finite Elements for Continua and Structures. John Wiley & Sons, Ltd, 2001.
[29] Cormier, J., Rickman, J.M. and Delph, T.J. Stress calculation in atomistic simulations of perfect and imperfect solids. Journal of Applied Physics 2001, 89, 99–104.
[30] Belytschko, T. and Bindeman, L.P. Assumed strain stabilization of the eight node hexahedral. Computer Methods in Applied Mechanics and Engineering 1993, 105, 225–260.
[31] Nurminen, L., Tavazza, F., Landau, D.P., Kuronen, A. and Kaski, K. Comparative study of Si(001) surface structure and interatomic potentials in finite-temperature simulations, Physical Review B. 2003, 67, 1–10.

[32] Bond, S.D., Leimkuhler, B.J. and Laird, B.B. The Nosé-Poincaré method for constant temperature molecular dynamics. Journal of Computational Physics 1999, 151, 114–134.
[33] Menon, M., Srivastava, D., Ponomareva, I. and Chernozatonskii, A.L. Nanomechanics of silicon nanowires. Physical Review B 2004, 70(12) (2004), 125313.
[34] Verlet, L. Computer "experiments" on classical fluids. I. Thermodynamical properties of Lennard-Jones molecules. Physical Review 1967, 159(1), 98–103.
[35] Swope, W.C., Anderson, H.C., Berens, P.H. and Wilson, K.R. A computer simulation method for the calculation of equilibrium constants for the formation of physical clusters of molecules: Application to small water clusters. The Journal of Chemical Physics 1982, 76(1), 637–649.
[36] Gear, C.W. Numerical Initial Value Problems in Ordinary Differential Equations. Prentice-Hall, 1971, chapter 9.
[37] Bennemann, C., Paul, W., Binder, K. and Dünweg, B. Molecular-dynamics simulations of the thermal glass transition in polymer melts: α-relaxation behavior. Physical Review E 1998, 57(1), 843–851.
[38] Rapaport, D.C. The Art of Molecular Dynamics Simulations. Cambridge University Press, 1995.
[39] Anderson, H.C. Rattle: a velocity version of the shake algorithm for molecular dynamics calculations. Journal of Computational Physics 1983, 52, 24–34.
[40] Tuckerman, M., Berne, B.J. and Martyna, G.J. Reversible multiple time scale molecular dynamics. The Journal of Chemical Physics 1992, 97(3), 1990–2001.
[41] Garcia-Archilla, B., Sanz-Serna, J.M. and Skeel, R.D. The mollified impulse method for oscillatory differential equations. SIAM Journal on Scientific Computing 1998, 20, 930–963.
[42] Leimkuhler, B.J., Reich, S. and Skeel, R.D. Integration methods for molecular dynamics. IMA Volumes in Mathematics and its Applications 1997, 82, 161–186.
[43] Janezic, D. and Orel, B. Implicit Runge-Kutta method for molecular dynamics integration. Journal of Chemical Information and Computer Sciences 1993, 33, 252–257.
[44] Waisman, H. and Fish, J. A heterogeneous space-time full approximation storage multilevel method for molecular dynamics simulations. International Journal for Numerical Methods in Engineering 2008, 73, 407–426.
[45] Brandt, A. Multi-level adaptive solutions to boundary-value problems. Mathematics of Computation 1977, 31(138), 333–390.
[46] Vandewalle, S. The parallel solution of parabolic partial differential equations by multigrid waveform relaxation methods. PhD thesis, Katholike Universiteit Leuven, 1992.
[47] Horton, G. and Vandewalle, S. A space-time multigrid method for parabolic partial differential equations. SIAM Journal on Scientific and Statistical Computing 1995, 16(4), 848–864.
[48] Leimkuhler, B. Timestep acceleration of waveform realxation. SIAM Journal on Numerical Analysis 1998, 35(1), 31–50.
[49] Saleh, R. and White, J. Accelerating relaxation algorithms for circuit simulation using waveform-newton and step-size refinement. IEEE Transactions on Computer-Aided Design 1990, 9(9), 951–958.
[50] Miekkala, U. and Nevanlinna, O. Convergence of dynamic iteration methods for initial value problems. SIAM Journal on Scientific and Statistical Computing 1987, 8(4), 459–482.
[51] Rudd, R.E. and Broughton, J.Q. Coarse-grained molecular dynamics and the atomic limit of finite elements. Physical Review B 1998, 58(10), R5893–R5896.

4

Reduced Order Homogenization

4.1 Introduction

The upscaling methods outlined in Chapter 2 for continua and in Chapter 3 for atomistic media have had little impact on nonlinear history-dependent problems due to the enormous computational complexity involved. To illustrate the computational challenge, consider a coarse-scale problem with n_{cells} Gauss points, n_{inc} load increments at the coarse scale, and I_{coarse} and I_{fine} number of average iterations at the coarse and fine scales, respectively. The total number of linear solves of the unit cell problem is thus $n_{\text{cells}} \cdot n_{\text{inc}} \cdot I_{\text{coarse}} \cdot I_{\text{fine}}$ — a formidable computational cost when dealing with large numbers of unit cells and when the size (degrees of freedom) of these unit cells is substantial. This tyranny of scales can be effectively addressed by using a combination of parallel methods and by introducing an intermediate reduced order (or so-called mesomechanical) model as shown in Figure 4.1.

Development of reduced order models for heterogeneous continua has been an active research area in the past decade. One of the oldest reduced order models is based on the purely kinematical Taylor's hypothesis (see the Voigt model in Chapter 2), which assumes uniform deformation at the fine scale; it satisfies compatibility but fails to account for equilibrium across the boundaries of fine-scale constituents. Major progress in mesomechanical modeling (at the expense of computational cost) has been made by utilizing the Voronoi cell method [1,2], the spectral method [3], the network approximation method [4], the fast Fourier transforms [5,6], the lattice discrete particle model (LDPM) [7,8], the meshfree reproducing kernel particle method (RKPM) [9,10], finite-volume direct averaging micromechanics (FVDAM) [11], the transformation field analysis (TFA) [12,13], the methods of cells [3] or its generalization [14] and methods based on control theory including balanced truncation [15,16], the optimal Hankel norm approximation [17], and proper orthogonal decomposition [18].

Practical Multiscaling, First Edition. Jacob Fish.

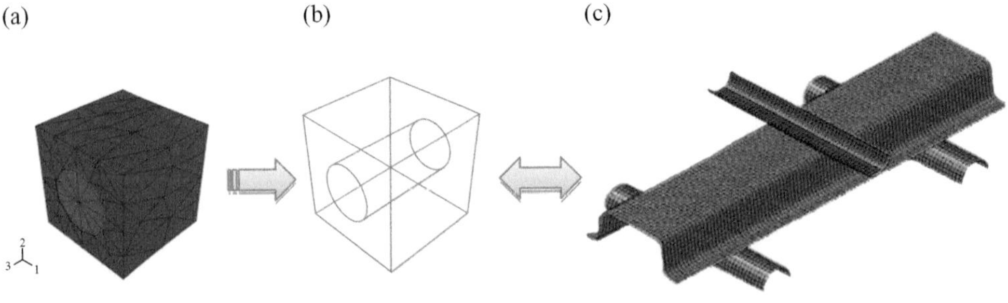

Figure 4.1 Linking fine-scale and coarse-scale models through a reduced order multiscale model: (a) fine-scale model; (b) meso-scale model; and (c) coarse-scale model

This chapter focuses on a variant of the TFA [19,20,21,22] that allows precomputing certain information (localization operators, concentration tensors, and transformation influence functions) in the preprocessing phase prior to nonlinear analysis, which can then be carried out with a small subset of unknowns. By using this approach, the effect of so-called eigenstrains (representing inelastic strains, thermal strains, or phase transformation strains) is accounted for by solving a sequence of linear elasticity problems. The salient feature of TFA is that the unit cell equilibrium equations, which have to be solved $n_{\text{cells}} \cdot n_{\text{inc}} \cdot I_{\text{coarse}} \cdot I_{\text{fine}}$ times by one of the upscaling methods discussed in Chapter 2, are satisfied a priori at the preprocessing stage.

Yet, despite its promise, TFA has found limited success in practice. There are several reasons for this:

1. It does not account for interface failure.
2. It is inaccurate for lower order approximation of eigenstrains, whereas for higher order approximation of eigenstrains its computational complexity is comparable with that of direct homogenization.
3. It is not hierarchical in the sense that model improvement is not possible when higher accuracy is needed, such as in the regions of boundary layers.
4. It is limited to two scales.

The above deficiencies have been circumvented by the so-called reduced order homogenization (ROH) method [23,24,25,26,27,28]. An early version of ROH that did not consider strong discontinuities at the material interfaces was developed in [29,30]. The focus of this chapter is on the ROH method developed in [28], which has the following characteristics:

1. It accounts for interface failure in addition to failure of fine-scale phases; interface failure is modeled using so-called *eigenseparations*, a concept similar to using eigenstrains for modeling inelastic deformation of phases.
2. It is equipped with a hierarchical model improvement capability, that is, it incorporates a hierarchical sequence of computational homogenization models where the most inexpensive member of the sequence is based on a hybrid impotent-incompatible eigenstrain approximation within each fine-scale phase (inclusion, matrix, and interface) [28], whereas the most comprehensive model of the hierarchical sequence coincides with the direct homogenization method considered in Chapter 2.

3. Like TFA, it constructs residual-free fields and thus de facto eliminates the need for costly solution of discrete equilibrium equations of the unit cell problem. This removes a major computational bottleneck and thus permits accounting for arbitrary fine-scale details at a cost comparable with phenomenological modeling of a nonlinear heterogeneous medium [31].

Based on the above characteristics, ROH falls into the subclass of coarse-graining methods (see Chapter 1 for classification) since complete fine-scale information is not revisited throughout the nonlinear analysis. Instead, the material database constructed prior to the analysis is accessed throughout the nonlinear analysis.

We will start with a formulation for small-deformation two-scale problems in section 4.2. Extensions to multiple scales, large deformations, wave dispersion, and multiple physical processes are given in subsequent sections.

4.2 Reduced Order Homogenization for Two-Scale Problems

4.2.1 Governing Equations

Consider a strong form of the boundary value problem stated on the composite domain Ω^ζ

$$\sigma_{ij,j}^\zeta(\boldsymbol{x})+b_i^\zeta(\boldsymbol{x})=0 \quad \boldsymbol{x}\in\Omega^\zeta \tag{4.1}$$

$$\sigma_{ij}^\zeta(\boldsymbol{x})=L_{ijkl}^\zeta(\boldsymbol{x})\left(\varepsilon_{kl}^\zeta(\boldsymbol{x})-\mu_{kl}^\zeta(\boldsymbol{x})\right) \quad \boldsymbol{x}\in\Omega^\zeta \tag{4.2}$$

$$\varepsilon_{ij}^\zeta(\boldsymbol{x})=u_{(i,j)}^\zeta(\boldsymbol{x})\equiv\frac{1}{2}\left(u_{i,j}^\zeta+u_{j,i}^\zeta\right) \quad \boldsymbol{x}\in\Omega^\zeta \tag{4.3}$$

$$\mu_{kl}^\zeta(\boldsymbol{x})=\sum_I \mu_{kl}^{I\zeta}(\boldsymbol{x}) \quad \boldsymbol{x}\in\Omega^\zeta \tag{4.4}$$

$$u_i^\zeta(\boldsymbol{x})=\bar{u}_i^\zeta(\boldsymbol{x}) \quad \boldsymbol{x}\in\partial\Omega^{u\zeta} \tag{4.5}$$

$$\sigma_{ij}^\zeta(\boldsymbol{x})n_j^\zeta(\boldsymbol{x})=\bar{t}_i^\zeta(\boldsymbol{x}) \quad \boldsymbol{x}\in\partial\Omega^{t\zeta} \tag{4.6}$$

where the total strain ε_{kl}^ζ is additively decomposed into elastic strain and inelastic strain, more generally referred to as eigenstrains μ_{kl}^ζ. The total eigenstrain is a sum of various eigenstrain types, denoted by $\mu_{kl}^{I\zeta}(\boldsymbol{x})$, including inelastic strain, thermal strain, moisture-induced strain, and so on. In the above, $\boldsymbol{n}^\zeta$ denotes the unit normal to the composite domain boundary $\partial\Omega^\zeta$.

In this section, small deformations are considered with no distinction made between the deformed and undeformed coordinates. Extension to large deformation is considered in section 4.7.

Following the linear elasticity formulation in Chapter 2, various fields are assumed to depend on the coarse-scale coordinate $\boldsymbol{x}$ and the unit cell coordinate $\boldsymbol{y}=\boldsymbol{x}/\zeta$. They are expressed in terms of the two-scale asymptotic expansion as follows

$$u_i^\zeta(\boldsymbol{x})\equiv u_i(\boldsymbol{x},\boldsymbol{y})=u_i^c(\boldsymbol{x})+\zeta u_i^{(1)}(\boldsymbol{x},\boldsymbol{y})+O(\zeta^2) \tag{4.7a}$$

$$\varepsilon_{ij}^{\zeta}(\boldsymbol{x}) \equiv \varepsilon_{ij}(\boldsymbol{x},\boldsymbol{y}) = \varepsilon_{ij}^{f}(\boldsymbol{x},\boldsymbol{y}) + \zeta\varepsilon_{ij}^{(1)}(\boldsymbol{x},\boldsymbol{y}) + O(\zeta^2) \tag{4.7b}$$

$$\sigma_{ij}^{\zeta}(\boldsymbol{x}) \equiv \sigma_{ij}(\boldsymbol{x},\boldsymbol{y}) = \sigma_{ij}^{f}(\boldsymbol{x},\boldsymbol{y}) + \zeta\sigma_{ij}^{(1)}(\boldsymbol{x},\boldsymbol{y}) + O(\zeta^2) \tag{4.7c}$$

where $u_i^c(\boldsymbol{x}) \equiv u_i^{(0)}(\boldsymbol{x})$, $\varepsilon_{kl}^{f}(\boldsymbol{x},\boldsymbol{y}) \equiv \varepsilon_{ij}^{(0)}(\boldsymbol{x},\boldsymbol{y})$, and $\sigma_{ij}^{f}(\boldsymbol{x},\boldsymbol{y}) \equiv \sigma_{ij}^{(0)}(\boldsymbol{x},\boldsymbol{y})$.

The eigenstrain $\mu_{ij}^{\zeta} \equiv \mu_{ij}^{\zeta}\left(\boldsymbol{\varepsilon}^{\zeta}, \boldsymbol{\sigma}^{\zeta}, \tilde{\boldsymbol{s}}^{\zeta}\right)$ depends on the constitutive behavior of fine-scale phases. It can be expressed in terms of the state variables denoted by $\tilde{s}_i^{\zeta}$, as well as by strain and/or stress. Expanding the eigenstrain in a Taylor series around the corresponding leading order field yields

$$\begin{aligned}\mu_{ij}^{\zeta} &\equiv \mu_{ij}\left(\boldsymbol{\varepsilon}^{f}, \boldsymbol{\sigma}^{f}, \tilde{\boldsymbol{s}}^{f}\right) + \zeta\left(\left.\frac{\partial \mu_{ij}^{\zeta}}{\partial \varepsilon_{kl}^{\zeta}}\right|_{\boldsymbol{\varepsilon}^{f}, \boldsymbol{\sigma}^{f}, \tilde{\boldsymbol{s}}^{f}} \varepsilon_{kl}^{(1)} + \left.\frac{\partial \mu_{ij}^{\zeta}}{\partial \sigma_{kl}^{\zeta}}\right|_{\boldsymbol{\varepsilon}^{f}, \boldsymbol{\sigma}^{f}, \tilde{\boldsymbol{s}}^{f}} \sigma_{kl}^{(1)} + \left.\frac{\partial \mu_{ij}^{\zeta}}{\partial \tilde{s}_{k}^{\zeta}}\right|_{\boldsymbol{\varepsilon}^{f}, \boldsymbol{\sigma}^{f}, \tilde{\boldsymbol{s}}^{f}} \tilde{s}_{k}^{(1)}\right) + O\left(\zeta^2\right) \\ &= \mu_{ij}^{f} + \zeta\mu_{ij}^{(1)} + O\left(\zeta^2\right)\end{aligned} \tag{4.8}$$

where $\mu_{ij}^{f} \equiv \mu_{ij}\left(\boldsymbol{\varepsilon}^{f}, \boldsymbol{\sigma}^{f}, \tilde{\boldsymbol{s}}^{f}\right)$.

Recalling the two-scale differential rule and inserting the asymptotic expansion (4.7) into (4.2) and (4.3) yields

$$\varepsilon_{kl}^{f}(\boldsymbol{x},\boldsymbol{y}) = \varepsilon_{kl}^{c}(\boldsymbol{x}) + u_{(k,y_l)}^{(1)}(\boldsymbol{x},\boldsymbol{y}) \tag{4.9a}$$

$$\varepsilon_{kl}^{c}(\boldsymbol{x}) = u_{(k,x_l)}^{c} \tag{4.9b}$$

$$\sigma_{ij}^{f}(\boldsymbol{x},\boldsymbol{y}) = L_{ijkl}(\boldsymbol{y})\left(\varepsilon_{kl}^{f}(\boldsymbol{x},\boldsymbol{y}) - \mu_{kl}^{f}(\boldsymbol{x},\boldsymbol{y})\right) \tag{4.9c}$$

Following the derivation in Chapter 2, the leading order equilibrium equations are

$$O\left(\zeta^{-1}\right): \quad \sigma_{ij,y_j}^{f} = 0 \tag{4.10a}$$

$$O(1): \quad \sigma_{ij,x_j}^{c} + b_i^c = 0 \tag{4.10b}$$

where the coarse-scale fields are given as

$$\begin{aligned}\sigma_{ij}^{c} &= \frac{1}{|\Theta|}\int_{\Theta} \sigma_{ij}^{f}\, d\Theta; \quad b_i^c = \frac{1}{|\Theta|}\int_{\Theta} b_i^{\zeta}\, d\Theta \\ \bar{t}_i^{c} &= \frac{1}{|\partial\omega|}\int_{\partial\omega} \bar{t}_i^{\zeta}\, ds; \quad \bar{u}_i^{c} = \frac{1}{|\partial\omega|}\int_{\partial\omega} \bar{u}_i^{\zeta}\, ds\end{aligned} \tag{4.11}$$

For the unit cell problem (4.10a), we will consider damage at the interfaces between phases by introducing displacement jump δ_i^{ζ} (hereafter referred to as an eigenseparation), defined as

$$\delta_i^{\zeta}(\boldsymbol{x}) \equiv \llbracket u_i^{\zeta}(\boldsymbol{x}) \rrbracket = u_i^{\zeta}\Big|_{S^-} - u_i^{\zeta}\Big|_{S^+} \qquad \boldsymbol{x} \in S \tag{4.12a}$$

and imposing traction continuity condition

$$\sigma_{ij}^{\zeta}\breve{n}_j\Big|_{S^+} + \sigma_{ij}^{\zeta}\breve{n}_j\Big|_{S^-} = t_i^{\zeta}\Big|_{S^+} + t_i^{\zeta}\Big|_{S^-} = 0 \quad \boldsymbol{x} \in S \tag{4.12b}$$

where $\llbracket \bullet \rrbracket$ denotes the jump operator. The +/− signs indicate the two sides of the interface. $\breve{\boldsymbol{n}}$ denotes the unit normal at the interface between fine-scale constituents.

The eigenseparation $\delta_i^{\zeta} \equiv \delta_i^{\zeta}(\boldsymbol{t}^{\zeta}, \breve{\boldsymbol{s}}^{\zeta})$ depends on the cohesive law of the interface. It can be expressed in terms of state variables $\breve{s}_i^{\zeta}$ and traction t_i^{ζ}. Expanding the eigenseparation $\delta_i^{\zeta}(\boldsymbol{t}^{\zeta}, \breve{\boldsymbol{s}}^{\zeta})$ in a Taylor series around the leading order fields yields

$$\begin{aligned} \delta_i^{\zeta}(\boldsymbol{t}^{\zeta}, \breve{\boldsymbol{s}}^{\zeta}) &\equiv \delta_i(\boldsymbol{t}^f, \breve{\boldsymbol{s}}^f) + \zeta\left(\frac{\partial \delta_i^{\zeta}}{\partial t_k^{\zeta}}\bigg|_{\boldsymbol{t}^f, \breve{\boldsymbol{s}}^f} t_k^{(1)} + \frac{\partial \delta_i^{\zeta}}{\partial \breve{s}_k^{\zeta}}\bigg|_{\boldsymbol{t}^f, \breve{\boldsymbol{s}}^f} \breve{s}_k^{(1)}\right) + \ldots \\ &= \breve{\delta}_i^f + \zeta\breve{\delta}_i^{(1)} + \ldots = \zeta\delta_i^f + \zeta^2\delta_i^{(1)} + O(\zeta^3) \end{aligned} \tag{4.13}$$

where $\breve{\delta}_i^f \equiv \delta_i(\boldsymbol{t}^f, \breve{\boldsymbol{s}}^f)$. Further inserting (4.7) and (4.13) into (4.12a) and (4.12b) yields

$$\breve{\delta}_i^f \equiv \zeta\delta_i^f = \zeta\llbracket u_i^f \rrbracket = \zeta\left(u_i^f\big|_{S^-} - u_i^f\big|_{S^+}\right) \tag{4.14a}$$

$$\sigma_{ij}^f\breve{n}_j^+\Big|_{S^+} + \sigma_{ij}^f\breve{n}_j^-\Big|_{S^-} = t_i^f\Big|_{S^+} + t_i^f\Big|_{S^-} = 0 \tag{4.14b}$$

where u_i^f is the translation-free fine-scale displacement defined in (2.182). It is instructive to point out that the leading order eigenseparation $\breve{\delta}_i^f$ is of order $O(\zeta)$ due to the highly stiff response of a cohesive law, especially in the elastic regime. As the damage at the interface accumulates, the magnitude of separation $\breve{\delta}_i^f$ can increase substantially.

Equation (4.10a), equation (4.9), equation (4.14a), and equation (4.14b), together with periodic boundary conditions, comprise a unit cell problem. Once the unit cell problem is solved, the coarse-scale stress follows from (410b) and (4.11), which yields

$$\sigma_{ij}^c(\boldsymbol{x}) = \frac{1}{|\Theta|}\int_{\Theta} L_{ijkl}(\boldsymbol{y})\left(\varepsilon_{kl}^c(\boldsymbol{x}) + u_{(k,y_l)}^{(1)}(\boldsymbol{x},\boldsymbol{y}) - \mu_{kl}^f(\boldsymbol{x},\boldsymbol{y})\right)d\Theta \tag{4.15}$$

where discontinuity at the interface will be accounted for in the construction of the $u_k^{(1)}$ term.

The focus of Chapter 4 is on the effective solution of the unit cell problem and the subsequent calculation of the coarse-scale stress.

4.2.2 Residual-Free Fields and Model Reduction

The salient feature of ROH is that the fine-scale displacement field correction $u_i^{(1)}(\boldsymbol{x},\boldsymbol{y})$ is formulated to automatically satisfy the fine-scale equilibrium equation (410a) for arbitrary coarse-scale strain ε_{kl}^{c}, eigenstrains μ_{ij}^{f}, and eigenseparations $\delta_{\breve{n}}^{f}$. $u_i^{(1)}(\boldsymbol{x},\boldsymbol{y})$ is constructed as follows

$$u_i^{(1)}(\boldsymbol{x},\boldsymbol{y}) = H_i^{kl}(\boldsymbol{y})\,\varepsilon_{kl}^{c}(\boldsymbol{x}) + \int_{\Theta} \tilde{h}_i^{kl}(\boldsymbol{y},\tilde{\boldsymbol{y}})\,\mu_{kl}^{f}(\boldsymbol{x},\tilde{\boldsymbol{y}})d\tilde{\Theta} + \int_{S} \breve{h}_i^{\breve{n}}(\boldsymbol{y},\breve{\boldsymbol{y}})\,\delta_{\breve{n}}^{f}(\boldsymbol{x},\breve{\boldsymbol{y}})d\breve{S} \tag{4.16}$$

where H_i^{kl}, $\tilde{h}_i^{kl}$, and $\breve{h}_i^{\breve{n}}$ are so-called transformation influence functions for the coarse-scale strain, the fine-scale eigenstrain, and the fine-scale eigenseparation, respectively. The physical meaning of (4.16) is that the eigenstrain (or so-called transformation strain) $\mu_{kl}^{f}(\boldsymbol{x},\tilde{\boldsymbol{y}})$ introduces elastic deformation in the magnitude of $\tilde{h}_i^{kl}(\boldsymbol{y},\tilde{\boldsymbol{y}})$ due to volume and/or shape changes at an infinitesimal neighborhood of a point $\tilde{\boldsymbol{y}} \in \Theta$ [12,19,22,32,33,34,35,36]. The volume integral represents the accumulative effect of all possible eigenstrains in the unit cell domain. Likewise, the eigenseparation $\delta_{\breve{n}}^{f}(\boldsymbol{x},\breve{\boldsymbol{y}})$ gives rise to elastic deformation equal to $\breve{h}_i^{\breve{n}}(\boldsymbol{y},\breve{\boldsymbol{y}})$ as a result of a debonding (or displacement jump) at an infinitesimal neighborhood of a point $\breve{\boldsymbol{y}} \in S$ at the interface [24,25,26]. The integral over all the interfaces in the unit cell represents an accumulative effect of all eigenseparations. Note that (4.16) holds for arbitrary eigenstrains as long as the strain follows the additive decomposition (4.9c).

Inserting (4.16) into (4.9) and (4.10a) results in a fine-scale equilibrium equation for the transformation influence functions

$$\left(L_{ijkl}(\boldsymbol{y}) \left(\begin{array}{l} \left(I_{klmn} + H_{(k,y_l)}^{mn}\right)\varepsilon_{mn}^{c}(\boldsymbol{x}) + \left(\int_{\Theta} \tilde{h}_{(k,y_l)}^{mn}(\boldsymbol{y},\tilde{\boldsymbol{y}})\,\mu_{mn}^{f}(\boldsymbol{x},\tilde{\boldsymbol{y}})d\tilde{\Theta} - \mu_{kl}^{f}(\boldsymbol{x},\boldsymbol{y})\right) \\ + \int_{S} \breve{h}_{(k,y_l)}^{\breve{n}}(\boldsymbol{y},\breve{\boldsymbol{y}})\,\delta_{\breve{n}}^{f}(\boldsymbol{x},\breve{\boldsymbol{y}})d\breve{S} \end{array} \right) \right)_{,y_j} = 0 \tag{4.17}$$

An analytical solution of (4.17) can be thought of by expressing eigenstrains and eigenseparations, $\mu_{kl}^{f}(\boldsymbol{x},\boldsymbol{y})$ and $\delta_{\breve{n}}^{f}(\boldsymbol{x},\boldsymbol{y})$, as

$$\begin{array}{ll} \mu_{kl}^{f}(\boldsymbol{x},\boldsymbol{y}) = a_{kl}(\boldsymbol{x})d(\boldsymbol{y}-\underset{\sim}{\boldsymbol{y}}) & \underset{\sim}{\boldsymbol{y}} \in \Theta \\ \delta_{\breve{n}}^{f}(\boldsymbol{x},\boldsymbol{y}) = b_{\breve{n}}(\boldsymbol{x})d(\boldsymbol{y}-\underset{\smile}{\boldsymbol{y}}) & \underset{\smile}{\boldsymbol{y}} \in S \end{array} \tag{4.18}$$

where $d(\boldsymbol{y}-\underset{\sim}{\boldsymbol{y}})$ is the Dirac delta function in the unit cell domain and $d(\boldsymbol{y}-\underset{\smile}{\boldsymbol{y}})$ is the Dirac delta function at the interfaces of fine-scale phases. The transformation influence functions can be computed by inserting (4.18) into (4.17) and requiring arbitrariness of $\varepsilon_{kl}^{c}(\boldsymbol{x})$, $a_{kl}(\boldsymbol{x})$, and $b_{\breve{n}}(\boldsymbol{x})$, which yields

$$\left\{L_{ijkl}(\boldsymbol{y})\left[I_{klmn} + H_{(k,y_l)}^{mn}(\boldsymbol{y})\right]\right\}_{,y_j} = 0 \tag{4.19a}$$

$$\left\{L_{ijkl}(\boldsymbol{y})\left[\tilde{h}_{(k,y_l)}^{mn}(\boldsymbol{y},\underset{\sim}{\boldsymbol{y}}) - I_{klmn}d(\boldsymbol{y}-\underset{\sim}{\boldsymbol{y}})\right]\right\}_{,y_j} = 0 \quad \forall \underset{\sim}{\boldsymbol{y}} \in \Theta \tag{4.19b}$$

$$\left\{L_{ijkl}(\boldsymbol{y})\breve{h}_{(k,y_l)}^{\breve{n}}(\boldsymbol{y},\underset{\smile}{\boldsymbol{y}})\right\}_{,y_j} = 0; \quad \left[\!\left[\breve{h}_k^{\breve{n}}(\boldsymbol{y},\underset{\smile}{\boldsymbol{y}})\right]\!\right] = d(\boldsymbol{y}-\underset{\smile}{\boldsymbol{y}}) \quad \forall \underset{\smile}{\boldsymbol{y}} \in S \tag{4.19c}$$

Equations (4.19) represent elastic boundary value problems that can be solved prior to the nonlinear coarse-scale analysis. An analytical solution of (4.19) for complex unit cells is usually unknown. Equation (4.19a) is identical to that considered for linear elasticity problems in Chapter 2 and can be solved using the finite element method.

Equation (4.19b) and equation (4.19c) comprise an infinite set of elastic boundary value problems, since (4.19b) and (4.19c) have to be satisfied for arbitrary $\boldsymbol{y} \in \Theta$ and $\boldsymbol{y} \in S$, respectively. To reduce the computational complexity involved, a reduced unit cell problem can be constructed by discretizing eigenstrains μ_{ij}^{f} and eigenseparations $\delta_{\breve{n}}^{f}$ in (4.16) as

$$\mu_{ij}^{f}(\boldsymbol{x},\boldsymbol{y}) = \sum_{\alpha=1}^{\tilde{M}} \tilde{N}^{(\alpha)}(\boldsymbol{y})\mu_{ij}^{(\alpha)}(\boldsymbol{x}) \tag{4.20a}$$

$$\delta_{\breve{n}}^{f}(\boldsymbol{x},\boldsymbol{y}) = \sum_{\xi=1}^{\breve{M}} \breve{N}^{(\xi)}(\boldsymbol{y})\delta_{\breve{n}}^{(\xi)}(\boldsymbol{x}) \tag{4.20b}$$

where $\tilde{M}$ and $\breve{M}$ are the number of phases and interface modes, respectively, and $\mu_{ij}^{(\alpha)}$ and $\delta_{\breve{n}}^{(\xi)}$ are the corresponding coefficients.

Since eigenstrains do not have to be continuous, the eigenstrain shape functions $\tilde{N}^{(\alpha)}(\boldsymbol{y})$ can be chosen to be $C^{-1}(\Theta)$ and thus satisfy partition of unity condition $\sum_{\alpha=1}^{\tilde{M}} \tilde{N}^{(\alpha)}(\boldsymbol{y}) = 1$. The eigenseparations, on the other hand, have to be $C^0(S)$ functions to allow for continuous crack patterns. Thus, the eigenseparation shape functions $\breve{N}^{(\xi)}(\boldsymbol{y})$ are chosen to be $C^0(S)$ in order to satisfy the partition of unity condition, that is, $\sum_{\xi=1}^{\breve{M}} \breve{N}^{(\xi)}(\boldsymbol{y}) = 1$.

The partition of unity condition for eigenstrain shape functions $\tilde{N}^{(\alpha)}(\boldsymbol{y})$ can be satisfied using a piecewise constant approximation

$$\tilde{N}^{(\alpha)}(\boldsymbol{y}) = \begin{cases} 1 & \boldsymbol{y} \in \Theta^{(\alpha)} \\ 0 & \boldsymbol{y} \in \Theta^{(\alpha)} \end{cases} \tag{4.21}$$

whereas the interface shape functions $\breve{N}^{(\xi)}(\boldsymbol{y})$ over the interface partition ξ can be constructed as follows

$$\breve{N}^{(\xi)}(\boldsymbol{y}) = \begin{cases} \sum\limits_{A \in S^{(\xi)}} N_A^{f}(\boldsymbol{y}) & \boldsymbol{y} \in S^{(\xi)} \\ 0 & \boldsymbol{y} \notin S^{(\xi)} \end{cases} \tag{4.22}$$

where $N_A^{f}(\boldsymbol{y})$ are the unit cell finite element shape functions $u_i^{f}(\boldsymbol{x},\boldsymbol{y}) = N_A^{f}(\boldsymbol{y})d_{iA}^{f}(\boldsymbol{x})$ with subscript A denoting the finite element node number.

For the piecewise constant approximation, $\mu_{ij}^{(\alpha)}$ and $\delta_{\breve{n}}^{(\xi)}$ represent the uniform eigenstrain and eigenseparation in phase partition $\Theta^{(\alpha)}$ and interface partition $S^{(\xi)}$, respectively. Each finite element (or node) in a unit cell domain can be defined as a partition, and any patch of elements within a single phase can serve as a partition. A partition consisting of a single finite element will have the highest computational complexity, whereas a partition made of all the elements within a phase will comprise the most inexpensive member of the hierarchy. We will see that the solution of the most inexpensive member of the hierarchy is often not sufficiently accurate. In section 4.3 we will consider approximations of eigenstrains other than (4.21), which will be more suitable for lower order approximations.

Figure 4.2 depicts an example of two interface partitions for a fibrous unit cell. Figure 4.2(a) shows the unit cell mesh; the two interface partitions are illustrated in Figure 4.2(b). The overlapping areas are shown in brown. Figure 4.2(c) and (d) depict the two corresponding shape functions.

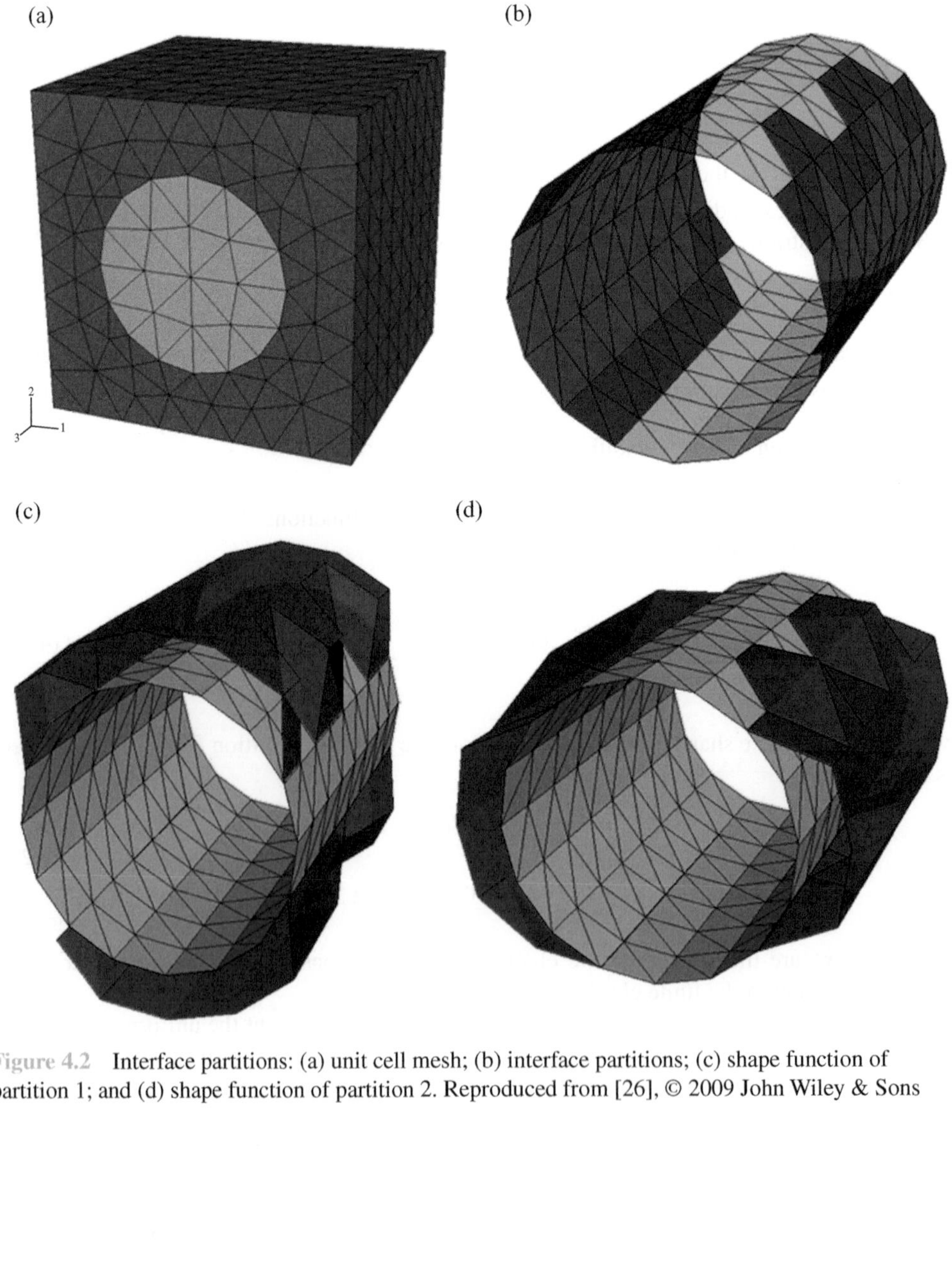

Figure 4.2 Interface partitions: (a) unit cell mesh; (b) interface partitions; (c) shape function of partition 1; and (d) shape function of partition 2. Reproduced from [26], © 2009 John Wiley & Sons

Inserting the piecewise discretizations (4.21) and (4.22) into (4.17) and requiring the unit cell equilibrium to be satisfied for arbitrary $\mu_{ij}^{(\alpha)}$ and $\delta_{\hat{n}}^{(\xi)}$ yields the following strong forms for the eigenstrains and eigenseparations influence functions:

(1) Find $\tilde{h}_i^{mn(\alpha)}(\boldsymbol{y})$ such that

$$\left\{L_{ijkl}(\boldsymbol{y})\left[P_{kl}^{mn(\alpha)}(\boldsymbol{y})-I_{klmn}^{(\alpha)}(\boldsymbol{y})\right]\right\}_{,y_j}=0 \quad \forall\alpha \tag{4.23}$$

where

$$P_{ij}^{mn(\alpha)}(\boldsymbol{y})=\left(\int_{\Theta^{(\alpha)}}\tilde{h}_i^{mn}(\boldsymbol{y},\tilde{\boldsymbol{y}})\,d\tilde{\Theta}\right)_{,y_j}\equiv\tilde{h}_{(i,y_j)}^{mn(\alpha)}$$

$$I_{klmn}^{(\alpha)}(\boldsymbol{y})=\begin{cases}I_{klmn} & \boldsymbol{y}\in\Theta^{(\alpha)}\\ 0 & \text{elsewhere}\end{cases} \tag{4.24}$$

subjected to periodic boundary conditions.

(2) Find $\breve{h}_i^{\hat{n}(\xi)}(\boldsymbol{y})$ such that

$$\left\{L_{ijkl}(\boldsymbol{y})\,Q_{kl}^{\hat{n}(\xi)}(\boldsymbol{y})\right\}_{,y_j}=0 \quad \forall\xi \tag{4.25a}$$

$$\left[\!\left[\breve{h}_i^{\hat{n}(\xi)}(\boldsymbol{y})\right]\!\right]=\breve{N}^{(\xi)}(\boldsymbol{y}) \quad \boldsymbol{y}\in S^{(\xi)} \tag{4.25b}$$

where

$$Q_{ij}^{\hat{n}(\xi)}(\boldsymbol{y})=\left(\int_S\breve{h}_i^{\hat{n}}(\boldsymbol{y},\tilde{\boldsymbol{y}})\;\breve{N}^{(\xi)}(\tilde{\boldsymbol{y}})\,d\tilde{S}\right)_{,y_j}\equiv\breve{h}_{(i,y_j)}^{\hat{n}(\xi)} \tag{4.26}$$

subjected to periodic boundary conditions.

The boundary value problems (4.23), (4.24), (4.25), and (4.26) are solved using a Galerkin approximation of the eigenstrains trial functions $\tilde{h}_i^{mn(\alpha)}$, the eigenseparations trial functions $\breve{h}_i^{\hat{n}(\xi)}$, and the test w_i^f functions

$$\tilde{h}_i^{mn(\alpha)}(\boldsymbol{y})=N_{i\beta}^f(\boldsymbol{y})\tilde{d}_\beta^{mn(\alpha)} \tag{4.27a}$$

$$\breve{h}_i^{\hat{n}(\xi)}(\boldsymbol{y})=N_{i\beta}^f(\boldsymbol{y})\breve{d}_\beta^{\hat{n}(\xi)} \tag{4.27b}$$

$$w_i^f=N_{i\alpha}^f(\boldsymbol{y})c_\alpha \tag{4.27c}$$

where $\tilde{d}_\beta^{mn(\alpha)}$ and $\breve{d}_\beta^{\hat{n}(\xi)}$ denote the nodal degrees of freedom of trial functions, c_α denotes the nodal degrees of freedom of test functions, $N_{i\beta}^f(\boldsymbol{y})$ is a $C^0(\Theta)$ continuous fine-scale shape function, and $B_{ij\alpha}^f(\boldsymbol{y})=N_{(i,y_j)\alpha}^f(\boldsymbol{y})$ is a corresponding symmetric gradient of the shape function.

Writing the weak form of (4.23) and employing Galerkin approximation (4.27a) and (4.27c) gives the discrete fine-scale problem for $\tilde{h}_i^{mn(\alpha)}$, which states:

Given L_{ijkl}, find $\tilde{d}_{\beta}^{mn(\alpha)}$ periodic such that

$$c_{\gamma}\left(\tilde{K}_{\gamma\beta}^{f}\tilde{d}_{\beta}^{mn(\alpha)}-\tilde{f}_{\gamma}^{mn(\alpha)}\right)=0 \quad \forall c_{\gamma}-\text{periodic} \tag{4.28}$$

where

$$\begin{aligned}\tilde{K}_{\gamma\beta}^{f} &= \int_{\Theta} B_{ij\gamma}^{f} L_{ijkl} B_{kl\beta}^{f} d\Theta \\ \tilde{f}_{\gamma}^{mn(\alpha)} &= \int_{\Theta^{(\alpha)}} B_{ij\gamma}^{f} L_{ijmn} d\Theta\end{aligned} \tag{4.29}$$

For 3D problems, equation (4.28) is solved $6\cdot\tilde{M}$ times corresponding to six eigenstrain modes applied at every volume partition. For implementation in a commercial finite element software package, the forcing term $\tilde{f}_{\gamma}^{mn(\alpha)}$ can be computed by subjecting the unit cell to six possible unit thermal strains $\varepsilon_{mn}^{\text{therm}}$ defined as

$$\varepsilon_{mn}^{\text{therm}} = \kappa_{mn}\cdot\Delta T \tag{4.30}$$

where thermal expansion coefficient κ_{mn} is defined as in Chapter 2 and

$$\Delta T = \begin{cases} 1 & y\in\Theta^{(\alpha)} \\ 0 & \text{elsewhere}\end{cases} \tag{4.31}$$

Likewise, writing the weak form of (4.25a) and employing Galerkin approximation (4.27b) and (4.27c) gives the discrete fine-scale problem for $\breve{h}_{i}^{\hat{n}(\xi)}$, which states:

Given L_{ijkl}, find $\breve{d}_{\beta}^{\hat{n}(\xi)}$ periodic such that

$$\begin{aligned} & c_{\gamma}\breve{K}_{\gamma\beta}^{f}\breve{d}_{\beta}^{\hat{n}(\xi)}=0 \quad \forall c_{\gamma}-\text{periodic} \\ & N_{i\beta}^{f}(\mathbf{y})\left[\!\left[\breve{d}_{\beta}^{\hat{n}(\xi)}\right]\!\right]=\breve{N}^{(\xi)}(\mathbf{y}) \quad \mathbf{y}\in S^{(\xi)}\end{aligned} \tag{4.32}$$

where

$$\breve{K}_{\gamma\beta}^{f} = \int_{\Theta} B_{ij\gamma}^{f} L_{ijkl} B_{kl\beta}^{f} d\Theta \tag{4.33}$$

For 3D problems, equation (4.32) is solved $3\cdot\breve{M}$ times corresponding to three eigenseparation modes applied at every interface partition. Equation (4.32b) can be enforced by inserting double nodes at the interface $S^{(\xi)}$. The stiffness matrix $\tilde{K}_{\gamma\beta}^{f}$ is different from $\breve{K}_{\gamma\beta}^{f}$ due to existence of the double nodes at the interface.

Remark 4.1 It is instructive to compare the governing equations for eigenstrains $\tilde{h}_{(k,y_l)}^{mn}(\mathbf{y},\tilde{\mathbf{y}})$ and $\tilde{h}_{(i,y_j)}^{mn(\alpha)}$

$$\left\{L_{ijkl}(\mathbf{y})\left[\tilde{h}_{(k,y_l)}^{mn}(\mathbf{y},\tilde{\mathbf{y}})-I_{klmn}d(\mathbf{y}-\tilde{\mathbf{y}})\right]\right\}_{,y_j}=0 \quad \forall\tilde{\mathbf{y}}\in\Theta \tag{4.34a}$$

$$\left\{L_{ijkl}(\boldsymbol{y})\left[\tilde{h}^{mn(\alpha)}_{(k,y_l)}(\boldsymbol{y})-I^{(\alpha)}_{klmn}(\boldsymbol{y})\right]\right\}_{,y_j}=0 \quad \forall\alpha \tag{4.34b}$$

It can be seen that if we replace Dirac delta function $d(\boldsymbol{y}-\tilde{\boldsymbol{y}})$ in (4.34a) by the numerical delta function $\tilde{d}_{\Theta^{(\alpha)}}(\boldsymbol{y})$

$$\tilde{d}_{\Theta^{(\alpha)}}(\boldsymbol{y})=\begin{cases}1 & \boldsymbol{y}\in\Theta^{(\alpha)}\\ 0 & \text{elsewhere}\end{cases} \tag{4.35}$$

then $\tilde{h}^{mn}_{(k,y_l)}(\boldsymbol{y},\tilde{\boldsymbol{y}})$ will coincide with $\tilde{h}^{mn(\alpha)}_{(k,y_l)}$. Likewise, comparing the eigenseparation influence functions $\breve{h}^{\breve{n}}_k(\boldsymbol{y},\breve{\boldsymbol{y}})$ and $\breve{h}^{\breve{n}(\xi)}_i(\breve{\boldsymbol{y}})$

$$\left[\!\left[\breve{h}^{\breve{n}}_k(\boldsymbol{y},\breve{\boldsymbol{y}})\right]\!\right]=d(\boldsymbol{y}-\breve{\boldsymbol{y}}) \quad \forall\breve{\boldsymbol{y}}\in S \tag{4.36a}$$

$$\left[\!\left[\breve{h}^{\breve{n}(\xi)}_i(\boldsymbol{y})\right]\!\right]=\breve{N}^{(\xi)}(\boldsymbol{y}) \quad \forall\xi \tag{4.36b}$$

suggests that if we replace the Dirac delta function in $d(\boldsymbol{y}-\breve{\boldsymbol{y}})$ (4.36a) by the numerical delta function $\breve{d}_{S^{(\xi)}}(\breve{\boldsymbol{y}})$

$$\breve{d}_{S^{(\xi)}}(\breve{\boldsymbol{y}})=\begin{cases}\breve{N}^{(\xi)}(\breve{\boldsymbol{y}}) & \breve{\boldsymbol{y}}\in S^{(\xi)}\\ 0 & \text{elsewhere}\end{cases} \tag{4.37}$$

then $\breve{h}^{\breve{n}}_k(\boldsymbol{y},\breve{\boldsymbol{y}})$ and $\breve{h}^{\breve{n}(\xi)}_i(\breve{\boldsymbol{y}})$ will coincide.

An example of the node-by-node interface partition is shown in Figure 4.3. The difference between the $d(\boldsymbol{y}-\breve{\boldsymbol{y}})$ and its numerical counterpart, $\breve{d}_{S^{(\xi)}}(\breve{\boldsymbol{y}})$, is that instead of imposing an eigenseparation over an infinitesimal interface region, it is imposed over a finite interface partition as shown in Figure 4.3.

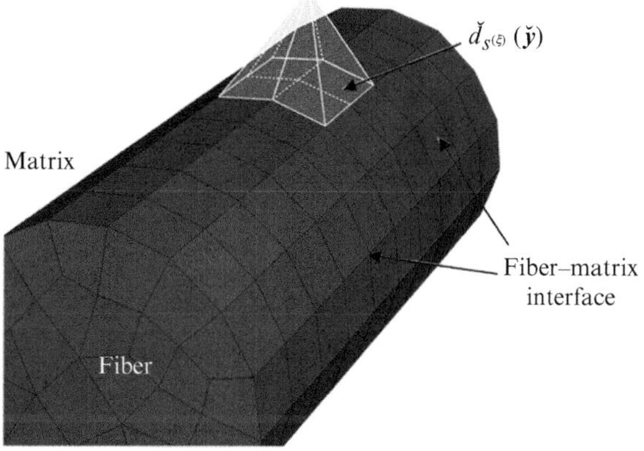

Figure 4.3 Node-by-node interface partition [24]

4.2.3 Reduced Order System of Equations

With the influence function problems solved in the preprocessing stage (prior to nonlinear analysis), the reduced order system of equations can be constructed to calculate the eigenstrains and eigenseparations at each iteration of the nonlinear coarse-scale analysis. The reduced order system consists of:

(a) reduced order transformation influence equations;
(b) reduced order traction continuity equations along the interfaces; and
(c) reduced order constitutive equations for fine-scale phases and interfaces (cohesive laws).

Consider the residual-free strain field obtained by combining (4.9) and (4.16)

$$\varepsilon_{ij}^{f}(\boldsymbol{x},\boldsymbol{y})=E_{ij}^{kl}(\boldsymbol{y})\varepsilon_{kl}^{c}(\boldsymbol{x})+\int_{\Theta}\tilde{h}_{(i,y_j)}^{kl}(\boldsymbol{y},\tilde{\boldsymbol{y}})\,\mu_{kl}^{f}(\boldsymbol{x},\tilde{\boldsymbol{y}})d\tilde{\Theta}+\int_{S}\breve{h}_{(i,y_j)}^{\breve{n}}(\boldsymbol{y},\breve{\boldsymbol{y}})\,\delta_{\breve{n}}^{f}(\boldsymbol{x},\breve{\boldsymbol{y}})d\breve{S} \tag{4.38}$$

where

$$E_{ij}^{kl}=I_{ijkl}+H_{(i,y_j)}^{kl} \tag{4.39}$$

Inserting the approximation of eigenstrains and eigenseparations (4.20) into (4.38) yields

$$\varepsilon_{ij}^{f}(\boldsymbol{x},\boldsymbol{y})=E_{ij}^{kl}(\boldsymbol{y})\varepsilon_{kl}^{c}(\boldsymbol{x})+\sum_{\alpha=1}^{\tilde{M}}P_{ij}^{kl(\alpha)}(\boldsymbol{y})\mu_{kl}^{(\alpha)}(\boldsymbol{x})+\sum_{\xi=1}^{\breve{M}}Q_{ij}^{\breve{n}(\xi)}(\boldsymbol{y})\delta_{\breve{n}}^{(\xi)}(\boldsymbol{x}) \tag{4.40}$$

Averaging above equation over partition domain $\Theta^{(\beta)}$ yields

$$\varepsilon_{ij}^{(\beta)}(\boldsymbol{x})-\sum_{\alpha=1}^{\tilde{M}}P_{ij}^{kl(\beta\alpha)}\,\mu_{kl}^{(\alpha)}(\boldsymbol{x})-\sum_{\xi=1}^{\breve{M}}Q_{ij}^{\breve{n}(\beta\xi)}\,\delta_{\breve{n}}^{(\xi)}(\boldsymbol{x})=E_{ij}^{kl(\beta)}\varepsilon_{kl}^{c}(\boldsymbol{x}) \tag{4.41}$$

where

$$\begin{aligned}
P_{ij}^{kl(\beta\alpha)}&=\frac{1}{|\Theta^{(\beta)}|}\int_{\Theta^{(\beta)}}P_{ij}^{kl(\alpha)}d\Theta=\frac{1}{|\Theta^{(\beta)}|}\int_{\Theta^{(\beta)}}\tilde{h}_{(i,y_j)}^{kl(\alpha)}d\Theta\\
Q_{ij}^{\breve{n}(\beta\xi)}&=\frac{1}{|\Theta^{(\beta)}|}\int_{\Theta^{(\beta)}}Q_{ij}^{\breve{n}(\xi)}d\Theta=\frac{1}{|\Theta^{(\beta)}|}\int_{\Theta^{(\beta)}}\breve{h}_{(i,y_j)}^{\breve{n}(\xi)}d\Theta\\
E_{ij}^{kl(\beta)}&=I_{ijkl}+\frac{1}{|\Theta^{(\beta)}|}\int_{\Theta^{(\beta)}}H_{(i,y_j)}^{kl}d\Theta
\end{aligned} \tag{4.42}$$

The reduced order traction $t_{\breve{m}}^{(\eta)}$ along the interface is obtained by averaging the fine-scale traction $t_{\breve{m}}^{f}=a_{\breve{m}i}\sigma_{ij}^{f}\breve{n}_j$ as defined in (4.40) and (4.9c) over the interface partition, where $a_{\breve{n}i}$ is the transformation matrix from the global coordinates system to the local interface coordinate system, which yields

$$-\sum_{\alpha=1}^{\tilde{M}} C_{\breve{m}}^{kl(\eta\alpha)} \mu_{kl}^{(\alpha)}(\boldsymbol{x}) + t_{\breve{m}}^{(\eta)}(\boldsymbol{x}) - \sum_{\xi=1}^{\breve{M}} D_{\breve{m}}^{\breve{n}(\eta\xi)} \delta_{\breve{n}}^{(\xi)}(\boldsymbol{x}) = T_{\breve{m}}^{kl(\eta)} \varepsilon_{kl}^{c}(\boldsymbol{x}) \tag{4.43}$$

where

$$\begin{aligned} C_{\breve{m}}^{kl(\alpha)} &= a_{\breve{m}i} L_{ijpq}(\boldsymbol{y}) \left[P_{pq}^{kl(\alpha)}(\boldsymbol{y}) - I_{pqkl}^{(\alpha)}(\boldsymbol{y}) \right] \breve{n}_j(\boldsymbol{y}) \\ D_{\breve{m}}^{\breve{n}(\xi)} &= a_{\breve{m}i} L_{ijpq}(\boldsymbol{y}) Q_{pq}^{\breve{n}(\xi)}(\boldsymbol{y}) \breve{n}_j(\boldsymbol{y}) \\ T_{\breve{m}}^{kl} &= a_{\breve{m}i} L_{ijpq}(\boldsymbol{y}) E_{pq}^{kl}(\boldsymbol{y}) \breve{n}_j(\boldsymbol{y}) \end{aligned} \tag{4.44}$$

and

$$C_{\breve{m}}^{kl(\eta\alpha)} \equiv \frac{1}{\left|S^{(\eta)}\right|} \int_{S^{(\eta)}} C_{\breve{m}}^{kl(\alpha)} dS; \quad D_{\breve{m}}^{\breve{n}(\eta\xi)} \equiv \frac{1}{\left|S^{(\eta)}\right|} \int_{S^{(\eta)}} D_{\breve{m}}^{\breve{n}(\xi)} dS; \quad T_{\breve{m}}^{kl(\eta)} \equiv \frac{1}{\left|S^{(\eta)}\right|} \int_{S^{(\eta)}} T_{\breve{m}}^{kl} dS \tag{4.45}$$

Consider, for simplicity, reduced order constitutive relations for eigenstrains and eigenseparations in the form of

$$\mu_{ij}^{(\alpha)}(\boldsymbol{x}) = f\left(\varepsilon_{ij}^{(\alpha)}(\boldsymbol{x})\right) \tag{4.46a}$$

$$t_{\breve{n}}^{(\eta)}(\boldsymbol{x}) = g\left(\delta_{\breve{n}}^{(\eta)}(\boldsymbol{x})\right) \tag{4.46b}$$

Equation (4.41), equation (4.43), and equation (4.46) comprise the reduced order system of equations for independent unknowns $\varepsilon_{ij}^{(\alpha)}(\boldsymbol{x})$ and $\delta_{\breve{n}}^{(\xi)}(\boldsymbol{x})$.

Finally, the reduced order form for the coarse-scale stress is obtained by integrating σ_{ij}^{f} over the unit cell domain, which yields

$$\sigma_{ij}^{c}(\boldsymbol{x}) = L_{ijkl}^{c} \varepsilon_{kl}^{c}(\boldsymbol{x}) + \sum_{\alpha=1}^{\tilde{M}} A_{ijkl}^{c(\alpha)} \mu_{kl}^{(\alpha)}(\boldsymbol{x}) + \sum_{\xi=1}^{\breve{M}} B_{ij\breve{n}}^{c(\xi)} \delta_{\breve{n}}^{(\xi)}(\boldsymbol{x}) \tag{4.47}$$

where

$$L_{ijkl}^{c} = \frac{1}{|\Theta|} \int_{\Theta} L_{ijmn}(\boldsymbol{y}) E_{mn}^{kl}(\boldsymbol{y})\, d\Theta \tag{4.48a}$$

$$A_{ijkl}^{c(\alpha)} = \frac{1}{|\Theta|} \int_{\Theta} L_{ijmn}(\boldsymbol{y}) \left[P_{mn}^{kl(\alpha)}(\boldsymbol{y}) - I_{mnkl}^{(\alpha)}(\boldsymbol{y}) \right] d\Theta \tag{4.48b}$$

$$B_{ij\breve{n}}^{c(\xi)} = \frac{1}{|\Theta|} \int_{\Theta} L_{ijmn}(\boldsymbol{y}) Q_{mn}^{\breve{n}(\xi)}(\boldsymbol{y})\, d\Theta \tag{4.48c}$$

Equation (4.47) can be rearranged by defining a coarse-scale eigenstrain $\mu_{kl}^{c}(\boldsymbol{x})$

$$\mu^{c}_{kl}(\boldsymbol{x}) = \sum_{\alpha=1}^{\tilde{M}} \left(-M^{c}_{klij} A^{c(\alpha)}_{ijmn}\right) \mu^{(\alpha)}_{mn}(\boldsymbol{x}) + \sum_{\xi=1}^{\breve{M}} \left(-M^{c}_{klij} B^{c(\xi)}_{ij\hat{n}}\right) \delta^{(\xi)}_{\hat{n}}(\boldsymbol{x}) \tag{4.49}$$

which yields a coarse-scale constitutive equation with a structure similar to that of the fine-scale constitutive equation

$$\sigma^{c}_{ij}(\boldsymbol{x}) = L^{c}_{ijkl}\left(\varepsilon^{c}_{kl}(\boldsymbol{x}) - \mu^{c}_{kl}(\boldsymbol{x})\right) \tag{4.50}$$

It can be seen from equation (4.49) that in the absence of eigenseparations, the coarse-scale eigenstrain is expressed in terms of a linear combination of partition eigenstrains. An equivalent relation to (4.49) was proposed by Mandel [32] and Levin [33] in the context of thermal strains and later generalized by Dvorak and Benveniste [19] to inelastic deformation.

Note that for uniform phase elastic properties, $A^{c(\alpha)}_{ijkl}$ in (4.48b) becomes

$$A^{c(\alpha)}_{ijkl} = \sum_{\beta=1}^{\tilde{M}} \phi^{(\beta)} L^{(\beta)}_{ijmn}\left[P^{kl(\beta\alpha)}_{mn} - \delta_{\alpha\beta} I_{mnkl}\right] \tag{4.51}$$

where $\phi^{(\beta)}$ is a volume fraction of phase β.

In summary, in the preprocessing stage prior to nonlinear coarse-scale analysis, the influence problems (4.23) and (4.25) are solved and the coefficient tensors (4.42), (4.45), and (4.48) are calculated. At each iteration of nonlinear coarse-scale analysis, the eigenstrains and eigenseparations are updated by solving the reduced order unit cell system of equation (4.41), equation (4.43), and equation (4.46), and finally, the coarse-scale stress is updated using (4.47).

4.2.4 One-Dimensional Model Problem

In this section, a 1D model problem is studied to verify the ROH approach.

The unit cell consists of two homogenous materials each occupying one-half of the unit cell domain schematically depicted in Figure 4.4. The Young's modulus of the two materials is denoted by $E^{(1)}$ and $E^{(2)}$, respectively.

We first consider Problem A in Figure 4.4(a) with inelastic phases and perfect interface. An analytical solution of Problem A subjected to monotonically increasing coarse-scale strain ε^c can be obtained by solving the following nonlinear system of equations for the fine-scale strains $\varepsilon^{(1)}$ and $\varepsilon^{(2)}$

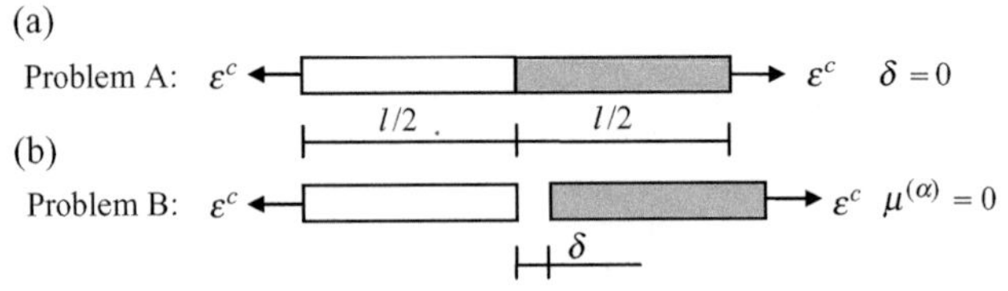

Figure 4.4 Two-scale 1D model problems: (a) inelastic phases with perfect interface δ=0; and (b) elastic phases with imperfect interface $\mu^{(\alpha)}$=0

$$\begin{aligned}&\sigma^{(1)}=\sigma^{(2)}\Rightarrow E^{(1)}(\varepsilon^{(1)}-\mu^{(1)})=E^{(2)}(\varepsilon^{(2)}-\mu^{(2)})\\&\varepsilon^{c}\equiv\int_0^1\varepsilon(x)dx=\frac{1}{2}\varepsilon^{(1)}+\frac{1}{2}\varepsilon^{(2)}\end{aligned}\tag{4.52}$$

where $\mu^{(1)}(\varepsilon^{(1)})$ and $\mu^{(2)}(\varepsilon^{(2)})$ are plastic strains in the two phases, which depend on evolution equations. A closed form solution of (4.52) in terms of plastic strains is given by

$$\begin{aligned}\varepsilon^{(1)}&=\frac{2E^{(2)}}{E^{(1)}+E^{(2)}}\varepsilon^{c}+\frac{E^{(1)}}{E^{(1)}+E^{(2)}}\mu^{(1)}-\frac{E^{(2)}}{E^{(1)}+E^{(2)}}\mu^{(2)}\\\varepsilon^{(2)}&=\frac{2E^{(1)}}{E^{(1)}+E^{(2)}}\varepsilon^{c}-\frac{E^{(1)}}{E^{(1)}+E^{(2)}}\mu^{(1)}+\frac{E^{(2)}}{E^{(1)}+E^{(2)}}\mu^{(2)}\end{aligned}\tag{4.53}$$

and

$$\begin{aligned}&\sigma^{(1)}=\sigma^{(2)}=\sigma^{c}=L^{c}(\varepsilon^{c}-\mu^{c})\\&\mu^{c}=\frac{\mu^{(1)}+\mu^{(2)}}{2}\end{aligned}\tag{4.54}$$

In the following, we prove that the solution obtained by the ROH with two partitions is identical to (4.52), (4.53), and (4.54). We start by solving for elastic and eigenstrain influence functions and calculating the relevant coefficient tensors.

The governing equation for the elastic influence function is given as

$$\{L(y)[1+H_{,y}(y)]\}_{,y}=0\quad y\in\Theta\tag{4.55}$$

where

$$\begin{aligned}&L=\begin{cases}E^{(1)} & y\in[0,0.5)\\E^{(2)} & y\in[0.5,1]\end{cases}\\&H(0)=H(1)=0\end{aligned}\tag{4.56}$$

The solution for the elastic influence function is given in equation (2.124) (see section 2.2.6.1 in Chapter 2)

$$H_{,y}=\begin{cases}\dfrac{E^{(2)}-E^{(1)}}{E^{(1)}+E^{(2)}} & y\in[0,0.5)\\[2ex]\dfrac{E^{(1)}-E^{(2)}}{E^{(1)}+E^{(2)}} & y\in[0.5,1]\end{cases}\tag{4.57}$$

and the corresponding coarse-scale elastic modulus is given by

$$L^{c}=\frac{2E^{(1)}E^{(2)}}{E^{(1)}+E^{(2)}}\tag{4.58}$$

The governing equation of the eigenstrain influence function $\tilde{h}^{(\alpha)}$ [see equation (4.24)] is given by (α=1,2)

$$\left\{L(y)\left[\tilde{h}^{(\alpha)}_{,y}(y)-1^{(\alpha)}(y)\right]\right\}_{,y}=0 \quad y\in\Theta \tag{4.59}$$

subjected to periodic boundary conditions that eliminate rigid body motion of the unit cell

$$\tilde{h}^{(\alpha)}(0)=\tilde{h}^{(\alpha)}(1)=0 \tag{4.60}$$

where $\tilde{h}^{(\alpha)}_{,y}(y)=P^{(\alpha)}$ and

$$1^{(\alpha)}(y)=\begin{cases}1 & y\in\Theta^{(\alpha)}\\ 0 & y\notin\Theta^{(\alpha)}\end{cases} \tag{4.61}$$

The solution of equation (4.59) and equation (4.60) is given by

$$P^{(1)}=\begin{cases}\dfrac{E^{(1)}}{E^{(1)}+E^{(2)}} & y\in[0,0.5)\\[2ex] \dfrac{-E^{(1)}}{E^{(1)}+E^{(2)}} & y\in[0.5,1]\end{cases};\qquad P^{(2)}=\begin{cases}\dfrac{-E^{(2)}}{E^{(1)}+E^{(2)}} & y\in[0,0.5)\\[2ex] \dfrac{E^{(2)}}{E^{(1)}+E^{(2)}} & y\in[0.5,1]\end{cases} \tag{4.62}$$

The resulting coefficient tensors are calculated from (4.42)

$$\begin{aligned}&P^{(1,1)}=\frac{E^{(1)}}{E^{(1)}+E^{(2)}};\qquad P^{(1,2)}=\frac{-E^{(2)}}{E^{(1)}+E^{(2)}};\\ &P^{(2,1)}=\frac{-E^{(1)}}{E^{(1)}+E^{(2)}};\qquad P^{(2,2)}=\frac{E^{(2)}}{E^{(1)}+E^{(2)}};\\ &\mathscr{E}^{(1)}=1+H_{,y}=\frac{2E^{(2)}}{E^{(1)}+E^{(2)}};\quad y\in[0,0.5)\\ &\mathscr{E}^{(2)}=1+H_{,y}=\frac{2E^{(1)}}{E^{(1)}+E^{(2)}};\quad y\in[0.5,1]\end{aligned} \tag{4.63}$$

The fine-scale strains $\varepsilon^{(1)}$ and $\varepsilon^{(2)}$ are given by

$$\begin{aligned}\varepsilon^{(1)}&=\mathscr{E}^{(1)}\varepsilon^{c}+P^{(1,1)}\mu^{(1)}+P^{(1,2)}\mu^{(2)}\\ \varepsilon^{(2)}&=\mathscr{E}^{(1)}\varepsilon^{c}+P^{(2,1)}\mu^{(1)}+P^{(2,2)}\mu^{(2)}\end{aligned} \tag{4.64}$$

Inserting (4.63) into (4.64) yields (4.53). Inserting (4.64) into constitutive equations of phases yields (4.54), which completes the proof.

We now consider Problem B with elastic phases and imperfect interface governed by traction separation law, $t(\delta)$. An analytical solution of Problem B subjected to monotonically increasing coarse-scale strain ε^c can be obtained by solving the following nonlinear system of equations for the eigenseparation

$$\begin{aligned} &E^{(1)}\varepsilon^{(1)}=E^{(2)}\varepsilon^{(2)}=t(\delta) \\ &\varepsilon^{(1)}\frac{l}{2}+\varepsilon^{(1)}\frac{l}{2}+\delta=\varepsilon^{c}l \end{aligned} \tag{4.65}$$

where the nonlinear equation for δ is given by

$$t(\delta)+\frac{2E^{(1)}E^{(2)}}{E^{(2)}+E^{(2)}}\delta=\frac{2E_1^{(1)}E^{(2)}}{E^{(1)}+E^{(2)}}\varepsilon^{c} \tag{4.66}$$

We now consider ROH. Since both phases are elastic, the governing equation for δ follows from (4.43)

$$t+D\delta=T\varepsilon^{c} \tag{4.67}$$

where

$$\begin{aligned} &D(y)=L(y)Q(y)=L\breve{h}_{,y} \\ &T(y)=L(y)\mathcal{E}(y)=L(1+H_{,y}) \end{aligned} \tag{4.68}$$

Here we omit the superscript denoting interface partition count due to the existence of one interface partition. From (4.57), it follows that

$$T(y)=\frac{2E^{(1)}E^{(2)}}{E^{(1)}+E^{(2)}} \tag{4.69}$$

$\breve{h}$ can be obtained by solving for the influence function problem (4.25)

$$\begin{aligned} &[L(y)\breve{h}_{,y}]_{,y}=0 \quad y\in[0,1] \\ &[\![\breve{h}]\!]_{y=0.5}=1 \\ &[\![L(y)\breve{h}_{,y}]\!]_{y=0.5}=0 \\ &\breve{h}(0)=\breve{h}(l)=0 \end{aligned} \tag{4.70}$$

A closed form solution of (4.70) gives

$$Q=\breve{h}_{,y}=\begin{cases} \dfrac{2E^{(2)}}{E^{(1)}+E^{(2)}} & y\in[0,0.5) \\ \dfrac{2E^{(1)}}{E^{(1)}+E^{(2)}} & y\in[0.5,1] \end{cases} \tag{4.71}$$

and

$$D=\frac{2E^{(1)}E^{(2)}}{E^{(1)}+E^{(2)}} \quad y\in[0,1] \tag{4.72}$$

Inserting (4.69) and (4.72) into (4.67) yields (4.66), which completes the proof.

4.2.5 Computational Aspects

In this section, we will focus on computational aspects of the ROH and integration of the ROH into a conventional finite element code architecture using a standard user-defined material interface. Most legacy finite element codes provide functionality to add user-defined material models. However, the ROH outlined in the previous section has a nonconventional data structure that complicates construction of a user-defined material model. In this section, we recast the ROH formulation into a form that can be easily added as a user-defined material model.

To solve the nonlinear system of equation (4.41) and equation (4.43), we define the following unit cell residual

$$\boldsymbol{r}^{(\beta,\eta)} \equiv \boldsymbol{r}\left(\Delta\varepsilon_{ij}^{(\beta)}, \Delta\delta_{\breve{n}}^{(\eta)}\right) = \begin{bmatrix} \Delta\varepsilon_{ij}^{(\beta)} - \sum_{\gamma=1}^{\tilde{M}} P_{ij}^{kl(\beta\gamma)} \Delta\mu_{kl}^{(\gamma)} - \sum_{\tau=1}^{\breve{M}} Q_{ij}^{\breve{n}(\beta\tau)} \Delta\delta_{\breve{n}}^{(\tau)} - E_{ij}^{kl(\beta)} \Delta\varepsilon_{kl}^{c} \\ -\sum_{\gamma=1}^{\tilde{M}} C_{\breve{n}}^{kl(\eta\gamma)} \Delta\mu_{kl}^{(\gamma)} + \Delta t_{\breve{n}}^{(\eta)} - \sum_{\tau=1}^{\breve{M}} D_{\breve{n}}^{\breve{m}(\eta\tau)} \Delta\delta_{\breve{m}}^{(\tau)} - T_{\breve{n}}^{kl(\eta)} \Delta\varepsilon_{kl}^{c} \end{bmatrix} = \boldsymbol{0} \tag{4.73}$$

The unknowns are the increments of phase strain $\Delta\varepsilon_{ij}^{(\beta)}$ (or eigenstrains) and phase separation $\Delta\delta_{\breve{n}}^{(\tau)}$. The coarse-scale strain ${}_{n+1}^{i+1}\Delta\varepsilon_{kl}^{c}$ is prescribed by the coarse-scale problem at every load increment n+1 and iteration i+1. Here we omit the left indices denoting the load increment and the iteration count. The nonlinear system of equations (4.73) is solved using the Newton method, which requires function derivatives with respect to the variable $\boldsymbol{\theta}^{(\alpha,\xi)} \equiv \left[\Delta\varepsilon_{kl}^{(\alpha)} \; \Delta\delta_{\breve{m}}^{(\xi)}\right]^{\mathrm{T}}$

$$\begin{aligned} \frac{\partial \boldsymbol{r}^{(\beta,\eta)}}{\partial \boldsymbol{\theta}^{(\alpha,\xi)}} &= \begin{pmatrix} \delta_{\beta\alpha} I_{ijmn} - \sum_{\gamma=1}^{\tilde{M}} P_{ij}^{kl(\beta\gamma)} \dfrac{\partial \Delta\mu_{kl}^{(\gamma)}}{\partial \Delta\varepsilon_{mn}^{(\alpha)}} \delta_{\alpha\gamma} & -\sum_{\tau=1}^{\breve{M}} Q_{ij}^{\breve{n}(\beta\tau)} \delta_{\tau\xi} \\ -\sum_{\gamma=1}^{\tilde{M}} C_{\breve{n}}^{kl(\eta\gamma)} \dfrac{\partial \Delta\mu_{kl}^{(\gamma)}}{\partial \Delta\varepsilon_{mn}^{(\alpha)}} \delta_{\alpha\gamma} & \delta_{\eta\xi} \dfrac{\partial \Delta t_{\breve{n}}^{(\eta)}}{\partial \Delta\delta_{\breve{m}}^{(\xi)}} - \sum_{\tau=1}^{\breve{M}} D_{\breve{n}}^{\breve{m}(\eta\tau)} \delta_{\tau\xi} \end{pmatrix} \\ &= \begin{pmatrix} \delta_{\beta\alpha} I_{ijmn} - P_{ij}^{kl(\beta\alpha)} \dfrac{\partial \Delta\mu_{kl}^{(\alpha)}}{\partial \Delta\varepsilon_{mn}^{(\alpha)}} & -Q_{ij}^{\breve{n}(\beta\xi)} \\ -C_{\breve{n}}^{kl(\eta\alpha)} \dfrac{\partial \Delta\mu_{kl}^{(\alpha)}}{\partial \Delta\varepsilon_{mn}^{(\alpha)}} & \delta_{\eta\xi} \dfrac{\partial \Delta t_{\breve{n}}^{(\eta)}}{\partial \Delta\delta_{\breve{m}}^{(\xi)}} - D_{\breve{n}}^{\breve{m}(\eta\xi)} \end{pmatrix} \end{aligned} \tag{4.74}$$

where δ_{ij} is the Kronecker delta. $\dfrac{\partial \Delta\mu_{mn}^{(\alpha)}}{\partial \Delta\varepsilon_{kl}^{(\alpha)}}$ can be computed from the constitutive equation for phase α

$$\Delta\mu_{ij}^{(\alpha)} = \Delta\varepsilon_{ij}^{(\alpha)} - M_{ijkl}^{(\alpha)} \Delta\sigma_{kl}^{(\alpha)} \tag{4.75}$$

where $M_{ijkl}^{(\alpha)}$ is the elastic compliance for phase partition α. Equation (4.75) follows from (4.2), assuming that the elastic properties are uniform within each phase.

The derivative of the eigenstrain is given by

$$\frac{\partial \Delta \mu_{ij}^{(\alpha)}}{\partial \Delta \varepsilon_{kl}^{(\alpha)}} = I_{ijkl} - M_{ijmn}^{(\alpha)} \frac{\partial \Delta \sigma_{mn}^{(\alpha)}}{\partial \Delta \varepsilon_{kl}^{(\alpha)}} \tag{4.76}$$

where $\frac{\partial \Delta \sigma_{mn}^{(\alpha)}}{\partial \Delta \varepsilon_{kl}^{(\alpha)}}$ is a single-scale consistent tangent stiffness for the corresponding phase. Formulation of the consistent tangent operator for each phase is a standard building block in any implicit finite element code, that is, given $\frac{\partial \Delta \sigma_{mn}^{(\alpha)}}{\partial \Delta \varepsilon_{kl}^{(\alpha)}}$, the eigenstrain derivative $\frac{\partial \Delta \mu_{ij}^{(\alpha)}}{\partial \Delta \varepsilon_{kl}^{(\alpha)}}$ follows from (4.76). Consequently, the derivatives $\frac{\partial \Delta \mu_{ij}^{(\alpha)}}{\partial \Delta \varepsilon_{kl}^{(\alpha)}}$ and $\frac{\partial \Delta t_{\breve{n}}^{(\xi)}}{\partial \Delta \delta_{\breve{m}}^{(\xi)}}$ appearing in (4.74) can be obtained from a single-scale constitutive model of phases and cohesive law.

Once the $\boldsymbol{\theta}^{(\alpha,\zeta)} \equiv \left[\Delta \varepsilon_{kl}^{(\alpha)} \Delta \delta_{\breve{m}}^{(\xi)} \right]^{\mathrm{T}}$ is solved for, the eigenstrain $\mu_{ij}^{(\alpha)}$ is subsequently computed from constitutive equation (4.46a), whereas the coarse-scale stress follows from equation (4.47).

For implicit computations, it remains to compute the consistent tangent operator $\frac{\partial \Delta \sigma_{ij}^{c}}{\partial \Delta \varepsilon_{kl}^{c}}$. Taking the derivative of $\Delta \sigma_{ij}^{c}$ in (4.49) with respect to $\Delta \varepsilon_{kl}^{c}$ yields

$$\frac{\partial \Delta \sigma_{ij}^{c}}{\partial \Delta \varepsilon_{kl}^{c}} = L_{ijkl}^{c} + \sum_{\alpha=1}^{\tilde{M}} A_{ijmn}^{c(\alpha)} \frac{\partial \Delta \mu_{mn}^{(\alpha)}}{\partial \Delta \varepsilon_{kl}^{c}} + \sum_{\xi=1}^{\breve{M}} B_{ij\breve{n}}^{c(\xi)} \frac{\partial \Delta \delta_{\breve{n}}^{(\xi)}}{\partial \Delta \varepsilon_{kl}^{c}} \tag{4.77}$$

Using the chain rule, equation (4.77) can be expressed as

$$\frac{\partial \Delta \sigma_{ij}^{c}}{\partial \Delta \varepsilon_{kl}^{c}} = L_{ijkl}^{c} + \sum_{\alpha=1}^{\tilde{M}} A_{ijmn}^{c(\alpha)} \frac{\partial \Delta \mu_{mn}^{(\alpha)}}{\partial \Delta \varepsilon_{pq}^{(\alpha)}} \boxed{\frac{\partial \Delta \varepsilon_{pq}^{(\alpha)}}{\partial \Delta \varepsilon_{kl}^{c}}} + \sum_{\xi=1}^{\breve{M}} B_{ij\breve{n}}^{c(\xi)} \boxed{\frac{\partial \Delta \delta_{\breve{n}}^{(\xi)}}{\partial \Delta \varepsilon_{kl}^{c}}} \tag{4.78}$$

The unknown quantities in the boxes are obtained as follows. Recall the reduced order unit cell problem

$$\begin{aligned} &\Delta \varepsilon_{ij}^{(\beta)} - \sum_{\gamma=1}^{\tilde{M}} P_{ij}^{kl(\beta\gamma)} \Delta \mu_{kl}^{(\gamma)} - \sum_{\tau=1}^{\breve{M}} Q_{ij}^{\breve{n}(\beta\tau)} \Delta \delta_{\breve{n}}^{(\tau)} = E_{ij}^{kl(\beta)} \Delta \varepsilon_{kl}^{c} \\ &-\sum_{\gamma=1}^{\tilde{M}} C_{\breve{n}}^{kl(\eta\gamma)} \Delta \mu_{kl}^{(\gamma)} + \Delta t_{\breve{n}}^{(\eta)} - \sum_{\tau=1}^{\breve{M}} D_{\breve{n}}^{\breve{m}(\eta\tau)} \Delta \delta_{\breve{m}}^{(\tau)} = T_{\breve{n}}^{kl(\eta)} \Delta \varepsilon_{kl}^{c} \end{aligned} \tag{4.79}$$

Taking the derivative of (4.79) with respect to $\Delta \varepsilon_{kl}^{c}$ gives

$$\begin{aligned} &\frac{\partial \Delta \varepsilon_{ij}^{(\beta)}}{\partial \Delta \varepsilon_{kl}^{c}} - \sum_{\gamma=1}^{\tilde{M}} P_{ij}^{mn(\beta\gamma)} \frac{\partial \Delta \mu_{mn}^{(\gamma)}}{\partial \Delta \varepsilon_{kl}^{c}} - \sum_{\tau=1}^{\breve{M}} Q_{ij}^{\breve{n}(\beta\tau)} \frac{\partial \Delta \delta_{\breve{n}}^{(\tau)}}{\partial \Delta \varepsilon_{kl}^{c}} = E_{ij}^{kl(\beta)} \\ &-\sum_{\gamma=1}^{\tilde{M}} C_{\breve{n}}^{mn(\eta\gamma)} \frac{\partial \Delta \mu_{mn}^{(\gamma)}}{\partial \Delta \varepsilon_{kl}^{c}} + \frac{\partial \Delta t_{\breve{n}}^{(\eta)}}{\partial \Delta \varepsilon_{kl}^{c}} - \sum_{\tau=1}^{\breve{M}} D_{\breve{n}}^{\breve{m}(\eta\tau)} \frac{\partial \Delta \delta_{\breve{m}}^{(\tau)}}{\partial \Delta \varepsilon_{kl}^{c}} = T_{\breve{n}}^{kl(\eta)} \end{aligned} \tag{4.80}$$

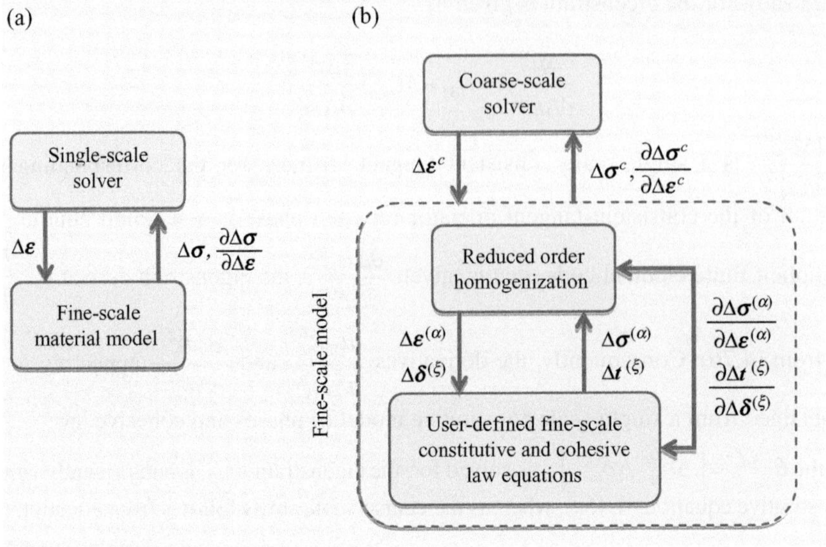

Figure 4.5 Information flow: (a) single-scale model; and (b) two-scale reduced order model

By applying the chain rule, we have

$$\begin{aligned}
&\left(\delta_{\beta\alpha} I_{ijpq} - P_{ij}^{mn(\beta\alpha)} \frac{\partial \Delta\mu_{mn}^{(\alpha)}}{\partial \Delta\varepsilon_{pq}^{(\alpha)}}\right) \boxed{\frac{\partial \Delta\varepsilon_{pq}^{(\alpha)}}{\partial \Delta\varepsilon_{kl}^{c}}} - Q_{ij}^{\breve{m}(\beta\xi)} \boxed{\frac{\partial \Delta\delta_{\breve{m}}^{(\xi)}}{\partial \Delta\varepsilon_{kl}^{c}}} = E_{ij}^{kl(\beta)} \\
&-C_{\breve{n}}^{mn(\eta\alpha)} \frac{\partial \Delta\mu_{mn}^{(\alpha)}}{\partial \Delta\varepsilon_{pq}^{(\alpha)}} \boxed{\frac{\partial \Delta\varepsilon_{pq}^{(\alpha)}}{\partial \Delta\varepsilon_{kl}^{c}}} + \left(\delta_{\eta\xi} \frac{\partial \Delta t_{\breve{n}}^{(\eta)}}{\partial \Delta\delta_{\breve{m}}^{(\xi)}} - D_{\breve{n}}^{\breve{m}(\eta\xi)}\right) \boxed{\frac{\partial \Delta\delta_{\breve{m}}^{(\xi)}}{\partial \Delta\varepsilon_{kl}^{c}}} = T_{\breve{n}}^{kl(\eta)}
\end{aligned} \tag{4.81}$$

The unknown quantities in the boxes can be found by solving the linear system of equations (4.81). Consequently, the coarse-scale consistent tangent $\dfrac{\partial \Delta\sigma_{ij}^{c}}{\partial \Delta\varepsilon_{kl}^{c}}$ follows from equation (4.78).

Figure 4.5 depicts the canonical structure of the ROH and its implementation in the conventional finite element code architecture.

4.3 Lower Order Approximation of Eigenstrains

It has been observed by a number of investigators [37,38,39] that lower order approximation of eigenstrains, and in particular a one-partition-per-phase scenario, gives rise to an inaccurate solution of overall properties, while subdividing each phase into numerous partitions, such as one partition per element, is often computationally prohibitive. The need for several partitions for each inelastic phase stems from the intrinsic nonuniformity of eigenstrains within a single material phase. In order to accurately reproduce the actual effective behavior of the composite, it is important to account *directly* or *indirectly* for the nonuniformity of eigenstrains. Several strategies are known to alleviate (at least partially) these shortcomings:

(a) *Variable and/or dynamic partitioning scheme* [24]. In the variable partitioning scheme, the inelastic coarse-scale domain is subdivided into several zones and various numbers of partitions are used in each zone. In the dynamic partitioning scheme, the number of partitions at a coarse-scale quadrature point is continuously adapted based on the severity of inelastic deformation during the deformation process.
(b) *Nonuniform eigenstrain field* [40]. The basic idea of this approach is to utilize the nonuniform eigenstrain shape functions that have their support entirely contained in a single material phase.
(c) *Hybrid compatible-incompatible eigenstrain field* [28]. In devising an optimal eigenstrain approximation, microscopic phase topology has to be taken into account. For single-phase dominated (matrix-dominated) modes of deformation with only one connected topology, the eigenstrains are discretized using C^0 continuous functions, whereas for the multiphase mode of deformation, the eigenstrains are approximated using the usual C^{-1} approximation.
(d) *Asymptotically consistent eigenstrain field* [39]. By using this approach, eigenstrain influence functions are constructed to satisfy consistency with the direct computational homogenization method considered in Chapter 2. Such a consistency adjustment requires continuous recalculation of the eigenstrain influence functions, unless it is imposed in the asymptotic limit of the instantaneous microphase properties only.

Methods (a) and (b) explicitly account for the nonuniform eigenstrains within each phase by either increasing the number of phase partitions or by constructing a piecewise constant approximation within each phase. The latter depends on how well the representative eigenstrain modes are a priori chosen. In the following section, we will focus on methods (c) and (d), which account for nonuniform eigenstrains indirectly and thus possess the lowest computational complexity, that is, one partition per phase. We will start with an exposition of ideas leading to the formulation of the indirect methods (c) and (d) by pinpointing the inability of the uniform one-partition-per-phase approach to resolve the highly nonlinear and heterogeneous behavior of composites.

4.3.1 The Pitfalls of a Piecewise Constant One-Partition-Per-Phase Model

Consider a two-phase model with a piecewise constant discretization of eigenstrains and one partition per phase. To demonstrate the shortcomings of this model, consider a matrix-dominated mode of deformation where the matrix (phase 2) is perfectly plastic but the inclusion (phase 1) remains elastic. The inclusion and matrix domains are denoted by $\Theta^{(1)}$ and $\Theta^{(2)}$, respectively. Let $\Delta\varepsilon^c_{kl}$ be the coarse-scale strain increment that the unit cell is subjected to. The resulting fine-scale stress increment $\Delta\sigma^f_{ij}$ should vanish for arbitrary eigenstrain in the matrix phase $\Delta\mu^{(2)}_{mn}$

$$\Delta\sigma^f_{ij} = 0 \text{ for } \forall\Delta\mu^{(2)}_{kl}; \quad \Delta\varepsilon^{(2)}_{kl} = \Delta\mu^{(2)}_{kl}; \quad \Delta\varepsilon^{(1)}_{kl} = \Delta\mu^{(1)}_{kl} = 0 \tag{4.82}$$

In the absence of eigenseparations we have

$$0=\Delta\sigma_{ij}^{f}=L_{ijkl}(y)E_{kl}^{mn}(y)\Delta\varepsilon_{mn}^{c}+L_{ijkl}(\boldsymbol{y})\left[P_{kl}^{mn(2)}(\boldsymbol{y})-I_{klmn}^{(2)}(\boldsymbol{y})\right]\Delta\mu_{mn}^{(2)} \tag{4.83}$$

and in absence of elastic strains we have

$$\Delta\varepsilon_{kl}^{c}=\frac{1}{|\Theta|}\int_{\Theta}\Delta\mu_{kl}^{f}\;d\Theta=\phi^{(2)}\Delta\mu_{kl}^{(2)};\quad \phi^{(2)}=\text{const} \tag{4.84}$$

Inserting (4.84) into (4.83) yields

$$0=\Delta\sigma_{ij}^{f}=L_{ijkl}(\boldsymbol{y})\left[P_{kl}^{mn(2)}(\boldsymbol{y})-I_{klmn}^{(2)}(\boldsymbol{y})+\phi^{(2)}E_{kl}^{mn}(\boldsymbol{y})\right]\Delta\mu_{mn}^{(2)}\quad \forall\Delta\mu_{mn}^{(2)} \tag{4.85}$$

In general, for arbitrary $\Delta\boldsymbol{\mu}^{(2)}$ and positive-definite $\boldsymbol{L}$, equation (4.85) is not satisfied unless we replace $\boldsymbol{P}^{(2)}$ by a newly constructed $\tilde{\boldsymbol{P}}^{(2)}$ that satisfies

$$\tilde{P}_{kl}^{mn(2)}(\boldsymbol{y})\equiv I_{klmn}^{(2)}(\boldsymbol{y})-\phi^{(2)}E_{kl}^{mn}(\boldsymbol{y}) \tag{4.86}$$

For partition β we have

$$\tilde{P}_{kl}^{mn(\beta 2)}=\delta_{\alpha\beta}I_{klmn}-\phi^{(2)}E_{kl}^{mn(\beta)} \tag{4.87}$$

where

$$\tilde{P}_{kl}^{mn(\beta\alpha)}=\frac{1}{\left|\Theta^{(\beta)}\right|}\int_{\Theta^{(\beta)}}\tilde{P}_{kl}^{mn(\alpha)}\;d\Theta \tag{4.88}$$

Figure 4.6 depicts an elastic inclusion phase that deforms as the result of a matrix-dominated mode of deformation with one partition per phase. It can be seen that if the matrix phase is perfectly plastic, the overall macroscopic stress will not vanish due to the elastic deformation of the inclusion phase. This so-called *inclusion-locking phenomenon* can be alleviated by increasing the number of partitions per phase, ultimately resulting in a nonuniform distribution of eigenstrains over the phase. This phenomenon is reminiscent of shear and pressure locking of lower order finite elements, since their kinematics are not rich enough to represent the correct solution.

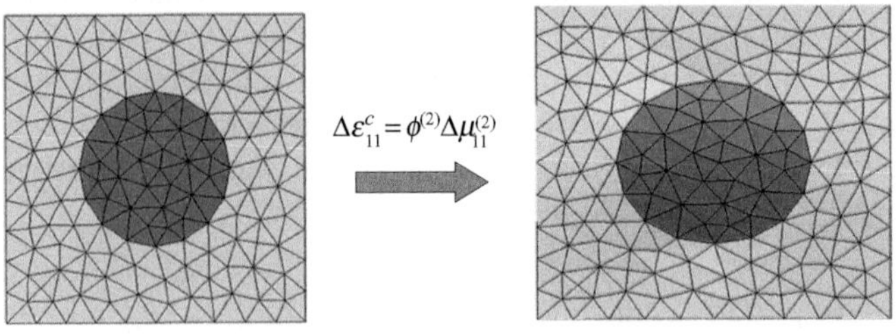

Figure 4.6 Inclusion locking in the matrix-dominated mode of deformation and uniform eigenstrain discretization with one partition per phase. Reproduced from [28], © 2013 John Wiley & Sons

4.3.2 Impotent Eigenstrain

One possible way to alleviate inclusion locking emanating from the lower order approximation of eigenstrains, and in particular for the case of one partition per phase, is to construct a compatible eigenstrain approximation. An inclusion-locking phenomenon can be alleviated if an eigenstrain in a matrix phase does not impose any eigenstrain in the inclusion phase. For this to happen, the compatible eigenstrain field at the matrix–fiber interface has to vanish.

Note that by definition an eigenstrain is an incompatible strain. The concept of an eigenstrain serving as the source of residual stress was developed by Mindlin and Cheng [34] with the terminology of nuclei of strain. Eshelby [35] referred to it as a stress-free transformation strain, while Ueda *et al.* [41] termed it an inherent strain. The term eigenstrain was originally coined by Mura [36], based on Reissner's paper [42]. The notion that an eigenstrain can be both compatible and incompatible was originally suggested by Fruhashi and Mura [43], who referred to the compatible eigenstrain as an *impotent* (or *harmless*) *eigenstrain*. Hereafter, we will refer to such an eigenstrain as either an impotent eigenstrain or a compatible eigenstrain.

4.3.2.1 Reduced Order Homogenization with Impotent (Compatible) Eigenstrain

Consider an additive decomposition of a total fine-scale strain $\boldsymbol{\varepsilon}^f$ into an elastic strain $\boldsymbol{e}^f$ and impotent eigenstrain $\mu^{f,\mathrm{comp}}$

$$\mu_{ij}^{f,\mathrm{comp}} = \varepsilon_{ij}^f - e_{ij}^f \tag{4.89}$$

Since the total and impotent fine-scale strains are both compatible, the elastic strain will be compatible as well. Thus, both the elastic strain and eigenstrain can be derived from the corresponding displacements as

$$u_i^{(0),\mathrm{comp}}(\boldsymbol{x}) = u_i^{e(0)}(\boldsymbol{x}) + u_i^{\mu(0)}(\boldsymbol{x}); \quad u_i^{(1),\mathrm{comp}}(\boldsymbol{x},\boldsymbol{y}) = u_i^{e(1)}(\boldsymbol{x},\boldsymbol{y}) + u_i^{\mu(1)}(\boldsymbol{x},\boldsymbol{y}) \tag{4.90}$$

where displacements $u_i^{e(I)}$ and $u_i^{\mu(I)}$ for $I=0,1$ correspond to the elastic strain and eigenstrain contributions of $u_i^{(I)}$, respectively, with $u_i^{e(1)}$ and $u_i^{\mu(1)}$ being $\boldsymbol{y}$-periodic functions. Following the decomposition of the total strain in (4.9a), the eigenstrain is decomposed into the coarse-scale eigenstrain $\mu_{ij}^{c,\mathrm{comp}}(\boldsymbol{x})$ and fine-scale correction $u_{(i,y_j)}^{\mu(1)}(\boldsymbol{x},\boldsymbol{y})$

$$\mu_{ij}^{f,\mathrm{comp}}(\boldsymbol{x},\boldsymbol{y}) = \mu_{ij}^{c,\mathrm{comp}}(\boldsymbol{x}) + u_{(i,y_j)}^{\mu(1)}(\boldsymbol{x},\boldsymbol{y}) \tag{4.91a}$$

$$\mu_{ij}^{c,\mathrm{comp}}(\boldsymbol{x}) \equiv \frac{1}{|\Theta|}\int_\Theta \mu_{ij}^{f,\mathrm{comp}}\, d\Theta = u_{(i,x_j)}^{\mu(0)}(\boldsymbol{x}) \tag{4.91b}$$

In the absence of eigenseparations, inserting the fine-scale eigenstrain into the residual-free strain field decomposition (4.16) gives

$$u_i^{(1),\mathrm{comp}}(\boldsymbol{x},\boldsymbol{y}) = H_i^{kl}(\boldsymbol{y})\varepsilon_{kl}^c(\boldsymbol{x}) + \mu_{kl}^{c,\mathrm{comp}}(\boldsymbol{x})\int_\Theta \tilde{h}_i^{kl}(\boldsymbol{y},\tilde{\boldsymbol{y}})d\tilde{\Theta} + \int_\Theta \tilde{h}_i^{kl}(\boldsymbol{y},\tilde{\boldsymbol{y}})u_{(k,\tilde{y}_l)}^{\mu(1)}(\boldsymbol{x},\tilde{\boldsymbol{y}})d\tilde{\Theta} \tag{4.92}$$

which further results in the following fine-scale strains and stresses

$$\varepsilon_{ij}^{f,\mathrm{comp}}(\boldsymbol{x},\boldsymbol{y}) = E_{ij}^{kl}(\boldsymbol{y})\varepsilon_{kl}^{c}(\boldsymbol{x}) + R_{ij}^{kl}(\boldsymbol{y})\mu_{kl}^{c,\mathrm{comp}}(\boldsymbol{x}) + \int_{\Theta} \tilde{h}_{(i,y_j)}^{kl}(\boldsymbol{y},\tilde{\boldsymbol{y}}) u_{(k,\tilde{y}_l)}^{\mu(1)}(\boldsymbol{x},\tilde{\boldsymbol{y}}) d\tilde{\Theta} \tag{4.93a}$$

$$\begin{aligned}\sigma_{ij}^{f,\mathrm{comp}}(\boldsymbol{x},\boldsymbol{y}) &= L_{ijkl}(\boldsymbol{y})E_{ij}^{mn}(\boldsymbol{y})\varepsilon_{mn}^{c}(\boldsymbol{x}) + L_{ijkl}(\boldsymbol{y})\left[R_{kl}^{mn}(\boldsymbol{y}) - I_{ijmn}\right]\mu_{mn}^{c,\mathrm{comp}}(\boldsymbol{x}) \\ &+ L_{ijkl}(\boldsymbol{y})\left[\int_{\Theta} \tilde{h}_{(k,y_l)}^{mn}(\boldsymbol{y},\tilde{\boldsymbol{y}}) u_{(m,\tilde{y}_n)}^{\mu(1)}(\boldsymbol{x},\tilde{\boldsymbol{y}}) d\tilde{\Theta} - u_{(k,y_l)}^{\mu(1)}(\boldsymbol{x},\boldsymbol{y})\right]\end{aligned} \tag{4.93b}$$

where

$$R_{ij}^{kl}(\boldsymbol{y}) \equiv \int_{\Theta} \tilde{h}_{(i,y_j)}^{kl}(\boldsymbol{y},\tilde{\boldsymbol{y}}) d\tilde{\Theta} \tag{4.94}$$

Consider now a unit cell problem, $\sigma_{ij,y_j}^{f,\mathrm{comp}} = 0$, where $\sigma_{ij}^{f,\mathrm{comp}}$ is given by (4.93b). As usual, assuming arbitrary $\varepsilon_{mn}^{c}(\boldsymbol{x})$, $\mu_{mn}^{c,\mathrm{comp}}(\boldsymbol{x})$, and $u_{(k,y_l)}^{\mu(1)}(\boldsymbol{x},\boldsymbol{y})$ yields the following influence function problems

$$\left\{L_{ijkl}(\boldsymbol{y})\left[H_{(k,y_l)}^{mn}(\boldsymbol{y}) + I_{klmn}\right]\right\}_{,y_j} = 0 \tag{4.95a}$$

$$\left\{L_{ijkl}(\boldsymbol{y})\left[R_{kl}^{mn}(\boldsymbol{y}) - I_{ijmn}\right]\right\}_{,y_j} = 0 \tag{4.95b}$$

$$\left\{L_{ijkl}(\boldsymbol{y})\left[\int_{\Theta} \tilde{h}_{(k,y_l)}^{mn}(\boldsymbol{y},\tilde{\boldsymbol{y}}) u_{(m,\tilde{y}_n)}^{\mu(1)}(\boldsymbol{x},\tilde{\boldsymbol{y}}) d\tilde{\Theta} - u_{(k,y_l)}^{\mu(1)}(\boldsymbol{x},\boldsymbol{y})\right]\right\}_{,y_j} = 0 \tag{4.95c}$$

subjected to periodic boundary conditions and utilizing (4.39). The boundary value problem (4.95a) coincides with (4.19a). It can be seen that (4.95a) and (4.95b) are equivalent in the sense that

$$R_{kl}^{mn}(\boldsymbol{y}) = -H_{(k,y_l)}^{mn}(\boldsymbol{y}) \tag{4.96}$$

The only new influence function problem that has to be solved is (4.95c).

The resulting fine-scale stress (4.93b) can be rewritten as

$$\begin{aligned}\sigma_{ij}^{f,\mathrm{comp}}(\boldsymbol{x},\boldsymbol{y}) &= L_{ijkl}(\boldsymbol{y})E_{kl}^{mn}(\boldsymbol{y})\left[\varepsilon_{mn}^{c}(\boldsymbol{x}) - \mu_{mn}^{c,\mathrm{comp}}(\boldsymbol{x})\right] \\ &+ L_{ijkl}(\boldsymbol{y})\left[\int_{\Theta} \tilde{h}_{(k,y_l)}^{mn}(\boldsymbol{y},\tilde{\boldsymbol{y}}) u_{(m,\tilde{y}_n)}^{\mu(1)}(\boldsymbol{x},\tilde{\boldsymbol{y}}) d\tilde{\Theta} - u_{(k,y_l)}^{\mu(1)}(\boldsymbol{x},\boldsymbol{y})\right]\end{aligned} \tag{4.97}$$

Consider the weak form of the unit cell problem (4.95c)

$$\begin{aligned}&\int_{\Theta} w_{(i,y_j)}(\boldsymbol{y}) L_{ijkl}(\boldsymbol{y}) v_{(k,y_l)}(\boldsymbol{x},\boldsymbol{y}) d\Theta = 0 \quad \text{in } \Theta \\ &\forall w_i \in W_{\Theta} = \left\{\boldsymbol{w}(\boldsymbol{y}) \text{ defined in } \Theta, C^0(\Theta), \boldsymbol{y}\text{-periodic}\right\} \\ &v_k(\boldsymbol{x},\boldsymbol{y}) = v_k(\boldsymbol{x},\boldsymbol{y}+\boldsymbol{l}) \text{ on } \partial\Theta\end{aligned} \tag{4.98}$$

where

$$v_{(k,y_l)}(\boldsymbol{x},\boldsymbol{y}) \equiv \int_{\Theta} \tilde{h}^{mn}_{(k,y_l)}(\boldsymbol{y},\tilde{\boldsymbol{y}})u^{\mu(1)}_{(m,\tilde{y}_n)}(\boldsymbol{x},\tilde{\boldsymbol{y}})d\tilde{\Theta} - u^{\mu(1)}_{(k,y_l)}(\boldsymbol{x},\boldsymbol{y}) \tag{4.99}$$

and $\boldsymbol{l}$ is a vector denoting the unit cell dimensions.

Due to the $\boldsymbol{y}$-periodicity of $v_k(\boldsymbol{x},\boldsymbol{y})$, we may choose the test function as $w_i = v_i(\boldsymbol{x}',\boldsymbol{y})$ for a given $\boldsymbol{x}=\boldsymbol{x}'$, which gives

$$\int_{\Theta} v_{(i,y_j)}(\boldsymbol{x}',\boldsymbol{y})L_{ijkl}(\boldsymbol{y})v_{(k,y_l)}(\boldsymbol{x}',\boldsymbol{y})d\Theta = 0 \tag{4.100}$$

Since the linear-elastic constitutive tensor $\boldsymbol{L}(\boldsymbol{y})$ is positive-definite, equation (4.100) can be satisfied if and only if $v_{(k,y_l)} = 0$, which yields

$$\int_{\Theta} \tilde{h}^{mn}_{(k,y_l)}(\boldsymbol{y},\tilde{\boldsymbol{y}})u^{\mu(1)}_{(m,\tilde{y}_n)}(\boldsymbol{x},\tilde{\boldsymbol{y}})d\tilde{\Theta} = u^{\mu(1)}_{(k,y_l)}(\boldsymbol{x},\boldsymbol{y}) \tag{4.101}$$

Inserting solution (4.101) into the fine-scale stress (4.97) yields

$$\sigma^{f,\text{comp}}_{ij}(\boldsymbol{x},\boldsymbol{y}) = L_{ijkl}(\boldsymbol{y})E^{mn}_{kl}(\boldsymbol{y})\left[\varepsilon^{c}_{mn}(\boldsymbol{x}) - \mu^{c,\text{comp}}_{mn}(\boldsymbol{x})\right] \tag{4.102}$$

and the fine-scale strain (4.93a) is given by

$$\varepsilon^{f,\text{comp}}_{ij}(\boldsymbol{x},\boldsymbol{y}) = E^{kl}_{ij}(\boldsymbol{y})\varepsilon^{c}_{kl}(\boldsymbol{x}) - H^{kl}_{(i,y_j)}(\boldsymbol{y})\mu^{c,\text{comp}}_{kl}(\boldsymbol{x}) + u^{\mu(1)}_{(i,y_j)}(\boldsymbol{x},\boldsymbol{y}) \tag{4.103}$$

Averaging (4.102) gives the coarse-scale stress–strain relation

$$\sigma^{c,\text{comp}}_{ij} = L^{c}_{ijkl}\left[\varepsilon^{c}_{kl} - \mu^{c,\text{comp}}_{kl}\right] \tag{4.104}$$

We will discuss the results obtained with the impotent and incompatible eigenstrain formulations in section 4.3.5.4.

4.3.2.2 The Impotent (Compatible) Eigenstrain Influence Function

Consider the following discretization of the impotent eigenstrain

$$\mu^{f,\text{comp}}_{ij}(\boldsymbol{x},\boldsymbol{y}) = \sum_{\alpha=1}^{\tilde{M}} \tilde{\mathcal{N}}^{(\alpha)}(\boldsymbol{y})\mu^{(\alpha),\text{comp}}_{ij}(\boldsymbol{x}) \tag{4.105a}$$

$$\mu^{(\alpha),\text{comp}}_{ij} \equiv \frac{1}{\left|\Theta^{(\alpha)}\right|}\int_{\Theta^{(\alpha)}} \mu^{f,\text{comp}}_{ij}\, d\Theta \tag{4.105b}$$

where the phase partition shape function $\tilde{\mathcal{N}}^{(\alpha)}(\boldsymbol{y})$ satisfies

$$\delta_{\alpha\beta}=\frac{1}{\left|\Theta^{(\beta)}\right|}\int_{\Theta^{(\beta)}}\tilde{\mathcal{N}}^{(\alpha)}(\boldsymbol{y})\;d\Theta \tag{4.106}$$

In the absence of eigenseparations, the resulting fine-scale strain follows from (4.40)

$$\varepsilon_{ij}^{f,\text{comp}}(\boldsymbol{x},\boldsymbol{y})=E_{ij}^{kl}(\boldsymbol{y})\varepsilon_{kl}^{c}(\boldsymbol{x})+\sum_{\alpha=1}^{\tilde{M}}\tilde{P}_{ij}^{kl(\alpha)}(\boldsymbol{y})\mu_{kl}^{(\alpha),\text{comp}}(\boldsymbol{x}) \tag{4.107}$$

where $\tilde{P}_{ij}^{kl(\alpha)}$ is an impotent eigenstrain influence function. Combining (4.103) and (4.107) gives

$$u_{(i,y_j)}^{\mu(1)}(\boldsymbol{x},\boldsymbol{y})=H_{(i,y_j)}^{kl}(\boldsymbol{y})\mu_{kl}^{c,\text{comp}}(\boldsymbol{x})+\sum_{\alpha=1}^{\tilde{M}}\tilde{P}_{ij}^{kl(\alpha)}(\boldsymbol{y})\mu_{kl}^{(\alpha),\text{comp}}(\boldsymbol{x}) \tag{4.108}$$

Using (4.108) and the definition (4.91a) yields

$$\mu_{ij}^{f,\text{comp}}(\boldsymbol{x},\boldsymbol{y})=E_{ij}^{kl}(\boldsymbol{y})\mu_{kl}^{c,\text{comp}}(\boldsymbol{x})+\sum_{\alpha=1}^{\tilde{M}}\tilde{P}_{ij}^{kl(\alpha)}(\boldsymbol{y})\mu_{kl}^{(\alpha),\text{comp}}(\boldsymbol{x}) \tag{4.109}$$

From (4.91b), (4.105), and (4.106), and utilizing $\Theta=\bigcup_{\beta=1}^{\tilde{M}}\Theta^{(\beta)}$, we have

$$\mu_{ij}^{c,\text{comp}}(\boldsymbol{x})=\sum_{\alpha=1}^{\tilde{M}}\phi^{(\alpha)}\mu_{ij}^{(\alpha),\text{comp}};\quad \phi^{(\alpha)}=\left|\Theta^{(\alpha)}\right|/\left|\Theta\right| \tag{4.110}$$

where $\phi^{(\alpha)}$ is a volume fraction of phase α. Inserting (4.105a) and (4.110) into (4.109) yields

$$\sum_{\alpha=1}^{\tilde{M}}\left[\tilde{P}_{ij}^{kl(\alpha)}(\boldsymbol{y})-\tilde{\mathcal{N}}^{(\alpha)}(\boldsymbol{y})I_{ijkl}+\phi^{(\alpha)}E_{ij}^{kl}(\boldsymbol{y})\right]\mu_{kl}^{(\alpha),\text{comp}}(\boldsymbol{x})=0 \tag{4.111}$$

Requiring arbitrariness of $\mu_{kl}^{(\alpha),\text{comp}}$ yields a closed form expression for $\tilde{P}_{ij}^{kl(\alpha)}$

$$\tilde{P}_{ij}^{kl(\alpha)}(\boldsymbol{y})=\tilde{\mathcal{N}}^{(\alpha)}(\boldsymbol{y})I_{ijkl}-\phi^{(\alpha)}E_{ij}^{kl}(\boldsymbol{y}) \tag{4.112a}$$

$$\tilde{P}_{ij}^{kl(\beta\alpha)}=\delta_{\alpha\beta}I_{ijkl}-\phi^{(\alpha)}E_{ij}^{kl(\beta)} \tag{4.112b}$$

Note that based on physical considerations, the impotent eigenstrain influence function in (4.112b) is identical to that obtained in (4.87), provided that $\phi^{(\alpha)}$ is a volume fraction.

For the coarse-scale stress expression in (4.47), we replace $A_{ijmn}^{c(\alpha)}$ by $\tilde{A}_{ijmn}^{c(\alpha)}$, given as

$$\tilde{A}_{ijmn}^{c(\alpha)}\equiv\frac{1}{|\Theta|}\int_{\Theta}L_{ijkl}\left[\tilde{P}_{kl}^{mn(\alpha)}-\tilde{\mathcal{N}}^{(\alpha)}I_{klmn}\right]d\Theta=-\phi^{(\alpha)}L_{ijmn}^{c} \tag{4.113}$$

Note that the fine-scale strain constructed from the impotent eigenstrain (4.107) remains residual-free, that is, it satisfies the unit cell equilibrium equation (4.95).

4.3.3 Hybrid Impotent-Incompatible Eigenstrain Mode Estimators

The impotent eigenstrain influence functions should typically be used for the matrix-dominated mode of deformation, whereas the usual incompatible eigenstrain piecewise constant approximation should be used for the inclusion-dominated mode of deformation. For simple microstructural geometries, the proper mode can be selected by careful observation of the unit cell geometry. However, for complex microstructures, such a determination is not obvious. In this section, we will describe a simple algorithm for an automatic selection of either the matrix- or inclusion-dominated mode.

Consider a multiphase unit cell with one of the phases being continuous. The continuous phase is referred to as a matrix, while the remaining (one or more) phases are termed inclusion phases. The mode selection algorithm is then as follows:

- *Step 1*: Assume isotropic elasticity for all phases with Young's modulus for the matrix and inclusion phases equal to 10^{-8} and 1.0, respectively. Assume all Poisson's ratios to be equal to zero. Compute the elastic influence functions and calculate six overall homogenized moduli: E_{11}, E_{22}, E_{33}, G_{23}, G_{13}, and G_{23}.
- *Step 2*: In the case of a matrix-dominated mode of deformation, the corresponding homogeneous modulus should be of order 10^{-8} (Reuss bound), while for an inclusion-dominated mode, it should be of order 1.0 (Voigt bound).

To this end, Figure 4.7 depicts various two-phase unit cells, and their mode dominance is summarized in Table 4.1. For the matrix-dominated mode, the impotent eigenstrain influence

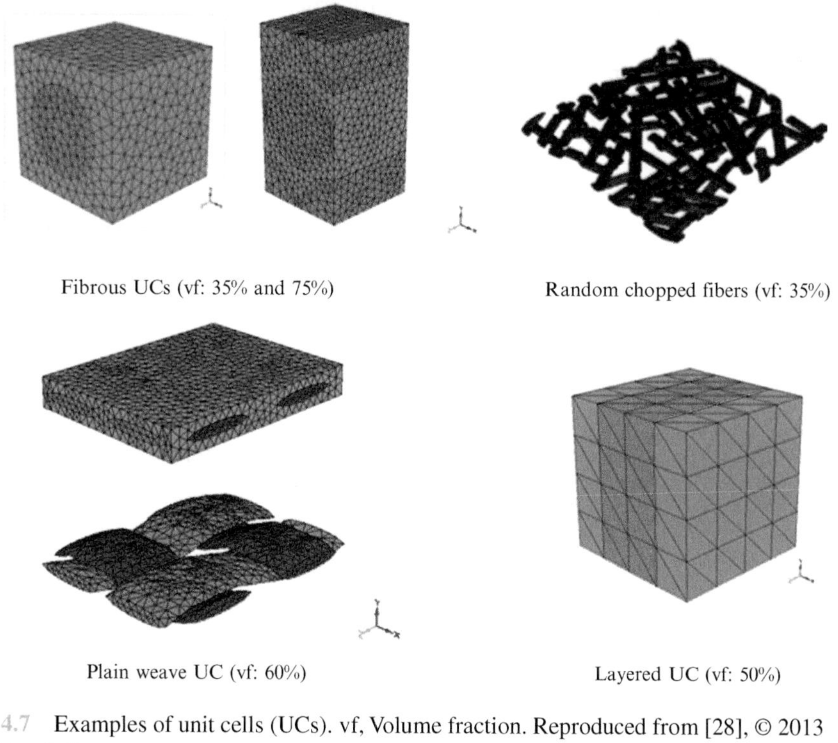

Figure 4.7 Examples of unit cells (UCs). vf, Volume fraction. Reproduced from [28], © 2013 John Wiley & Sons

Table 4.1 Eigenmode dominance for various unit cells in Figure 4.7

Unit cell	E_{11}	E_{22}	E_{33}	G_{23}	G_{13}	G_{12}
Fibrous	Matrix	Matrix	Inclusion	Matrix	Matrix	Matrix
Plain (or five harness) weave	Inclusion	Matrix	Inclusion	Matrix	Matrix	Matrix
Random chopped fibers	Matrix	Matrix	Matrix	Matrix	Matrix	Matrix
Layered	Matrix	Inclusion	Inclusion	Inclusion	Matrix	Matrix

functions $\tilde{P}_{mn}^{kl(\alpha\beta)}$ are employed, whereas for the inclusion-dominated mode, the incompatible eigenstrain influence functions $P_{mn}^{kl(\alpha\beta)}$ are used instead.

4.3.4 Chaboche Modification

Chaboche *et al.* [39] proposed to derive the eigenstrain influence functions from the instantaneous tangent stiffness. While such an approach would make the cost of the reduced order formulation comparable with that of direct homogenization, matching the two formulations only in the asymptotic range of the instantaneous inelastic properties makes the reduced order formulation computationally viable.

Let $\hat{\boldsymbol{L}}^{(\alpha)}$ and $\hat{\boldsymbol{E}}^{(\alpha)}$ be the instantaneous material properties such that

$$\Delta\hat{\sigma}_{ij}^{(\beta)} = \hat{L}_{ijkl}^{(\beta)}\Delta\hat{\varepsilon}_{kl}^{(\beta)} \equiv L_{ijkl}^{(\beta)}\left[\Delta\hat{\varepsilon}_{kl}^{(\beta)} - \Delta\mu_{kl}^{(\beta)}\right] \tag{4.114a}$$

$$\Delta\hat{\varepsilon}_{ij}^{(\beta)} = \hat{E}_{ij}^{kl(\beta)}\Delta\varepsilon_{kl}^{c} \tag{4.114b}$$

$$\hat{E}_{ij}^{kl} = I_{ijkl} + \hat{H}_{(i,y_j)}^{kl} \tag{4.114c}$$

$$\hat{E}_{ij}^{kl(\beta)} = \frac{1}{\left|\Theta^{(\beta)}\right|}\int_{\Theta^{(\beta)}} \hat{E}_{ij}^{kl}\; d\Theta \tag{4.114d}$$

The instantaneous displacement influence function $\hat{H}_i^{kl}(\boldsymbol{x},\boldsymbol{y})$, which is related to the incremental fields by $\Delta u_i^{(1)}(\boldsymbol{x},\boldsymbol{y}) = \hat{H}_i^{kl}(\boldsymbol{x},\boldsymbol{y})\Delta\varepsilon_{kl}^{c}$, can be computed from the incremental unit cell equilibrium $\Delta\hat{\sigma}_{ij,y_j} = 0$. The resulting boundary value problem for the $\boldsymbol{y}$-periodic function $\hat{H}_i^{kl}(\boldsymbol{x},\boldsymbol{y})$ is given by

$$\left(\hat{L}_{ijkl}\left(I_{klmn} + \hat{H}_{(k,y_l)}^{mn}\right)\right)_{,y_j} = 0 \tag{4.115}$$

where $\hat{L}_{ijkl}$ is an instantaneous constitutive tensor equal to $\hat{L}_{ijkl}^{(\beta)}$ for each phase β.

From (4.114a) we have

$$\Delta\mu_{ij}^{(\beta)} = \left[I_{ijkl} - M_{ijmn}^{(\beta)}\hat{L}_{mnkl}^{(\beta)}\right]\Delta\hat{\varepsilon}_{kl}^{(\beta)} = \left[I_{ijkl} - M_{ijmn}^{(\beta)}\hat{L}_{mnkl}^{(\beta)}\right]\hat{E}_{kl}^{pq(\beta)}\Delta\hat{\varepsilon}_{pq}^{c} \tag{4.116}$$

We further introduce a modified eigenstrain influence function $\widehat{P}_{ij}^{mn(\beta\alpha)}$

$$\Delta\widehat{\varepsilon}_{ij}^{(\beta)} = E_{ij}^{kl(\beta)}\Delta\varepsilon_{kl}^{c} + \sum_{\alpha=1}^{\tilde{M}} \widehat{P}_{ij}^{mn(\beta\alpha)}\Delta\mu_{mn}^{(\alpha)} \tag{4.117}$$

that, in absence of eigenseparations, satisfies a relation similar to (4.41).

Combining (4.116), (4.114b), and (4.117) and requiring arbitrariness of $\Delta\varepsilon_{kl}^{c}$ yields an equation for $\widehat{P}_{ij}^{mn(\beta\alpha)}$

$$\widehat{E}_{ij}^{pq(\beta)} - E_{ij}^{pq(\beta)} = \sum_{\alpha=1}^{\tilde{M}} \widehat{P}_{ij}^{mn(\beta\alpha)}\left[I_{mnst} - M_{mnrv}^{(\alpha)}\widehat{L}_{rvst}^{(\alpha)}\right]\widehat{E}_{st}^{pq(\alpha)} \tag{4.118}$$

In [39], $\widehat{P}_{ij}^{mn(\beta\alpha)}$ in (4.118) is replaced by $P_{ij}^{kl(\beta\alpha)}K_{kl}^{mn(\alpha)}$, where $K_{kl}^{mn(\alpha)}$ is a so-called correction tensor.

In general, the instantaneous properties $\widehat{L}_{ijkl}^{(\alpha)}$ and $\widehat{E}_{ij}^{pq(\alpha)}$ should be computed at each increment from the phase constitutive relation $\widehat{\sigma}_{ij}^{(\beta)}\left(\widehat{\varepsilon}_{ij}^{(\beta)}\right)$ and the unit cell problem (4.115). However, in the limit, when $\widehat{L}_{ijkl}^{(\alpha)}$ is known a priori, instantaneous properties $\widehat{E}_{ij}^{pq(\alpha)}$ can be precomputed a priori.

Note that $\widehat{A}_{ijmn}^{c(\alpha)}$ is computed from $\widehat{P}_{ij}^{mn(\beta\alpha)}$ similarly to (4.113).

4.3.5 Analytical Relations for Various Approximations of Eigenstrain Influence Functions

Consider a multiphase media with uniform elastic stiffness $L_{ijkl}^{(\alpha)}$ for each phase partition $\Theta^{(\alpha)}$. In section 4.3.5.1 we start by establishing a general relation between coefficients $P_{ij}^{kl(\alpha)}$ and E_{ij}^{kl} that will be valid for arbitrary unit cell geometry and material properties. Then in section 4.3.5.2 we derive the relation for $P_{ij}^{kl(\alpha)}$, E_{ij}^{kl} [19, 39] and $\widehat{P}_{ij}^{kl(\alpha)}$, $\tilde{P}_{ij}^{kl(\alpha)}$ for two-phase media. In section 4.3.5.3 we establish the relation between $P_{ij}^{kl(\alpha\beta)}$, $\tilde{P}_{ij}^{kl(\alpha\beta)}$, $\widehat{P}_{ij}^{kl(\alpha\beta)}$ for the two-phase model. Finally, in section 4.3.5.4. we compare incompatible and hybrid impotent-incompatible eigenstrain formulations with the Chaboche approximation and direct numerical simulation.

4.3.5.1 Multiphase Media Relations

To establish the relations between various eigenstrain influence functions, we start by proving the following relations [19]

$$L_{ijkl}^{c} = \sum_{\beta=1}^{\tilde{M}} \phi^{(\beta)} L_{ijmn}^{(\beta)} E_{mn}^{kl(\beta)} \tag{4.119a}$$

$$\sum_{\beta=1}^{\tilde{M}} \phi^{(\beta)} E_{ij}^{kl(\beta)} = I_{ijkl} \tag{4.119b}$$

$$\sum_{\beta=1}^{\tilde{M}} P_{ij}^{kl(\alpha\beta)} = I_{ijkl} - E_{ij}^{kl(\alpha)} \tag{4.119c}$$

$$\sum_{\beta=1}^{\tilde{M}} P_{ij}^{kl(\alpha\beta)} M_{klmn}^{(\beta)} = 0 \tag{4.119d}$$

$$\left(L_{ijmn}^{(\alpha)} - L_{ijmn}^{(\beta)}\right)^{-1} L_{mnkl}^{(\alpha)} = -M_{ijmn}^{(\beta)}\left(M_{mnkl}^{(\alpha)} - M_{mnkl}^{(\beta)}\right)^{-1} \tag{4.119e}$$

From the definitions of L_{ijkl}^{c} and $E_{ij}^{kl(\beta)}$ and from the assumption that elastic properties are constant within each phase, we obtain (4.119a)

$$L_{ijkl}^{c} = \frac{1}{|\Theta|}\sum_{\beta=1}^{\tilde{M}} L_{ijmn}^{(\beta)} \int_{\Theta^{(\beta)}} E_{mn}^{kl} d\Theta^{(\beta)} = \sum_{\beta=1}^{\tilde{M}} \phi^{(\beta)} L_{ijmn}^{(\beta)} E_{mn}^{kl(\beta)} \tag{4.120}$$

Note that for compatible eigenstrain formulation and (4.120), we have

$$\tilde{A}_{ijmn}^{c(\alpha)} = \sum_{\beta=1}^{\tilde{M}} \phi^{(\beta)} L_{ijkl}^{(\beta)} \left[\tilde{P}_{kl}^{mn(\beta\alpha)} - \delta_{\alpha\beta} I_{klmn}\right] = -\phi^{(\alpha)} \sum_{\beta=1}^{\tilde{M}} \phi^{(\beta)} L_{ijkl}^{(\beta)} E_{kl}^{mn(\beta)} = -\phi^{(\alpha)} L_{ijmn}^{c} \tag{4.121}$$

Equation (4119b) follows from the homogenization of elastic properties. In the absence of eigenstrains and eigenseparations, it follows from (4.41) that

$$\varepsilon_{ij}^{(\beta)} = E_{ij}^{kl(\beta)} \varepsilon_{kl}^{c} \tag{4.122}$$

Combining (4.122) with the definition of coarse-scale strain

$$\varepsilon_{ij}^{c} = \frac{1}{|\Theta|}\int_{\Theta} \varepsilon_{ij}^{f} d\Theta = \sum_{\beta=1}^{\tilde{M}} \phi^{(\beta)} \varepsilon_{ij}^{(\beta)} \tag{4.123}$$

gives

$$\varepsilon_{ij}^{c} = \sum_{\beta=1}^{\tilde{M}} \phi^{(\beta)} E_{ij}^{kl(\beta)} \varepsilon_{kl}^{c} \tag{4.124}$$

which proves (4.119b).

Note that for instantaneous fields (4.114) and (4.117), a similar relation to (4.119b) can be derived

$$\sum_{\beta=1}^{\tilde{M}} \phi^{(\beta)} \hat{E}_{ij}^{kl(\beta)} = I_{ijkl} \tag{4.125}$$

To derive (4.119c) and (4.119d), we consider the coarse-scale strain and phase strain from (4.50) and (4.75)

$$\varepsilon_{ij}^{c} = M_{ijkl}^{c}\sigma_{kl}^{c} + \mu_{ij}^{c}; \quad \varepsilon_{ij}^{(\alpha)} = M_{ijkl}^{(\alpha)}\sigma_{kl}^{(\alpha)} + \mu_{ij}^{(\alpha)} \tag{4.126}$$

We define a piecewise uniform eigenstrain $\mu_{ij}^{(\alpha)}$ [19] that gives rise to a uniform stress σ_{ij}^{c} and uniform strain ε_{ij}^{c} over a unit cell domain

$$\begin{aligned}
&\sigma_{ij}^{(\alpha)} = \sigma_{ij}^{c}\\
&\varepsilon_{ij}^{c} = \varepsilon_{ij}^{(\alpha)} = M_{ijkl}^{(1)}\sigma_{kl}^{c} + \mu_{ij}^{(1)} = \ldots = M_{ijkl}^{(\tilde{M})}\sigma_{kl}^{c} + \mu_{ij}^{(\tilde{M})}; \quad \alpha = 1,2\ldots\tilde{M}\\
&\mu_{ij}^{(\alpha)} = \mu_{ij}^{c} - \left[M_{ijkl}^{(\alpha)} - M_{ijkl}^{c}\right]\sigma_{kl}^{c}
\end{aligned} \tag{4.127}$$

where the residual-free strain (4.41) satisfies

$$\varepsilon_{ij}^{(\beta)} = E_{ij}^{kl(\beta)}\varepsilon_{kl}^{c} + \sum_{\alpha=1}^{\tilde{M}} P_{ij}^{kl(\beta\alpha)}\mu_{kl}^{(\alpha)} \tag{4.128}$$

Inserting (4.126) and (4.127) into (4.128) gives

$$\left[E_{ij}^{kl(\beta)} + \sum_{\alpha=1}^{\tilde{M}} P_{ij}^{kl(\beta\alpha)} - I_{ijkl}\right]\varepsilon_{kl}^{c} = \sum_{\alpha=1}^{\tilde{M}} P_{ij}^{kl(\beta\alpha)} M_{klmn}^{(\alpha)}\sigma_{mn}^{c} \tag{4.129}$$

Since (4.129) should hold for arbitrary σ_{ij}^{c} (and thus for arbitrary ε_{ij}^{c}), we arrive at (4.119c) and (4.119d).

Equation (4.119e) is obtained by expressing $\left(L_{ijmn}^{(\alpha)}M_{mnkl}^{(\beta)} - I_{ijkl}\right)$ in two different ways

$$\begin{aligned}
L_{ijmn}^{(\alpha)}M_{mnkl}^{(\beta)} - I_{ijkl} &= L_{ijmn}^{(\alpha)}\left(M_{mnkl}^{(\beta)} - M_{mnkl}^{(\alpha)}\right)\\
L_{ijmn}^{(\alpha)}M_{mnkl}^{(\beta)} - I_{ijkl} &= \left(L_{ijmn}^{(\alpha)} - L_{ijmn}^{(\beta)}\right)M_{mnkl}^{(\beta)}
\end{aligned} \tag{4.130}$$

which gives

$$L_{ijmn}^{(\alpha)}\left(M_{mnkl}^{(\alpha)} - M_{mnkl}^{(\beta)}\right) = -\left(L_{ijmn}^{(\alpha)} - L_{ijmn}^{(\beta)}\right)M_{mnkl}^{(\beta)} \tag{4.131}$$

and from (4.131) we have (4.119e).

4.3.5.2 Two-Phase Media Relations

Consider now two-phase media (one partition per phase) with uniform elastic stiffness $L_{ijkl}^{(\alpha)}$ over each phase. In the following, we establish analytical relations between tensors $E_{ij}^{kl(\alpha)}$ and $P_{ij}^{kl(\alpha\beta)}$ (see also [19,39])

$$P_{ij}^{kl(\alpha\alpha)} = \left(I_{ijmn} - E_{ij}^{mn(\alpha)}\right)\left(L_{mnst}^{(\alpha)} - L_{mnst}^{(\beta)}\right)^{-1} L_{stkl}^{(\alpha)}; \quad \alpha \neq \beta \tag{4.132a}$$

$$P_{ij}^{kl(\alpha\beta)} = \left(I_{ijmn} - E_{ij}^{mn(\alpha)}\right)\left(L_{mnst}^{(\beta)} - L_{mnst}^{(\alpha)}\right)^{-1} L_{stkl}^{(\beta)} \tag{4.132b}$$

$$E_{ij}^{kl(\alpha)} = \frac{1}{\phi^{(\alpha)}}\left(L_{ijmn}^{(\alpha)} - L_{ijmn}^{(\beta)}\right)^{-1}\left(L_{mnkl}^{c} - L_{mnkl}^{(\beta)}\right) \tag{4.132c}$$

To prove (4.132), we utilize the multiphase relations derived in section 4.3.5.1. Combining (411.9a) and (4.119b) for two phases α, β gives

$$\left(L_{ijmn}^{(\alpha)} - L_{ijmn}^{(\beta)}\right)\phi^{(\alpha)} E_{mn}^{kl(\alpha)} = L_{ijkl}^{c} - L_{ijkl}^{(\beta)} \tag{4.133}$$

From (4.133) we obtain (4.132c).

Rewriting (4.119c) and (4.119d) for two phases α, β yields

$$\begin{aligned} P_{ij}^{kl(\alpha\alpha)} &= -\left(I_{ijmn} - E_{ij}^{mn(\alpha)}\right) M_{mnst}^{(\beta)}\left(M_{stkl}^{(\alpha)} - M_{stkl}^{(\beta)}\right)^{-1} \\ P_{ij}^{kl(\alpha\beta)} &= \left(I_{ijmn} - E_{ij}^{mn(\alpha)}\right) M_{mnst}^{(\alpha)}\left(M_{stkl}^{(\alpha)} - M_{stkl}^{(\beta)}\right)^{-1} \end{aligned} \tag{4.134}$$

Combining (4.134) with (4.119e) gives (4.132a) and (4.132b).

Consider now a Chaboche approximation for two-phase material with $\Theta^{(1)}$ phase elastic and $\Theta^{(2)}$ phase inelastic. For the elastic phase, we have $\widehat{L}_{ijkl}^{(1)} = L_{ijkl}^{(1)}$ and $I_{ijkl} - M_{ijmn}^{(1)}\widehat{L}_{mnkl}^{(1)} = 0$, which removes the summation from equation (4.118). The resulting eigenstrain influence function $\widehat{P}_{ij}^{kl(\beta 2)}$ follows from

$$\widehat{P}_{ij}^{mn(\beta 2)}\left[I_{mnst} - M_{mnrv}^{(2)}\widehat{L}_{rvst}^{(2)}\right]\widehat{E}_{st}^{pq(2)} = \widehat{E}_{ij}^{pq(\beta)} - E_{ij}^{pq(\beta)} \tag{4.135}$$

The remaining eigenstrain influence function components are computed using a piecewise constant approximation of eigenstrains, $\widehat{P}_{ij}^{mn(\beta 1)} = P_{ij}^{mn(\beta 1)}$. Further manipulation of (4.135) gives

$$\begin{aligned} \widehat{P}_{ij}^{kl(\beta 2)} &= \left[\widehat{E}_{ij}^{ab(\beta)} - E_{ij}^{ab(\beta)}\right]\left(\widehat{E}_{ab}^{rv(2)}\right)^{-1}\left[I_{rvkl} - M_{rvcd}^{(2)}\widehat{L}_{cdkl}^{(2)}\right]^{-1} \\ \widehat{P}_{ij}^{mn(\beta 1)} &= P_{ij}^{mn(\beta 1)} \end{aligned} \tag{4.136}$$

The impotent approximation $\tilde{P}_{ij}^{kl(\alpha\beta)}$ for the multiphase media was given in (4.112). For the two-phase medium, it can be simplified as

$$\begin{aligned} \tilde{P}_{ij}^{kl(\alpha\alpha)} &= I_{ijkl} - \phi^{(\alpha)} E_{ij}^{kl(\alpha)} = \phi^{(\beta)} E_{ij}^{kl(\beta)} \\ \tilde{P}_{ij}^{kl(\alpha\beta)} &= -\phi^{(\beta)} E_{ij}^{kl(\alpha)}; \quad \alpha \neq \beta \end{aligned} \tag{4.137}$$

4.3.5.3 Relation Between $P^{(\beta\alpha)}$, $\tilde{P}^{(\beta\alpha)}$, $\hat{P}^{(\beta\alpha)}$ for the Two-Phase Media

Combining (4.125) and (4.119b) for the two-phase media, and exploiting that $I_{ijkl}=(\phi^{(1)}+\phi^{(2)})I_{ijkl}$, gives

$$\hat{E}_{ij}^{kl(1)}-E_{ij}^{kl(1)}=-\frac{\phi^{(2)}}{\phi^{(1)}}\left[\hat{E}_{ij}^{kl(2)}-E_{ij}^{kl(2)}\right] \tag{4.138a}$$

$$I_{ijkl}-E_{ij}^{kl(1)}=-\frac{\phi^{(2)}}{\phi^{(1)}}\left[I_{ijkl}-E_{ij}^{kl(2)}\right] \tag{4.138b}$$

from which follows

$$\left(I_{stij}-E_{st}^{ij(1)}\right)^{-1}\left[\hat{E}_{ij}^{ab(1)}-E_{ij}^{ab(1)}\right]=\left(I_{stij}-E_{st}^{ij(2)}\right)^{-1}\left[\hat{E}_{ij}^{ab(2)}-E_{ij}^{ab(2)}\right] \tag{4.139}$$

Using (4.132), (4.136), and (4.139), it can be shown that

$$\hat{P}_{mn}^{kl\,(\alpha\beta)}=P_{mn}^{ij\,(\alpha\beta)}K_{ij}^{kl(\beta)} \tag{4.140a}$$

$$K_{ij}^{kl(1)}=I_{ijkl} \tag{4.140b}$$

$$\begin{aligned}K_{ij}^{kl(2)}=&\left(I_{ijst}-M_{ijmn}^{(2)}L_{mnst}^{(1)}\right)\left(I_{stpq}-E_{st}^{pq(2)}\right)^{-1}\\&\left[\hat{E}_{pq}^{ab(2)}-E_{pq}^{ab(2)}\right]\left(\hat{E}_{ab}^{rv(2)}\right)^{-1}\left[I_{rvkl}-M_{rvcd}^{(2)}\hat{L}_{cdkl}^{(2)}\right]^{-1}\end{aligned} \tag{4.140c}$$

which recovers the Chaboche correction matrix $\boldsymbol{K}^{(\alpha)}$ [39].

To this end, it is convenient to define the correction tensor $\mathscr{K}_{mn}^{kl(\beta)}$ as

$$\tilde{P}_{mn}^{kl(\alpha\beta)}=\hat{P}_{mn}^{ij(\alpha\beta)}\mathscr{K}_{ij}^{kl(\beta)} \tag{4.141}$$

Then using (4.112b), and applying (4.136) for inelastic phase $\beta=2$ and (4.132a) and (4.132b) and (4.138b) for elastic phase $\beta=1$, gives

$$\begin{aligned}\mathscr{K}_{ij}^{kl(2)}&=\left[I_{ijmn}-M_{ijst}^{(2)}\hat{L}_{stmn}^{(2)}\right]\hat{E}_{mn}^{st(2)}\left[\hat{E}_{st}^{rv(\alpha)}-E_{st}^{rv(\alpha)}\right]^{-1}\left[I_{rvkl}\delta_{\alpha 2}-\phi^{(2)}E_{rv}^{kl(\alpha)}\right]\\ \mathscr{K}_{ij}^{kl(1)}&=\left(I_{ijst}-M_{ijmn}^{(1)}L_{mnst}^{(2)}\right)\left(I_{strv}-E_{st}^{rv(1)}\right)^{-1}\phi^{(2)}E_{rv}^{kl(2)}\end{aligned} \tag{4.142}$$

Note that $\mathscr{K}_{ij}^{kl(2)}$ is the same for $\alpha=1,2$ due to (4.137) and (4.138a), which gives

$$\left[\hat{E}_{rv}^{st(1)}-E_{rv}^{st(1)}\right]^{-1}\left[-\phi^{(2)}E_{st}^{kl(1)}\right]=\left[\hat{E}_{rv}^{st(2)}-E_{rv}^{st(2)}\right]^{-1}\left[I_{stkl}-\phi^{(2)}E_{st}^{kl(2)}\right] \tag{4.143}$$

Combining (4.140a) and (4.141) yields

$$\tilde{P}_{mn}^{kl(\alpha\beta)}=P_{mn}^{ij\,(\alpha\beta)}K_{ij}^{st(\beta)}\mathscr{K}_{st}^{kl(\beta)} \tag{4.144}$$

where

$$K_{ij}^{st(2)}\mathscr{K}_{st}^{kl(2)} = -\left(I_{ijst} - M_{ijmn}^{(2)}L_{mnst}^{(1)}\right)\left(I_{stpq} - E_{st}^{pq(1)}\right)^{-1}\phi^{(2)}E_{pq}^{kl\,(1)}$$
$$K_{ij}^{st(1)}\mathscr{K}_{st}^{kl(1)} = \mathscr{K}_{ij}^{kl(1)} \tag{4.145}$$

Equation (4.140),equation (4.141), and equation (4.144) provide the relations between $\hat{P}_{mn}^{kl\,(\alpha\beta)}$, $\tilde{P}_{mn}^{ij(\alpha\beta)}$, and $P_{mn}^{ij(\alpha\beta)}$. It can be shown that for a 1D case, $K^{(\beta)}\mathscr{K}^{(\beta)}=1$ and thus $\tilde{P}^{(\alpha\beta)} = P^{(\alpha\beta)}$ (see Problem 4.6).

4.3.5.4 Verification of the Incompatible Eigenstrain Formulation, the Hybrid Impotent-Incompatible Eigenstrain Formulation, and the Chaboche Approximation

In this section, we compare various reduced order formulations including the incompatible and hybrid impotent-incompatible eigenstrain formulations with the Chaboche approximation and the direct numerical simulation (DNS), which serves as a reference solution. For DNS, we use the same finite element mesh from which the influence functions are computed in the reduced order methods. We consider a two-phase fibrous unit cell (Figure 4.7) with a 50% fiber volume fraction. The fiber is aligned in the x_3=Z direction. The unit cell is subjected to monotonic uniaxial or shear loading. The deformation modes are shown in Figure 4.8. We consider a stiff elastic inclusion embedded in either ductile (von Mises plasticity) or brittle (continuum damage law) matrix phase. The nondimensional elastic and inelastic properties are given in Table 4.2 and Table 4.3.

In Table 4.3, σ^0 is the stress at the damage initiation, ε^1 is an equivalent strain at a fully damaged state, and $C = 1$ denotes identical behavior in tension and compression [44].

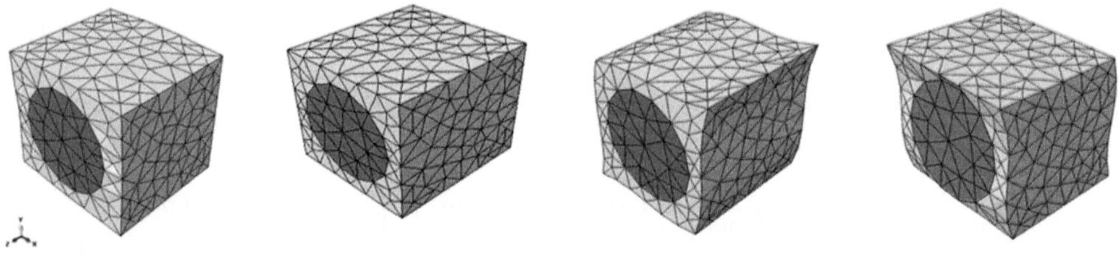

XX=11 mode   ZZ=33 mode   XY=12 mode   YZ=23 mode

Figure 4.8 Types of loading modes for the fibrous unit cell. Reproduced from [28], © 2013 John Wiley & Sons

Table 4.2 Nondimensional isotropic elastic properties of the fibrous composite unit cell

	Young's modulus	Poisson's ratio
Matrix	1	0
Inclusion	100	0

Table 4.3 Nondimensional matrix inelastic material constants for the isotropic bilinear damage law [44]

σ^0	ε^1	C
0.015	0.02	1

Table 4.4 Nondimensional matrix inelastic material constants for von Mises plasticity law with exponential isotropic hardening [45]

K_0	K_∞	δ	H
0.01	0.02	100	0.1

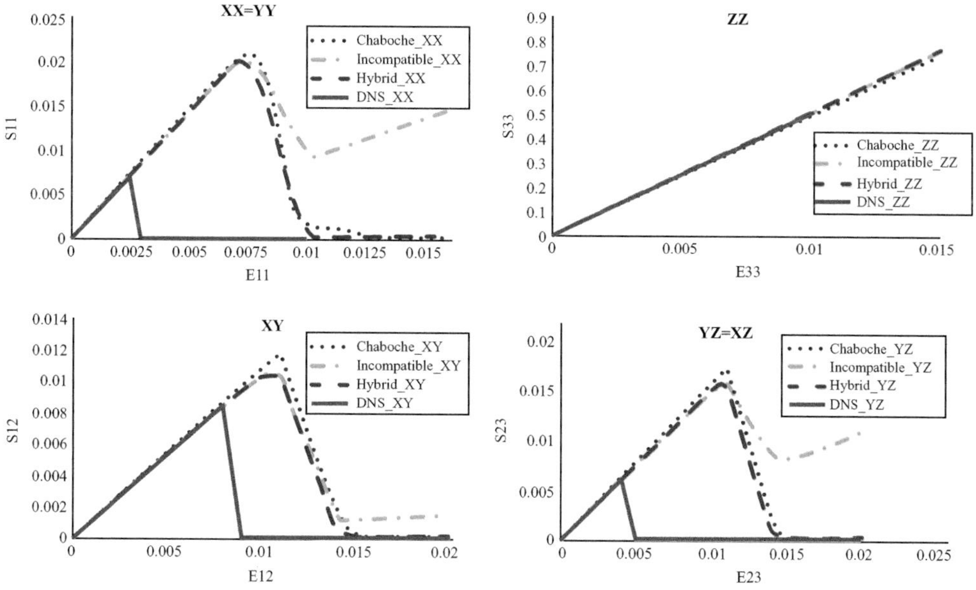

Figure 4.9 Fibrous unit cell (50%). Isotropic bilinear damage law for the matrix phase. Comparison of the incompatible and hybrid impotent-incompatible eigenstrain formulations with the Chaboche approximation and the reference solution. Material parameters are not calibrated. Reproduced from [28], © 2013 John Wiley & Sons

In Table 4.4, K_0 is the yield stress, K_∞ is the ultimate strength, δ is the exponent for the evolution law, and H is the linear portion of the hardening law [45].

Figure 4.9 and Figure 4.10 compare the simulation results obtained with the incompatible and hybrid impotent-incompatible eigenstrain formulations, the Chaboche approximation, and the DNS for a unit cell subjected to overall unit strain modes. The same material parameters are used for all reduced order formulations, as well as for DNS. The most notable result is the erroneous post-failure behavior of the incompatible eigenstrain formulation, which shows increasing stress even after the matrix phase (modeled by continuum damage law) has been fully damaged. On the other hand, the impotent eigenstrain

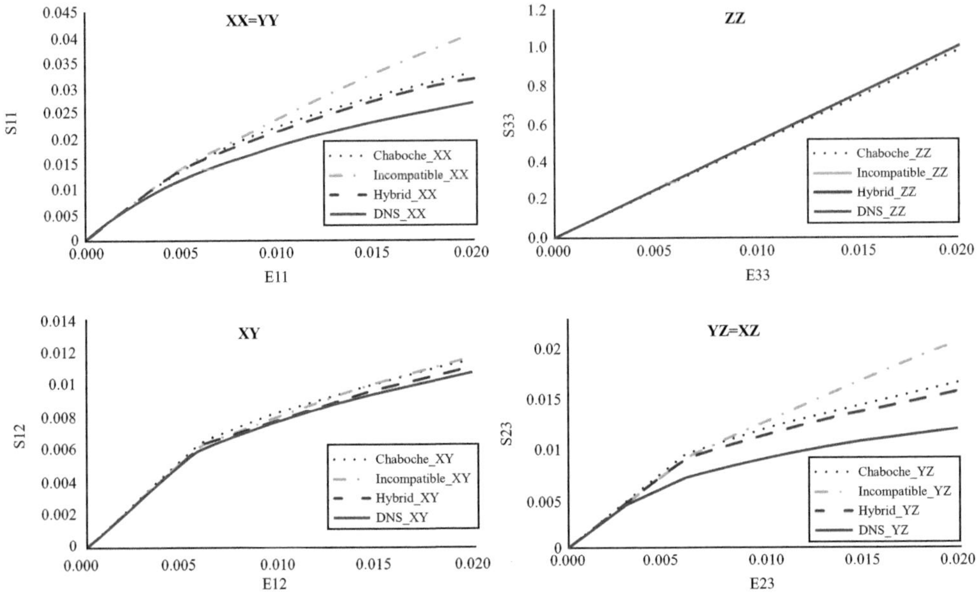

Figure 4.10 Fibrous unit cell (50%). Von Mises plasticity with isotropic exponential hardening for the matrix phase. Comparison of the incompatible and hybrid impotent-incompatible eigenstrain formulations with the Chaboche approximation and the reference solution. Material parameters are not calibrated. Reproduced from [28], © 2013 John Wiley & Sons

formulation correctly predicts zero stress in the post-failure regime. The Chaboche formulation seems to closely follow the hybrid impotent-incompatible eigenstrain formulation in all cases considered. Hereafter, we will only study the hybrid incompatible-impotent formulation, whose computational complexity is substantially lower than that of the Chaboche approximation.

We now investigate whether calibration (identification, see section 4.10) of inelastic material constants in the reduced order formulation improves the quality of the overall stress–strain response. Indeed, Figure 4.11 shows that such a calibration significantly improves the overall response in some but not all modes of loading. It can be seen from Figure 4.11 that with the calibrated material parameters depicted in Table 4.5, the results obtained in the axial and transverse tension are in good agreement with the reference solution. Yet the results in shear, and in particular in-plane shear, are far from satisfactory.

The erroneous shear response as predicted by the reduced order method can be circumvented by introducing the eigenstrain upwinding scheme described in the next section.

4.3.6 Eigenstrain Upwinding

It can be seen from Figure 4.9 and Figure 4.10 that the onset and subsequent inelastic behavior predicted by the ROH is somewhat delayed in comparison with the reference solution obtained by DNS. This is because the inelastic deformation in the ROH is driven by average phase fields rather than by localized deformation. Typically in brittle materials, once the critical strain at a

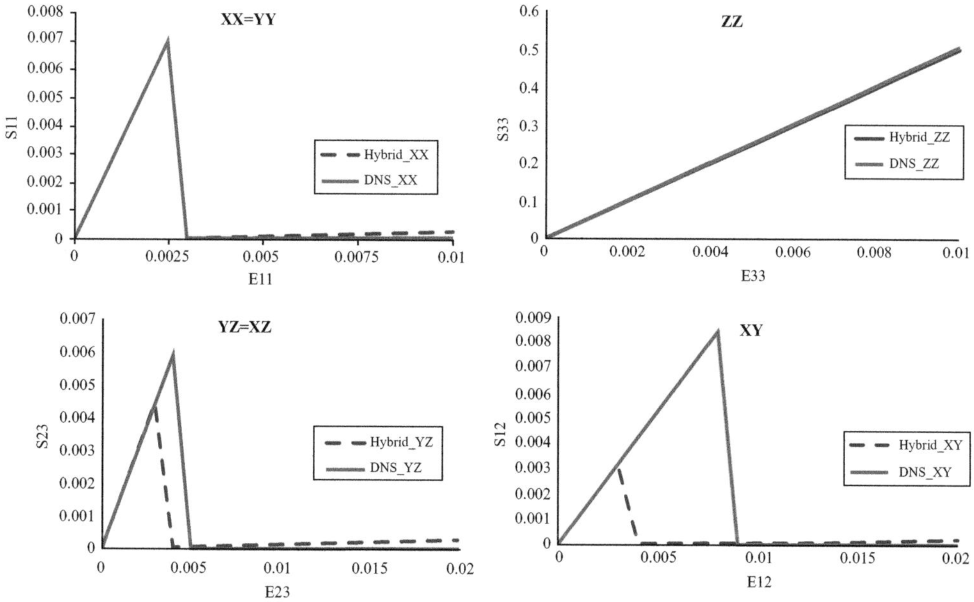

Figure 4.11 Fibrous unit cell (50%). Isotropic bilinear damage law for the matrix phase. Comparison of the calibrated hybrid impotent-incompatible eigenstrain formulation with the reference solution. Reproduced from [28], © 2013 John Wiley & Sons

Table 4.5 Calibrated inelastic parameters for isotropic bilinear damage law. Notably, the failure initiation stress and strain at complete failure are both reduced (see Table 4.3)

σ^0	ε^1	C
0.005	0.005	1

point in a unit cell is reached, the damage zone propagates rapidly, having an almost immediate effect on the overall stress–strain response. In ductile materials, on the other hand, there is a redistribution of stresses following the onset of inelastic deformation, and the overall inelastic response is less pronounced in comparison with the evolution of inelastic response locally.

To compensate for the delay in the onset and subsequent evolution of inelastic response, we introduce an *eigenstrain upwind parameter* $\alpha^{(\beta)} \geq 1$ for each phase, such that the phase constitutive equation (4.46) is restated as

$$\mu_{ij}^{(\beta)} = f\left(\alpha^{(\beta)}\varepsilon_{ij}^{(\beta)}\right) \tag{4.146}$$

The resulting stress update procedure consists of two steps. First, an update procedure is carried out given the upwind strains $\alpha^{(\beta)}\varepsilon_{ij}^{(\beta)}$, from which the eigenstrains are computed using (4.146). Subsequently, stresses are obtained from $\sigma_{ij}^{(\beta)} = L_{ijkl}^{(\beta)}\left(\varepsilon_{kl}^{(\beta)} - \mu_{kl}^{(\beta)}\right)$ using the original strain $\varepsilon_{kl}^{(\beta)}$ and the upwind eigenstrains that were computed from (4.146).

The eigenstrain upwind parameter for each phase is defined as

$$\alpha^{(\beta)} \equiv 1 + \vartheta^{(\beta)}(\eta^{(\beta)} - 1) \tag{4.147}$$

where $\vartheta^{(\beta)}$ is the material *ductility constant*. For highly ductile materials, $\vartheta^{(\beta)} \approx 0$, whereas for highly brittle materials, $\vartheta^{(\beta)} > 1$. Thus, for highly ductile materials such as metal matrix composites, $\alpha^{(\beta)} \approx 1$. In general, the ductility constant can be identified together with other material parameters to fit the experimental data at a coupon level, as discussed in section 4.3.8.

In (4.147), $\eta^{(\beta)}$ is a load-dependent phase *localization parameter*. It represents the magnitude of the localized deformation in each material phase in comparison with the phase averages as discussed below.

Consider a unit cell subjected to six unit overall strain modes, $I \in [1,6]$. For each phase β and mode I we compute an equivalent stress or strain, denoted by $\varphi_{\text{eqv}}^{I,(\beta)}(\boldsymbol{y})$. For brittle materials typically governed by continuum damage mechanics law, $\varphi_{\text{eqv}}^{I,(\beta)}(\boldsymbol{y})$ denotes the equivalent strain defined by [44]

$$\varphi_{\text{eqv}}^{I,(\beta)}(\boldsymbol{y}) \equiv \text{Eqv}\left\{\varphi_K^{I,(\beta)}(\boldsymbol{y})\right\} = \text{Eqv}\left\{\varepsilon_K^{I,(\beta)}(\boldsymbol{y})\right\} \equiv \sqrt{\sum_{i=1}^{3}\left[\hat{\varepsilon}_i^{I,(\beta)}(\boldsymbol{y})\right]^2} \tag{4.148}$$

where subscript $K \in [1,6]$ denotes six second-order tensor components and $\hat{\varepsilon}_i^{I,(\beta)}(\boldsymbol{y})$ is the principle strain i in phase β.

For ductile materials governed by von Mises plasticity, $\varphi_{\text{eqv}}^{I,(\beta)}(\boldsymbol{y})$ denotes the von Mises stress in phase β and mode I, defined by

$$\varphi_{\text{eqv}}^{I,(\beta)}(\boldsymbol{y}) \equiv \text{Eqv}\left\{\varphi_K^{I,(\beta)}(\boldsymbol{y})\right\} = \text{Eqv}\left\{\sigma_K^{I,(\beta)}(\boldsymbol{y})\right\} \equiv \sqrt{\frac{3}{2} s_K^{I,(\beta)}(\boldsymbol{y}) s_K^{I,(\beta)}(\boldsymbol{y})} \tag{4.149}$$

where $s_K^{I,(\beta)}(\boldsymbol{y})$ is a deviatoric stress in phase β and mode I.

For each phase β and each mode I, we identify a critical element e for which the equivalent stress/strain has a maximum value over each phase

$$\varphi_{\text{eqv}}^{e,I,(\beta)} = \max_{y \in \Theta^{(\beta)}} \text{Eqv}\left\{\varphi_K^{I,(\beta)}(\boldsymbol{y})\right\} = \text{Eqv}\left\{\varphi_K^{e,I,(\beta)}\right\} \tag{4.150}$$

and we compute the average stress/strain value over each phase and for each loading mode, denoted here as $\text{avr}\left(\varphi_J^{I,(\beta)}\right)$.

Consider now a unit cell subjected to a coarse-scale strain $\varepsilon_I^c(\boldsymbol{x})$ that gives rise to phase strains $\varepsilon_I^{(\beta)}(\boldsymbol{x})$. Since phase strains $\varepsilon_I^{(\beta)}(\boldsymbol{x})$ can be arbitrary, the location of the maximal equivalent stress/strain may vary throughout the loading history. For a large unit cell, it is convenient to assume that the maximum equivalent stress/strain will be in one of the six elements identified in (4.150). Thus, for each critical element e we compute the value of the stress/strain $\varphi_J^{e,(\beta)}$ using the linear combination as

$$\varphi_J^{e,(\beta)}(\boldsymbol{x}) \equiv \sum_{I=1}^{6} \varphi_J^{e,I,(\beta)} \varepsilon_I^{(\beta)}(\boldsymbol{x}) \tag{4.151}$$

and the corresponding equivalent value $\varphi_{\text{eqv}}^{e,(\beta)}(\boldsymbol{x}) = \text{Eqv}\left\{\varphi_J^{e,(\beta)}(\boldsymbol{x})\right\}$. Likewise, we calculate an average stress/strain field using a linear combination of average fields

$$\varphi_J^{(\beta)} \equiv \sum_{I=1}^{6} \text{avr}\left(\varphi_J^{I,(\beta)}\right)\varepsilon_I^{(\beta)}(\boldsymbol{x}) \tag{4.152}$$

and the corresponding equivalent value of these averages, $\varphi_{\text{eqv}}^{(\beta)} \equiv \text{Eqv}\left\{\varphi_J^{(\beta)}\right\}$. Finally, the phase localization parameter $\eta^{(\beta)}$ is defined as the ratio between the maximum and the average equivalent stress/strain values in each phase

$$\eta^{(\beta)} = \frac{\max\limits_{e}\left\{\varphi_{\text{eqv}}^{e,(\beta)}\right\}}{\varphi_{\text{eqv}}^{(\beta)}} \tag{4.153}$$

Consequently, the eigenstrain upwind parameter follows from (4.147).

In the remainder of this section, we will study the performance of the upwinding scheme for the test problems described in the previous section. Figure 4.12 and Figure 4.13 show the results of the hybrid impotent-incompatible egenstrain formulation with identified material parameters and eigenstrain upwinding for the continuum damage and von Mises plasticity models, respectively. The calibrated material parameters are depicted in Table 4.6 and Table 4.7 for the two material models, respectively. The residual errors in the L_2 norm between the ROH and the reference solution are 9.41×10^{-3} and 2.88×10^{-3}, respectively.

We use the calibrated material parameters to predict the overall response of the unit cell subjected to the uniform and nonuniform biaxial transverse tension. The results for damage and plastiticy models are shown in Figure 4.14 and Figure 4.15.

It can be seen that for the continuum damage model, the results predicted in the uniform biaxial test are in good agreement with the reference solution [Figure 4.14(a)], and yet the peak load in the nonuniform biaxial tension shows more than 100% error in the Y direction [Figure 4.14(c)]. For the plasticity model, a substantially stiffer response is observed in comparison with the reference solution (Figure 4.15). Further model improvements aimed at adressing these discrepancies are discussed in the next section.

4.3.7 Enhancing Constitutive Laws of Phases

Inelastic properties of phases are typically identified using inverse methods (see section 4.10) that minimize the error between the reduced order model and the reference solution. The reference solution can be obtained either from experimental data or from DNS with the microstructural geometry and constitutive equations at the material scale, or from a combination of the two when limited experimental data are available. Here we consider using a DNS to serve as a reference solution. In the numerical experiments conducted in the previous section, the constitutive models at the microscale used in the DNS and the ROH method were assumed to be identical, but the inelastic material constants of phases were calibrated to find the best fit between the DNS and the reduced model for all the experiments considered. The question is: can a handful of calibrated material parameters compensate for the complex inelastic deformation occurring at the microscale? Two simple examples considered in the previous section reveal the limitation of this approach.

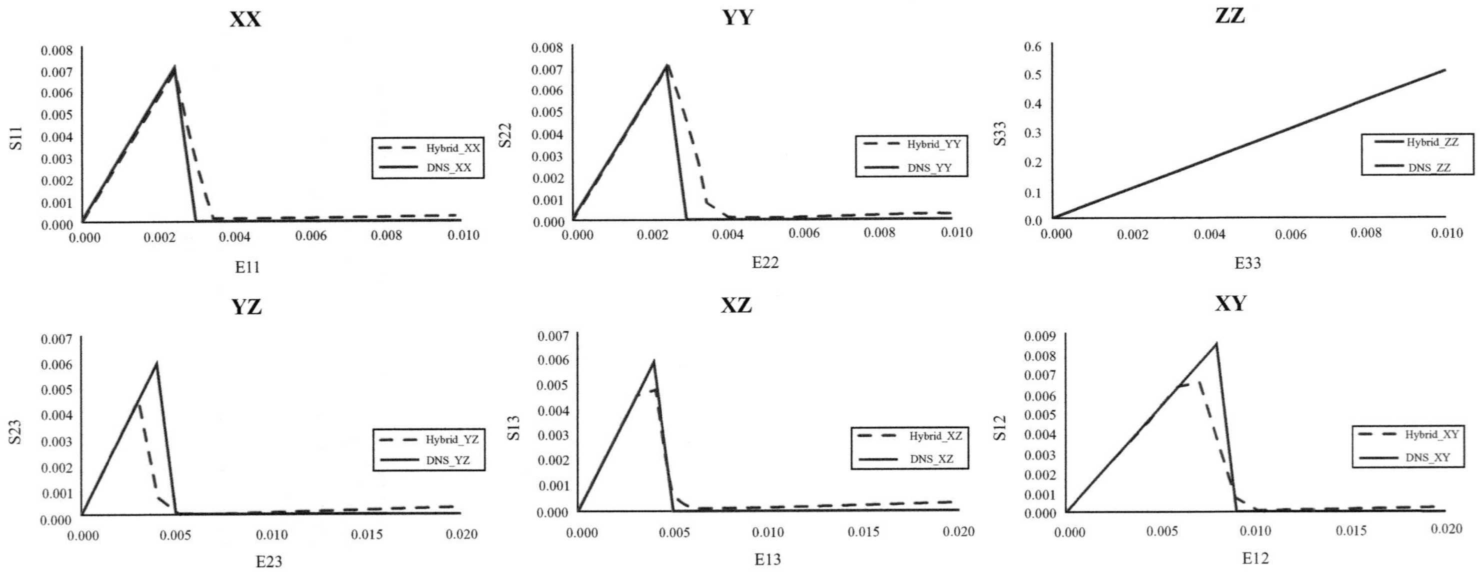

Figure 4.12 Fibrous unit cell (50%). Isotropic bilinear damage law for the matrix phase. Comparison of the hybrid impotent-incompatible eigenstrain formulation with calibrated material parameters and eigenstrain upwinding with the reference solution. Nonsymmetric YZ and XZ response arises due to nonsymmetry in the finite element mesh. Reproduced from [28], © 2013 John Wiley & Sons

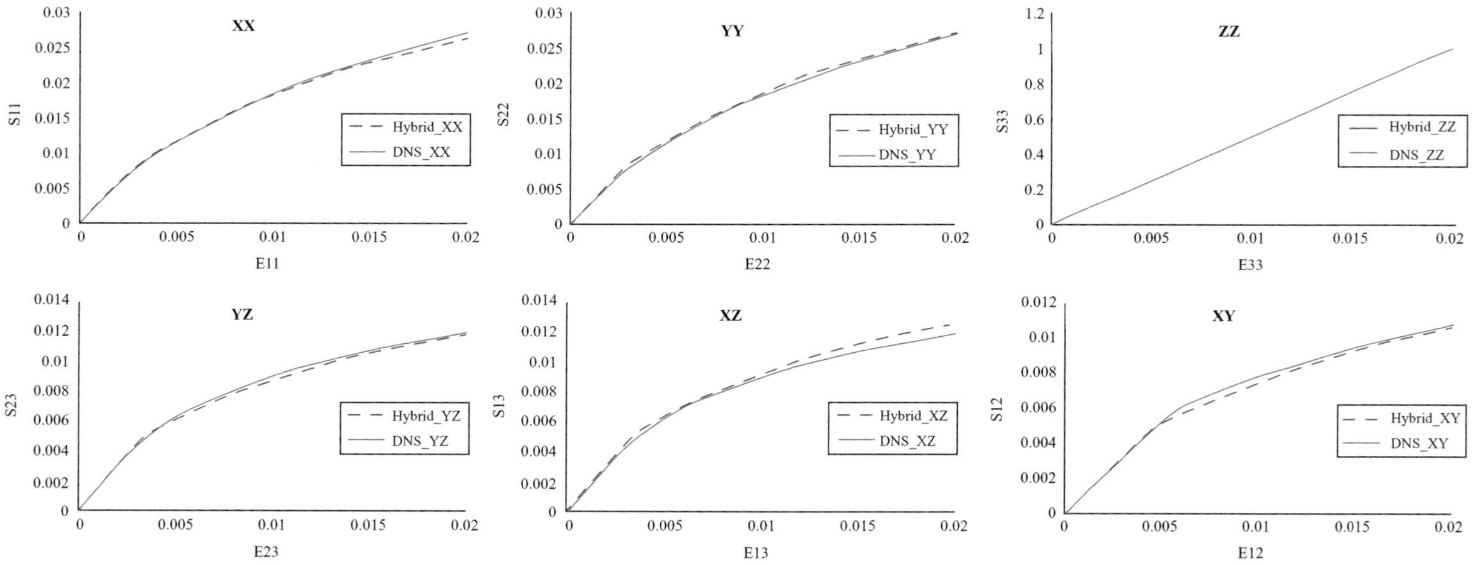

Figure 4.13 Fibrous unit cell (50%). Von Mises plasticity with isotropic exponential hardening for the matrix. Comparison of the hybrid impotent-incompatible eigenstrain formulation with calibrated material parameters and eigenstrain upwinding with the reference solution. Reproduced from [28], © 2013 John Wiley & Sons

Table 4.6 Calibrated inelastic material parameters for isotropic bilinear damage law

σ^0	ε^1	C	$\vartheta^{(\text{matrix})}$
0.0229	0.0315	1	2

Table 4.7 Calibrated inelastic material parameters for plasticity with exponential isotropic hardening

K_0	K_∞	δ	H	$\vartheta^{(matrix)}$
0.01	0.0216	99.9385	0.1366	0.3914

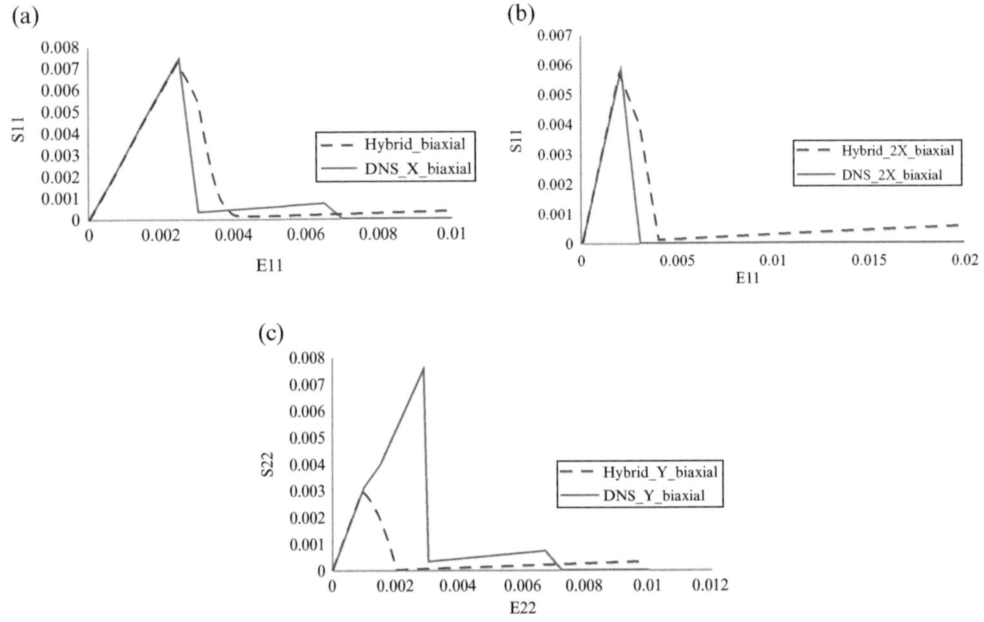

Figure 4.14 Fibrous unit cell (50%). Isotropic bilinear damage law for the matrix phase. Comparison of the hybrid impotent-incompatible eigenstrain formulation with calibrated material parameters and eigenstrain upwinding with the reference solution for (a) uniform and (b, c) nonuniform biaxial loading. Reproduced from [28], © 2013 John Wiley & Sons

Figure 4.14(c) depicts a fibrous composite with elastic fiber and a brittle matrix (governed by scalar damage mechanics law) subjected to biaxial transverse tension. It can be seen that the ROH predicts substantially different post-failure responses than the reference solution. This is because the reduced order approach with scalar damage law for the matrix phase predicts isotropic damage formation in the matrix phase, which gives rise to an isotropic coarse-scale

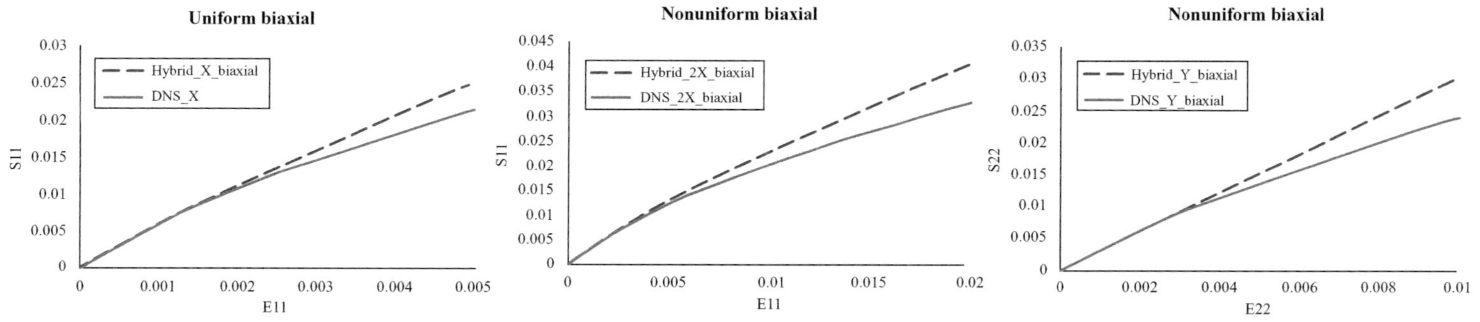

Figure 4.15 Fibrous unit cell (50%). Von Mises plasticity with isotropic exponential hardening for the matrix. Comparison of the hybrid impotent-incompatible eigenstrain formulation with calibrated material parameters and eigenstrain upwinding with the reference solution for biaxial loadings. Reproduced from [28], © 2013 John Wiley & Sons

response. On the other hand, the reference solution shows localized damage formation in the plane perpendicular to the loading in the matrix phase emanating from the fiber. The overall response is clearly anisotropic. The reduced order formulation is unable to fit the multiple experimental data despite parameter calibration. This can be circumvented by introducing an anisotropic (or transverse orthotropic, in this case) damage model [46,47,48] for the matrix phase, as illustrated in the examples below.

Figure 4.15 depicts a fibrous composite with elastic fiber and a ductile matrix (modeled by von Mises plasticity) subjected to biaxial tension in the transverse direction. The reduced order model predicts a nearly hydrostatic stress state in the matrix phase and, since a deviatoric plasticity model is employed, the overall response is close to linear in the post-failure regime. The DNS, on the other hand, shows localized deformation in the vicinity of the fiber and in the formation of localized deviatoric stress at that location. Thus, the overall response shows a considerably softer behavior in the inelastic regime. This phenomenon can be accounted for by a pressure-sensitive matrix response using a cap plasticity model [49]. In the numerical examples considered below, we used identical caps for hydrostatic tension and compression as an extension of von Mises yield surface.

A detailed formulation of the anisotropic damage and cap plasticity models is given in [50].

4.3.8 *Validation of the Hybrid Impotent-Incompatible Reduced Order Model with Eigenstrain Upwinding and Enhanced Constitutive Model of Phases*

In this section, we study the performance of the hybrid impotent-incompatible reduced order model with eigenstrain upwinding and an enhanced constitutive model of phases. We consider both fibrous and woven composite unit cells, with the matrix phase modeled by either anisotropic damage or cap plasticity models in the reduced order model.

We start with a damage model. The calibrated nondimensional transverse orthotropic elastic material constants are summarized in Table 4.8, and the corresponding inelastic material constants are given in Table 4.9. The predicted overall responses of the unit cells subjected to the overall unit strain and the nonuniform biaxial mode of loading are shown in Figure 4.16(a) and (b), respectively.

In Table 4.9, σ_I^0, σ_{II}^0, and σ_{III}^0 are the uniaxial tensile strengths in three principal directions, and ε_I^1, ε_{II}^1, and ε_{III}^1 are the corresponding principal strains in a fully damaged state.

Finally, we study the model predictability for arbitrary loading history. We choose a simultaneous loading in shear YZ and tensile X directions. The results given in Figure 4.17 show a reasonable agreement with the reference solution.

We now focus on the fibrous unit cell where the matrix phase is ductile, obeying the von Mises plasticity model. The reduced order model employs the cap plasticity model for the

Table 4.8 Nondimensional transverse orthotropic elastic material constants

E_1	E_2	E_3	G_{12}	G_{13}	G_{23}	ν_{12}	ν_{13}	ν_{23}
1	1	1	0.5	0.5	0.5	0	0	0

Table 4.9 Nondimensional calibrated inelastic material constants for the transverse orthotropic bilinear damage law

σ_I^0	ε_I^1	σ_{II}^0	ε_{II}^1	σ_{III}^0	ε_{III}^1	C	$\vartheta^{(\text{matrix})}$
0.0148	0.016	0.0148	0.016	0.0141	1.5813	1	1.0156

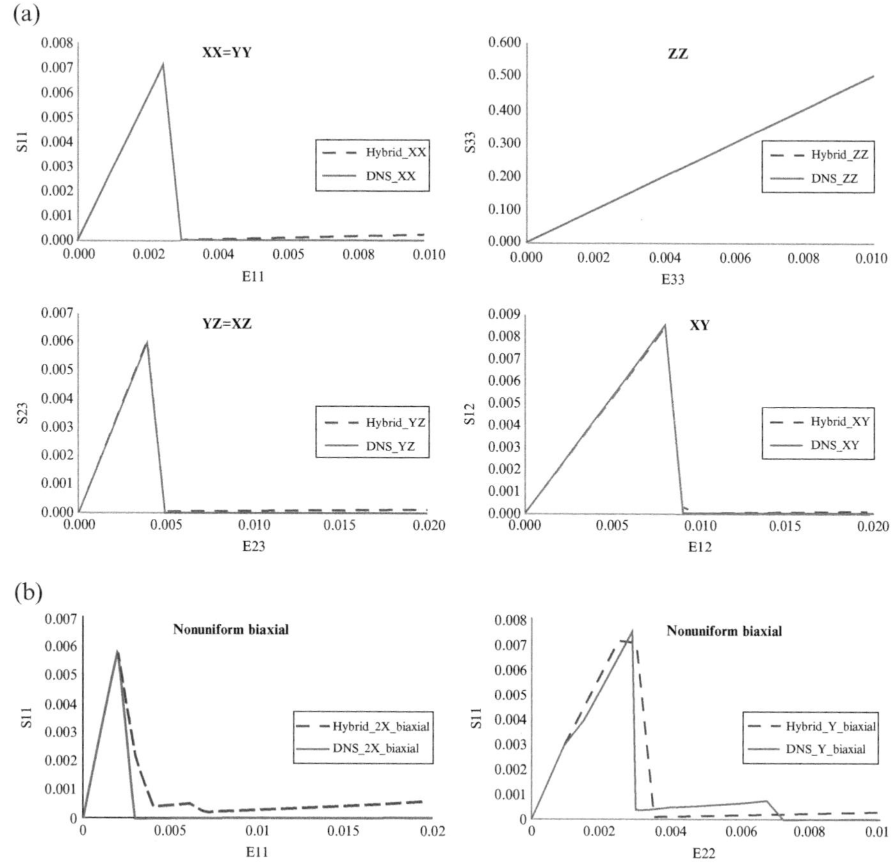

Figure 4.16 Fibrous unit cell (50%). Transverse orthotropic bilinear damage law for the matrix phase. Comparison of the hybrid impotent-incompatible correction with calibrated material parameters and eigenstrain upwinding with the reference solution for (a) unit strain modes and (b) nonuniform biaxial loading. Reproduced from [28], © 2013 John Wiley & Sons

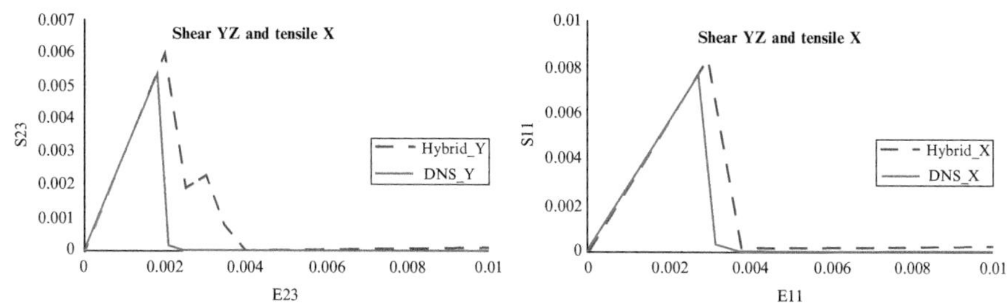

Figure 4.17 Fibrous unit cell (50%). Transverse orthotropic bilinear damage law for the matrix. Verification of the hybrid impotent-incompatible eigenstrain formulation with calibrated material parameters and eigenstrain upwinding subjected to simultaneous shear and tensile loading. Reproduced from [28], © 2013 John Wiley & Sons

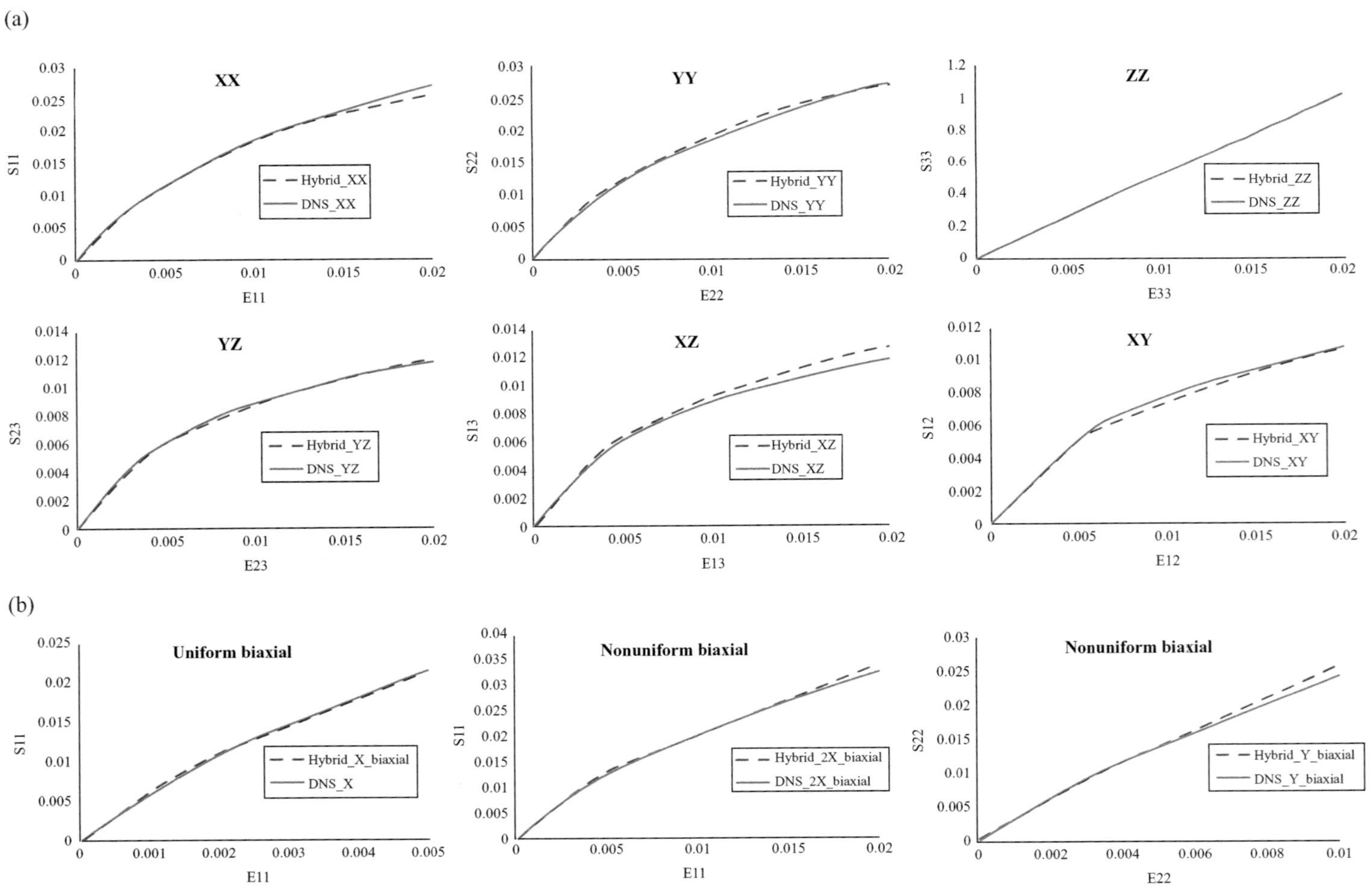

Figure 4.18 Fibrous unit cell (50%). Cap plasticity for the matrix phase. Comparison of the hybrid impotent-incompatible eigenstrain formulation with calibrated material parameters and eigenstrain upwinding with the reference solution for (a) six overall strain modes and (b) biaxial uniform and nonuniform loading. Reproduced from [28], © 2013 John Wiley & Sons

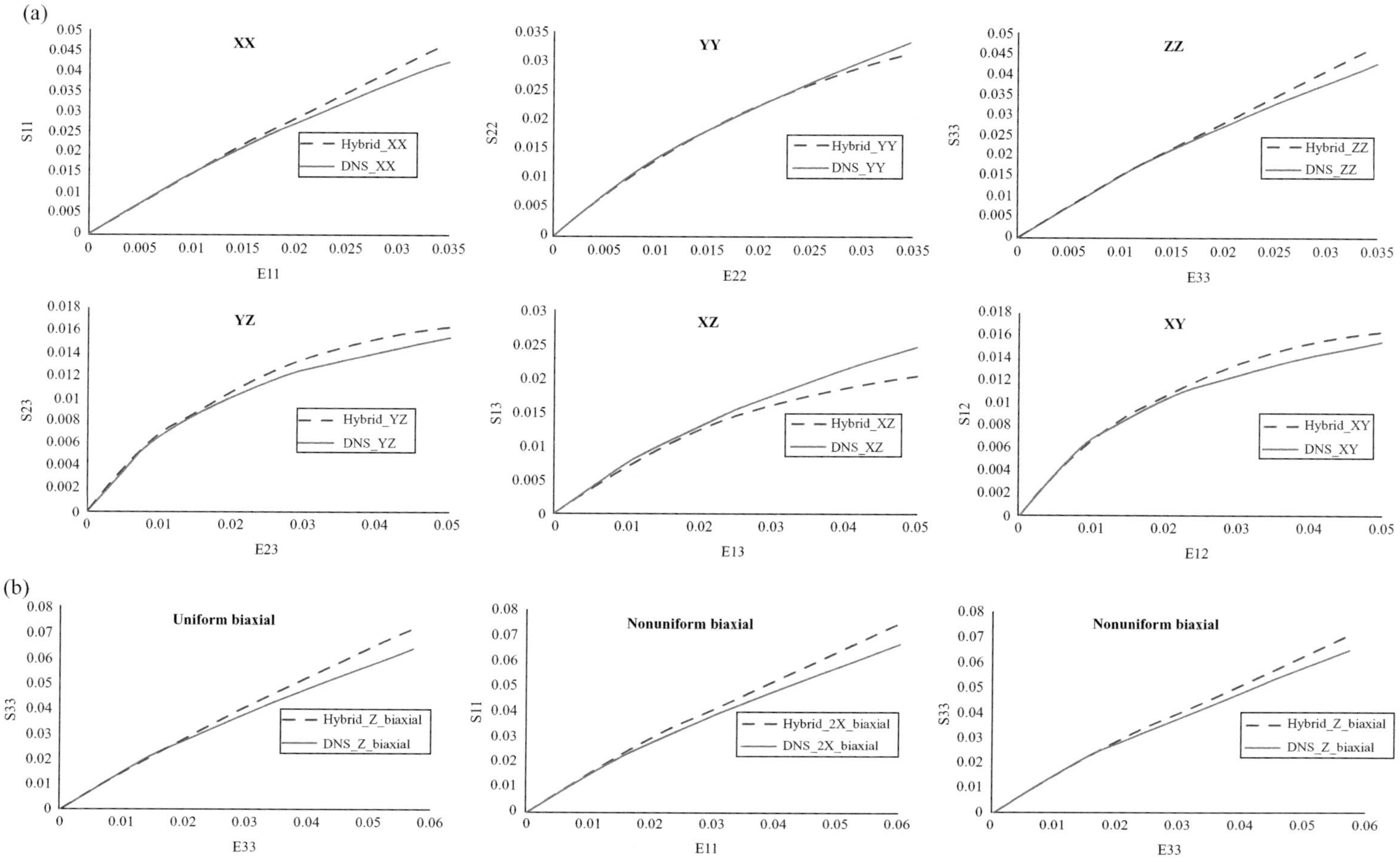

Figure 4.19 Plain weave unit cell (60%). Cap plasticity for the matrix phase. Comparison of the hybrid impotent-incompatible eigenstrain formulation with calibrated material parameters and eigenstrain upwinding with the reference solution for (a) six overall strain modes and (b) biaxial uniform and nonuniform loading in X and Z directions. Reproduced from [28], © 2013 John Wiley & Sons

Table 4.10 Nondimensional calibrated inelastic material parameters for the cap plasticity model. Fibrous unit cell

K_0	K_∞	δ	H	R	W	D	κ_0	$\vartheta^{(matrix)}$
0.01	0.0216	99.9385	0.1366	1	1	0.3	0.0268	0.3721

Table 4.11 Calibrated inelastic material parameters for the cap plasticity model. Plain weave unit cell

K_0	K_∞	δ	H	R	W	D	κ_0	$\vartheta^{(matrix)}$
0.0094	0.0231	99.9987	0.0883	1	1	0.5	0.024	0

matrix phase to compensate for the nonuniform fields developing in a unit cell subjected to hydrostatic loading, as discussed in the previous section. For model identification, we consider a unit cell subjected to six overall strain modes [Figure 4.18(a)] along with the uniform biaxial tension [Figure 4.18(b), left] for calibration of cap parameters. The identified model parameters are summarized in Table 4.10. The response in nonuniform biaxial loading [Fig. 4.18(b), middle and right] is then predicted using calibrated model parameters. The error in the L_2 norm between the reference solution and the reduced order model is 3.46×10^{-3}.

In Table 4.10, *R*, *W*, and *D* are the material parameters of the plasticity cap [49,50], and $\kappa_0 \geq K_\infty$ is the initial value of a cap hardening.

We consider a plain weave unit cell (Figure 4.7) with a matrix phase obeying von Mises plasticity. We consider elastic tows with nondimensional Young's modulus E=2 and zero Poisson's ratio. Simulation results are shown in Figure 4.19, with calibrated material constants listed in Table 4.11. The error in the L_2 norm between the reference solution and the reduced order model is 3.97×10^{-2}.

4.4 Extension to Nonlocal Heterogeneous Media

Results obtained using the formulation presented in the previous sections are independent of the unit cell size. In practice, however, strong unit cell size dependence has been observed in a variety of heterogeneous materials. For instance, the yield stress in metal matrix composites has been found to decrease with an increase in inclusion size [51]. In thin crystalline solids, such as aluminum films [52], the yield stress and strain hardening are strongly dependent on film thickness *L*, grain size *l*, and plasticity length scale *d*, characterizing geometrically necessary dislocations. When $L \approx l \approx d$, the hardening decreases with the increase of *l*. A comprehensive treatise on the dependence of strength on size effect has been given in a monograph by Bazant [53]. From the modeling point of view, these types of effects can be accounted for within a rich framework provided by nonlocal continuum theory and its various variants.

Generally speaking, the nonlocal approach consists of replacing a certain variable by its nonlocal counterpart, which is obtained by weighted averaging over a spatial neighborhood of each point under consideration. If $f(\boldsymbol{x})$ is a *local field* in a domain *V*, the corresponding nonlocal field $\tilde{f}(\boldsymbol{x})$ is defined by

$$\tilde{f}(\boldsymbol{x}) = \int_V \alpha'(\boldsymbol{x},\xi) f(\xi) d\xi \tag{4.154}$$

where $\alpha'(\boldsymbol{x},\boldsymbol{\xi})$ is a nonlocal weight function (or kernel) that depends on the distance between the *source point* $\boldsymbol{\xi}$ and the *effect point* $\boldsymbol{x}$. Since the influence of microstructure only becomes noticeable upon localization primarily governed by inelasticity, the nonlocal formulation (4.154) is typically applied to the inelastic part of the constitutive equation, causing elasticity to remain local. This type of formulation was originated by Pijaudier-Cabot and Bazant [54] and Bazant [55]. For heterogeneous materials, nonlocal theories present several challenges, namely:

(a) *Computational*: Since nonlocal field $\tilde{f}(\boldsymbol{x})$ depends on the information in a characteristic domain V, which could be larger than a finite element size in a coarse-scale mesh, this gives rise to coupled stress updates at various quadrature points in the coarse-scale domain.
(b) *Conceptual*: For heterogeneous media with continuous phases, the definition or even the existence of the nonlocal kernel function $\alpha'(\boldsymbol{x},\boldsymbol{\xi})$ is questionable.

The computational complexity issue can be partially addressed by means of higher grade continua, which are characterized by higher order spatial derivatives of the displacement field [56,57,58,59,60,61,62]. Higher grade continuum models can be derived as an approximation of nonlocal models by means of a truncated Taylor series [63]. These types of models are often referred to as explicit gradient models. However, by virtue of truncation, higher grade models assume that nonlocal interactions are limited to an infinitesimal neighborhood and, therefore, they are considered local in nature. Nonlocal interactions can be accounted for by means of a second-order implicit higher grade continuum model [64], which can be constructed by applying a Laplacian operator to the explicit gradient model. While the computational complexity of both gradient models (compared with nonlocal models) is reduced, substantial overhead is required to deal with the higher order continuity (or mixed methods), and the boundary conditions remain.

The conceptual issue of consistent formulation of kernels, or even the existence of nonlocal kernels for heterogeneous media with continuous fine-scale phases, has received little attention. For example, should the kernel function $\alpha'(\boldsymbol{x},\boldsymbol{\xi})$ be smooth over the characteristic domain V, or should it be oscillatory, somehow mimicking the distribution of material heterogeneity, or should the nonlocality of each phase be accounted for independently by using different kernel functions for each phase?

The primary objective of this section is to present a simple approach that addresses the aforementioned issues [65]. The computational issue is resolved by the so-called *staggered nonlocal model*, where the information from the source point $\boldsymbol{\xi}\neq\boldsymbol{x}$ is taken from the previous load increment (or time step). We will show that for typical load increments, the loss in accuracy as compared with the classical nonlocal methods (hereafter referred to as implicit nonlocal methods) is inconsequential.

In the staggered nonlocal model, the nonlocal kernels are defined separately for each microphase, that is, $\alpha'^{(\beta)}(\boldsymbol{x},\boldsymbol{\xi})$ where β denotes a material phase assumed to have nonlocal interactions. Local fields are equilibrated by constructing residual-free fields that satisfy the fine-scale equilibrium equation for arbitrary inelastic deformation (or eigenstrains). Only phase eigenstrains that account for inelastic deformation are assumed to be nonlocal, while elastic deformation remains local. Attention is restricted to macroscopically homogeneous heterogeneous media. Besides algorithmic convenience, the staggered nonlocal model has physical merit. For instance, in unidirectional composites subjected to tension in the fiber direction, failure of one of the fibers will transfer the load to the surrounding fibers; and therefore, averaging should be carried out over the fiber phase only. In a shear-dominated mode of deformation, the matrix is responsible for the load transfer; and therefore, averaging should be performed over the matrix phase only. In the case of elastic inclusions, averaging in the matrix should exclude fibers since there are typically no eigenstrains in the inclusion phase.

4.4.1 Staggered Nonlocal Model for Homogeneous Materials

In the following, we consider an inelastic homogeneous material system undergoing a small deformation. The local constitutive equation is given by

$$\sigma_{ij}(\boldsymbol{x}) = L_{ijkl}(\boldsymbol{x})\left[\varepsilon_{kl}(\boldsymbol{x}) - \mu_{kl}(\boldsymbol{\varepsilon}(\boldsymbol{x}), \text{history})\right] \tag{4.155}$$

In the staggered nonlocal model, the nonlocal strain $\tilde{\boldsymbol{\varepsilon}}(\boldsymbol{x})$ at every point is computed as a weighted average of strains $\boldsymbol{\varepsilon}(\boldsymbol{\xi})$ in a spatial neighborhood V

$$\tilde{\boldsymbol{\varepsilon}}(\boldsymbol{x}) = \int_V \alpha'(\boldsymbol{x},\boldsymbol{\xi})\boldsymbol{\varepsilon}(\boldsymbol{\xi})d\boldsymbol{\xi} \tag{4.156}$$

where, neglecting boundary effects, the kernel function $\alpha'(\boldsymbol{x},\boldsymbol{\xi})$ is defined as

$$\alpha'(\boldsymbol{x},\boldsymbol{\xi}) = \frac{\alpha(|\boldsymbol{x}-\boldsymbol{\xi}|)}{\int_V \alpha(|\boldsymbol{x}-\boldsymbol{\xi}|)d\boldsymbol{\xi}} \tag{4.157}$$

In (4.157), $\alpha(r)$ is a monotonically decreasing nonnegative function of the distance $r=|\boldsymbol{x}-\boldsymbol{\xi}|$, typically described by a Gauss or Bell-shaped function. Here, the Bell-shaped weight function

$$\alpha(r) = \begin{cases} \left(1 - \dfrac{r^2}{R^2}\right)^2 & \text{if } 0 \le r \le R \\ 0 & \text{if } r \ge R \end{cases} \tag{4.158}$$

is employed due to its simplicity. In (4.158), R denotes the interaction radius, which is assumed to be an intrinsic material parameter.

In the nonlocal model, the local eigenstrain $\mu_{kl}(\boldsymbol{\varepsilon}(\boldsymbol{x}), \text{history})$ is replaced by a nonlocal eigenstrain $\tilde{\mu}_{kl}(\tilde{\boldsymbol{\varepsilon}}(\boldsymbol{x}), \text{history})$, and the corresponding local constitutive model (4.155) is replaced by its nonlocal counterpart

$$\sigma_{ij}(\boldsymbol{x}) = L_{ijkl}(\boldsymbol{x})\left[\varepsilon_{kl}(\boldsymbol{x}) - \tilde{\mu}_{kl}(\tilde{\boldsymbol{\varepsilon}}(\boldsymbol{x}), \text{history})\right] \tag{4.159}$$

Note that nonlocality in elastic deformation, which is not considered here, can be accounted for by replacing local strain $\varepsilon_{kl}(\boldsymbol{x})$ with nonlocal strain $\tilde{\varepsilon}_{kl}(\boldsymbol{x})$ in (4.159).

In the staggered nonlocal algorithm (Box 4.1), the nonlocal strain $\tilde{\varepsilon}_{kl}$ driving the inelastic deformation is constructed from the past history.

Box 4.1 Staggered nonlocal model algorithm for a homogeneous medium

(A) Initialization

(A1) For each quadrature point $\boldsymbol{x}_I$, generate an adjacent set of quadrature points $\boldsymbol{\xi}_J \in Q_I$, $\{Q_I \,|\, |\boldsymbol{x}_I - \boldsymbol{\xi}_J| \le R,\ \forall \boldsymbol{\xi}_J \in Q_I\}$

(A2) For each set Q_I, renormalize the weights: $\alpha^*(\boldsymbol{x}_I, \boldsymbol{\xi}_J) = \dfrac{\alpha'(\boldsymbol{x}_I, \boldsymbol{\xi}_J)}{\sum_{\boldsymbol{\xi}_J \in Q_I} \alpha'(\boldsymbol{x}_I, \boldsymbol{\xi}_J)}$

(A3) For each set Q_I, initialize the strain "backup" database: $\boldsymbol{E}_I^{\text{back}} = \{\boldsymbol{\varepsilon}_J^{\text{back}}\} = \mathbf{0}$

(A4) For each set Q_I, initialize the "in-use" database: $\boldsymbol{E}_I^{\text{use}} = \{\boldsymbol{\varepsilon}_J^{\text{use}}\} = \mathbf{0}$

(B) For each load increment $n+1$ and iteration $i+1$, compute:

LOOP over all sets Q_I:

(B1) Compute the strain at quadrature point ${}_{n+1}^{i+1}\boldsymbol{\varepsilon}_I$ belonging to set Q_I

(B2) LOOP over all quadrature points $\boldsymbol{\xi}_J \in Q_I$

(B3) Approximate nonlocal strain as: ${}_{n+1}^{i+1}\tilde{\boldsymbol{\varepsilon}}_I = \sum_{\boldsymbol{\xi}_J \in Q_I} \alpha^*(\boldsymbol{x}_I, \boldsymbol{\xi}_J)\, {}_{n+1}^{i+1}\hat{\boldsymbol{\varepsilon}}(\boldsymbol{\xi}_J)$,

where $${}_{n+1}^{i+1}\hat{\boldsymbol{\varepsilon}}_I = \begin{cases} {}_{n+1}^{i+1}\boldsymbol{\varepsilon}_I & \text{if } \boldsymbol{\xi}_J = \boldsymbol{x}_I \\ \boldsymbol{\varepsilon}_J^{\text{back}} & \text{if } \boldsymbol{\xi}_J \ne \boldsymbol{x}_I \end{cases}$$

END LOOP

(B4) Update the nonlocal eigenstrains: ${}_{n+1}^{i+1}\tilde{\boldsymbol{\mu}}_I = \tilde{\boldsymbol{\mu}}\left({}_{n+1}^{i+1}\tilde{\boldsymbol{\varepsilon}}_I\right)$

(B5) Calculate the stress: ${}_{n+1}^{i+1}\boldsymbol{\sigma}_{n+1} = \boldsymbol{\sigma}\left({}_{n+1}^{i+1}\tilde{\boldsymbol{\mu}}_I, {}_{n+1}^{i+1}\boldsymbol{\varepsilon}_I\right)$

(B6) Update the "in-use" database: $\boldsymbol{\varepsilon}_I^{\text{use}} = {}_{n+1}^{i+1}\boldsymbol{\varepsilon}_I$

END LOOP

(C) Update the "backup" database

IF the load increment converged, THEN

Update the "backup" database $\boldsymbol{E}_I^{\text{back}} = \boldsymbol{E}_I^{\text{use}}$ for each set Q_I

Go to the next load increment

ELSE

Restore the "in-use" database $\boldsymbol{E}_I^{\text{use}} = \boldsymbol{E}_I^{\text{back}}$ for each set Q_I

Decrease the load increment, and return to Step B

END IF

It is noteworthy to point out that since stress updates are carried out at the quadrature points, it is convenient to express nonlocal strain integrals (4.156) in terms of the information available in the quadrature points only. This requires renormalization of the weights [equation (A2) in Box 4.1] and approximation of the integral [equation (B3) in Box 4.1].

Equation (A2) and equation (B3) in Box 4.1 can be rationalized by the existence of so-called numerical kernel function $\alpha^c(\boldsymbol{x},\xi)$, defined as

$$\tilde{f}(\boldsymbol{x}_I)=\int_V \alpha^c(\boldsymbol{x}_I,\xi)f(\xi)d\xi=\sum_{\xi_J\in Q_I} W_J J_J \alpha^c(\boldsymbol{x}_I,\xi_J)f(\xi_J) \tag{4.160}$$

where J_J is the Jacobian that maps a sphere of radius R into a unit sphere and W_J is the quadrature weight. By comparing equation (4.160) and equation (B3), $\alpha^*(\boldsymbol{x}_I,\xi_J)$ is defined as

$$\alpha^*(\boldsymbol{x}_I,\xi_J)=W_J J_J \alpha^c(\boldsymbol{x}_I,\xi_J) \tag{4.161}$$

Note that equation (4.161) is used for interpretation of kernel $\alpha^*(\boldsymbol{x}_I,\xi_J)$ only – the actual computation is carried out based on equation (A2) and (B3) in Box 4.1.

4.4.2 Staggered Nonlocal Multiscale Model

In this section, we describe a two-scale staggered nonlocal model for a heterogeneous medium. For simplicity, we consider a small deformation problem where inelastic deformation is represented by eigenstrains. In the spirit of the single-scale formulation, only the eigenstrains are assumed to be nonlocal.

We consider a strong form of boundary value problems (4.1), (4.2), (4.3), (4.4), (4.5), and (4.6), with the exception that eigenstrains are considered to be nonlocal

$$\sigma_{ij}^{\zeta}(\boldsymbol{x})=L_{ijkl}^{\zeta}(\boldsymbol{x})\left(\varepsilon_{kl}^{\zeta}(\boldsymbol{x})-\tilde{\mu}_{kl}^{\zeta}(\tilde{\varepsilon}(\boldsymbol{x}))\right)\quad \boldsymbol{x}\in\Omega^{\zeta} \tag{4.162}$$

where $\tilde{\varepsilon}_{ij}$ and $\tilde{\mu}_{ij}$ are nonlocal strain and nonlocal eigenstrain, respectively, to be subsequently defined. We consider the unit cell and coarse-scale problems defined in equation (4.10) and equation (4.11).

In the following derivation, existence of strain gradients in the fine-scale problem will be accounted for by means of nonlocal formulation of phase fields in the physical domain in combination with ROH (see section 4.2). Note that averaging of fine-scale fields in the unit cell domain does not capture nonlocal effects because the unit cell in $O(1)$ homogenization has no physical size. Furthermore, even higher order homogenization theories cannot fully account for nonlocality since the characteristic length may be significantly larger than the coarse-scale element, the size of which is dictated by the coarse-scale solution gradients developing in the post-failure regime.

Reduced order homogenization is employed here not as a localization limiter, but in order to reduce computational cost as compared with direct homogenization.

Following ROH (section 4.1, section 4.2, and section 4.3), the fine-scale strain and stress are constructed to satisfy local equilibrium equations for arbitrary coarse-scale strains and eigenstrains as

$$\varepsilon_{ij}^{f}(\boldsymbol{x},\boldsymbol{y}) = E_{ij}^{kl}(\boldsymbol{y})\varepsilon_{kl}^{c}(\boldsymbol{x}) + \sum_{\alpha=1}^{\tilde{M}} P_{ij}^{kl(\alpha)}(\boldsymbol{y})\,\tilde{\mu}_{kl}^{(\alpha)}\left(\tilde{\boldsymbol{\varepsilon}}^{(\alpha)}(\boldsymbol{x}),\text{history}\right) \tag{4.163}$$

$$\sigma_{ij}^{f}(\boldsymbol{x},\boldsymbol{y}) = L_{ijkl}(\boldsymbol{y})\left[E_{kl}^{mn}(\boldsymbol{y})\varepsilon_{mn}^{c}(\boldsymbol{x}) + \sum_{\alpha=1}^{\tilde{M}}\left(P_{kl}^{mn(\alpha)}(\boldsymbol{y}) - I_{klmn}^{(\alpha)}(\boldsymbol{y})\right)\tilde{\mu}_{mn}^{(\alpha)}\left(\tilde{\boldsymbol{\varepsilon}}^{(\alpha)}(\boldsymbol{x}),\text{history}\right)\right] \tag{4.164}$$

with $I_{ijkl}^{(\alpha)}(\boldsymbol{y})$, $E_{ij}^{kl}(\boldsymbol{y})$, $P_{ij}^{kl(\alpha)}(\boldsymbol{y})$ defined as in section 4.2.

The salient feature of the staggered nonlocal multiscale approach is that phase eigenstrains in equation (4.163) and equation (4.164) are nonlocal. The nonlocal phase eigenstrain $\tilde{\mu}_{mn}^{(\alpha)}\left(\tilde{\boldsymbol{\varepsilon}}^{(\alpha)}(\boldsymbol{x}),\text{history}\right)$ depends on nonlocal phase strain $\tilde{\boldsymbol{\varepsilon}}^{(\alpha)}(\boldsymbol{x})$ and on deformation history. The nonlocal phase strain is evaluated (see also Box 4.1) as

$$\tilde{\boldsymbol{\varepsilon}}^{(\beta)}(\boldsymbol{x}_I) \equiv \tilde{\boldsymbol{\varepsilon}}_I^{(\beta)} = \sum_{\xi_J \in Q_I} \alpha^{(\beta)} * (\boldsymbol{x}_I, \xi_J)\,\hat{\boldsymbol{\varepsilon}}^{(\beta)}(\xi_J) \tag{4.165}$$

where

$$\alpha^{(\beta)} * (\boldsymbol{x}_I, \xi_J) = \frac{\alpha^{(\beta)\prime}(\boldsymbol{x}_I, \xi_J)}{\sum_{\xi_J \in Q_I} \alpha^{(\beta)\prime}(\boldsymbol{x}_I, \xi_J)} \tag{4.166}$$

Note that various fine-scale phases may have different kernel functions $\alpha^{(\beta)\prime}$ governed by the radius of kernel function $R^{(\beta)}$ [see (4.158)]. Furthermore, nonlocality in elastic deformation can be accounted for by replacing the coarse-scale strain $\varepsilon_{kl}^{c}(\boldsymbol{x})$ in equation (4.163) and equation (4.164) with nonlocal coarse-scale strain $\tilde{\varepsilon}_{kl}^{c}(\boldsymbol{x})$ [defined similarly to equation (4.165) and equation (4.166)], but that case is not considered in this book. Other alternatives are considered in Chapter 5.

The resulting coarse-scale stress is expressed in terms of nonlocal phase eigenstrain

$$\sigma_{ij}^{c}(\boldsymbol{x}) = L_{ijkl}^{c}(\boldsymbol{x})\varepsilon_{kl}^{c}(\boldsymbol{x}) + \sum_{\alpha=1}^{\tilde{M}} A_{ijkl}^{c(\alpha)}(\boldsymbol{x})\;\tilde{\mu}_{kl}^{(\alpha)}\left(\tilde{\boldsymbol{\varepsilon}}^{(\alpha)}(\boldsymbol{x}),\text{history}\right) \tag{4.167}$$

where $L_{ijkl}^{c}(\boldsymbol{x})$ and $A_{ijkl}^{c(\alpha)}(\boldsymbol{x})$ are defined as in section 4.2. The solution of the reduced order nonlocal problem is identical to that presented in section 4.2. The results are presented in section 4.4.3.

4.4.3 Validation of the Nonlocal Model

Consider a steel wire reinforced cement beam subjected to four-point bending. The four-point bending beam test results were reported by Jiang *et al.* [66]. The dimensions of the specimen and loading configuration and material microstructure are illustrated in Figure 4.20(a). The volume fraction of steel wire reinforcement is 0.86%. The length of the steel wire is 28 mm, and the diameter of the steel wire is 0.84 mm.

The reinforcement wires are commercial grade utility steel with a yield strength of 260 MPa and a failure strain of 12.6%. Matrix material is assumed to obey continuum damage mechanics

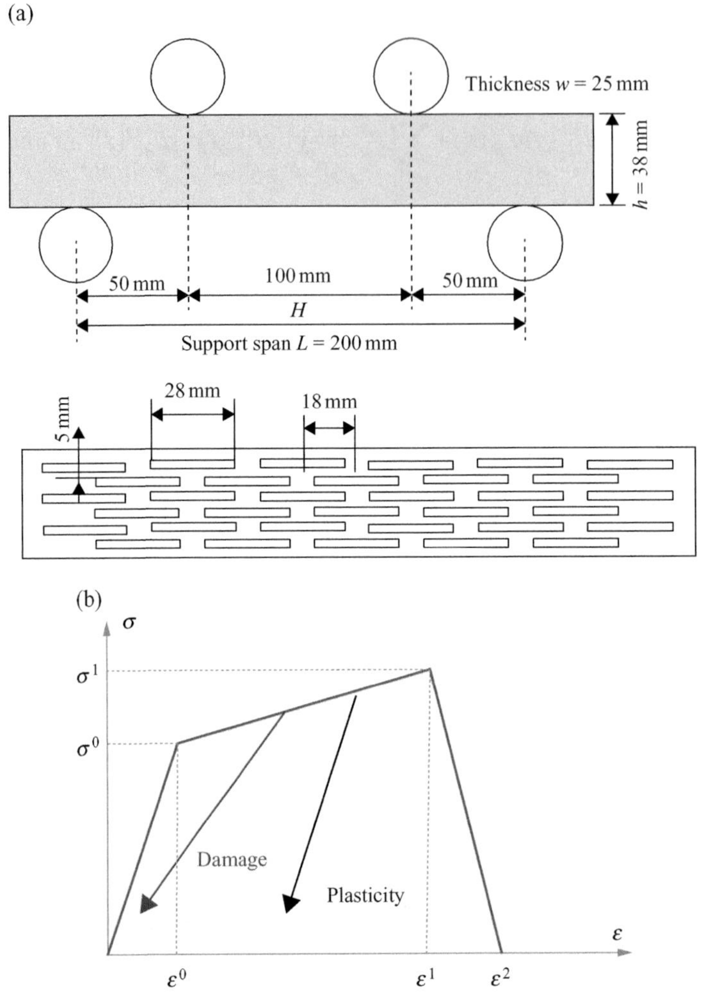

Figure 4.20 (a) Schematics of the steel wire reinforced cement beam; and (b) damage and plasticity bilinear model

law; fibers are modeled as perfectly plastic. Model parameters for the matrix and fibers are summarized in Figure 4.20(b), Table 4.12 and Table 4.13.

The microstructure of the steel wire reinforced cement is idealized by a periodic unit cell with dimensions of 10 mm × 10 mm × 32 mm as shown in Figure 4.21. The three finite element meshes of the beam considered in the present study are shown in Figure 4.22. For the staggered nonlocal multiscale model, a characteristic length for the matrix material has been chosen by inverse calibration to the experimental data, which gives 15 mm. Fibers are assumed to obey the local model. Coincidently, a cube of length of 15 mm has almost the same volume as a unit cell. Note also that the coarse-scale element size has been chosen to be considerably smaller than the unit cell size in order to capture coarse-scale solution gradients in a post-failure

Table 4.12 Model parameters for steel wire

Young's modulus (GPa)	Poisson's ratio	σ^0 (MPa)	σ^1 (MPa)	ε^1
9.0	0.17	260	350	0.126

Table 4.13 Model parameters for cement matrix

Young's modulus (GPa)	Poisson's ratio	σ^0 (MPa)	σ^1 (MPa)	ε^1	ε^2	C
8.5	0.20	6.5	2.2	3.1×10^{-3}	1.0×10^{-2}	0.1

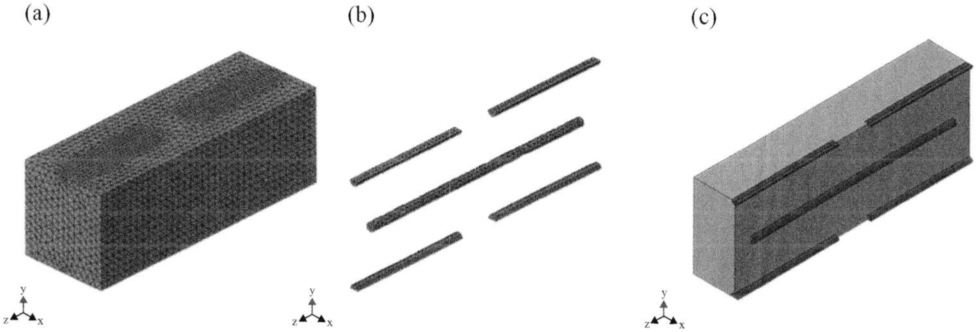

Figure 4.21 Unit cell model of the wire reinforced beam: (a) unit cell finite element model; (b) unit cell finite element model without matrix; and (c) one-half of the unit cell CAD model. Reproduced with permission [65], © 2012 John Wiley & Sons

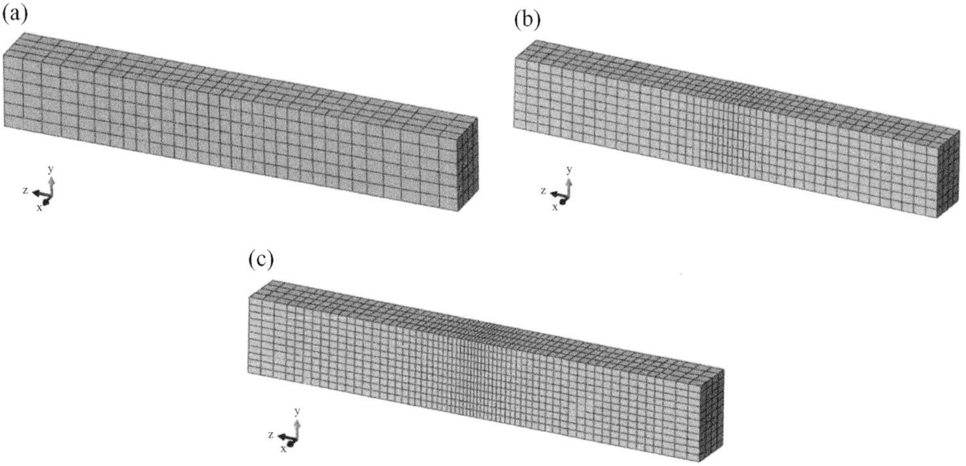

Figure 4.22 Three finite element meshes considered in the study: (a) mesh-1, 768 elements; (b) mesh-2, 1472 elements; and (c) mesh-3, 2816 elements. Reproduced with permission [65], © 2012 John Wiley & Sons

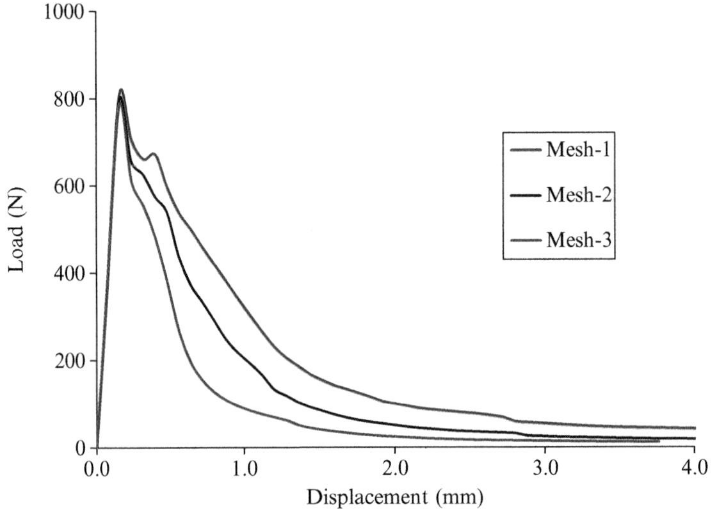

Figure 4.23 Numerical simulation results of load versus displacement for the four-point bending test using a local multiscale model. Reproduced with permission [65], © 2012 John Wiley & Sons

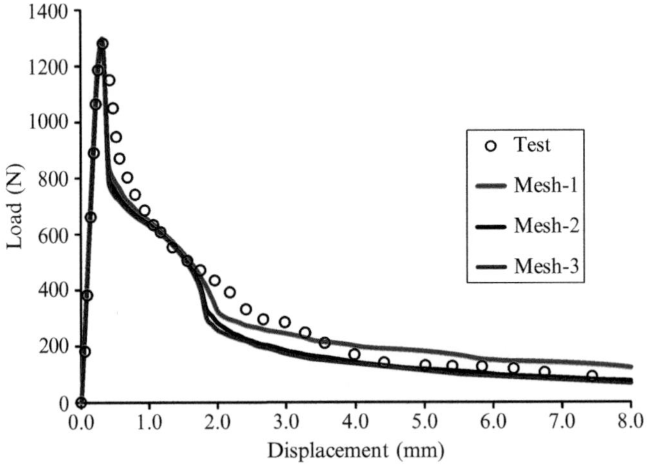

Figure 4.24 Numerical simulation results of load versus displacement for the four-point bending test using a staggered nonlocal multiscale model with a characteristic length of 15 mm. Reproduced with permission [65], © 2012 John Wiley & Sons

regime. Thus, a simple localization limiter that limits the size of the smallest coarse-scale element to a size equal to or larger than the unit cell size is inappropriate.

The results of numerical simulations for the four-point bending beam are illustrated in Figure 4.23 and Figure 4.24 for local and staggered nonlocal multiscale models, respectively. The local multiscale model shows obvious mesh size dependence and becomes unstable

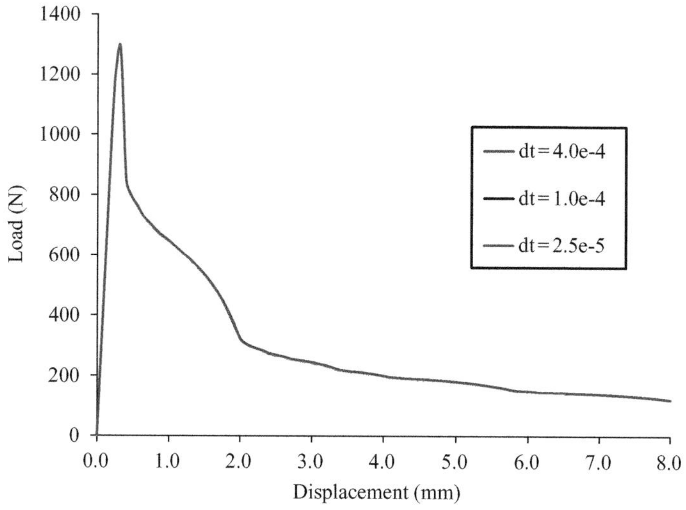

Figure 4.25 The influence of load increment size on the convergence of a staggered nonlocal multiscale model. Reproduced with permission [65], © 2012 John Wiley & Sons

and/or difficult to converge past the peak load, whereas the staggered nonlocal multiscale model is practically mesh size independent and shows reasonable agreement with the test data. Perhaps surprisingly, and despite the overhead, the staggered nonlocal multiscale model requires less CPU time due to the considerably faster convergence of the Newton method.

Finally, we study the convergence of the staggered nonlocal multiscale model to the implicit nonlocal multiscale model. Load increments ranging from 1 to 0.04% have been considered. Figure 4.25 shows that the results obtained by using different load increments are almost identical. It seems that as long as the solution converges, the results of the staggered nonlocal model should be very close to those of the implicit nonlocal model.

4.4.4 Rescaling Constitutive Equations

An alternative to nonlocal formulation is the rescaling of constitutive equations. The notion of rescaling the constitutive equations was introduced by Bazant [67] (see also [68]) in the context of a blunt crack model. To illustrate the basic idea, we first consider a single-scale model with an effective stress–strain relation that exhibits softening behavior as shown in Figure 4.26(a).

$$\sigma = f(\varepsilon) \tag{4.168}$$

The *softening strain*, denoted by κ, is measured from the unloading branch emanating from the peak stress $\sigma_{\max}$ to the softening branch [Figure 4.26(b)]

$$\kappa = \varepsilon^s - \varepsilon^u \tag{4.169}$$

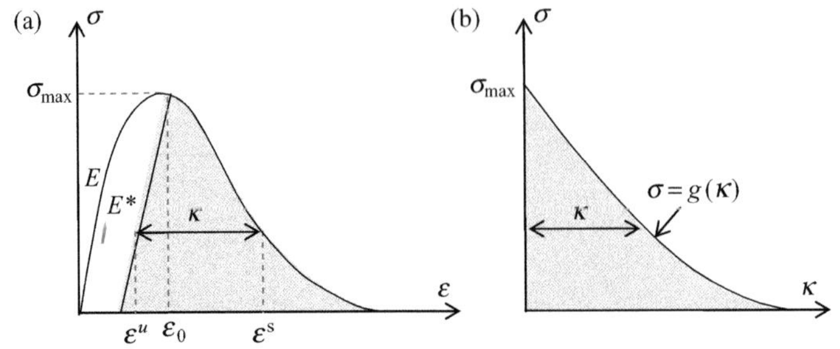

Figure 4.26 (a) Effective stress–strain relation with definition of unloading and softening branches. (b) Effective stress-softening strain curve corresponding to the effective stress–strain curve in (a)

where ε^s and ε^u are effective strains on the softening and the unloading branches corresponding to the same stress value, respectively. The unloading modulus E^* for plasticity and continuum damage mechanics is defined as

$$E^* = \begin{cases} E & \text{for plasticity} \\ (1-\omega)E & \text{for continuum damage mechanics} \end{cases}$$

where $\omega \in [0, 1)$ is the damage parameter. With the above definition of the softening strain κ, we can now construct the softening branch of stress as a function of softening strain

$$\sigma = f(\varepsilon(\kappa)) = g(\kappa) \tag{4.170}$$

The shaded area under the $\sigma - \varepsilon$ curve in Figure 4.26(a), which is identical to the area under the $\sigma - \kappa$ curve in Figure 4.26(b), is the portion of internal work required to fully damage a unit material volume from its initial (possibly) damaged state corresponding to peak stress.

We now focus on computing the so-called post-peak fracture energy G_c, which is the amount of post-peak energy removed from the finite element mesh $\int_0^\infty g(\kappa) d\kappa\, \Omega_c$ divided by the new surface area introduced by S_c

$$\left(\int_0^\infty g(\kappa) d\kappa\ \Omega_c \right) \Big/ S_c = G_c \tag{4.171}$$

where Ω_c is a characteristic material volume. Note that the emphasis here is on the softening branch of the constitutive equation since only the softening branch of the constitutive equation is mesh dependent and therefore rescaled.

Consider a uniform finite element mesh, or so-called *reference mesh*, where the element size is equal to the characteristic material length scale h_c. In explicit finite element codes, when a one-point quadrature element completely fails, that is, stress reduces to zero due to softening, the element is removed from the mesh due to stability, whereas in implicit codes, it is most often left to preserve element connectivity. Whether the element is removed or left is a matter of computational convenience, but de facto it no longer exists, and thus effectively introduces a new free surface, S_c. While the exact free surface area S_c created by element erosion depends on the element geometry and fracture pattern, it is convenient to introduce an approximation, $\Omega_c / S_c \approx \sqrt[3]{\Omega_c} \equiv h_c$, and thus simplify equation (4.171) as

$$\int_0^\infty g(\kappa)d\kappa \; h_c = G_c \tag{4.172}$$

Consider now crack propagation in a uniform finite element mesh with a mesh size equal to h^e. We require the relation (4.172) to hold for arbitrary finite element mesh, that is,

$$\int_0^\infty g^e(\kappa^e)d\kappa^e \; h^e = G_c \tag{4.173}$$

where the superscript e denotes a finite element count and $g^e(\kappa^e)$ is an unknown function at this point.

Combining equation (4.172) and equation (4.173) yields

$$\int_0^\infty g^e(\kappa^e)\left(\frac{h^e}{h_c}\right)d\kappa^e = \int_0^\infty g(\kappa)d\kappa \tag{4.174}$$

It can be seen that a linear map in the form of $g^e(\kappa^e(\kappa)) = g(\kappa)$ with

$$\kappa^e = \frac{h_c}{h^e}\kappa \tag{4.175}$$

satisfies equation (4.174). Equation (4.175) suggests that an element erosion criterion should depend on the element size h^e. The simplest approach to make the element erosion criterion independent of element size is to rescale the softening strain κ^e so that the shaded area under the stress–strain curve in Figure 4.27(b) is scaled by h_c/h^e with respect to the shaded area under the curve in the reference mesh [Figure 4.27(a)]. Likewise, the area under the $\sigma - \kappa^e$

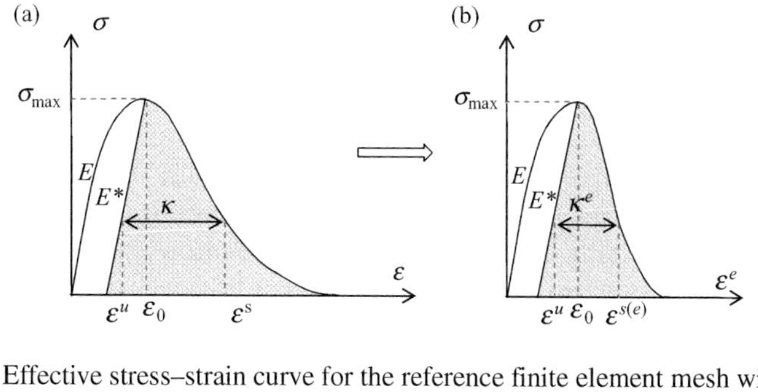

Figure 4.27 (a) Effective stress–strain curve for the reference finite element mesh with element size h_c. (b) Effective stress–strain curve for the arbitrary uniform finite element mesh with element size h^e. The shaded area corresponding to the softening portion of the stress–strain curve is rescaled to keep the post-peak fracture energy G_c invariant with respect to the mesh size. The shaded area in (a) is rescaled by rescaling the softening strain axis while keeping the hardening and unloading portions unchanged

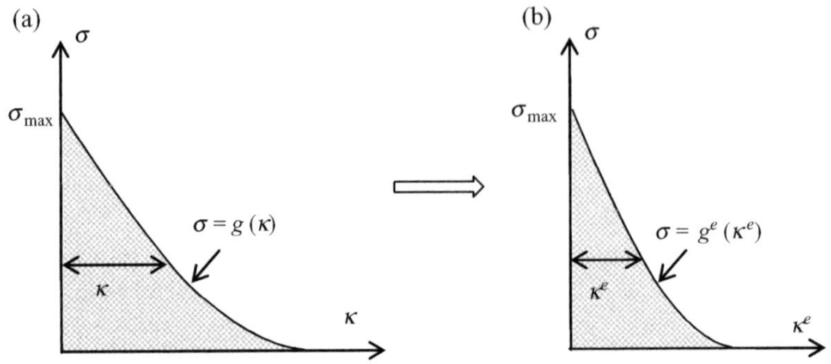

Figure 4.28 Effective stress-softening strain curve: (a) reference mesh; and (b) arbitrary uniform mesh of size h^e

curve in Figure 4.28(b) is consequently scaled by h_c/h^e with respect to the corresponding area in the reference mesh [Figure 4.28(a)].

Note that in practice, however, the characteristic length h_c may be considerably smaller than a structural component size, and therefore a uniform finite element mesh with h_c as a mesh size might be computationally prohibitive. Instead, we may choose a coarser mesh size $h' > h_c$ as a reference mesh and find (or calibrate) a $g'(\kappa')$ that will reproduce the experimental data in terms of crack length. Then the rescaling process will be similar to equation (4.175)

$$\kappa^e = \frac{h'}{h^e}\kappa' \tag{4.176}$$

Consider now a heterogeneous nonlinear medium. The energy released from a uniform coarse-scale finite element mesh by removing a single finite element is given by

$$G_c = h^e \sum_{\beta} \phi^{(\beta)} \int_0^{\infty} g^{e(\beta)}\left(\kappa^{e(\beta)}\right) d\kappa^{e(\beta)} \tag{4.177}$$

where $\phi^{(\beta)}$ is a volume fraction of the phase β.

To extend the single-scale approach outlined earlier to the two-scale problems, the following issues have to be addressed:

(1) For single-scale problems, only the *softening strain* κ has to be rescaled to preserve invariance of post-peak fracture toughness G_c with respect to mesh size. For more than one phase, various rescaling parameters can be used for each phase.
(2) There are various ways to trigger element erosion. For instance, an element can be marked for erosion if either one or all phases fail.

To fix ideas, consider a fibrous composite with elastic fibers throughout the deformation. With loading orthogonal to the fiber axis, matrix cracking will eventually cause the overall failure of the unit cell. Once the matrix fails, the coarse-scale finite element can be eroded even though the inclusion remains elastic. The fracture energy balance equation can be written as

$$G_c = \left(\int_0^\infty g^{e(m)}\left(\kappa^{e(m)}\right) d\kappa^{e(m)} \phi^{(m)} \right) \frac{\Omega^e}{S^{e(m)}} = \left(\int_0^\infty g'^{(m)}(\kappa'^{(m)}) d\kappa'^{(m)} \phi^{(m)} \right) h' \quad (4.178)$$

where the superscript m denotes the matrix phase and $S^{e(m)}$ is a new surface area within the matrix phase. It is again convenient to assume that $\Omega^e / S^{e(m)} \approx \sqrt[3]{\Omega^e} \equiv h^e$, and consequently, scaling laws (4.175) and (4.176) can be extended to the matrix phase as

$$\kappa^{e(m)} = \frac{h'}{h^e} \kappa'^{(m)} \quad (4.179)$$

Consider now a two-phase material where both phases fail. Using the same considerations as before yields

$$\begin{aligned} &\int_0^\infty \left(g^{e(m)}\left(\frac{h'}{h^e} \kappa'^{(m)} \right) - g'^{(m)}\left(\kappa'^{(m)}\right) \right) d\kappa'^{(m)} \phi^{(m)} \\ &+ \int_0^\infty \left(g^{e(f)}\left(\frac{h'}{h^e} \kappa'^{(f)} \right) - g'^{(f)}\left(\kappa'^{(f)}\right) \right) d\kappa'^{(f)} \phi^{(f)} = 0 \end{aligned} \quad (4.180)$$

which gives one equation with two unknowns. While there is no unique solution, one possible solution is to rescale each phase by the same factor as

$$\kappa^{e(m)} = \frac{h'}{h^e} \kappa'^{(m)}; \quad \kappa^{e(f)} = \frac{h'}{h^e} \kappa'^{(f)} \quad (4.181)$$

where superscripts m and f denote matrix and fiber phases, respectively.

In the context of the ROH, rescaling can be used to improve the initial guess of the damage parameters in the calibration procedure.

4.5 Extension to Dispersive Heterogeneous Media

The inertia term is most commonly accounted for in the coarse-scale equation of motion, whereas the unit cell problem remains quasi-static (see Remark 2.1 in Chapter 2 for details). This approach, however, is adequate for low rates of loading and for short observation times. For high rates of loading and long observation times, the dominant feature that has been observed in the experiments and simulations is that of dispersion, which cannot be predicted by the effective medium theories derived by homogenization. This is because internal material interfaces in a heterogeneous material cause reflection and refraction of stress waves, giving rise to dispersion and attenuation of waves within the material microstructure [69,70].

How to account for this fine-scale phenomenon without exhausting computational resources has been a topic of considerable interest in the scientific and engineering communities. The most common approach to accounting for the dispersion phenomenon is to enrich the continuum description in some way [71,72]. The effective stiffness theory developed by Achenbach and Herrmann [73] was one of the first dispersive continuum models for composites. Subsequently, several higher order homogenization-based theories were proposed [74,75,76,77].

Boutin and Auriault [76] demonstrated that the terms of higher order successively introduce effects of polarization, dispersion, and attenuation. Various self-consistent schemes were employed for the purposes of dynamic homogenization by Sabina and Willis [78] and Kanaun and Levin [79,80]. Santosa and Symes [81] developed a dispersive effective medium theory using a Bloch wave expansion, which is in essence a representation of the solution in terms of the eigenmodes. For an excellent survey of various higher order continuum models in the elastodynamics of composites, we refer the reader to [82].

Most of the higher order homogenization theories introduce multiple spatial scales and higher order terms in the asymptotic expansion. However, while higher order terms in the asymptotic expansion are capable of capturing dispersion effects, they introduce secular terms which grow unbounded in time [83,84,85]. When the observation time is small, higher order terms introduce the necessary correction to the leading order term capable of resolving the dispersion effect. However, as the time window increases, the higher order terms become close to or larger than the leading order term, owing to the existence of secularity. Consequently, the asymptotic expansion breaks down as it ceases to be valid. The secularity of the asymptotic expansion can be eliminated by introducing slow timescale(s) aimed at capturing the long-time effect of dispersion, in addition to the fast spatial scale aimed at spatial resolution of the microstructure [83,84,85]. As a variant to the space–time homogenization theory [83,84,85], a nonlocal dispersive model has been developed by adding various order equilibrium equations [86,87]. The resulting equation is independent of slow timescales, but it contains a fourth-order spatial derivative and thus requires a C^1 continuous finite element formulation. For a case of constant mass density, the fourth-order spatial derivative term can be approximated by a mixed second-order derivative in space and time.

The mathematical framework developed for the dispersive continuum theories [76,77,78,79, 80,81,82,83,84,85,86,87] has been limited to linear problems. Nonlinear dispersive formulations are very rare. Among the very few exceptions are the works of Molinari and Mercier [88], Wang and Jiang [89], Leveque and Yong [90],Wang and Sun [91], and more recently, Fish *et al.* [92].

As the primary goal of this book is to explore the practical aspects of multiscale modeling, this section focuses on a general purpose computational framework possessing the following characteristics:

(i) *Generality*: valid for nonlinear problems.
(ii) *Compatibility with standard finite element code architecture*: use of standard C^0 continuity formalism with no higher order boundary conditions.
(iii) *Computational efficiency*: capable of systematically reducing the computational complexity of solving nonlinear unit cell problems.

The fine-scale inertia effect is accounted for by formulating a *quasi-dynamic unit cell problem* where the fine-scale inertia effect is represented by the so-called *inertia-induced eigenstrain*. Solution of the nonlinear quasi-dynamic unit cell problem gives rise to either modification of the coarse-scale mass matrix in the implicit solvers or modification of the internal force in the explicit solvers. Scale separation is assumed, just as in classical homogenization theory, but higher order homogenization is not pursued in order to avoid higher order coarse-scale gradients, higher order continuity, and higher order boundary conditions.

The outline of section 4.5 is as follows. Asymptotic expansions of displacements, inertia, and test functions are introduced in section 4.5.1, which, when combined with the governing equations stated on the fine scale, will lead to the derivation of the coarse-scale problem. The formulation of the quasi-dynamic unit cell problem is given in section 4.5.2. A closed-form solution for a linear elastic periodic heterogeneous medium, which coincides with the space–time solution obtained in [83,84,85], is given in section 4.5.3. Solution of the nonlinear model problem is given in section 4.5.4. Implicit and explicit integration schemes are formulated in section 4.5.5.

4.5.1 Dispersive Coarse-scale Problem

Similar to the expansion of displacements in section 4.2.1, the inertia force $\rho^\zeta \ddot{u}_i^\zeta$ is expanded in the asymptotic Taylor series expansion

$$\rho^\zeta \ddot{u}_i^\zeta(\boldsymbol{x}, \boldsymbol{y}, t) = \rho^c \ddot{u}_i^c(\boldsymbol{x}, t) + \zeta \rho^{(1)} \ddot{u}_i^{(1)}(\boldsymbol{x}, \boldsymbol{y}, t) + O(\zeta^2) \tag{4.182}$$

where ρ^c is the coarse-scale density defined by $\rho^c = \frac{1}{|\Theta|}\int_\Theta \rho^\zeta \, d\Theta$ and $\rho^{(1)}\ddot{u}_i^{(1)}$ is the perturbation from the average inertia, which assumes the following decomposition

$$\rho^{(1)} \ddot{u}_i^{(1)}(\boldsymbol{x}, \boldsymbol{y}, t) = \rho^c \widehat{h}_i^{mn}(\boldsymbol{x}, \boldsymbol{y}, t) \ddot{\varepsilon}_{mn}^c(\boldsymbol{x}, t) \tag{4.183}$$

where $\widehat{h}_i^{mn}(\boldsymbol{x}, \boldsymbol{y}, t)$ is the $\boldsymbol{y}$-periodic tensor function normalized as

$$\int_\Theta \widehat{h}_i^{mn}(\boldsymbol{x}, \boldsymbol{y}, t) d\Theta = 0 \tag{4.184}$$

The weight (or test) function in the weak form is expanded similarly to the trial function in equation (4.7a) as

$$w_i^\zeta(\boldsymbol{x}, t) = w_i^c(\boldsymbol{x}, t) + \zeta w_i^{(1)}(\boldsymbol{x}, \boldsymbol{y}, t) + O(\zeta^2) \tag{4.185}$$

where $w_i^{(1)}$ is the $\boldsymbol{y}$-periodic function satisfying the following normalization condition

$$\int_\Theta w_i^{(1)} \, d\Theta = 0 \tag{4.186}$$

Appling the two-scale spatial differentiation rule [see (2.11) in Chapter 2] to the weight function asymptotic expansion yields

$$w_{(i,j)}^\zeta(\boldsymbol{x}, t) = w_{(i,x_j)}^c(\boldsymbol{x}, t) + w_{(i,y_j)}^{(1)}(\boldsymbol{x}, \boldsymbol{y}, t) + \zeta w_{(i,x_j)}^{(1)}(\boldsymbol{x}, \boldsymbol{y}, t) + O(\zeta^2) \tag{4.187}$$

Further integrating the weak form over the composite domain using the two-scale integration scheme [see section 2.2.2 in Chapter 2], then inserting (4.184), (4.186), and (4.187), exploiting the periodicity of $w_i^{(1)}$, and neglecting terms of the order $O(\zeta^3)$ and higher yields

$$\int_{\Omega} w^c_{(i,x_j)} \sigma^c_{ij}\, d\Omega + \int_{\Omega} w^c_i \rho^c \ddot{u}^c_i\, d\Omega + \int_{\Omega} \left[\frac{1}{|\Theta|} \int_{\Theta} w^{(1)}_{(i,y_j)} \sigma^f_{ij}\, d\Theta + \zeta^2 \rho^c \ddot{\varepsilon}^c_{mn} \frac{1}{|\Theta|} \int_{\Theta} w^{(1)}_i \hat{h}^{mn}_i\, d\Theta \right] d\Omega$$
$$= \int_{\partial\Omega^t} w^c_i \bar{t}^c_i\, d\Gamma \tag{4.188}$$
$$\forall \boldsymbol{w}^c \in W_{\Omega} = \left\{ \boldsymbol{w}^c \text{ defined in } \Omega, C^0(\Omega), \boldsymbol{w}^c = \boldsymbol{0} \text{ on } \partial\Omega^u \right\}$$
$$\forall \boldsymbol{w}^{(1)} \in W_{\Theta} = \left\{ \boldsymbol{w}^{(1)} \text{ defined in } \Theta, C^0(\Theta), \boldsymbol{y}\text{-periodic}, \int_{\Theta} \boldsymbol{w}^{(1)} d\Theta = \boldsymbol{0} \right\}$$

where the fine-scale traction is assumed to be constant over $\partial\Theta$ and $\boldsymbol{t}^{\zeta} = \bar{\boldsymbol{t}}^c$.

Integrating (4.188) by parts gives

$$\int_{\Omega} w^c_i \left[\sigma^c_{ij,x_j} - \rho^c \ddot{u}^c_i \right] d\Omega + \int_{\Omega} \left[\frac{1}{|\Theta|} \int_{\Theta} w^{(1)}_i \sigma^f_{ij,y_j}\, d\Theta - \zeta^2 \rho^c \ddot{\varepsilon}^c_{mn} \frac{1}{|\Theta|} \int_{\Theta} w^{(1)}_i \hat{h}^{mn}_i\, d\Theta \right] d\Omega$$
$$= \int_{\partial\Omega^t} w^c_i \left[t^c_i - \bar{t}^c_i \right] d\Gamma; \qquad \forall \boldsymbol{w}^c \in W_{\Omega}, \forall \boldsymbol{w}^{(1)} \in W_{\Theta} \tag{4.189}$$

where $t^c_i = \sigma^c_{ij} n^c_j$.

The first and second terms in (4.189) correspond to the coarse-scale and unit cell problems, respectively. Here we have two ways to proceed. One approach is to account for the inertia term in both the unit cell and coarse-scale problems. Assuming arbitrariness of $\boldsymbol{w}^{(1)}(\boldsymbol{y}) \in W_{\Theta}$ yields the dynamic unit cell problem and the classical coarse-scale problem

$$\sigma^f_{ij,y_j} = \zeta^2 \rho^c \hat{h}^{mn}_i \ddot{\varepsilon}^c_{mn} \tag{4.190a}$$

$$\sigma^c_{ij,x_j} = \rho^c \ddot{u}^c_i \tag{4.190b}$$

$$\bar{t}^c_i = t^c_i \tag{4.190c}$$

This variant introduces two-way coupling even for linear problems where the unit cell problem (4.190a) has to be continuously reanalyzed for different coarse-scale acceleration gradients.

An alternative approach is to seek an approximation of the unit cell problem that does not depend on the coarse-scale solution for linear problems. This is accomplished by eliminating the inertia term from the fine-scale problem, resulting in the classical quasi-static unit cell problem

$$\sigma^f_{ij,y_j} = 0 \tag{4.191}$$

and by constructing the test function $w^{(1)}_i$ as

$$w^{(1)}_i(\boldsymbol{x}, \boldsymbol{y}, t) = \hat{h}^{kl}_i(\boldsymbol{x}, \boldsymbol{y}, t)\, w^c_{(k,x_l)}(\boldsymbol{x}, t) \tag{4.192}$$

where the inertia influence function $\hat{h}_t^{ij} \in W_\Theta$ is constructed from the solution of the *quasi-dynamic* unit cell problem (see section 4.5.2 for details).

Thus, inserting (4.191) and (4.192) into (4.188) and (4.189) yields the weak form of the coarse-scale problem

$$\int_\Omega w^c_{(i,x_j)}\left(\sigma^c_{ij} + \zeta^2 D_{ijmn}\ddot{\varepsilon}^c_{mn}\right)d\Omega + \int_\Omega w^c_i \rho^c \ddot{u}^c_i \, d\Omega = \int_{\partial\Omega^t} w^c_i \bar{t}^c_i d\Gamma \qquad (4.193)$$
$$\forall w^c_i \in W_\Omega$$

where the so-called dispersion tensor $\boldsymbol{D}$ is defined by

$$D_{ijmn}(\boldsymbol{x},t) = \rho^c \frac{1}{|\Theta|}\int_\Theta \hat{h}^{ij}_s(\boldsymbol{x},\boldsymbol{y},t)\hat{h}^{mn}_s(\boldsymbol{x},\boldsymbol{y},t)d\Theta \qquad (4.194)$$

Integrating by parts the first term in (4.193) and imposing arbitrariness of $\boldsymbol{w}^c$ yields the following strong form of the coarse-scale problem

$$\frac{\partial}{\partial x_j}\left(\sigma^c_{ij} + \zeta^2 D_{ijkl}\ddot{\varepsilon}^c_{kl}\right) - \rho^c \ddot{u}^c_i = 0 \quad \text{on} \quad \Omega$$
$$\left(\sigma^c_{ij} + \zeta^2 D_{ijkl}\ddot{\varepsilon}^c_{kl}\right)n^c_j = \bar{t}^c_i \quad \text{on} \quad \partial\Omega^t \qquad (4.195)$$

Equation (4.195) suggests redefinition of the coarse-scale stress for high frequency dynamic problems as

$$\bar{\sigma}^c_{ij} = \sigma^c_{ij} + \zeta^2 D_{ijkl}\ddot{\varepsilon}^c_{kl} \qquad (4.196)$$

4.5.2 The Quasi-Dynamic Unit Cell Problem

The effect of a stress wave propagating in a unit cell is accounted for via inertia-induced strain ε^I_{kl}, which is assumed to be proportional to the material density

$$\varepsilon^I_{kl}(\boldsymbol{x},\boldsymbol{y},t) = \frac{\rho(\boldsymbol{y})}{\rho^c} f^c_{kl}(\boldsymbol{x},t) \qquad (4.197)$$

where the coarse-scale function $f^c_{mn}(\boldsymbol{x},t)$ is assumed to depend on $\ddot{\varepsilon}^c_{kl}$ so that in absence of $\ddot{\varepsilon}^c_{kl}$, $f^c_{mn} = 0$. In other words, inertia acts as an eigenstrain, which changes the shape and/or volume of the material element.

Consider constitutive relation (4.2) expressed in terms of instantaneous constitutive tensor $\hat{\boldsymbol{L}}^\zeta(\boldsymbol{x},\boldsymbol{y},t) = \hat{\boldsymbol{L}}(\boldsymbol{y},t)$ as

$$\dot{\sigma}^f_{ij}(\boldsymbol{x},t) = \hat{L}_{ijkl}(\boldsymbol{y},t)\dot{\varepsilon}^f_{kl}(\boldsymbol{x},t) \qquad (4.198)$$

Assuming small deformations (or corotational frame), the rate form of the fine-scale problem (4.191) is given by

$$\dot{\sigma}^f_{ij,y_j} = 0 \qquad (4.199)$$

The symmetric fine-scale velocity gradient $\dot{u}^{(1)}_{(i,y_j)}(\boldsymbol{x},\boldsymbol{y},t)$ is decomposed into two parts. One is driven by the coarse-scale strain; the other is governed by the function that depends on the coarse-scale acceleration $f^c_{mn}(\boldsymbol{x},t)$

$$\dot{u}^{(1)}_{(i,y_j)}(\boldsymbol{x},\boldsymbol{y},t) = \widehat{H}^{kl}_{(i,y_j)}(\boldsymbol{x},\boldsymbol{y},t)\dot{\varepsilon}^c_{kl}(\boldsymbol{x},t) + \dot{\varepsilon}^I_{ij}(\boldsymbol{x},\boldsymbol{y},t) + \zeta\widehat{h}^{kl}_{(i,y_j)}(\boldsymbol{x},\boldsymbol{y},t)\dot{f}^c_{kl}(\boldsymbol{x},t) \tag{4.200}$$

where $\widehat{H}^{kl}_i(\boldsymbol{x},\boldsymbol{y},t)$ is a $\boldsymbol{y}$-periodic tensor that depends on both the linear and nonlinear material behavior.

Inserting (4.200) into (4.9a), the fine-scale strain rate $\dot{\varepsilon}^f_{kl}(\boldsymbol{x},\boldsymbol{y},t)$ can be expressed as

$$\begin{aligned}\dot{\varepsilon}^f_{kl}(\boldsymbol{x},\boldsymbol{y},t) &= \left[I_{klmn} + \widehat{H}^{mn}_{(k,y_l)}(\boldsymbol{x},\boldsymbol{y},t)\right]\dot{\varepsilon}^c_{mn}(\boldsymbol{x},t) \\ &\quad + \left(\frac{\rho(\boldsymbol{y})}{\rho^c}I_{klmn} + \widehat{h}^{mn}_{(k,y_l)}(\boldsymbol{x},\boldsymbol{y},t)\right)\dot{f}^c_{mn}(\boldsymbol{x},t)\end{aligned} \tag{4.201}$$

Further inserting (4.201) into (4.198) and (4.199) and requiring the resulting equation to be satisfied for arbitrary coarse-scale fields $\dot{\varepsilon}^c_{kl}$ and $\dot{f}^c_{mn}$ yields the two influence functions problems

$$\left[\widehat{L}_{ijkl}(\boldsymbol{y},t)\left[I_{klmn} + \widehat{H}^{mn}_{(k,y_l)}(\boldsymbol{x},\boldsymbol{y},t)\right]\right]_{,y_j} = 0 \tag{4.202a}$$

$$\left[\widehat{L}_{ijkl}(\boldsymbol{y},t)\left(\frac{\rho(\boldsymbol{y})}{\rho^c}I_{klmn} + \widehat{h}^{mn}_{(k,y_l)}(\boldsymbol{x},\boldsymbol{y},t)\right)\right]_{,y_j} = 0 \tag{4.202b}$$

Equation (4.202a) is a classical nonlinear quasi-static unit cell problem (see, for instance, [93]) that will be further reformulated. We will refer to equation (4.202b) as a quasi-dynamic unit cell problem to reflect the fact that the inertia is represented implicitly via the eigenstrain. From (4.202a) and (4.202b), it can be seen that the influence functions $\widehat{H}^{kl}_i(\boldsymbol{x},\boldsymbol{y},t)$ and $\widehat{h}^{mn}_k$ are related by

$$\widehat{h}^{mn}_{(k,y_l)}(\boldsymbol{x},\boldsymbol{y},t) = \widehat{H}^{mn}_{(k,y_l)}(\boldsymbol{x},\boldsymbol{y},t) + \frac{\Delta\rho(\boldsymbol{y})}{\rho^c}I_{klmn} \tag{4.203a}$$

$$\Delta\rho(\boldsymbol{y}) = \rho^c - \rho(\boldsymbol{y}) \tag{4.203b}$$

and therefore only the quasi-static unit cell boundary value problem has to be solved, whereas $\widehat{h}^{mn}_k$ can be subsequently computed from (4.203a).

We now focus on reformulation and solution of the quasi-static unit cell problem (4.202a). For the quasi-static unit cell problem, the fine-scale velocity and displacements are given by

$$\dot{u}^{(1)}_i(\boldsymbol{x},\boldsymbol{y},t) = \widehat{H}^{kl}_i(\boldsymbol{x},\boldsymbol{y},t)\dot{\varepsilon}^c_{kl}(\boldsymbol{x},t) \tag{4.204}$$

$$u^{(1)}_i(\boldsymbol{x},\boldsymbol{y},t) = H^{kl}_i(\boldsymbol{y})\varepsilon^c_{kl}(\boldsymbol{x},t) + f^\mu_i(\boldsymbol{x},\boldsymbol{y},t) \tag{4.205}$$

$$f^\mu_i(\boldsymbol{x},\boldsymbol{y},t) = \int_\Theta \tilde{h}^{kl}_i(\boldsymbol{y},\tilde{\boldsymbol{y}})\mu^f_{kl}(\boldsymbol{x},\tilde{\boldsymbol{y}},t)d\tilde{\Theta} \tag{4.206}$$

where f_i^{μ} is the $\boldsymbol{y}$-periodic tensor function corresponding to the eigenstrain $\boldsymbol{\mu}^f$, such that $\boldsymbol{f}^{\mu}=\boldsymbol{0}$ for $\boldsymbol{\mu}^f=\boldsymbol{0}$. Comparing the incremental form of (4.205) with the rate form (4.204), $\widehat{H}_i^{kl}$ can be approximated as

$$\widehat{H}_i^{kl}(\boldsymbol{x},\boldsymbol{y},t) \approx H_i^{kl}(\boldsymbol{x},\boldsymbol{y}) + \frac{\partial \Delta f_i^{\mu}(\boldsymbol{x},\boldsymbol{y},t)}{\partial \Delta \varepsilon_{kl}^{c}(\boldsymbol{x},t)} \tag{4.207}$$

where, for the second-order accuracy, the derivative in (4.207) is computed at the mid-step. The solution of the quasi-static unit cell problem is identical to that discussed in section 4.2.

For the quasi-dynamic unit cell problem, consider an additive decomposition of $\widehat{\boldsymbol{h}}$ into the constant and variable functions in $(\boldsymbol{x},t)$, denoted by $\boldsymbol{h}(\boldsymbol{y})$ and $\boldsymbol{k}(\boldsymbol{x},\boldsymbol{y},t)$, respectively

$$\begin{aligned} &\widehat{h}_t^{mn}(\boldsymbol{x},\boldsymbol{y},t) = h_t^{mn}(\boldsymbol{y}) + k_t^{mn}(\boldsymbol{x},\boldsymbol{y},t) \\ &\int_{\Theta} k_t^{mn}(\boldsymbol{x},\boldsymbol{y},t)\,d\Theta = 0; \quad \int_{\Theta} h_t^{mn}(\boldsymbol{y})\,d\Theta = 0 \end{aligned} \tag{4.208}$$

Note that $\boldsymbol{h}$, $\boldsymbol{k}$, and $\widehat{\boldsymbol{h}}$ are $O(1)$.

Inserting (4.208) into (4.203) and (4.207) and assuming linear material, that is, $\boldsymbol{f}^{\mu}\equiv\boldsymbol{0}$, yields the relation between $\boldsymbol{h}$ and $\boldsymbol{H}$

$$h_{(i,y_j)}^{kl}(\boldsymbol{y}) = H_{(i,y_j)}^{kl}(\boldsymbol{y}) + \frac{\Delta\rho(\boldsymbol{y})}{\rho^c} I_{ijkl} \tag{4.209}$$

The tensor function $\boldsymbol{k}$ follows from

$$k_i^{mn} = \frac{\partial \Delta f_i^{\mu}}{\partial \Delta \varepsilon_{mn}^{c}} = \frac{\partial \Delta f_i^{\mu}}{\partial \Delta \mu_{kl}^{f}} \frac{\partial \Delta \mu_{kl}^{f}}{\partial \Delta \varepsilon_{mn}^{c}} \tag{4.210}$$

Note that the derivative $\partial \Delta \mu_{kl}^{f} / \partial \Delta \varepsilon_{mn}^{c}$ can be computed from the constitutive relation as discussed below.

The discretized influence function $\boldsymbol{k}$ in (4.210) can be specified from the incompatible eigenstrain discretization (4.20) and expression (4.206)

$$\begin{aligned} &k_t^{mn}(\boldsymbol{x},\boldsymbol{y},t) = \sum_{\alpha=1}^{\tilde{M}} \tilde{h}_i^{kl(\alpha)}(\boldsymbol{y}) \frac{\partial \Delta \mu_{kl}^{(\alpha)}(\boldsymbol{x},t)}{\partial \Delta \varepsilon_{mn}^{c}(\boldsymbol{x},t)} \\ &\tilde{h}_i^{kl(\alpha)}(\boldsymbol{y}) = \int_{\Theta} \tilde{h}_i^{kl}(\boldsymbol{y},\tilde{\boldsymbol{y}}) \tilde{N}^{(\alpha)}(\tilde{\boldsymbol{y}})\,d\tilde{\Theta} \end{aligned} \tag{4.211}$$

The dispersion coefficient (4.194) can be decomposed into linear and nonlinear parts, $\boldsymbol{D}=\boldsymbol{D}^{\mathrm{lin}}+\boldsymbol{D}^{\mathrm{nonlin}}$, defined as

$$(D_{ijmn})^{\mathrm{lin}} = \rho^c \frac{1}{|\Theta|} \int_{\Theta} h_s^{ij} h_s^{mn}\,d\Theta; \; (D_{ijmn})^{\mathrm{nonlin}} = \rho^c \frac{1}{|\Theta|} \int_{\Theta} \left[k_s^{ij} k_s^{mn} + h_s^{ij} k_s^{mn} + k_s^{ij} h_s^{mn} \right] d\Theta \tag{4.212}$$

For implementation, it is convenient to express the nonlinear dispersion coefficient as

$$(D_{ijmn})^{\text{nonlin}} = \sum_{\alpha=1}^{\tilde{M}}\sum_{\beta=1}^{\tilde{M}} \frac{\partial \Delta\mu_{st}^{(\alpha)}}{\partial \Delta\varepsilon_{ij}^{c}} R_{stpq}^{(\alpha\beta)} \frac{\partial \Delta\mu_{pq}^{(\beta)}}{\partial \Delta\varepsilon_{mn}^{c}} + \sum_{\alpha=1}^{\tilde{M}} Q_{ijpq}^{(\alpha)} \frac{\partial \Delta\mu_{pq}^{(\alpha)}}{\partial \Delta\varepsilon_{mn}^{c}} + \sum_{\alpha=1}^{\tilde{M}} Q_{mnpq}^{(\alpha)} \frac{\partial \Delta\mu_{pq}^{(\alpha)}}{\partial \Delta\varepsilon_{ij}^{c}} \tag{4.213}$$

$$R_{stpq}^{(\alpha\beta)} = \rho^{c}\frac{1}{|\Theta|}\int_{\Theta} \tilde{h}_{r}^{st(\alpha)}\tilde{h}_{r}^{pq(\beta)} d\Theta; \quad Q_{ijpq}^{(\alpha)} = \rho^{c}\frac{1}{|\Theta|}\int_{\Theta} h_{r}^{ij}\tilde{h}_{r}^{pq(\alpha)} d\Theta$$

Note that tensor coefficients $\boldsymbol{D}^{\text{lin}}, \boldsymbol{R}^{(\alpha\beta)}, \boldsymbol{Q}^{(\alpha)}$ are computed prior to the coarse-scale analysis.

Finally, the nonlinear dispersion coefficient in (4.213) is computed from

$$\frac{\partial \Delta\mu_{kl}^{(\gamma)}}{\partial \Delta\varepsilon_{mn}^{c}} = \frac{\partial \Delta\mu_{kl}^{(\gamma)}}{\partial \Delta\varepsilon_{ij}^{(\gamma)}}\frac{\partial \Delta\varepsilon_{ij}^{(\gamma)}}{\partial \Delta\varepsilon_{mn}^{c}} \tag{4.214}$$

where the first term $\partial\Delta\mu_{kl}^{(\gamma)}/\partial\Delta\varepsilon_{ij}^{(\gamma)}$ is obtained from the consistent tangent stiffness [section 4.2.5, (4.76)], and the second term $\partial\Delta\varepsilon_{ij}^{(\gamma)}/\partial\Delta\varepsilon_{mn}^{c}$ follows from the solution of the residual-free unit cell problem (4.81).

4.5.3 Linear Model Problem

Consider a 1D unit cell model problem consisting of two elastic phases, $(\rho^{(1)}, E^{(1)})$ and $(\rho^{(2)}, E^{(2)})$, where $\phi^{(1)}$ and $\phi^{(2)}$ are volume fractions such that $\phi^{(1)}+\phi^{(2)}=1$. Phase 2 is placed in the center of the unit cell.

The quasi-dynamic unit cell problem (4.202b) for the linear model problem is given by

$$\left[E(y)\left(\frac{\rho(y)}{\rho^{c}} + h_{,y}(y)\right)\right]_{,y} = 0 \tag{4.215}$$

A closed-form solution of (4.215) is

$$h(y) = \begin{cases} -\phi^{(1)}\lambda(y+l) & -l \le y < -\phi^{(1)}l \\ \phi^{(2)}\lambda y & -\phi^{(1)}l \le y \le \phi^{(1)}l; \\ -\phi^{(1)}\lambda(y-l) & \phi^{(1)}l < y \le l \end{cases} \qquad l = \frac{l_x}{2\zeta} \tag{4.216}$$

where λ is the *normalized acoustic impedance variation* given by

$$\lambda = \frac{E^{(2)}\rho^{(2)} - E^{(1)}\rho^{(1)}}{\rho^{c}\left(\phi^{(1)}E^{(2)} + \phi^{(2)}E^{(1)}\right)} \tag{4.217}$$

The coarse-scale dispersion coefficient (4.194) for the model problem is given by

$$D^{c} = \zeta^{2} D = \frac{1}{12}\rho^{c}\left(\phi^{(1)}\phi^{(2)}\right)^{2}\lambda^{2}(l_x)^{2} \tag{4.218}$$

It can be seen that there are two major factors affecting the dispersion coefficient: (i) the square of the unit cell size l_x in the physical domain; and (ii) the square of the normalized impedance variation λ. Incidentally, the dispersion coefficient in (4.218) is identical to the one obtained in [77,83,84,85,86,87].

The resulting strong form of the coarse-scale problem is given by

$$E^c \frac{d^2u^c}{dx^2} + D^c \frac{d^2\ddot{u}^c}{dx^2} = \rho^c \ddot{u}^c \tag{4.219}$$

where $E^c = (\phi^{(1)}/E^{(1)} + \phi^{(2)}/E^{(2)})^{-1}$.

To quantify the dispersion D^c, consider harmonic wave $u^c = u_0 e^{ik(x-ct)}$ and insert it into (4.219), which yields

$$-E^c + D^c k^2 c^2 = -\rho^c c^2 \tag{4.220}$$

from which the phase velocity speed c and angular frequency $\omega = kc$ can be determined

$$c = \sqrt{\frac{E^c}{\rho^c + k^2 D^c}}; \quad \omega = \sqrt{\frac{k^2 E^c}{\rho^c + k^2 D^c}} \tag{4.221}$$

Note that due to the appearance of the dispersive coefficient, the wave speed depends on wave number k. It can be seen that dispersion effectively increases material density for high wave numbers and thus slows down propagation of high frequency waves.

4.5.4 Nonlinear Model Problem

Consider the following nonlinear 1D model problem where the two phases obey exponential law [90]

$$\sigma^{(\beta)} = e^{K^{(\beta)}\varepsilon^{(\beta)}} - 1 \tag{4.222}$$

where $K^{(1)}$ and $K^{(2)}$ are material parameters of the two phases. The volume fractions of the two phases are equal, and the unit cell length is $l_x = 2l\zeta$. Generalization of the exponential law to multidimensions and corresponding results are given in [92].

In the following, we derive a closed-form dispersion coefficient (4.212) for the model problem by using direct homogenization and model reduction. Since for the 1D model problem stress is uniform within each phase, the model reduction and direct homogenization approaches are expected to provide identical results.

The instantaneous constitutive tensor (4.198) follows from (4.222)

$$\hat{L}^{(\beta)} = K^{(\beta)}\left(\sigma^{(\beta)} + 1\right) \tag{4.223}$$

Similar to the linear model (see section 4.5.3), solving the nonlinear quasi-dynamic unit cell (4.202b) gives

$$\hat{h}(y,t) = \begin{cases} -\hat{\lambda}(y+l)/2 & -l \le y < -l/2 \\ \hat{\lambda} y/2 & -l/2 \le y \le l/2 \\ -\hat{\lambda}(y-l)/2 & l/2 < y \le l \end{cases} \tag{4.224}$$

where

$$\hat{\lambda}=2\frac{\hat{L}^{(2)}\rho^{(2)}-\hat{L}^{(1)}\rho^{(1)}}{\rho^{c}(\hat{L}^{(1)}+\hat{L}^{(2)})} \tag{4.225}$$

Exploiting the traction continuity condition $\sigma^{(1)}=\sigma^{(2)}$ yields

$$\hat{\lambda}=2\frac{K^{(2)}\rho^{(2)}-K^{(1)}\rho^{(1)}}{\rho^{c}(K^{(1)}+K^{(2)})} \tag{4.226}$$

The resulting dispersion coefficient is given by

$$D^{c}=\zeta^{2}D=\frac{\rho^{c}}{12}\left(\frac{\hat{\lambda}}{4}\right)^{2}(l_x)^{2}=\frac{(l_x)^{2}}{48\rho^{c}}\left(\frac{K^{(2)}\rho^{(2)}-K^{(1)}\rho^{(1)}}{K^{(1)}+K^{(2)}}\right)^{2} \tag{4.227}$$

We now consider the model reduction approach. The initial elastic module is $L^{(\beta)}=K^{(\beta)}$. The 1D counterpart of the influence functions problems is given by

$$\begin{aligned}&\{L(y)[1+H_{,y}(y)]\}_{,y}=0\\&\{L(y)(\tilde{h}_{,y}^{(\alpha)}(y)-1^{(\alpha)}(y))\}_{,y}=0\end{aligned} \tag{4.228}$$

Solution of (4.228) is given by

$$H=\begin{cases}-a(y+l) & -l\le y<-l/2\\ ay & -l/2\le y\le l/2\\ -a(y-l) & l/2<y\le l\end{cases}\;;\quad a=\frac{K^{(2)}-K^{(1)}}{K^{(1)}+K^{(2)}} \tag{4.229a}$$

$$\tilde{h}^{(\beta)}=\begin{cases}-(-1)^{\beta+1}b^{(\beta)}(y+l) & -l\le y<-l/2\\ (-1)^{\beta+1}b^{(\beta)}y & -l/2\le y\le l/2\\ -(-1)^{\beta+1}b^{(\beta)}(y-l) & l/2<y\le l\end{cases}\;;\quad b^{(\beta)}=\frac{K^{(\beta)}}{K^{(1)}+K^{(2)}} \tag{4.229b}$$

and

$$\begin{aligned}&P^{(11)}=b^{(1)};\quad P^{(21)}=-b^{(1)};\quad P^{(22)}=b^{(2)};\quad P^{(12)}=-b^{(2)}\\&E^{(1)}=a+1=2b^{(2)};\quad E^{(2)}=1-a=2b^{(1)}\end{aligned} \tag{4.230}$$

The linear influence function $h(y)$ was computed in (4.216) and (4.217) and is given by

$$h(y)=\begin{cases}-\lambda(y+l)/2 & -l\le y<-l/2\\ \lambda y/2 & -l/2\le y\le l/2\\ -\lambda(y-l)/2 & l/2<y\le l\end{cases}\;;\quad \lambda=2\frac{K^{(2)}\rho^{(2)}-K^{(1)}\rho^{(1)}}{\rho^{c}(K^{(1)}+K^{(2)})} \tag{4.231}$$

To compute the nonlinear influence function $k(y)$, consider the eigenstrain field defined as

$$\begin{aligned} &\mu^{(\beta)} = \varepsilon^{(\beta)} - \frac{1}{K^{(\beta)}}\left[e^{K^{(\beta)}\varepsilon^{(\beta)}} - 1\right] \\ &\frac{\partial \Delta\mu^{(\beta)}}{\partial \Delta\varepsilon^{(\beta)}} \approx \frac{\partial \mu^{(\beta)}}{\partial \varepsilon^{(\beta)}} = -\sigma^{(\beta)} \end{aligned} \tag{4.232}$$

The resulting residual-free problem is given by

$$\begin{aligned} &\{1 + b^{(1)}\sigma^{(1)}\}\frac{\partial \Delta\varepsilon^{(1)}}{\partial \Delta\varepsilon^{c}} - b^{(2)}\sigma^{(2)}\frac{\partial \Delta\varepsilon^{(2)}}{\partial \Delta\varepsilon^{c}} = 2b^{(2)} \\ &-b^{(1)}\sigma^{(1)}\frac{\partial \Delta\varepsilon^{(1)}}{\partial \Delta\varepsilon^{c}} + \{1 + b^{(2)}\sigma^{(2)}\}\frac{\partial \Delta\varepsilon^{(2)}}{\partial \Delta\varepsilon^{c}} = 2b^{(1)} \end{aligned} \tag{4.233}$$

Solving (4.233) for $b^{(1)}\sigma^{(1)} + b^{(2)}\sigma^{(2)} + 1 \neq 0$ gives

$$\frac{\partial \Delta\varepsilon^{(1)}}{\partial \Delta\varepsilon^{c}} = 2\frac{b^{(2)}(\sigma^{(2)}+1)}{b^{(1)}\sigma^{(1)} + b^{(2)}\sigma^{(2)} + 1}; \qquad \frac{\partial \Delta\varepsilon^{(2)}}{\partial \Delta\varepsilon^{c}} = 2\frac{b^{(1)}(\sigma^{(1)}+1)}{b^{(1)}\sigma^{(1)} + b^{(2)}\sigma^{(2)} + 1} \tag{4.234}$$

The 1D counterpart of equation (4.214) is

$$\frac{\partial \Delta\mu^{(\beta)}}{\partial \Delta\varepsilon^{c}} = -\sigma^{(\beta)}\frac{\partial \Delta\varepsilon^{(\beta)}}{\partial \Delta\varepsilon^{c}} \tag{4.235}$$

From (4.229b) and (4.234), the influence function $k(y)$ (4.211) is given by

$$k = \sum_{\alpha=1}^{\tilde{M}} \tilde{h}^{(\alpha)}\frac{\partial \Delta\mu^{(\alpha)}}{\partial \Delta\varepsilon^{c}}; \quad k(y) = \begin{cases} -C(y+l) & -l \le y < -l/2 \\ Cy & -l/2 \le y \le l/2 \\ -C(y-l) & l/2 < y \le l \end{cases} \tag{4.236a}$$

$$C = b^{(1)}\frac{\partial \Delta\mu^{(1)}}{\partial \Delta\varepsilon^{c}} - b^{(2)}\frac{\partial \Delta\mu^{(2)}}{\partial \Delta\varepsilon^{c}} \tag{4.236b}$$

Combining (4.234), (4.235), and (4.236b) gives

$$C = \frac{2b^{(1)}b^{(2)}\left(\sigma^{(2)} - \sigma^{(1)}\right)}{b^{(1)}\sigma^{(1)} + b^{(2)}\sigma^{(2)} + 1} \tag{4.237}$$

Applying the traction continuity condition $\sigma^{(1)}=\sigma^{(2)}$ yields $C=0$. Consequently, $k=0$, which yields $\widehat{h}=h$. Thus, the dispersion coefficient (4.212) obtained by model reduction coincides with (4.227).

4.5.5 Implicit and Explicit Formulations

Consider finite element discretization of the coarse-scale displacements, test functions, and strains

$$\boldsymbol{u}^c = \boldsymbol{N}\boldsymbol{d}; \quad \boldsymbol{w}^c = \boldsymbol{N}\boldsymbol{c}; \quad \boldsymbol{\varepsilon}^c = \boldsymbol{B}\boldsymbol{d} \tag{4.238}$$

where $\boldsymbol{N}$, $\boldsymbol{B}$, $\boldsymbol{d}$, and $\boldsymbol{c}$ are the shape functions, their symmetric gradients, the nodal displacements, and the nodal test functions, respectively.

Discretizing the weak form of the coarse-scale problem (4.193) using finite element discretization (4.238) yields

$$\boldsymbol{f}^{\text{int}} + \boldsymbol{m}^D\ddot{\boldsymbol{d}} + \boldsymbol{M}\ddot{\boldsymbol{d}} = \boldsymbol{f}^{\text{ext}} \tag{4.239}$$

where internal force $\boldsymbol{f}^{\text{int}}$, external force $\boldsymbol{f}^{\text{ext}}$, and consistent mass matrix $\boldsymbol{M}$ are given as

$$\boldsymbol{f}^{\text{int}} = \int_\Omega \boldsymbol{B}^T\boldsymbol{\sigma}^c d\Omega; \quad \boldsymbol{f}^{\text{ext}} = \int_{\partial\Omega^t} \boldsymbol{N}^T\bar{\boldsymbol{t}}^c d\Gamma; \quad \boldsymbol{M} = \rho^c\int_\Omega \boldsymbol{N}^T\boldsymbol{N} d\Omega \tag{4.240}$$

and the dispersion matrix is

$$\boldsymbol{m}^D = \int_\Omega \boldsymbol{B}^T\boldsymbol{D}\boldsymbol{B} d\Omega \tag{4.241}$$

It is instructive to point out that the lumping of dispersion matrix $\boldsymbol{m}^D$ yields a zero matrix and therefore the discrete system of equations

$$\begin{aligned} \bar{\boldsymbol{M}}\ddot{\boldsymbol{d}} + \boldsymbol{f}^{\text{int}} &= \boldsymbol{f}^{\text{ext}} \\ \bar{\boldsymbol{M}} &\equiv \boldsymbol{M} + \boldsymbol{m}^D \end{aligned} \tag{4.242}$$

can be employed in the context of the implicit integration method only.

For explicit methods, the effect of dispersion has to be accounted for in the stress field [see (4.196)]

$$\bar{\boldsymbol{f}}^{\text{int}} = \int_\Omega \boldsymbol{B}^T\bar{\boldsymbol{\sigma}}^c d\Omega \tag{4.243a}$$

$$\bar{\boldsymbol{\sigma}}^c = \boldsymbol{\sigma}^c + \boldsymbol{D}\boldsymbol{B}\ddot{\boldsymbol{d}} \tag{4.243b}$$

The resulting acceleration in the explicit method follows from

$$\ddot{d} = \tilde{\tilde{M}}^{-1}(f^{\text{ext}} - \bar{f}^{\text{int}}) \tag{4.244}$$

where $\tilde{\tilde{M}}$ is a lumped (or diagonal) matrix. Note that the acceleration $\ddot{d}$ appearing in (4.243b) is taken from the previous increment.

4.6 Extension to Multiple Spatial Scales

In this section, we outline the extension of the ROH to nonlinear problems where material heterogeneity exists at more than one scale. The governing equations for multiple-scale homogenization of a linear elastic heterogeneous solid were introduced in Chapter 2, section 2.6.

In this section, the left superscript in parentheses denotes the scale. The position vector ${}^{(I)}\boldsymbol{x}$ at scale I is related to the position vector at scale I–1 as

$${}^{(I)}\boldsymbol{x} \equiv {}^{(I-1)}\boldsymbol{x}/\zeta \quad \text{for } I = 1,\ldots,n_{sc}-1 \tag{4.245}$$

where n_{sc} is the number of scales. As a prelude, we summarize the multiple-scale constitutive equations

$$\sigma_{ij}^{f}\left({}^{(0)}\boldsymbol{x},\ldots,{}^{(n_{sc}-1)}\boldsymbol{x}\right) = L_{ijkl}\left({}^{(n_{sc}-1)}\boldsymbol{x}\right)\left[\varepsilon_{kl}^{f}\left({}^{(0)}\boldsymbol{x},\ldots,{}^{(n_{sc}-1)}\boldsymbol{x}\right) - \mu_{kl}^{f}\left({}^{(0)}\boldsymbol{x},\ldots,{}^{(n_{sc}-1)}\boldsymbol{x}\right)\right] \tag{4.246a}$$

$${}^{(I-1)}\sigma_{ij}\left({}^{(0)}\boldsymbol{x},\ldots,{}^{(I-1)}\boldsymbol{x}\right) \equiv \frac{1}{\left|{}^{(I)}\Theta\right|}\int_{I_\Theta} {}^{(I)}\sigma_{ij}\left({}^{(0)}\boldsymbol{x},\ldots,{}^{(I)}\boldsymbol{x}\right)d\Theta \quad \text{for } I = 1,\ldots,n_{sc}-1 \tag{4.246b}$$

$${}^{(n_{sc}-1)}\sigma_{ij}\left({}^{(0)}\boldsymbol{x},\ldots,{}^{(n_{sc}-1)}\boldsymbol{x}\right) \equiv \sigma_{ij}^{f}\left({}^{(0)}\boldsymbol{x},\ldots,{}^{(n_{sc}-1)}\boldsymbol{x}\right) \tag{4.246c}$$

$${}^{(n_{sc}-1)}\varepsilon_{ij}\left({}^{(0)}\boldsymbol{x},\ldots,{}^{(n_{sc}-1)}\boldsymbol{x}\right) \equiv \varepsilon_{ij}^{f}\left({}^{(0)}\boldsymbol{x},\ldots,{}^{(n_{sc}-1)}\boldsymbol{x}\right) \tag{4.246d}$$

$${}^{(n_{sc}-1)}\mu_{ij}\left({}^{(0)}\boldsymbol{x},\ldots,{}^{(n_{sc}-1)}\boldsymbol{x}\right) \equiv \mu_{ij}^{f}\left({}^{(0)}\boldsymbol{x},\ldots,{}^{(n_{sc}-1)}\boldsymbol{x}\right) \tag{4.246e}$$

where the left superscript in parentheses, $I = n_{sc} - 1$, denotes the finest scale of interest. Equation (4.246a) denotes the constitutive equation at the finest scale of interest; the constitutive equation at the coarser scales is obtained by reduced order homogenization, as described in the next section.

The multiple-scale equilibrium equations (see Chapter 2) are

$$\begin{aligned} &O\left(\zeta^{-I}\right): {}^{(I)}\sigma_{ij,{}^{(I)}x_j}\left({}^{(0)}\boldsymbol{x},\ldots,{}^{(I)}\boldsymbol{x}\right)=0 \quad \text{for } I=1,\ldots,n_{sc}-1 \\ &O\left(\zeta^{0}\right): {}^{(0)}\sigma_{ij,{}^{(0)}x_j}\left({}^{(0)}\boldsymbol{x}\right)+{}^{(0)}b_i=0 \end{aligned} \tag{4.247}$$

where ${}^{(0)}b_i$ is the average body force at the coarsest scale. As for the two-scale nonlinear problems, the leading order displacement is assumed to be a function of the coarse-scale coordinate. In (4.247) it is also assumed that the body force is of order $O(1)$. If this is not the case, then oscillatory body force terms, such as ${}^{(-1)}b_i=O\left(\zeta^{-1}\right)$, will appear in one or more fine-scale equations (4.247a).

4.6.1 Residual-Free Governing Equations at Multiple Scales

In this section, we will formulate equilibrium equations in terms of eigenstrain and eigenseparation modes that a priori satisfy equilibrium equations at multiple scales, except for the coarsest scale. We start by expressing the equilibrium equations (4.247) at scale $I>0$ in terms of constitutive relation (4.246), which gives

$$\left\{{}^{(I)}L_{ijkl}\left({}^{(0)}\boldsymbol{x},\ldots,{}^{(I)}\boldsymbol{x}\right)\left[{}^{(I-1)}\varepsilon_{kl}\left({}^{(0)}\boldsymbol{x},\ldots,{}^{(I-1)}\boldsymbol{x}\right)+u^{(I)}_{(k,{}^{(I)}x_l)}\left({}^{(0)}\boldsymbol{x},\ldots,{}^{(I)}\boldsymbol{x}\right)-{}^{(I)}\mu_{kl}\left({}^{(0)}\boldsymbol{x},\ldots,{}^{(I)}\boldsymbol{x}\right)\right]\right\}_{,{}^{(I)}x_j}=0 \tag{4.248}$$

At the finest scale ($I=n_{sc}-1$), the eigenstrain ${}^{(n_{sc}-1)}\mu_{kl}\equiv\mu^f_{kl}$ is defined by the constitutive relation; at all other scales, the eigenstrain is upscaled from the finer scale quantities as described below.

Extending the two-scale formulation (see section 4.2) to multiple scales, the residual-free displacement field is constructed using an additive decomposition of the elastic influence function ${}^{(I)}H^{kl}_i$, the eigenstrain influence function ${}^{(I)}\tilde{h}^{kl}_i$, and the eigenseparation influence function ${}^{(I)}\breve{h}^{\breve{n}}_i$ as

$$\begin{aligned} u_i^{(I)}\left({}^{(0)}\boldsymbol{x},\ldots,{}^{(I)}\boldsymbol{x}\right) &= {}^{(I)}H^{kl}_i\left({}^{(0)}\boldsymbol{x},\ldots,{}^{(I)}\boldsymbol{x}\right){}^{(I-1)}\varepsilon_{kl}\left({}^{(0)}\boldsymbol{x},\ldots,{}^{(I-1)}\boldsymbol{x}\right) \\ &+\int_{{}^{(I)}\Theta}{}^{(I)}\tilde{h}^{kl}_i\left({}^{(0)}\boldsymbol{x},\ldots,{}^{(I)}\boldsymbol{x},{}^{(I)}\tilde{\boldsymbol{x}}\right){}^{(I)}\mu_{kl}\left({}^{(0)}\boldsymbol{x},\ldots,{}^{(I)}\tilde{\boldsymbol{x}}\right)d\tilde{\Theta} \\ &+\int_{{}^{(I)}S}{}^{(I)}\breve{h}^{\breve{n}}_i\left({}^{(0)}\boldsymbol{x},\ldots,{}^{(I)}\boldsymbol{x},{}^{(I)}\breve{\boldsymbol{x}}\right){}^{(I)}\delta_{\breve{n}}\left({}^{(0)}\boldsymbol{x},\ldots,{}^{(I)}\breve{\boldsymbol{x}}\right)d\breve{S} \end{aligned} \tag{4.249}$$

where ${}^{(I)}H^{kl}_i$, ${}^{(I)}\tilde{h}^{kl}_i$, and ${}^{(I)}\breve{h}^{\breve{n}}_i$ are computed by solving a sequence of elastic boundary value problems independent of and prior to solving the nonlinear coarse-scale problem. These influence functions, which are similar to those in the two-scale problems considered in section 4.2, are chosen to satisfy the equilibrium equations at multiple scales (4.248).

Similar to eigenstrains, the eigenseparations at the finest scale of interest are denoted as

$$ {}^{(n_{sc}-1)}\delta_{\breve{n}}\left({}^{(0)}\boldsymbol{x},\ldots,{}^{(n_{sc}-1)}\boldsymbol{x}\right) \equiv \delta_{\breve{n}}^{f}\left({}^{(0)}\boldsymbol{x},\ldots,{}^{(n_{sc}-1)}\boldsymbol{x}\right) \tag{4.250} $$

The residual-free stress at scale I follows from (4.249) and is given by

$$ {}^{(I)}\sigma_{ij} = {}^{(I)}L_{ijmn}\left({}^{(0)}\boldsymbol{x},\ldots,{}^{(I)}\boldsymbol{x}\right)\left[\begin{array}{l} \left(I_{mnkl} + {}^{(I)}H^{kl}_{\left(m,{}^{(I)}x_n\right)}\left({}^{(0)}\boldsymbol{x},\ldots,{}^{(I)}\boldsymbol{x}\right)\right)^{(I-1)} \varepsilon_{kl}\left({}^{(0)}\boldsymbol{x},\ldots,{}^{(I-1)}\boldsymbol{x}\right) \\ +\int_{{}^{(I)}S} {}^{(I)}\breve{h}^{\breve{n}}_{\left(m,{}^{(I)}x_n\right)}\left({}^{(0)}\boldsymbol{x},\ldots,{}^{(I)}\boldsymbol{x},{}^{(I)}\breve{\boldsymbol{x}}\right)^{(I)} \delta_{\breve{n}}\left({}^{(0)}\boldsymbol{x},\ldots,{}^{(I)}\breve{\boldsymbol{x}}\right) d\breve{S} \\ +\int_{{}^{(I)}\Theta} {}^{(I)}\tilde{h}^{kl}_{\left(m,{}^{(I)}x_n\right)}\left({}^{(0)}\boldsymbol{x},\ldots,{}^{(I)}\boldsymbol{x},{}^{(I)}\tilde{\boldsymbol{x}}\right){}^{(I)}\mu_{kl}\left({}^{(0)}\boldsymbol{x},\ldots,{}^{(I)}\tilde{\boldsymbol{x}}\right) d\tilde{\Theta} \\ -{}^{(I)}\mu_{mn}\left({}^{(0)}\boldsymbol{x},\ldots,{}^{(I)}\boldsymbol{x}\right) \end{array}\right] \tag{4.251} $$

Averaging (4.251) and exploiting the constitutive equation at the I–1 scale, the eigenstrain at scale I–1 ($I \geq 2$) can be expressed in terms of the eigenstrains and eigenseparations at scale I

$$ \begin{aligned} &{}^{(I-1)}\mu_{ij}\left({}^{(0)}\boldsymbol{x},\ldots,{}^{(I-1)}\boldsymbol{x}\right) \\ &= -{}^{(I-1)}M_{ijst}\left({}^{(0)}\boldsymbol{x},\ldots,{}^{(I-1)}\boldsymbol{x}\right)\cdot\frac{1}{\left|{}^{(I)}\Theta\right|} \\ &\int_{{}^{(I)}\Theta}\left\{{}^{(I)}L_{stmn}\left({}^{(0)}\boldsymbol{x},\ldots,{}^{(1)}\boldsymbol{x}\right)\left[\begin{array}{l} \int_{{}^{(I)}S} {}^{(I)}\breve{h}^{\breve{n}}_{\left(m,{}^{(I)}x_n\right)}\left({}^{(0)}\boldsymbol{x},\ldots,{}^{(I)}\boldsymbol{x},{}^{(I)}\breve{\boldsymbol{x}}\right){}^{(I)}\delta_{\breve{n}}\left({}^{(0)}\boldsymbol{x},\ldots,{}^{(I)}\breve{\boldsymbol{x}}\right) d\breve{S} \\ +\int_{{}^{(I)}\Theta} {}^{(I)}\tilde{h}^{kl}_{\left(m,{}^{(I)}x_n\right)}\left({}^{(0)}\boldsymbol{x},\ldots,{}^{(I)}\boldsymbol{x},{}^{(I)}\,{}^{I}\tilde{\boldsymbol{x}}\right){}^{(I)}\mu_{kl}\left({}^{(0)}\boldsymbol{x},\ldots,{}^{I}\tilde{\boldsymbol{x}}\right) d\tilde{\Theta} \\ -{}^{(I)}\mu_{mn}\left({}^{(0)}\boldsymbol{x},\ldots,{}^{(I)}\boldsymbol{x}\right) \end{array}\right]\right\} d\Theta \end{aligned} \tag{4.252} $$

4.6.2 Multiple-Scale Reduced Order Model

As in the case of two scales, model reduction is accomplished through discretization of eigenstrains and eigenseparations at multiple scales and formulating the resulting residual-free reduced order unit cell equations.

Consider an incompatible formulation for eigenstrains. The eigenstrains at each scale are discretized in terms of the piecewise constant function $\tilde{N}^{(\alpha_I)}({}^{(I)}\boldsymbol{x})$ as

$$ {}^{(I)}\mu_{ij}\left({}^{(0)}\boldsymbol{x},\ldots,{}^{(I)}\boldsymbol{x}\right) = \sum_{\alpha_I=1}^{\tilde{M}_I} \tilde{N}^{(\alpha_I)}\left({}^{(I)}\boldsymbol{x}\right)\mu_{ij}^{(\alpha_I)}\left({}^{(0)}\boldsymbol{x},\ldots,{}^{(I-1)}\boldsymbol{x}\right) \tag{4.253} $$

where

$$\tilde{N}^{(\alpha_I)}\left({}^{(I)}\boldsymbol{x}\right)=\begin{cases}1 & {}^{(I)}\boldsymbol{x}\in\Theta^{(\alpha_I)}\\ 0 & {}^{(I)}\boldsymbol{x}\notin\Theta^{(\alpha_I)}\end{cases} \tag{4.254}$$

in which the total volume of the unit cell at scale I is partitioned into $\tilde{M}_I$ nonoverlapping subdomains denoted by $\Theta^{(\alpha_I)}$. In the case of one partition per phase, the C^{-1} continuous approximation of eigenstrains has to be replaced by the hybrid impotent-incompatible approximation described in section 4.3.

The eigenseparation ${}^{(I)}\delta_{\breve{n}}({}^{(0)}\boldsymbol{x},\ldots,{}^{(I)}\boldsymbol{x})$ at scale I is discretized in terms of the C^0 continuous shape function $\breve{N}^{(\xi_I)}({}^{(I)}\boldsymbol{x})$ as

$${}^{(I)}\delta_{\breve{n}}\left({}^{(0)}\boldsymbol{x},\ldots,{}^{(I)}\boldsymbol{x}\right)=\sum_{\xi_I=1}^{\breve{M}_I}\breve{N}^{(\xi_I)}\left({}^{(I)}\boldsymbol{x}\right)\delta_{\breve{n}}^{(\xi_I)}\left({}^{(0)}\boldsymbol{x},\ldots,{}^{(I-1)}\boldsymbol{x}\right) \tag{4.255}$$

where $\breve{N}^{(\xi_I)}({}^{(I)}\boldsymbol{x})$ is defined by a linear combination of the piecewise linear finite element shape functions over partition ξ_I

$$\breve{N}^{(\xi_I)}\left({}^{(I)}\boldsymbol{x}\right)=\begin{cases}\sum\limits_{A\in{}^{(I)}S^{(\xi_I)}} N_A^f\left({}^{(I)}\boldsymbol{x}\right) & {}^{(I)}\boldsymbol{x}\in S^{(\xi_I)}\\ 0 & {}^{(I)}\boldsymbol{x}\notin S^{(\xi_I)}\end{cases} \tag{4.256}$$

in which the total interface at scale I is divided into $\breve{M}_I$ partitions denoted by $S^{(\xi_I)}$, and $N_A^f({}^{(I)}\boldsymbol{x})$ is a linear shape function associated with finite element mesh node A along the interface of scale I.

Partitions at various scales are denoted by a right superscript enclosed in parentheses. For instance, $(\alpha_I),(\beta_I)$ denote phase (volume) partitions at scale I, whereas $(\xi_I),(\eta_I)$ are reserved for the interface (surface) partitions at scale I.

It is instructive to note that the reduced order eigenstrain at scale I-1 (for $I\geq 2$) is a function of eigenstrains and eigenseparations at a finer scale. It can be obtained by discretizing (4.252) and then averaging over the phase partition as $\frac{1}{\left|\Theta^{(\alpha_{I-1})}\right|}\int_{\Theta^{(\alpha_{I-1})}}\cdot d\Theta$. This yields the residual unit cell equations denoted by $\boldsymbol{r}^{(\alpha_{I-1})}\left(\varepsilon_{ij}^{(\alpha_I)},\delta_{\breve{n}}^{(\xi_I)}\right)=\boldsymbol{0}$ at multiple scales. For details, we refer the reader to [25,26]. In the following flowchart, we will briefly outline the hierarchical [26] solution strategy depicted in Figure 4.29. A similar strategy has been employed for the multiscale modeling of bones [94].

At scale I=1 (the coarsest scale unit cell):

1. Given $\Delta\varepsilon_{kl}^c$, find $\varepsilon_{ij}^{(\alpha_1)}$ and $\delta_{\breve{n}}^{(\xi_1)}$ by solving the reduced order unit cell problem at scale I=1

$$\boldsymbol{r}^c\left(\varepsilon_{ij}^{(\alpha_1)},\ \delta_{\breve{n}}^{(\xi_1)}\right)=\boldsymbol{0}$$

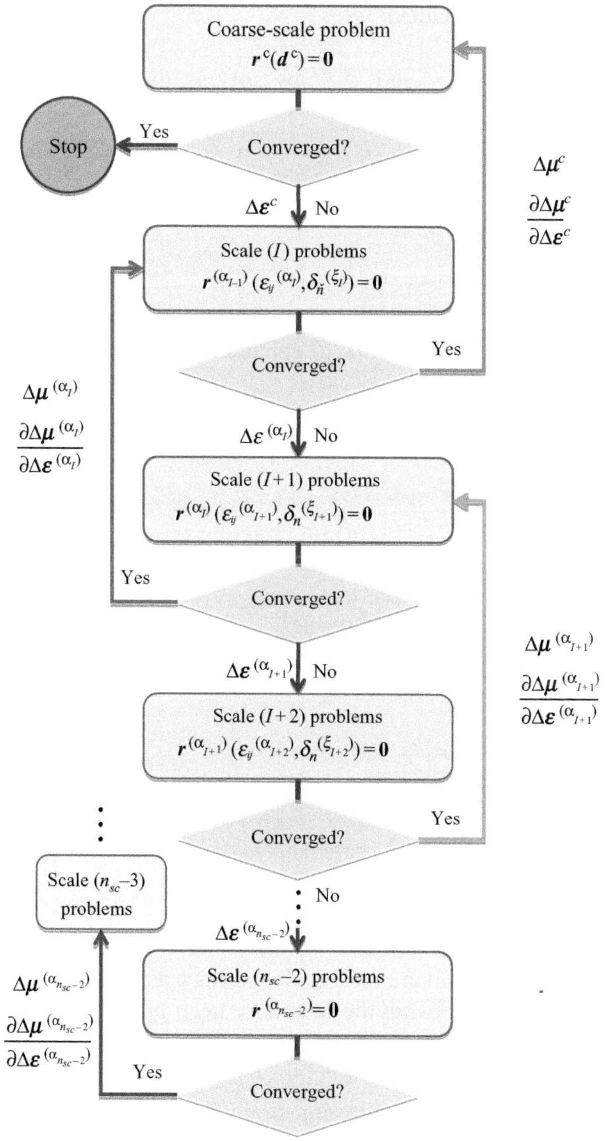

Figure 4.29 Information flow in a multiple-scale reduced order model

with specified cohesive law

$$t_{\breve{n}}^{(\xi_1)} = g\left({}^{(1)}\delta_{\breve{n}}^{(\xi_1)}\right)$$

2. At each iteration, $\mu_{ij}^{(\alpha_1)}$ is determined by solving a unit cell problem at scale I=2 using equation (4.252).
3. Given $\mu_{ij}^{(\alpha_1)}$ and $\delta_{\breve{n}}^{(\xi_1)}$ at each partition α_1, ξ_1, calculate the coarse-scale stress σ_{ij}^c by averaging the stress at scale I=1.

At scale I=2,..., n_{sc}−2 (all mesoscales):

1. Given $\Delta\varepsilon_{kl}^{(\beta_{I-1})} \forall \beta_{I-1}$, find $\varepsilon_{ij}^{(\alpha_I)}$ and $\delta_{\breve{n}}^{(\xi_I)}$ by solving the mesoscale reduced residual equation

$$\boldsymbol{r}^{(\alpha_{I-1})}\left(\varepsilon_{ij}^{(\alpha_I)}, \delta_{\breve{n}}^{(\xi_I)}\right) = \boldsymbol{0}$$

with the specified cohesive law

$$t_{\breve{n}}^{(\xi_I)} = g\left(\delta_{\breve{n}}^{(\xi_I)}\right)$$

2. Given $\mu_{ij}^{(\alpha_I)}$ and $\delta_{\breve{n}}^{(\xi_I)}$ at each partition α_I, ξ_I, calculate $\mu_{ij}^{(\alpha_{I-1})}$ using (4.252).
3. Go to the (I–1) unit cell problem.

At scale I=n_{sc}−1 (the finest scale):

1. Given $\Delta\varepsilon_{kl}^{(\beta_{n_{sc}-2})}$ at each partition $\beta_{n_{sc}-2}$, find $\varepsilon_{ij}^{(\alpha_{n_{sc}-1})}$ and $\delta_{\breve{n}}^{(\xi_{n_{sc}-1})}$ using the fine-scale reduced order residual equation

$$\boldsymbol{r}^{(\alpha_{n_{sc}-2})}\left(\varepsilon_{ij}^{(\alpha_{n_{sc}-1})}, \delta_{\breve{n}}^{(\xi_{n_{sc}-1})}\right) = \boldsymbol{0}$$

with

$$\mu_{ij}^{(\alpha_{n_{sc}-1})} = f\left(\varepsilon_{ij}^{(\alpha_{n_{sc}-1})}, \sigma_{ij}^{(\alpha_{n_{sc}-1})}\right); \quad t_{\breve{n}}^{(\xi_{n_{sc}-1})} = g\left(\delta_{\breve{n}}^{(\xi_{n_{sc}-1})}\right)$$

2. With the updated $\mu_{ij}^{(\alpha_{n_{sc}-1})}$ and $\delta_{\breve{n}}^{(\xi_{n_{sc}-1})}$, calculate $\mu_{ij}^{(\alpha_{n_{sc}-2})}$ using (4.252).
3. Go to the (n_{sc}−2) scale unit cell problem.

Remark 4.2 In the above algorithm, cohesive laws were a priori postulated at all mesoscales. Alternatively, we can consider a mesoscale interface as both a boundary layer and as a way to upscale cohesive laws by using the boundary layer approach considered in Chapter 2 (section 2.2.5) or by using any other upscaling method that makes no periodicity assumption in the direction normal to the interface. Alternatively, we can model the interface partition as a phase partition (or interphase) where partition strain $\varepsilon_{\breve{n}}^{(\xi_I)}$ is expressed in terms of the partition eigenseparation $\delta_{\breve{n}}^{(\xi_I)}$ and the characteristic interphase thickness ${}^{I}l$ at a mesoscale I.

4.7 Extension to Large Deformations

In this section, we focus on extending the ROH to large deformation problems. The general framework of ROH for large deformation problems in a periodic inelastic heterogeneous medium has been given in [95]. In the following, we describe a simplified variant based on the corotational formulation. The corotational framework allows us to (i) take advantage of the precomputed influence functions prior to nonlinear analysis and (ii) incorporate nonlinear geometrical information using conventional corotational finite element code architecture.

Development of a corotational formulation was actively pursued in the early 1980s. Important papers are those by Argyris [96], Belytschko and Hsieh [97], and Bergan and Horrigmoe [98]. In the following, we focus on the specifics of the corotational approach as applied to two-scale ROH.

In the corotational formulation, the motion of a heterogeneous body is decomposed into rigid body motion (rotation and translation) followed by deformational displacements. In the spatially discretized domain, this decomposition can be accomplished by attaching a local corotational coordinate frame either to each finite element or to each unit cell and then assuming that the deformational displacements are small. Here, the latter choice is pursued.

Consider large deformation-governing equations in the current (deformed) configuration

$$\sigma_{ij,j}^{\zeta}(\boldsymbol{x},t)+b_i^{\zeta}(\boldsymbol{x},t)=0 \tag{4.257a}$$

$$\sigma_{ij}^{\Re\zeta}(\boldsymbol{x},t)=\Re_{ki}^{\zeta}(\boldsymbol{x},t)\ \Re_{lj}^{\zeta}(\boldsymbol{x},t)\ \sigma_{kl}^{\zeta}(\boldsymbol{x},t) \tag{4.257b}$$

$$\dot{\sigma}_{ij}^{\Re\zeta}=L_{ijkl}^{\Re\zeta}\left(\boldsymbol{x}^{\Re}/\zeta\right)\left(\dot{\varepsilon}_{kl}^{\Re\zeta}(\boldsymbol{x},t)-\dot{\mu}_{kl}^{\Re\zeta}(\boldsymbol{x},t)\right) \tag{4.257c}$$

$$\dot{\varepsilon}_{kl}^{\Re\zeta}(\boldsymbol{x},t)=\frac{1}{2}\left(\frac{\partial v_k^{\Re\zeta}}{\partial x_l^{\Re}}+\frac{\partial v_l^{\Re\zeta}}{\partial x_k^{\Re}}\right) \tag{4.257d}$$

where the right superscript $\Re$ denotes various fields defined in the corotational coordinates $\boldsymbol{x}^{\Re}$ that are attached to the material microstructure and are related to the global coordinate system $\boldsymbol{x}$ by

$$x_i=\Re_{ij}^{\zeta}x_j^{\Re}+\hat{x}_i \tag{4.258}$$

$\Re_{ij}^{\zeta}$ denotes rigid body rotation of the corotational frame, and $\hat{x}_i$ is a centroid of corotational frame, subsequently to be used as the unit cell centroid; the constitutive tensor $L_{ijkl}^{\Re\zeta}\left(\boldsymbol{x}^{\Re}/\zeta\right)$ is assumed to be periodic in the corotational frame; σ_{ij}^{ζ} is Cauchy stress; $v_i^{\Re\zeta}=\dot{x}_i^{\Re\zeta}$ denotes velocity in the corotational frame; and the time t is used to track the load level. A superimposed dot denotes the material time derivative.

The velocity field in the global Cartesian coordinates is expanded in the asymptotic expansion as

$$v_k^{\zeta}(\boldsymbol{x},t)\equiv v_k(\boldsymbol{x},\boldsymbol{y},t)=v_k^{(0)}(\boldsymbol{x},t)+\zeta\, v_k^{(1)}(\boldsymbol{x},\boldsymbol{y},t)+O(\zeta^2) \tag{4.259}$$

Various terms in the asymptotic expansion of the velocity are expanded around the unit cell centroid $\hat{\boldsymbol{x}}$ in the current configuration where $\boldsymbol{x}-\hat{\boldsymbol{x}}=\zeta\boldsymbol{y}$

$$v_k^{\zeta}(\boldsymbol{x},t)\equiv v_k(\boldsymbol{x},\boldsymbol{y},t)=v_k^{(0)}(\hat{\boldsymbol{x}},t)+\zeta\left(v_k^{(1)}(\hat{\boldsymbol{x}},\boldsymbol{y},t)+v_{k,x_l}^{(0)}(\hat{\boldsymbol{x}},t)\Big|_{\hat{\boldsymbol{x}}}y_l\right)+O(\zeta^2) \tag{4.260}$$

We now define the unit cell corotational coordinate system $y_j^{\Re}$ placed at the unit cell centroid as

$$y_l = \Re_{lj}^{\zeta} y_j^{\Re} \tag{4.261}$$

It is convenient to express the velocity field in the corotational coordinate system as

$$v_i^{\Re\zeta}(\boldsymbol{x},t) \equiv v_i^{\Re}(\boldsymbol{x},\boldsymbol{y},t) = \Re_{ki}^{\zeta} v_k^{(0)}(\hat{\boldsymbol{x}},t) + \zeta\left(\Re_{ki}^{\zeta} v_k^{(1)}(\hat{\boldsymbol{x}},\boldsymbol{y},t) + \Re_{ki}^{\zeta} v_{k,x_l}^{(0)}(\hat{\boldsymbol{x}},t)\Big|_{\hat{\boldsymbol{x}}} y_l\right) + O(\zeta^2) \tag{4.262}$$

We now proceed by approximating $\Re_{ki}^{\zeta}$ as

$$\Re_{ki}^{\zeta}(\boldsymbol{x},t) \equiv \Re_{ki}(\boldsymbol{x},\boldsymbol{y},t) = \Re_{ki}^{c}(\hat{\boldsymbol{x}},t) + O(\zeta^2) \tag{4.263}$$

where $\Re_{ki}^{c}(\boldsymbol{x},t)$ denotes the coarse-scale rotation. Neglecting $O(\zeta^2)$ in (4.263) implies that all points in the unit cell have the same rotation (see Remark 4.3).

Inserting (4.263) and (4.261) into (4.262) yields

$$v_i^{\Re\zeta}(\boldsymbol{x},t) = \Re_{ki}^{c}(\hat{\boldsymbol{x}},t) v_k^{(0)}(\hat{\boldsymbol{x}},t) + \zeta\left(v_i^{\Re(1)}(\hat{\boldsymbol{x}},\boldsymbol{y},t) + \Re_{ki}^{c}(\hat{\boldsymbol{x}},t)\Re_{lj}^{c}(\hat{\boldsymbol{x}},t) v_{k,x_l}^{(0)}(\hat{\boldsymbol{x}},t)\Big|_{\hat{\boldsymbol{x}}} y_j^{\Re}\right) + O(\zeta^2) \tag{4.264}$$

where $v_i^{\Re(1)}(\hat{\boldsymbol{x}},\boldsymbol{y},t) = \Re_{ki}^{c}(\hat{\boldsymbol{x}},t) v_k^{(1)}(\hat{\boldsymbol{x}},\boldsymbol{y},t)$ is the fine-scale correction in the corotational frame. The velocity gradient in the corotational frame follows from (4.264)

$$\frac{\partial v_i^{\Re\zeta}}{\partial x_j^{\Re}} = \frac{\partial v_i^{\Re(1)}(\hat{\boldsymbol{x}},\boldsymbol{y},t)}{\partial y_j^{\Re}} + \Re_{ki}^{c}(\hat{\boldsymbol{x}},t)\Re_{lj}^{c}(\hat{\boldsymbol{x}},t) v_{k,x_l}^{(0)}(\hat{\boldsymbol{x}},t)\Big|_{\hat{\boldsymbol{x}}} + O(\zeta) \tag{4.265}$$

The rate of deformation is obtained by taking the symmetric part of the velocity gradient

$$\dot{\varepsilon}_{ij}^{\Re\zeta}(\boldsymbol{x}^{\Re},t) = v_{\left(i,y_j^{\Re}\right)}^{\Re(1)}(\hat{\boldsymbol{x}},\boldsymbol{y},t) + \Re_{ki}^{c}(\hat{\boldsymbol{x}},t)\Re_{lj}^{c}(\hat{\boldsymbol{x}},t) v_{(k,x_l)}^{(0)}(\hat{\boldsymbol{x}},t)\big|_{\hat{\boldsymbol{x}}} + O(\zeta) \tag{4.266}$$

Let ${}_n u_i^{\Re}$ be the converged displacement in the corotational frame at load increment n. We wish to find the displacement increment $\Delta u_i^{\Re}$ in the corotational frame by integrating (4.266) using the midpoint integration rule to obtain second-order accuracy

$$\int_{t_n}^{t_{n+1}} \dot{\varepsilon}_{ij}^{\Re\zeta}(\boldsymbol{x}^{\Re},t)dt \equiv \Delta\varepsilon_{ij}^{\Re\zeta} = \Delta\varepsilon_{ij}^{\Re f} + \zeta\Delta\varepsilon_{ij}^{\Re(1)} + O(\zeta^2) \tag{4.267}$$

where

$$\Delta\varepsilon_{ij}^{\Re f} = \Delta u_{\left(i,y_j^{\Re}\right)}^{\Re(1)}\left({}_{n+1/2}\hat{\boldsymbol{x}},\boldsymbol{y}\right) + \Re_{ki}^{c}\left({}_{n+1/2}\hat{\boldsymbol{x}}\right)\Re_{lj}^{c}\left({}_{n+1/2}\hat{\boldsymbol{x}}\right)\operatorname{sym}\left(\frac{\partial \Delta u_k^{(0)}}{\partial_{n+1/2} x_l}\right)\Bigg|_{{}_{n+1/2}\hat{\boldsymbol{x}}} \tag{4.268}$$

The first term in (4.268) denotes the fine-scale perturbation of the displacement increment, which is assumed to be small; the second term in (4.268) denotes the coarse-scale strain increment rotated into the corotational frame. The unit cell coordinate of the centroid at the mid-step ${}_{n+1/2}\hat{\boldsymbol{x}}$ is defined as

$$ {}_{n+1/2}\hat{\boldsymbol{x}} = \frac{1}{2}\left({}_{n}\hat{\boldsymbol{x}} + {}^{i+1}_{n+1}\hat{\boldsymbol{x}} \right) \tag{4.269} $$

where the left superscript denotes a Newton iteration count at the coarse scale. The coarse-scale displacement increment $\Delta u_k^{(0)}$ and the fine-scale perturbation increment in the corotational frame $\Delta u_i^{\Re(1)}$ are defined as

$$ \begin{aligned} \Delta u_k^{(0)} &= {}^{i+1}_{n+1}u_k^{(0)} - {}_{n}u_k^{(0)} \\ \Delta u_i^{\Re(1)} &= {}^{i+1}_{n+1}u_i^{\Re(1)} - {}_{n}u_i^{\Re(1)} \end{aligned} \tag{4.270} $$

Integrating (4.268) over the unit cell domain and exploiting solution periodicity in the corotational frame yields the coarse-scale strain increment

$$ \Delta\varepsilon_{ij}^{\Re c} = \Re_{ki}^{c}\left({}_{n+1/2}\hat{\boldsymbol{x}}\right)\Re_{lj}^{c}\left({}_{n+1/2}\hat{\boldsymbol{x}}\right)\,\mathrm{sym}\left(\left.\frac{\partial\Delta u_k^{(0)}}{\partial_{n+1/2}x_l}\right)\right|_{{}_{n+1/2}\hat{\boldsymbol{x}}} \tag{4.271} $$

It can be seen that the second term in (4.268) represents the average fine-scale strain increment.

We assume that the Cauchy stress is a function of the previously converged stress ${}_{n}\sigma_{ij}^{\zeta}$, rotation $\Re_{ij}^{\zeta}$, and the strain increment in the corotational frame $\Delta\varepsilon_{ij}^{\Re f}$, that is, $\sigma_{ij}^{\zeta}\left({}_{n}\sigma_{ij}^{\zeta},\Re_{ij}^{\zeta},\Delta\varepsilon_{ij}^{\Re\zeta}\right)$. To derive the equilibrium equation, we start by expanding Cauchy stress around $\left({}_{n}\sigma_{ij}^{\zeta},\Re_{kl}^{c},\Delta\varepsilon_{ij}^{\Re f}\right)$, which yields

$$ \begin{aligned} \sigma_{ij}^{\zeta}\left({}_{n}\boldsymbol{\sigma}^{\zeta},\boldsymbol{\Re}^{\zeta},\Delta\boldsymbol{\varepsilon}^{\Re\zeta}\right) &= \sigma_{ij}^{\zeta}\left({}_{n}\boldsymbol{\sigma}^{\zeta},\boldsymbol{\Re}^{c},\Delta\boldsymbol{\varepsilon}^{\Re f}\right) + \zeta\left(\left.\frac{\partial\sigma_{ij}^{\zeta}}{\partial\Re_{kl}^{\zeta}}\right|_{\boldsymbol{\Re}^{c}} O(\zeta) + \left.\frac{\partial\sigma_{ij}^{\zeta}}{\partial\Delta\varepsilon_{kl}^{\Re\zeta}}\right|_{\Delta\boldsymbol{\varepsilon}^{\Re c}} \Delta\varepsilon_{kl}^{\Re(1)}\right) \\ &= \sigma_{ij}^{(0)}\left({}_{n}\boldsymbol{\sigma}^{\zeta},\boldsymbol{\Re}^{c},\Delta\boldsymbol{\varepsilon}^{\Re f}\right) + \zeta\sigma_{ij}^{(1)} + O\left(\zeta^{2}\right) \end{aligned} \tag{4.272} $$

Further expanding (4.272) in a Taylor series around the unit cell centroid in the current configuration yields

$$ \begin{aligned} \sigma_{ij}^{\zeta}(\boldsymbol{x},\boldsymbol{y},t) &= \sigma_{ij}^{(0)}(\hat{\boldsymbol{x}},\boldsymbol{y},t) + \left.\frac{\partial\sigma_{ij}^{(0)}}{\partial x_k}\right|_{\hat{\boldsymbol{x}}}(x_k-\hat{x}_k) + \zeta\sigma_{ij}^{(1)}(\hat{\boldsymbol{x}},\boldsymbol{y},t) + \zeta\left.\frac{\partial\sigma_{ij}^{(1)}}{\partial x_k}\right|_{\hat{\boldsymbol{x}}}(x_k-\hat{x}_k)\cdots \\ &= \underbrace{\sigma_{ij}^{0}\left(\hat{\boldsymbol{x}},\boldsymbol{y},t\right)}_{\sigma_{ij}^{f}(\hat{\boldsymbol{x}},\boldsymbol{y},t)} + \zeta\left(\left.\frac{\partial\sigma_{ij}^{(0)}}{\partial x_k}\right|_{\hat{\boldsymbol{x}}} y_k + \sigma_{ij}^{(1)}(\hat{\boldsymbol{x}},\boldsymbol{y},t)\right) + O\left(\zeta^{2}\right) \end{aligned} \tag{4.273} $$

Inserting (4.273) into equilibrium equation (4.257a) yields the two-scale equilibrium equations

$$ O(\zeta^{-1}):\ \ \sigma_{ij,y_j}^{f}\left(\hat{\boldsymbol{x}},\boldsymbol{y},t\right) = 0 \tag{4.274a} $$

$$O(1): \quad \left.\frac{\partial \sigma_{ij}^{f}}{\partial x_j}\right|_{\hat{x}} + \sigma_{ij,y_j}^{(1)}(\hat{\boldsymbol{x}}, \boldsymbol{y}, t) + b_i^{\zeta}(\hat{\boldsymbol{x}}, t) = 0 \tag{4.274b}$$

Averaging (4.274b) over the deformed unit cell domain Θ_y gives the coarse-scale equilibrium equation

$$\left.\frac{\partial \sigma_{ij}^{c}}{\partial x_j}\right|_{\hat{x}} + b_i^{c}(\hat{\boldsymbol{x}}, t) = 0$$

$$\sigma_{ij}^{c}(\hat{\boldsymbol{x}}, t) = \frac{1}{|\Theta_y|}\int_{\Theta_y} \sigma_{ij}^{f}(\hat{\boldsymbol{x}}, \boldsymbol{y}, t)\, d\Theta_y \tag{4.275}$$

$$b_i^{c}(\hat{\boldsymbol{x}}, t) = \frac{1}{|\Theta_y|}\int_{\Theta_y} b_i(\hat{\boldsymbol{x}}, \boldsymbol{y}, t)\, d\Theta_y$$

It is convenient to express the unit cell equilibrium equation (4.274a) in the corotational frame. Rewriting (4.274a) gives

$$\frac{\partial \sigma_{ij}^{f}}{\partial y_k^{\Re}} \frac{\partial y_k^{\Re}}{\partial y_j} = \frac{\partial \sigma_{ij}^{f}}{\partial y_k^{\Re}} \Re_{jk}^{c}(\hat{\boldsymbol{x}}, t) = 0$$

Multiplying the above by $\Re_{il}^{c}(\hat{\boldsymbol{x}}, t)$ yields

$$\left(\Re_{il}^{c}(\hat{\boldsymbol{x}}, t)\, \Re_{jk}^{c}(\hat{\boldsymbol{x}}, t) \sigma_{ij}^{f}\right)_{,y_k^{\Re}} \equiv \sigma_{lk,y_k^{\Re}}^{\Re f} = 0 \tag{4.276}$$

The Cauchy stress in the corotational frame $\sigma_{ij}^{\Re f}$ can be expressed in terms of converged value ${}_n\sigma_{ij}^{\Re f}$ and the new increment $\Delta\sigma_{ij}^{\Re f}$, which yields the following unit cell governing equations

$$\Delta\sigma_{ij,y_j^{\Re}}^{\Re f} = 0$$

$$\Delta\sigma_{ij}^{\Re f} = L_{ijkl}^{\Re}\left(\Delta\,\varepsilon_{kl}^{\Re f} - \Delta\mu_{kl}^{\Re f}\right) \tag{4.277}$$

$$\Delta\varepsilon_{ij}^{\Re f} = \Delta u_{\left(i,y_j^{\Re}\right)}^{\Re(1)} + \Delta\varepsilon_{ij}^{\Re c}$$

The incremental unit cell problem in the corotational frame (4.277) is solved by constructing a residual-free field similar to that considered in section 4.1, section 4.2, and section 4.3

$$\Delta u_i^{\Re(1)}(\boldsymbol{x}, \boldsymbol{y}) = H_i^{kl}(\boldsymbol{y}) \Delta\varepsilon_{kl}^{\Re c}(\boldsymbol{x}) + \int_{\Theta_y} \tilde{h}_i^{kl}(\boldsymbol{y}, \tilde{\boldsymbol{y}})\; \Delta\mu_{kl}^{\Re f}(\boldsymbol{x}, \tilde{\boldsymbol{y}}) d\tilde{\Theta}$$
$$+ \int_{S_y} \breve{h}_i^{\hat{n}}(\boldsymbol{y}, \breve{\boldsymbol{y}})\; \Delta\delta_{\hat{n}}^{\Re f}(\boldsymbol{x}, \breve{\boldsymbol{y}}) d\breve{S} \tag{4.278}$$

with the only exception being that all the fields are expressed in the corotational frame. $\mu_{kl}^{\Re f}$ and $\delta_{\hat{n}}^{\Re f}$ denote the leading order (fine-scale) eigenstrain and eigenseparation, respectively, in the corotational frame.

Remark 4.3 In the case of a large unit cell distortion, the influence functions H_i^{kl}, $\tilde{h}_i^{kl}$, and $\breve{h}_i^{\hat{n}}$ have to be recomputed from time to time. It is, however, instructive to point out that if the unit cell geometry is changing considerably from one load increment to another, the ROH outlined in section 4.1, section 4.2, and section 4.3 will offer little advantage over the nonlinear direct homogenization approach considered in Chapter 2. Moreover, the approximation in (4.263) may no longer be accurate. Instead, a phase rotation $\mathfrak{R}_{ki}^{(\alpha)}(\hat{\boldsymbol{x}},t)$ that tracks an average rotation of each phase has to be defined.

Given equation (4.277) and equation (4.278), the process of constructing and solving the residual-free equations in the corotational frame is identical to that considered for small deformation problems in section 4.1, section 4.2, and section 4.3. Once the fine-scale stress in the corotational frame $\sigma_{ij}^{\mathfrak{R}f}$ has been computed, the coarse-scale stress $\sigma_{ij}^{\mathfrak{R}c}$ in the corotational frame is computed by averaging $\sigma_{ij}^{\mathfrak{R}f}$. The resulting coarse-scale stress in the global Cartesian coordinates is computed by rotating the coarse-scale stress from the corotational frame

$$\sigma_{ij}^{c}\left({}_{n+1}^{i+1}\hat{\boldsymbol{x}}\right) = \mathfrak{R}_{il}^{c}\left({}_{n+1}^{i+1}\hat{\boldsymbol{x}}\right)\,\mathfrak{R}_{jk}^{c}\left({}_{n+1}^{i+1}\hat{\boldsymbol{x}}\right)\,\sigma_{lk}^{\mathfrak{R}c}\left({}_{n+1}^{i+1}\hat{\boldsymbol{x}}\right) \tag{4.279}$$

Remark 4.4 Construction of the tangent stiffness matrix requires linearization of the rotational matrix $\mathfrak{R}_{jk}^{c}$ computed at the mid-step and at the end of the increment at each Newton iteration. Since $\mathfrak{R}_{jk}^{c}$ is assumed not to depend on the fine-scale problem, its construction and linearization are governed by the deformation of the coarse-scale problem only. The interested reader is referred to [99] for more details.

4.8 Extension to Multiple Temporal Scales with Application to Fatigue

Fatigue of heterogeneous material systems is a multiscale phenomenon in space and time. It is multiscale in space because defects and cracks can be several orders of magnitude smaller than those of a structural component. It is multiscale in time because the cyclic load period can be of the order of seconds, while component life may span years. Owing to this tremendous disparity of spatial and temporal scales, component fatigue life prediction poses a tremendous challenge to structural component and engine designers.

Fatigue life prediction methods range from sole experimentation to modeling and computational resolution of spatial scales. The basic design tool today is primarily experimental, based on so-called S-N curves, which provide a relation between specimen life and cyclic stress/strain level. Due to considerable scatter in fatigue life data, a family of S-N curves with a probability of failure known as S-N-P plots is often used. Fatigue experiments are generally limited to specimens or small structural components, and therefore, predictions of boundary and initial conditions between the components or interconnect of interest and the remaining structure involve some sort of modeling. Typically, finite element analysis is carried out to predict the *far fields* acting on the critical component or interconnect. These types of calculations do not take into account the force redistribution caused by the accumulation of damage taking place in the critical components or interconnects.

Paris law [100] represents one of the most common approaches used to empirically model the multiple temporal scales associated with fatigue life. It states that under ideal conditions of high cycle fatigue (or small-scale yielding) and constant amplitude loading, the growth rate of long cracks depends on the amplitude of stress intensity factors. Models departing from these ideal conditions have been widely used (e.g., [101,102,103,104]). Various crack growth "laws" have been used in conjunction with multiple spatial scales methods to propagate arbitrary discontinuities.

An alternative to using Paris-like fatigue models is to carry out a direct cycle-by-cycle simulation. These simulations often employ a cohesive law to model fatigue crack growth based on unloading–reloading hysteresis [105]. However, the cycle-by-cycle approach is not feasible for high cycle fatigue life prediction of large-scale heterogeneous systems. Nevertheless, an attractive feature of this approach is that it provides a unified treatment of long and short cracks and the ability to account for overloads.

To circumvent the computational challenges of the cycle-by-cycle simulation, several temporal multiscale approaches have been proposed. The first is known as a *cycle jump* technique [106,107,108], where a single load cycle is used to compute the rate of fatigue damage growth at each spatial integration point $\hat{x}_I$ and to construct the ordinary differential equation

$$\frac{d\omega(\hat{x}_I)}{dt} = \omega(\hat{x}_I)\Big|_K - \omega(\hat{x}_I)\Big|_{K-1} \tag{4.280}$$

where $\omega(\hat{x}_I)$ is a damage variable at integration point $\hat{x}_I$, t is the cycle variable, and K is the cycle count. The block cycle jump is closely related to the unified brittle–fatigue damage model [109]. The second approach [110,111,112,113,114] is based on a multiple temporal scale asymptotic analysis where loads and response fields are assumed to depend on the slow time coordinate t due to slow degradation of material properties during fatigue, as well as on the fast time coordinate τ due to locally periodic loading.

In section 4.8 we detail a unified formulation for temporal multiscale modeling of fatigue [115]. We show that the two temporal multiscale approaches are closely related and that the unified framework can be effectively utilized in practice for arbitrary material architectures and constitutive equations of microphases. We start by introducing multiple temporal scales in the context of a single spatial scale and discussing the application of the unified temporal homogenization method to fatigue. Consideration of multiple spatial and temporal scales will be subsequently discussed.

4.8.1 *Temporal Homogenization*

We consider two pseudo temporal scales aimed at resolving the deformation within a single load cycle and capturing the slow degradation of material properties due to multiple load cycles. The slow timescale is denoted by t, and the fast timescale is denoted by τ. The pseudo slow timescale is assumed to range from 0 to N, that is, $t \in [0, N]$, where N is the number of cycles until failure. The local periodicity (τ-periodicity) assumption is made similarly to the fine-scale periodicity assumption made in the spatial homogenization theory.

The aforementioned two scales are related by a small positive scaling parameter, η

$$\tau = \frac{t}{\eta}; \quad 0 < \eta \ll 1 \tag{4.281}$$

Various response fields ϕ are assumed to depend on the two temporal scales

$$\phi^{\eta}(\boldsymbol{x},t)=\phi(\boldsymbol{x},t,\tau) \tag{4.282}$$

The time differentiation of the response fields with respect to multiple temporal scales is given by the chain rule

$$\frac{d\phi^{\eta}(\boldsymbol{x},t)}{dt}=\frac{\partial\phi(\boldsymbol{x},t,\tau)}{\partial t}+\frac{1}{\eta}\frac{\partial\phi(\boldsymbol{x},t,\tau)}{\partial\tau}=\dot{\phi}(\boldsymbol{x},t,\tau)+\frac{1}{\eta}\phi'(\boldsymbol{x},t,\tau) \tag{4.283}$$

where $d\phi^{\eta}/dt$, $\dot{\phi}$, and ϕ' denote the total time derivative, the partial time derivative with respect to slow time variable t, and the partial time derivation with respect to fast time variable τ, respectively. Here we adopt the notation proposed in [114].

Consider a homogeneous inelastic solid subjected to periodic loads and/or boundary conditions in a time domain. Assuming, for simplicity, small deformations, the governing equations are

$$\begin{aligned}
&\sigma_{ij,j}(\boldsymbol{x},t,\tau)+b_i(\boldsymbol{x},t,\tau)=0 \quad \text{on} \quad \Omega\times[0,N]\times[0,\tau_0]\\
&\sigma_{ij}(\boldsymbol{x},t,\tau)=L_{ijkl}(\boldsymbol{x})\left(\varepsilon_{kl}(\boldsymbol{x},t,\tau)-\mu_{kl}(\boldsymbol{x},t,\tau)\right) \quad \text{on} \quad \Omega\times[0,N]\times[0,\tau_0]\\
&\varepsilon_{ij}(\boldsymbol{x},t,\tau)=u_{(i,j)}(\boldsymbol{x},t,\tau) \quad \text{on} \quad \Omega\times[0,N]\times[0,\tau_0]\\
&u_i(\boldsymbol{x},t,\tau)=\overline{u}_i(\boldsymbol{x},t,\tau) \quad \text{on} \quad \partial\Omega^u\times[0,N]\times[0,\tau_0]\\
&\sigma_{ij}(\boldsymbol{x},t,\tau)n_j=\overline{t}_i(\boldsymbol{x},t,\tau) \quad \text{on} \quad \partial\Omega^t\times[0,N]\times[0,\tau_0]
\end{aligned} \tag{4.284}$$

The eigenstrain evolution equation can be schematically expressed as

$$\frac{d\mu_{kl}^{\eta}}{dt}=f_{kl}(\boldsymbol{\sigma},\boldsymbol{\varepsilon},\boldsymbol{s}) \tag{4.285}$$

where s_i denotes the state variables. Applying the temporal differentiation rule (4.283) to (4.285) yields

$$\frac{d\mu_{kl}^{\eta}}{dt}=\dot{\mu}_{kl}+\frac{1}{\eta}\mu_{kl}'=f_{kl}(\boldsymbol{\sigma},\boldsymbol{\varepsilon},\boldsymbol{s}) \tag{4.286}$$

Consider the leading order term in the asymptotic expansion of eigenstrains

$$\mu_{kl}^{\eta}(\boldsymbol{x},t)\equiv\mu_{kl}(\boldsymbol{x},t,\tau)=\hat{\mu}_{kl}(\boldsymbol{x},t,\tau)+O(\zeta) \tag{4.287}$$

Inserting (4.287) into (4.286) yields the leading order equation

$$\frac{\partial\hat{\mu}_{kl}(\boldsymbol{x},t,\tau)}{\partial\tau}=0\Rightarrow\hat{\mu}_{kl}=\hat{\mu}_{kl}(\boldsymbol{x},t) \tag{4.288}$$

Thus, the leading order eigenstrain is only a function of slow time

$$\mu_{kl}(\boldsymbol{x},t,\tau)=\hat{\mu}_{kl}(\boldsymbol{x},t)+O(\eta) \tag{4.289}$$

Equation (4.289) suggests that inelastic deformation evolves slowly in time, that is, it is little affected by what happens in a single load cycle.

Applying the time-averaging operator

$$\langle \phi(\boldsymbol{x},t,\tau)\rangle \equiv \frac{1}{\tau_0}\int_0^{\tau_0} \phi(\boldsymbol{x},t,\tau)d\tau \tag{4.290}$$

to (4.289) yields

$$\langle \mu_{kl}(\boldsymbol{x},t,\tau)\rangle = \hat{\mu}_{kl}(\boldsymbol{x},t)+O(\eta) \tag{4.291}$$

Given the above time-averaging operator, we can define the leading order slow-evolving (or time-homogenized) governing equations

$$\begin{aligned}
&\langle \sigma_{ij}\rangle_{,j}(\boldsymbol{x},t)+\langle b_i\rangle(\boldsymbol{x},t)=0 \quad \text{on} \quad \Omega\times[0,N]\\
&\langle \sigma_{ij}\rangle(\boldsymbol{x},t)=L_{ijkl}(\boldsymbol{x})\left(\langle \varepsilon_{kl}\rangle(\boldsymbol{x},t)-\hat{\mu}_{kl}(\boldsymbol{x},t)\right) \quad \text{on} \quad \Omega\times[0,N]\\
&\langle \varepsilon_{ij}\rangle(\boldsymbol{x},t)=\langle u_{(i,j)}\rangle(\boldsymbol{x},t) \quad \text{on} \quad \Omega\times[0,N]\\
&\langle u_i\rangle(\boldsymbol{x},t)=\langle \bar{u}_i\rangle(\boldsymbol{x},t) \quad \text{on} \quad \partial\Omega^u\times[0,N]\\
&\langle \sigma_{ij}\rangle(\boldsymbol{x},t)n_j=\langle \bar{t}_i\rangle(\boldsymbol{x},t) \quad \text{on} \quad \partial\Omega^t\times[0,N]
\end{aligned} \tag{4.292}$$

To complete the definition of the initial-boundary value problem for the time averages, it remains to construct the evolution equation for the leading order eigenstrain $\hat{\mu}_{kl}(\boldsymbol{x},t)$ with respect to slow timescale. The time derivative $\dot{\hat{\mu}}_{kl}(\boldsymbol{x},t)$ after $t=K$ cycles can be evaluated using the finite difference in two subsequent cycles

$$\dot{\hat{\mu}}_{kl}(\boldsymbol{x},t)\Big|_{t=K} = \hat{\mu}_{kl}(\boldsymbol{x},t)\Big|_{t=K} - \hat{\mu}_{kl}(\boldsymbol{x},t)\Big|_{t=K-1} \tag{4.293}$$

Inserting (4.289) into (4.293) and denoting $(t=K-1, \tau=\tau_0)\equiv(t=K, \tau=0)$ yields the evolution equation for the leading order eigenstrain with respect to the slow timescale

$$\dot{\hat{\mu}}_{kl}(\boldsymbol{x},t)\Big|_{t=K} = \mu_{kl}(\boldsymbol{x},t=K-1,\tau=\tau_0)-\mu_{kl}(\boldsymbol{x},t=K-1,\tau=0) \tag{4.294}$$

Equation (4.294) states that the eigenstrain growth rate with respect to the slow timescale can be evaluated by computing the difference between the values of eigenstrains in the beginning and the end of the previous load cycle.

Using a forward Euler integration, the eigenstrain after Δt_K cycles of the current load cycle K can be approximated by

$$\hat{\mu}_{kl}(\boldsymbol{x},t)\Big|_{t=K+\Delta t} = \hat{\mu}_{kl}(\boldsymbol{x},t)\Big|_{t=K} + \dot{\hat{\mu}}_{kl}(\boldsymbol{x},t)\Big|_{t=K}\Delta t_K \tag{4.295}$$

It is important to note that evolving eigenstrains (4.295) while keeping the rest of the fields unchanged could violate the governing equations (4.284). This inconsistency can be alleviated by equilibrating discrete equilibrium equations. We will refer to this process as *consistency adjustment.*

The block size Δt_K is selected to ensure accuracy based on the following criteria:

1. Let $\Delta\hat{\mu}_a$ be the user-defined allowable eigenstrain increment for Δt_K cycles, and let $\max_I \|\Delta\hat{\mu}\|$ be the largest eigenstrain increment obtained in a single cycle among all the quadrature points I. Then the initial value of Δt_K can be evaluated as $\Delta t_K = \text{int}\left\{\Delta\hat{\mu}_a / \max_I \|\Delta\hat{\mu}\|\right\}$ where $\text{int}\{\cdot\}$ denotes truncation to the decimal part.
2. Given the above initial value of Δt_K, correct the predictor value in (4.295) as

$$\hat{\mu}_{kl}(\boldsymbol{x},t)\Big|_{t=K+\Delta t_K}^{cor} = \hat{\mu}_{kl}(\boldsymbol{x},t)\Big|_{t=K} + \frac{\Delta t_K}{2}\left(\dot{\hat{\mu}}_{kl}(\boldsymbol{x},t)\Big|_{t=K} + \dot{\hat{\mu}}_{kl}(\boldsymbol{x},t)\Big|_{t=K+\Delta t_K}\right) \tag{4.296}$$

3. If the difference in some norm between the predictor (4.295) and corrector (4.296) is sufficiently small, then the step is accepted and the aforementioned consistency adjustment is performed. Otherwise, the block size is halved and the predictor-corrector step is repeated.

We now focus on solution postprocessing, which is similar to the postprocessing in spatial homogenization. Let ϕ^* be the fast time correction from the average, defined by

$$\phi^* = \phi^{\zeta} - \langle\phi\rangle \tag{4.297}$$

The governing equations for the fast-scale correction $\overset{*}{u}_i$ can then be obtained by subtracting the governing equation for the temporal averages (4.292) from (4.284), which yields

$$\begin{aligned}
&\overset{*}{\sigma}_{ij,j}(\boldsymbol{x},\tau) + \overset{*}{b}_i(\boldsymbol{x},\tau) = 0 \quad \text{on} \quad \Omega\times[0,\tau_0] \\
&\overset{*}{\sigma}_{ij}(\boldsymbol{x},\tau) = L_{ijkl}(\boldsymbol{x})\overset{*}{\varepsilon}_{kl}(\boldsymbol{x},\tau) \quad \text{on} \quad \Omega\times[0,\tau_0] \\
&\overset{*}{\varepsilon}_{ij}(\boldsymbol{x},\tau) = \overset{*}{u}_{(i,j)}(\boldsymbol{x},\tau) \quad \text{on} \quad \Omega\times[0,\tau_0] \\
&\overset{*}{u}_i(\boldsymbol{x},\tau) = \overline{u}(\boldsymbol{x},\tau) \quad \text{on} \quad \partial\Omega^u\times[0,\tau_0] \\
&\overset{*}{\sigma}_{ij}(\boldsymbol{x},\tau)n_j = \overline{t}_i^{\,*}(\boldsymbol{x},\tau) \quad \text{on} \quad \partial\Omega^t\times[0,\tau_0]
\end{aligned} \tag{4.298}$$

It is instructive to point out that the above equations are linear. Moreover, if the prescribed boundary conditions and body forces have the same τ_0-periodicity, then $\overset{*}{b}_i(\boldsymbol{x},\tau)$, $\overline{u}_i^{\,*}(\boldsymbol{x},\tau)$, $\overline{t}_i^{\,*}(\boldsymbol{x},\tau)$, and the resulting fast-scale correction $\overset{*}{u}_i(\boldsymbol{x},\tau)$ can be computed by analyzing the linear problem (4.298) over a single load cycle. Finally, if $\overset{*}{b}_i(\boldsymbol{x},0)$, $\overline{u}_i^{\,*}(\boldsymbol{x},0)$, and $\overline{t}_i^{\,*}(\boldsymbol{x},0)$ vanish at τ=0, then for quasi-static loading, the fast-scale correction vanishes at the beginning and at the end of the cycle.

4.8.2 Multiple Temporal and Spatial Scales

For a heterogeneous inelastic solid subjected to cyclic loading, the spatial and temporal upscaling procedures outlined in the previous sections can be sequentially applied, starting with the spatial upscaling. For simplicity, we will consider a small deformation problem. Spatial upscaling based on ROH gives rise to the following two-scale equations

$$\begin{aligned}
&\sigma^c_{ij,x_j}(\boldsymbol{x},t,\tau)+b^c_i(\boldsymbol{x},t,\tau)=0 \quad \text{on} \quad \Omega\times[0,N]\times[0,\tau_0]\\
&\sigma^c_{ij}(\boldsymbol{x},t,\tau)=L^c_{ijkl}\left(\varepsilon^c_{kl}(\boldsymbol{x},t,\tau)-\mu^c_{kl}(\boldsymbol{x},t,\tau)\right) \quad \text{on} \quad \Omega\times[0,N]\times[0,\tau_0]\\
&\varepsilon^c_{ij}(\boldsymbol{x},t,\tau)=u^c_{(i,j)}(\boldsymbol{x},t,\tau) \quad \text{on} \quad \Omega\times[0,N]\times[0,\tau_0]\\
&u^c_i(\boldsymbol{x},t,\tau)=\overline{u}^c_i(\boldsymbol{x},t,\tau) \quad \text{on} \quad \partial\Omega^u\times[0,N]\times[0,\tau_0]\\
&\sigma^c_{ij}(\boldsymbol{x},t,\tau)n^c_j=\overline{t}^c_i(\boldsymbol{x},t,\tau) \quad \text{on} \quad \partial\Omega^t\times[0,N]\times[0,\tau_0]
\end{aligned} \tag{4.299}$$

with the coarse-scale eigenstrain evolution equations computed from the fine scale

$$\begin{aligned}
&\frac{d\mu^c_{kl}}{dt}=\sum_{\alpha=1}^{\tilde{M}}\left(-M^c_{klij}A^{c(\alpha)}_{ijmn}\right)\frac{d\mu^{(\alpha)}_{mn}}{dt}+\sum_{\xi=1}^{\breve{M}}\left(-M^c_{klij}B^{c(\xi)}_{ij\breve{n}}\right)\frac{d\delta^{(\xi)}_{\breve{n}}}{dt}\\
&\frac{d\mu^{(\alpha)}_{mn}}{dt}=f^{(\alpha)}_{mn}\left(\boldsymbol{\varepsilon}^{(\alpha)},\boldsymbol{s}^{(\alpha)}_{\mu}\right); \quad \frac{d\delta^{(\xi)}_{\breve{n}}}{dt}=f^{(\xi)}_{\breve{n}}\left(\boldsymbol{t}^{(\xi)}_{\breve{n}},\boldsymbol{s}^{(\xi)}_{\delta}\right)\\
&\boldsymbol{r}\left(\frac{d\boldsymbol{\mu}^{(\alpha)}}{dt},\frac{d\boldsymbol{\delta}^{(\xi)}}{dt},\frac{d\boldsymbol{\varepsilon}^c}{dt}\right)=\boldsymbol{0}
\end{aligned} \tag{4.300}$$

where $\boldsymbol{s}^{(\alpha)}_{\mu}$ and $\boldsymbol{s}^{(\xi)}_{\delta}$ are partitioned state variables. Applying the time differentiation rule (4.286) to the partitioned eigenstrains and eigenseparations, it follows that the leading order partitioned eigenstrains and eigenseparations do not depend on the fast time coordinates

$$\begin{aligned}
&\mu^{(\alpha)}_{kl}(\boldsymbol{x},t,\tau)=\widehat{\mu}^{(\alpha)}_{kl}(\boldsymbol{x},t)+O(\eta)\\
&\delta^{(\xi)}_{\breve{n}}(\boldsymbol{x},t,\tau)=\widehat{\delta}^{(\xi)}_{\breve{n}}(\boldsymbol{x},t)+O(\eta)
\end{aligned} \tag{4.301}$$

Likewise, the leading order coarse-scale eigenstrain can be expressed as a function of the slow timescale coordinate t

$$\mu^c_{kl}(\boldsymbol{x},t,\tau)=\widehat{\mu}^c_{kl}(\boldsymbol{x},t)+O(\eta) \tag{4.302}$$

Applying the time-averaging operator to coarse-scale equations (4.299) yields the coarse-scale slow time governing equations

$$
\begin{aligned}
&\left\langle \sigma_{ij,j}^{c} \right\rangle(\boldsymbol{x},t) + \left\langle b_i^c \right\rangle(\boldsymbol{x},t) = 0 \quad \text{on} \quad \Omega\times[0,N] \\
&\left\langle \sigma_{ij}^{c} \right\rangle(\boldsymbol{x},t) = L_{ijkl}^{c}(\boldsymbol{x})\left(\left\langle \varepsilon_{kl}^{c} \right\rangle(\boldsymbol{x},t) - \bar{\mu}_{kl}^{c}(\boldsymbol{x},t)\right) \quad \text{on} \quad \Omega\times[0,N] \\
&\left\langle \varepsilon_{ij}^{c} \right\rangle(\boldsymbol{x},t) = \left\langle u_{(i,x_j)}^{c} \right\rangle(\boldsymbol{x},t) \quad \text{on} \quad \Omega\times[0,N] \\
&\left\langle u_i^c \right\rangle(\boldsymbol{x},t) = \left\langle \bar{u}_i^c \right\rangle(\boldsymbol{x},t) \quad \text{on} \quad \partial\Omega^u\times[0,N] \\
&\left\langle \sigma_{ij}^{c} \right\rangle(\boldsymbol{x},t) n_j^c = \left\langle \bar{t}_i^c \right\rangle(\boldsymbol{x},t) \quad \text{on} \quad \partial\Omega^t\times[0,N]
\end{aligned} \tag{4.303}
$$

To complete the definition of the initial-boundary value problem for the coarse-scale time averages (4.303), it is necessary to evaluate the slow time derivative of the partitioned eigenstrain $\dot{\mu}_{kl}^{(\alpha)}(\boldsymbol{x},t)\Big|_K$ and the eigenseparation $\dot{\delta}_{\breve{n}}^{(\xi)}(\boldsymbol{x},t)\Big|_K$ using the finite difference in two subsequent cycles

$$
\begin{aligned}
\dot{\mu}_{kl}^{(\alpha)}(\boldsymbol{x},t)\Big|_{t=K} &= \mu_{kl}^{(\alpha)}(\boldsymbol{x}, t=K-1, \tau=\tau_0) - \mu_{kl}^{(\alpha)}(\boldsymbol{x}, t=K-1, \tau=0) \\
\dot{\delta}_{\breve{n}}^{(\xi)}(\boldsymbol{x},t)\Big|_{t=K} &= \hat{\delta}_{\breve{n}}^{(\xi)}(\boldsymbol{x}, t=K-1, \tau=\tau_0) - \hat{\delta}_{\breve{n}}^{(\xi)}(\boldsymbol{x}, t=K-1, \tau=0)
\end{aligned} \tag{4.304}
$$

The predictor-corrector Euler integrator algorithm outlined in the previous section can be applied to integrate (4.304). The value of the block size Δt_K is selected to ensure that (i) partitioned eigenstrain and eigenseparation increments do not exceed the user-defined tolerances and (ii) $\dot{\mu}_{kl}^{(\alpha)}(\boldsymbol{x},t)$ and $\dot{\hat{\delta}}_{\breve{n}}^{(\xi)}(\boldsymbol{x},t)$ vary little from the beginning to the end of cycle $K-1$.

Finally, the governing equations for the coarse-scale fast time correction u_i^{c*} are defined by subtracting (4.303) from (4.299), which yields the following linear strong form over $\Omega\times[0,\tau_0]$

$$
\begin{aligned}
&\sigma_{ij,j}^{c*}(\boldsymbol{x},\tau) + b_i^{c*}(\boldsymbol{x},\tau) = 0 \quad \text{on} \quad \Omega\times[0,\tau_0] \\
&\sigma_{ij}^{c*}(\boldsymbol{x},\tau) = L_{ijkl}^{c}(\boldsymbol{x})\varepsilon_{kl}^{c*}(x,\tau) \quad \text{on} \quad \Omega\times[0,\tau_0] \\
&\varepsilon_{ij}^{c*}(\boldsymbol{x},\tau) = u_{(i,j)}^{c*}(\boldsymbol{x},\tau) \quad \text{on} \quad \Omega\times[0,\tau_0] \\
&u_i^{c*}(\boldsymbol{x},\tau) = \bar{u}_i^{c*}(\boldsymbol{x},\tau) \quad \text{on} \quad \partial\Omega^u\times[0,\tau_0] \\
&\sigma_{ij}^{c*}(\boldsymbol{x},\tau) n_j^c = \bar{t}_i^{c*}(\boldsymbol{x},\tau) \quad \text{on} \quad \partial\Omega^t\times[0,\tau_0]
\end{aligned} \tag{4.305}
$$

4.8.3 Fatigue Constitutive Equation

Here we describe a continuum damage mechanics fatigue damage cumulative law originally proposed in [108], where the fatigue damage accumulation law in its rate form is given by

$$
\dot{\omega}^{(\eta)}(\boldsymbol{x},t) = \begin{cases} 0 & \hat{\varepsilon}^{(\eta)} < \hat{\varepsilon}_{ini}^{(\eta)} \\ \left(\dfrac{\Phi^{(\eta)}}{\omega^{(\eta)}}\right)^{\gamma^{(\eta)}} \dfrac{\partial\Phi^{(\eta)}}{\partial\hat{\varepsilon}^{(\eta)}} \left\langle \dot{\hat{\varepsilon}}^{(\eta)} \right\rangle_+ & \hat{\varepsilon}^{(\eta)} \geq \hat{\varepsilon}_{ini}^{(\eta)} \end{cases} \tag{4.306}
$$

For each phase, η, $\Phi^{(\eta)}$ is the damage evolution law, $\omega^{(\eta)}$ is the current state of damage, and $\gamma^{(\eta)}$ is the fatigue law material parameter, such that when $\gamma^{(\eta)} \to \infty$ the fatigue damage power law reduces to the static damage law. $\frac{\partial \Phi^{(\eta)}}{\partial \hat{\varepsilon}^{(\eta)}} \left\langle \dot{\hat{\varepsilon}}^{(\eta)} \right\rangle_+$ represents the instantaneous static damage accumulation due to the equivalent strain $\hat{\varepsilon}^{(\eta)}$. $\hat{\varepsilon}_{ini}^{(\eta)}$ is a material parameter denoting the equivalent strain at which the damage process initiates.

Starting from the definition of the *pseudo damage parameter*,

$$\hat{\omega}^{(\eta)}(\boldsymbol{x}, t) = \Phi\left(< \hat{\varepsilon}^{(\eta)}(\boldsymbol{x}, t) - \hat{\varepsilon}_{ini}^{(\eta)} >_+\right); \quad \frac{\partial \Phi\left(< \hat{\varepsilon}^{(\eta)}(\boldsymbol{x}, t) - \hat{\varepsilon}_{ini}^{(\eta)} >_+\right)}{\partial \hat{\varepsilon}^{(\eta)}} \geq 0 \tag{4.307}$$

where $\hat{\omega}^{(\eta)} \in [0,1]$ and the operator $< >_+$ denote the positive part, that is, $<\cdot>_+ = \sup\{0,\cdot\}$, the damage parameter $\omega^{(\eta)}$ is defined as

$$\omega^{(\eta)}(\boldsymbol{x},t) = \max\left\{\hat{\omega}^{(\eta)}(\boldsymbol{x},t) \middle| (\tau \leq t)\right\} \tag{4.308}$$

In the numerical examples in section 4.8.4, the damage evolution law $\Phi^{(\eta)} = \Phi\left(< \hat{\varepsilon}^{(\eta)}(\boldsymbol{x},t) - \hat{\varepsilon}_{ini}^{(\eta)} >_+\right)$ is chosen in the form of

$$\Phi^{(\eta)} = \frac{\operatorname{atan}\left[\alpha^{(\eta)} \frac{\left(< \hat{\varepsilon}^{(\eta)}(\boldsymbol{x},t) - \hat{\varepsilon}_{ini}^{(\eta)} >_+\right)}{\hat{\varepsilon}_0^{(\eta)}} - \beta^{(\eta)}\right] + \operatorname{atan}\left(\beta^{(\eta)}\right)}{\frac{\pi}{2} + \operatorname{atan}\left(\beta^{(\eta)}\right)} \tag{4.309}$$

where $\alpha^{(\eta)}, \beta^{(\eta)}, \hat{\varepsilon}_0^{(\eta)}$ are material constants.

The equivalent phase strain is defined as

$$\hat{\varepsilon}^{(\eta)} = \sqrt{\sum_{I=1}^{3} \left\{\varepsilon_I^{(\eta)}\right\}^2} \tag{4.310}$$

where

$$\{x\} = \begin{cases} x & x \geq 0 \\ Cx & x < 0 \end{cases} \tag{4.311}$$

Principal strains are denoted by $\varepsilon_I^{(\eta)}$, and C is a material parameter indicating sensitivity to compression.

4.8.4 Verfication of the Multiscale Fatigue Model

In this section, the multiscale fatigue model is verified against cycle-by-cycle fatigue. For model validation, we refer to [108,115].

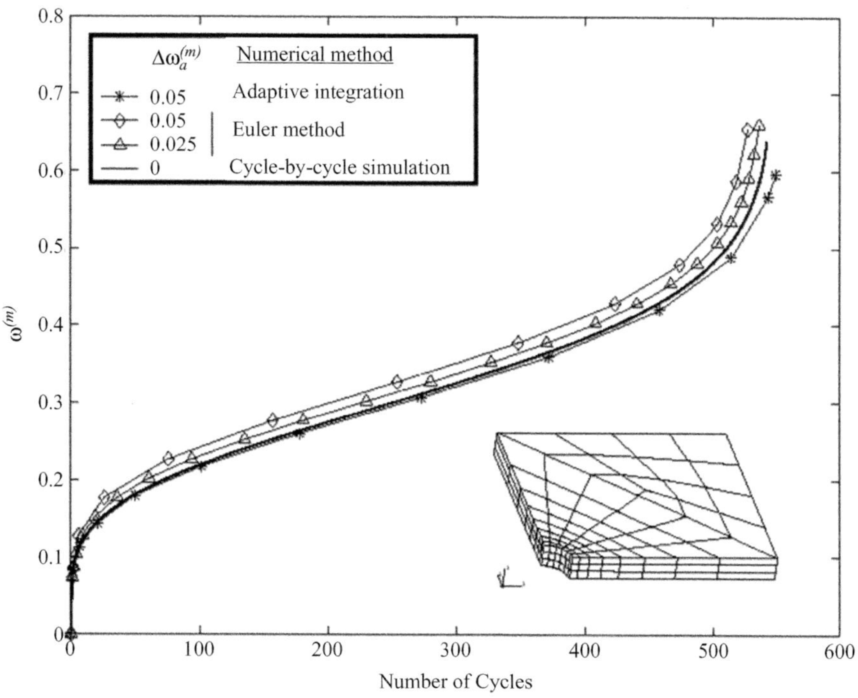

Figure 4.30 Fatigue damage accumulation for low-cycle fatigue for different maximum allowable damage parameters in a cycle $\Delta\omega_a^{(m)}$

The microstructure is that of fibrous composites. We consider a plate with a circular hole subjected to uniaxial tension perpendicular to the fiber direction. The materials properties are taken from [108]: fiber volume fraction $\phi^{(f)}=0.267$, Young's moduli $E^{(f)}=379$ GPa, $E^{(m)}=69$ GPa, Poisson's ratio $\nu^{(f)}=0.21$, and $\nu^{(m)}=0.33$. Parameters for the damage evolution law (4.307) are $\alpha^{(m)}=8.2, \beta^{(m)}=10.2, \hat{\varepsilon}_0^{(m)}=0.05\ (\text{MPa})^{1/2}$. For low cycle fatigue, $\gamma^{(m)}=4.5$. The static load capacity of the plate is 103.6 N. The cyclic loading is designed as a tension-to-zero loading with amplitude of 90 N. The accuracy is controlled by the maximum allowable damage parameter in a cycle $\Delta\omega_a^{(m)}$. The evolution of the maximum damage parameter in the matrix versus the number of cycles is shown in Figure 4.30.

4.9 Extension to Multiphysics Problems

Analysis of multiple physical processes at multiple scales has been an active area of research in recent years. Terada and Kurumatani developed a two-scale diffusion–deformation coupling model for material deterioration involving microcrack propagation [116], Ozdemir *et al.* studied the thermo-mechanical behavior of heterogeneous solids at multiple scales [117], Yu and Fish analyzed coupled thermo-viscoelastic solids at multiple spatial and temporal scales [118], and Kuznetsov and Fish developed a homogenization theory for electroactive continua [119]. Various staggered and monolithic solution techniques for multiscale-multiphysics problems have been studied in [120,121].

This section is concerned with the formulation and model reduction of a coupled vector-scalar field problem at multiple scales. As an example, we choose a coupled mechano-diffusion-reaction problem applied to high temperature polymer-matrix composites (PMCs) (such as carbon fiber/polyimide) and nonoxide ceramic-matrix composites (CMCs) (such as a SiC/SiC material system). The CMC is used in the hottest sections of advanced aircraft engines and land turbines, such as combustor liners and jet exhaust vanes, whereas high-temperature PMC is used in propulsion systems, such as turbine engines and engine exhaust washed structures.

4.9.1 Reduced Order Coupled Vector-Scalar Field Model at Multiple Scales

We start by stating a strong form of the governing equations of the coupled mechano-diffusion-reaction problem on the composite domain Ω^ζ

$$q^\zeta_{i,j} - r^\zeta(c^\zeta) = \dot{c}^\zeta \quad \text{on} \quad \Omega^\zeta \tag{4.312a}$$

$$\sigma^\zeta_{ij,j} + b^\zeta_i = 0 \quad \text{on} \quad \Omega^\zeta \tag{4.312b}$$

$$q^\zeta_i = D'^\zeta_{ij}\left(\tilde{s}^\zeta_D, \tilde{s}^\zeta_M\right) c^\zeta_{,j} \quad \text{on} \quad \Omega^\zeta \tag{4.312c}$$

$$\sigma^\zeta_{ij} = L^\zeta_{ijkl}\left(\varepsilon^\zeta_{kl} - \mu^\zeta_{kl}\left(\tilde{s}^\zeta_D, \tilde{s}^\zeta_M\right)\right) \quad \text{on} \quad \Omega^\zeta \tag{4.312d}$$

$$\varepsilon^\zeta_{ij}(\boldsymbol{x}) = u^\zeta_{(i,j)}(\boldsymbol{x}) \equiv \frac{1}{2}\left(u^\zeta_{i,j}(\boldsymbol{x}) + u^\zeta_{j,i}(\boldsymbol{x})\right) \quad \boldsymbol{x} \in \Omega^\zeta \tag{4.312e}$$

with initial and boundary conditions assumed to be functions of the coarse-scale coordinates

$$\begin{aligned}
u^\zeta_i(\boldsymbol{x},t) &= \hat{u}^\zeta_i(\boldsymbol{x}) && \boldsymbol{x} \in \Omega^\zeta, \quad t=0 \\
c^\zeta(\boldsymbol{x},t) &= \hat{c}^\zeta(\boldsymbol{x}) && \boldsymbol{x} \in \Omega^\zeta, \quad t=0 \\
u^\zeta_i(\boldsymbol{x},t) &= \bar{u}_i(\boldsymbol{x},t) && \boldsymbol{x} \in \partial\Omega^{u\zeta}, \quad t \in [0,t_o] \\
\sigma^\zeta_{ij}(\boldsymbol{x},t) n^\zeta_j &= \bar{t}^\zeta_i(\boldsymbol{x},t) && \boldsymbol{x} \in \partial\Omega^{t\zeta}, \quad t \in [0,t_o] \\
c^\zeta(\boldsymbol{x},t) &= \bar{c}^\zeta(\boldsymbol{x},t) && \boldsymbol{x} \in \partial\Omega^{c\zeta}, \quad t \in [0,t_o] \\
q^\zeta_n(\boldsymbol{x},t) &= \bar{q}_n(\boldsymbol{x},t) && \boldsymbol{x} \in \partial\Omega^{q\zeta}, \quad t \in [0,t_o]
\end{aligned} \tag{4.313}$$

where c^ζ is the fluid (or gas, such as oxygen) concentration, q^ζ_i is the diffusion flux, r^ζ is the fluid (oxygen) consumption rate, and D'^ζ_{ij} is the nonlinear diffusivity. The mechanical and diffusion-reaction problems are coupled by the dependence of μ^ζ_{kl} and D^ζ_{ij} on mechanical state variables $\tilde{s}^\zeta_M$ and diffusion state variables $\tilde{s}^\zeta_D$. For instance, eigenstrain can be a function of oxidation-induced degradation or shrinkage, which is quantified by the diffusion state variables $\tilde{s}_D^{\zeta}$. Similarly, crack formation, which is a function of mechanical state variables $\tilde{s}_M^{\zeta}$, may create new pathways for fluid penetration and, in turn, increase diffusivity. $\bar{c}^{\zeta}$ is the concentration specified on the Dirichlet boundary $\partial\Omega^{c\zeta}$, and $\bar{q}_n$ is the flux defined on the Neumann boundary $\partial\Omega^{q\zeta}$. We assume that the environmental conditions are specified directly on the boundary, that is, $\partial\Omega^{c\zeta} \cup \partial\Omega^{q\zeta} = \partial\Omega^{\zeta}$, $\partial\Omega^{c\zeta} \cap \partial\Omega^{q\zeta} = 0$, or otherwise the convection boundary condition must be specified.

For the nonlinear diffusion problem, it is convenient to restate Fick's first law in terms of damage-free diffusivity D_{ik}^{ζ} by additively decomposing the oxygen concentration gradient $c_{,j}^{\zeta}$ into the damage-free concentration $c_{,j}^{\zeta df}$ and so-called eigenconcentration gradient η_j^{ζ}

$$c_{,j}^{\zeta} = c_{,j}^{\zeta df} + \eta_j^{\zeta} \tag{4.314}$$

Consequently, Fick's first law in (4.312c) can be rewritten in terms of damage-free diffusivity D_{ik}^{ζ}

$$\begin{aligned} q_i^{\zeta} &= D_{ij}'^{\zeta} c_{,j}^{\zeta} \\ &= D_{ik}^{\zeta} (D_{lk}^{\zeta})^{-1} D_{lj}'^{\zeta} c_{,j}^{\zeta} \\ &= D_{ik}^{\zeta} [\delta_{kj} + (D_{lk}^{\zeta})^{-1} D_{lj}'^{\zeta} - \delta_{kj}] c_{,j}^{\zeta} \\ &= D_{ik}^{\zeta} (c_{,k}^{\zeta} - \eta_k^{\zeta}) \end{aligned} \tag{4.315}$$

with eigenconcentration gradient η_k^{ζ} defined as

$$\eta_k^{\zeta} \equiv [(D_{lk}^{\zeta})^{-1} D'^{\zeta}_{lj} - \delta_{kj}] c_{,j}^{\zeta} \tag{4.316}$$

Both the displacements and oxygen concentration are expanded in a two-scale asymptotic expansion as

$$\begin{aligned} u_i^{\zeta} &= u_i(\boldsymbol{x}, \boldsymbol{y}) = u_i^c(\boldsymbol{x}) + \zeta u_i^{(1)}(\boldsymbol{x}, \boldsymbol{y}) + O(\zeta^2) \\ c^{\zeta} &= c(\boldsymbol{x}, \boldsymbol{y}) = c^c(\boldsymbol{x}) + \zeta c^{(1)}(\boldsymbol{x}, \boldsymbol{y}) + O(\zeta^2) \end{aligned} \tag{4.317}$$

and the two eigenfields are expanded in an asymptotic expansion around the leading order fields

$$\begin{aligned} \mu_{ij}^{\zeta} &= \mu_{ij}^f + O(\zeta) \\ \eta_i^{\zeta} &= \eta_i^f + O(\zeta) \end{aligned} \tag{4.318}$$

Inserting the above asymptotic expansions (4.317) into the governing equations (4.312) yields the two-scale coupled mechanical diffusion problem

$$
O\left(\zeta^{-1}\right):\ \begin{cases} q^f_{i,y_i}=0 \\ \sigma^f_{ij,y_j}=0 \\ u_i^{(1)}(\boldsymbol{x},\boldsymbol{y}),c^{(1)}(\boldsymbol{x},\boldsymbol{y}) \quad \boldsymbol{y}-\text{periodic} \end{cases}
$$

$$
O\left(\zeta^{0}\right):\ \begin{cases} q^c_{i,x_i}-r^c=\dot{c}^c \\ \sigma^c_{ij,x_j}+b^c_i=\rho^c\ddot{u}^c_i \\ u^c_i(\boldsymbol{x})=\overline{u}^c_i(\boldsymbol{x}) \quad \boldsymbol{x}\in\partial\Omega^u \\ c^c(\boldsymbol{x})=\overline{c}^c(\boldsymbol{x}) \quad \boldsymbol{x}\in\partial\Omega^c \\ \sigma^c_{ij}(\boldsymbol{x})n^c_j(\boldsymbol{x})=\overline{t}^c_i(\boldsymbol{x}) \quad \boldsymbol{x}\in\partial\Omega^t \\ q^c_i(\boldsymbol{x})=\overline{q}^c_i(\boldsymbol{x}) \quad \boldsymbol{x}\in\partial\Omega^q \end{cases} \tag{4.319}
$$

where coarse-scale mechanical fields are defined as in Chapter 2 and

$$
\begin{aligned}
c^c(\boldsymbol{x})&=\frac{1}{|\Theta|}\int_\Theta c^f(\boldsymbol{x},\boldsymbol{y})\ d\Theta \\
q^c_i(\boldsymbol{x})&=\frac{1}{|\Theta|}\int_\Theta q^f_i(\boldsymbol{x},\boldsymbol{y})\ d\Theta \\
r^c(\boldsymbol{x})&=\frac{1}{|\Theta|}\int_\Theta r(\boldsymbol{x},\boldsymbol{y})\ d\Theta \\
\overline{c}^c(\boldsymbol{x})&=\frac{1}{|\partial\omega|}\int_{\partial\omega}\overline{c}(\boldsymbol{x},\boldsymbol{y})\ ds \\
\overline{q}^c&=\frac{1}{|\partial\omega|}\int_{\partial\omega}\overline{q}(\boldsymbol{x},\boldsymbol{y})\ ds
\end{aligned} \tag{4.320}
$$

where for simplicty we omitted initial conditions and dependence on time. For the mechanical problem, $u_i^{(1)}$ decomposition in (4.16) and the model reduction procedure remain the same as decribed in section 4.2. For simplicity, in the following we will focus on eigenstrains.

For the diffusion problem, $c^{(1)}(\boldsymbol{x},\boldsymbol{y})$ is approximated in terms of the residual-free concentration gradient and eigenconcentration gradient influence functions

$$
c^{(1)}(\boldsymbol{x},\boldsymbol{y})=H^k(\boldsymbol{y})c^c_{,k}(\boldsymbol{x})+\int_\Theta h^k(\boldsymbol{y},\tilde{\boldsymbol{y}})\ \eta^f_k\left(\boldsymbol{x},\tilde{\boldsymbol{y}}\right)d\tilde{\Theta} \tag{4.321}
$$

By utilizing the piecewise constant (or hybrid impotent-incompatible) approximation of the eigenconcentration gradient $\eta_k^{(\alpha)}$ over $\Theta^{(\alpha)}$, the residual-free diffusion flux $q^f_i(\boldsymbol{x},\boldsymbol{y})$ can be expressed in terms of the partitioned eigenconcentration gradient $\eta_k^{(\alpha)}(\boldsymbol{x})$ at the fine scale as follows

$$q_i^f(\boldsymbol{x},\boldsymbol{y}) = D_{ij}(\boldsymbol{y})\left[E_j^k(\boldsymbol{y})c_{,k}^c(\boldsymbol{x}) + \sum_{\alpha=1}^{\tilde{M}} S_j^{k(\alpha)}(\boldsymbol{y})\eta_k^{(\alpha)}(\boldsymbol{x})\right] \tag{4.322}$$

with

$$\begin{aligned} E_i^j(\boldsymbol{y}) &= \delta_{ij} + H_{,y_i}^j(\boldsymbol{y}) \\ S_i^{j(\alpha)}(\boldsymbol{y}) &= P_i^{j(\alpha)}(\boldsymbol{y}) - \delta_{ij}^{(\alpha)}(\boldsymbol{y}) = \tilde{h}_{,y_i}^{j(\alpha)}(\boldsymbol{y}) - \delta_{ij}^{(\alpha)}(\boldsymbol{y}) \end{aligned} \tag{4.323}$$

The coefficient tensors $(H^j, \tilde{h}^{j(\alpha)})$ are governed by the following equations

$$\begin{aligned} &\left\{D_{ij}(\boldsymbol{y})\left[\delta_{jk}(\boldsymbol{y}) + H_{,y_j}^k(\boldsymbol{y})\right]\right\}_{,y_i} = 0 \quad \boldsymbol{y} \in \Theta \\ &\left\{D_{ij}(\boldsymbol{y})\left[\tilde{h}_{,y_j}^{k(\alpha)}(\boldsymbol{y}) - \delta_{jk}^{(\alpha)}(\boldsymbol{y})\right]\right\}_{,y_i} = 0 \quad \boldsymbol{y} \in \Theta \end{aligned} \tag{4.324}$$

subjected to periodicity conditions.

Note the similarity of these equations to those used for the mechanical problem. The main difference is that for the diffusion problem, the solution variables are scalar fields.

Again, the salient feature of the reduced order method for multiple physical processes is that the computation of the influence functions and the coefficient tensors is independent of, and conducted prior to, the nonlinear coarse-scale analysis. These computations are performed in the preprocessing stage. The resulting residual-free diffusion flux at the coarse-scale is given as

$$q_i^c = D_{ik}^c c_{,k}^c(\boldsymbol{x}) + \sum_{\alpha=1}^{\tilde{M}} A_{ik}^{c(\alpha)} \eta_k^{(\alpha)}(\boldsymbol{x}) \tag{4.325}$$

where

$$\begin{aligned} D_{ik}^c &= \frac{1}{|\Theta|}\int_\Theta D_{ij}(\boldsymbol{y})E_j^k(\boldsymbol{y})\,d\Theta \\ A_{ik}^{c(\alpha)} &= \frac{1}{|\Theta|}\int_\Theta D_{ij}(\boldsymbol{y})S_j^{k(\alpha)}(\boldsymbol{y})\,d\Theta \end{aligned} \tag{4.326}$$

Since the coefficient tensors $(D_{ik}^c, A_{ik}^{c(\alpha)})$ can be precomputed, the only quantity that needs to be updated at each increment of the coarse-scale analysis is the partitioned eigenconcentration gradient $\eta_k^{(\alpha)}$, which has a limited number of degrees of freedom. The governing equations for the reduced order system are given by

$$\Delta c_{,i}^{(\beta)} = E_i^{j(\beta)} \Delta c_{,j}^c + \sum_{\alpha=1}^{\tilde{M}} P_i^{j(\beta\alpha)} \Delta \eta_j^{(\alpha)} \tag{4.327}$$

At each iteration of the nonlinear coarse-scale analysis, the following reduced order coupled system is solved

$$\begin{cases} \Delta c_{,i}^{(\beta)} = E_i^{j(\beta)} \Delta c_{,j}^c + \sum_{\alpha=1}^{\tilde{M}} P_i^{j(\beta\alpha)} \Delta \eta_j^{(\alpha)}\left(s_D^\zeta, s_M^\zeta\right) \\ \Delta \varepsilon_{ij}^{(\beta)} = E_{ij}^{kl(\beta)} \Delta \varepsilon_{kl}^c + \sum_{\alpha=1}^{\tilde{M}} P_{ij}^{kl(\beta\alpha)} \Delta \mu_{kl}^{(\alpha)}\left(s_D^\zeta, s_M^\zeta\right) \end{cases} \tag{4.328}$$

The coarse-scale diffusion flux and stress can be updated as follows

$$\begin{aligned} q_i^c &= D_{ik}^c \, c_{,k}^c + \sum_{\alpha=1}^{\tilde{M}} A_{ik}^{c(\alpha)} \eta_k^{(\alpha)} \\ \sigma_{ij}^c &= L_{ijkl}^c \, \varepsilon_{kl}^c + \sum_{\alpha=1}^{\tilde{M}} A_{ijkl}^{c(\alpha)} \mu_{kl}^{(\alpha)} \end{aligned} \tag{4.329}$$

4.9.2 Environmental Degradation of PMC

As an example of the coupled vector-scalar field problem at multiple scales, we consider environmental degradation of high-temperature PMCs, such as PMR-15 polymer resins reinforced with carbon fibers, which are often employed in the aerospace industry. The experimental results suggest that prior thermal aging in air and in argon significantly affect mechanical properties as well as the time-dependent creep response of the PMR-15 neat resin at 288 °C [122].

As a starting point of the coupled mechano-diffusion-reaction model of the polymer oxidation process under stress, we introduce the polymer availability state variable φ [123] to distinguish between three material regions with different levels of oxidation. The state variable φ is parameterized to range in $\varphi_{ox} \le \varphi \le 1$, where φ_{ox} denotes the completely oxidized polymer and $\varphi = 1$ denotes the unoxidized polymer. In a region close to the external surface, $\varphi = \varphi_{ox}$; in the interior of the material, $\varphi = 1$, whereas in the transition region, $\varphi_{ox} < \varphi < 1$.

Coupling of physical processes at multiple scales is introduced, assuming that (i) diffusivity is a function of temperature T, oxidation state φ, eigenconcentration η_i, and strain ε_{ij}, that is, $D(T,\varphi,\eta_i,\varepsilon_{ij})$, and (ii) stress is a function temperature T, strain ε_{ij}, inelastic strain ε_{ij}^{in}, temperature T, and oxidation state φ, that is, $\sigma\left(T,\varepsilon_{ij},\varepsilon_{ij}^{in},\varphi\right)$. In the following, we will discuss the specific forms of these relations.

4.9.2.1 Deformation-Dependent Diffusion-Reaction Model

Following [124], we assume that diffusivity depends on the oxidation state via the rule of mixtures

$$
\begin{aligned}
&\boldsymbol{D}(T,\varphi,\varepsilon_{ij}) = \boldsymbol{D}^{\mathrm{un}}(T,\varepsilon_{ij})\left(\frac{\varphi-\varphi_{\mathrm{ox}}}{1-\varphi_{\mathrm{ox}}}\right) + \boldsymbol{D}^{\mathrm{ox}}(T)\left(\frac{1-\varphi}{1-\varphi_{\mathrm{ox}}}\right) \\
&\boldsymbol{D}^{\mathrm{un}}(T,\varepsilon_{ij}) = \boldsymbol{D}^{*}_{\mathrm{un}}(\varepsilon_{ij})e^{\frac{-E_a^{\mathrm{un}}}{RT}};\quad \boldsymbol{D}^{\mathrm{ox}}(T) = \boldsymbol{D}^{*}_{\mathrm{ox}}e^{\frac{-E_a^{\mathrm{ox}}}{RT}} \\
&\boldsymbol{D}^{*}_{\mathrm{un}}(\varepsilon_{ij}) = \boldsymbol{D}^{0}_{\mathrm{un}} + \left(\boldsymbol{D}^{\infty}_{\mathrm{un}} - \boldsymbol{D}^{0}_{\mathrm{un}}\right)\left(1-e^{-\lambda\|\varepsilon_{ij}\|}\right)
\end{aligned}
\tag{4.330}
$$

where the temperature dependence of diffusivity for unoxidized and oxidized regions is expressed in the usual Arrhenius form, in which $\boldsymbol{D}^*$ is the pre-exponential factor and E_a denotes the activation energy. In (4.328), $\boldsymbol{D}^{0}_{\mathrm{un}}, \boldsymbol{D}^{\infty}_{\mathrm{un}}$ denote the diffusivities under the strain-free state and the fully deformed state, respectively, with $\|\varepsilon_{ij}\|$ being the effective strain.

The reaction rate $r(c,T)$ depends on the available oxygen concentration, temperature, and oxidation state

$$
\begin{aligned}
&r(c,T) = \begin{cases} r_0(T)\dfrac{2\beta c}{1+\beta c}\left[1-\dfrac{\beta c}{2(1+\beta c)}\right] & \varphi > \varphi_{\mathrm{ox}} \\ 0 & \varphi = \varphi_{\mathrm{ox}} \end{cases} \\
&r_0(T) = r^{*}e^{\frac{-r_a}{RT}}
\end{aligned}
\tag{4.331}
$$

where β is a material parameter, $r_0(T)$ denotes the saturated reaction rate governed by the Arrhenius-type kinetics model where the rate constant r^* and the activation parameter r_a govern the temperature dependence of the reaction rate, and R is the universal gas constant.

The evolution of the oxidation state variable φ depends on the reaction rate through the weight loss model, assuming that the rate of change of weight is proportional to the reaction rate, that is, $d\varphi/dt \sim dW/dt \sim -r(c,T)$. The weight loss model is formulated [123] as

$$
\begin{aligned}
&d\varphi/dt = -\alpha r(c,T) \qquad \text{for } \varphi_{\mathrm{ox}} < \varphi < 1 \\
&\alpha = \begin{cases} \alpha^1 - \left(\alpha^1 - \alpha^2\right)\left(\dfrac{t}{t^*}\right) & t < t^* \\ \alpha^2 & t \geq t^* \end{cases}
\end{aligned}
\tag{4.332}
$$

where α^1, α^2 are material parameters and t^* denotes the time when the weight loss rate reaches the constant value for longer aging time.

4.9.2.2 Oxidation-Dependent Mechanical Model

To model the effect of oxidation on the mechanical behavior of a polymer, we consider the viscoplasticity based on the overstress (VBO) model. The VBO model was originally developed by Krempl and coworkers for modeling the rate-dependent response of metals and polymers [125]. The framework was derived from the standard linear solid (SLS) model, which is a combination of the classical Maxwell and Kelvin models and is capable of predicting both creep and relaxation.

The oxidation is assumed to induce additional strain on the mechanical field, such as shrinkage caused by the weight loss of the material. This effect is modeled by adding the additional term $\varepsilon_{kl}^{\text{ox}}(\varphi)$ to the inelastic mechanical constitutive relation

$$\sigma_{ij} = L_{ijkl}\left(\varepsilon_{kl} - \varepsilon_{kl}^{in} - \varepsilon_{kl}^{\text{ox}}(\varphi)\right) \tag{4.333}$$

The oxidation-induced strain is employed to model shrinkage as follows

$$\varepsilon_{ij}^{\text{ox}} = \begin{cases} -\dfrac{1-\varphi}{1-\varphi^{\text{ox}}} \cdot \left|\varepsilon_{\text{shrink}}^{ub}\right| & \text{for } i = j \\ 0 & \text{for } i \neq j \end{cases} \tag{4.334}$$

where $\varepsilon_{\text{shrink}}^{ub}$ is the material parameter denoting the upper bound of shrinkage strain.

The inelastic strain rate is assumed to be volume preserving as

$$\begin{aligned} \dot{\varepsilon}_{ij}^{in} &= \dot{\eta}_{ij}^{in} = \sqrt{\frac{3}{2}}\,\bar{\dot{\eta}}^{in} n_{ij} \\ \bar{\dot{\eta}}^{in} &= \frac{\Gamma}{Ek(\Gamma)};\; n_{ij} = \sqrt{\frac{3}{2}}\frac{o_{ij}}{\Gamma} \end{aligned} \tag{4.335}$$

where E is the Young's modulus, k is the viscosity coefficient, $\dot{\eta}_{ij}^{in}$ is the deviatoric inelastic strain, and Γ is the scalar invariant of overstress deviator defined by

$$\begin{aligned} \Gamma &= \sqrt{\frac{3}{2} o_{ij} o_{ij}} \\ o_{ij} &= s_{ij} - g_{ij} \end{aligned} \tag{4.336}$$

where o_{ij} is the overstress deviator, s_{ij} is the deviatoric stress, and g_{ij} is the equilibrium stress deviator. The evolution equation of the equilibrium stress deviator is defined as

$$\dot{g}_{ij} = \frac{\psi}{E}\left[\dot{s}_{ij} + \frac{o_{ij}}{k} - \frac{\Gamma}{k}\frac{\left(g_{ij} - f_{ij}\right)}{A}\right] + \left(1 - \frac{\psi}{E}\right)\dot{f}_{ij} \tag{4.337}$$

where ψ is the positive shape function that models the transition from a quasi-linear elastic behavior to inelastic flow, given by

$$\psi = a_1 + (a_2 - a_1)e^{-a_3\Gamma} \tag{4.338}$$

A is the isotropic stress introduced to model the cyclic hardening or softening behavior. The evolution form is given by

$$\dot{A} = A_c\left(A_f - A\right)\dot{\bar{\eta}}^{in} \tag{4.339}$$

with the initial condition

$$A(t=0) = A_0 \tag{4.340}$$

where a_1, a_2, a_3, A_c, A_f, and A_0 are model parameters. f_{ij} is the kinematic stress introduced to model the work hardening.

The evolution of the kinematic stress is given as

$$\dot{f}_{ij} = \bar{E}_t \dot{\eta}_{ij}^{in} \tag{4.341}$$

with $\bar{E}_t$ defined by

$$\bar{E}_t \equiv \frac{2}{3}\frac{E_t}{(1 - E_t/E)} \tag{4.342}$$

where E_t is the terminating slope of the uniaxial stress–strain curve at the maximum strain of interest. For more details, we refer to [126].

4.9.3 Validation of the Multiphysics Model

The model parameters for the VBO model of PMR-15 are given in [127]. We first consider a diffusion-reaction model of a fibrous composite unit cell with a carbon fiber volume fraction of 56%. The carbon fiber T650-35 is assumed to be a transversely orthotropic elastic with material properties for each phase from [128]. In the present study, each microphase is modeled as a single partition.

For validation of the diffusion-reaction model of a fibrous composite, we consider the experimental data reported in [129] for unidirectional G30-500/PMR-15 composites aged in air at 288 °C. An anisotropic oxidative response is observed where materials degrade preferentially from the specimen surface perpendicular to the fibers. The oxidation growth in the transverse direction is observed to be the same as the oxidation growth in the neat resin, whereas the growth in the axial direction is much higher. Figure 4.31 shows that oxidation substantially increases in both the axial and transverse directions with aging time.

The diffusivity of the fiber itself is typically negligible compared with the matrix diffusivity. It is believed that the higher oxidation growth in the axial direction is caused by fiber–matrix interface (or interphase) degradation. A fibrous composite unit cell model is employed with an

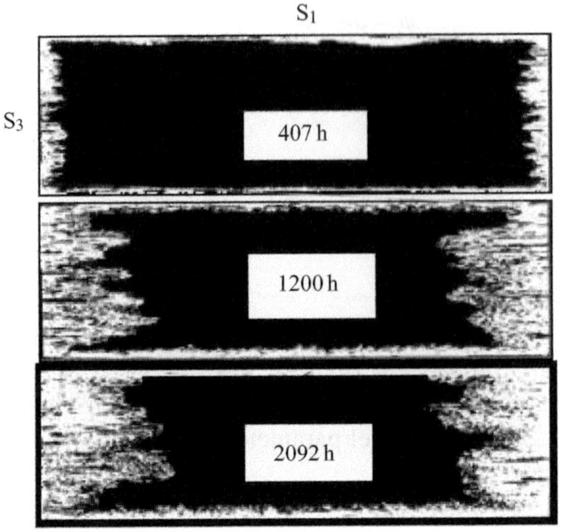

Figure 4.31 Unidirectional G30-500/PMR-15 at 407, 1200, and 2092 h of oxidation at 288 °C. Reproduced from [129], © 2007 Elsevier

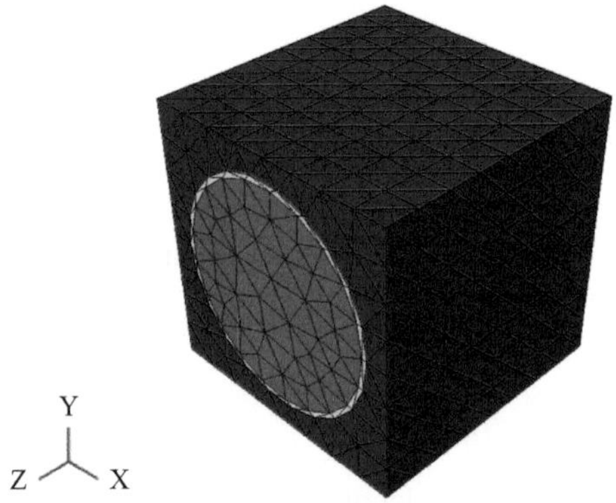

Figure 4.32 The unit cell model for the unidirectional G30-500/PMR-15 composite

explicit consideration of a thin layer of an interphase surrounding the fiber phase, as shown in Figure 4.32. The volume fraction of the fiber is 50% and the thickness of the interphase is set to be 5% of the radius of the fiber. The physical length of the unit cell is 12.5 μm.

The diffusivity of the fiber phase is assumed to be isotropic, 0.1% of the diffusivity of the matrix phase. The diffusivity of the interphase is assumed to be transversely orthotropic, the value of which is derived from the macro oxidation response. The identified diffusivities for each microphase and the corresponding effective diffusivities for the composite are summarized in Table 4.14.

Due to symmetry, one-quarter of the specimen of size 5250 μm × 1600 μm × 25 μm (Figure 4.31) is considered in the finite element analysis. The left (S_3) and top (S_1) surfaces are the outer boundaries exposed to oxygen.

Table 4.14 Micro and macro diffusivities (mm²/min⁻¹)

PMR-15 matrix (isotropic)	3.66×10^{-3}
G30-500 fiber (isotropic)	3.66×10^{-6}
Interphase (the transverse direction)	7.32×10^{-2}
Interphase (the axial direction)	3.66
The corresponding composite model	
Effective diffusivity (the transverse direction)	3.70×10^{-3}
Effective diffusivity (the axial direction)	1.90×10^{-1}

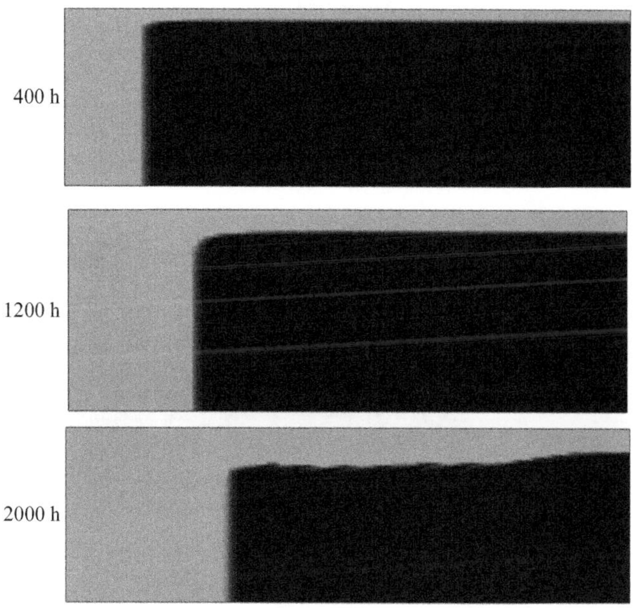

Figure 4.33 The quarter model after 400, 1200, and 2000 h of oxidation at 288 °C

Figure 4.33 shows the oxidation growth in the axial and transverse directions as a function of aging time. Figure 4.34 compares the numerical results of the oxidation growth in the axial and transverse directions with the experimental data.

To this end, we consider a coupled mechano-diffusion-reaction problem. A fully oxidized state variable is given as $\varphi^{\text{ox}}=0.18$. The upper bound of shrinkage strain $\left|\varepsilon_{\text{shrink}}^{ub}\right|$ is set to be 1%. All other material constants are the same as in the previous example. We consider the same macro problem as in Figure 4.31. Figure 4.35 shows the oxidation state after 2000 h.

Figure 4.36 compares the transverse stress as obtained without [Figure 4.36(a)] and with [Figure 4.36(b)] coupling of mechanical and diffusion reaction fields. The influence of oxidation can be clearly seen.

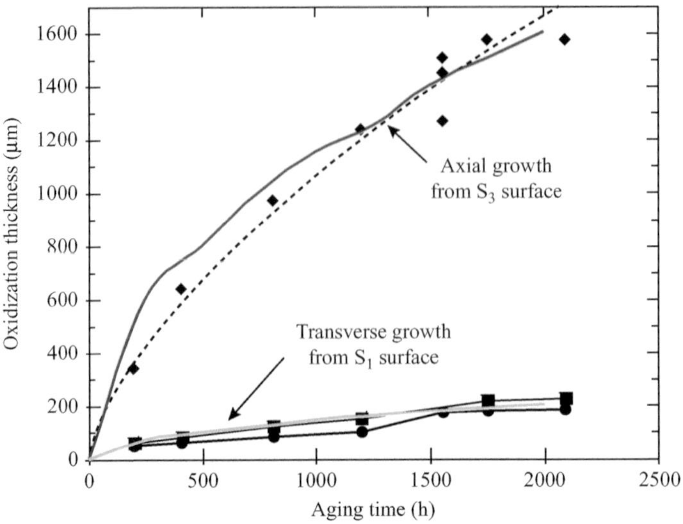

Figure 4.34 Comparison of the oxidation growth in the axial and transverse directions: experimental results shown in black and simulation results shown in color

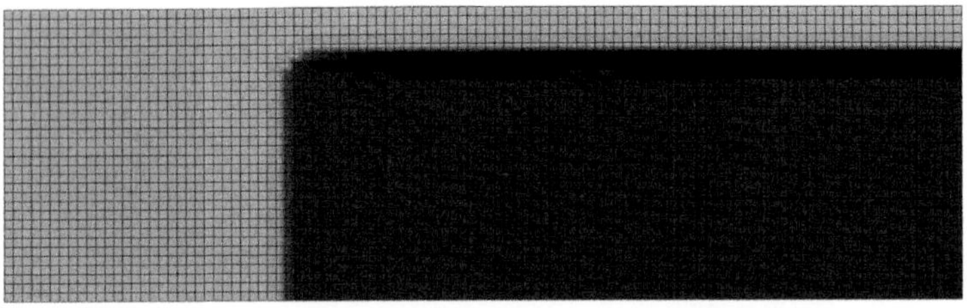

Figure 4.35 Oxidation at 288 °C after 2000 h (gray: oxidized; black: unoxidized)

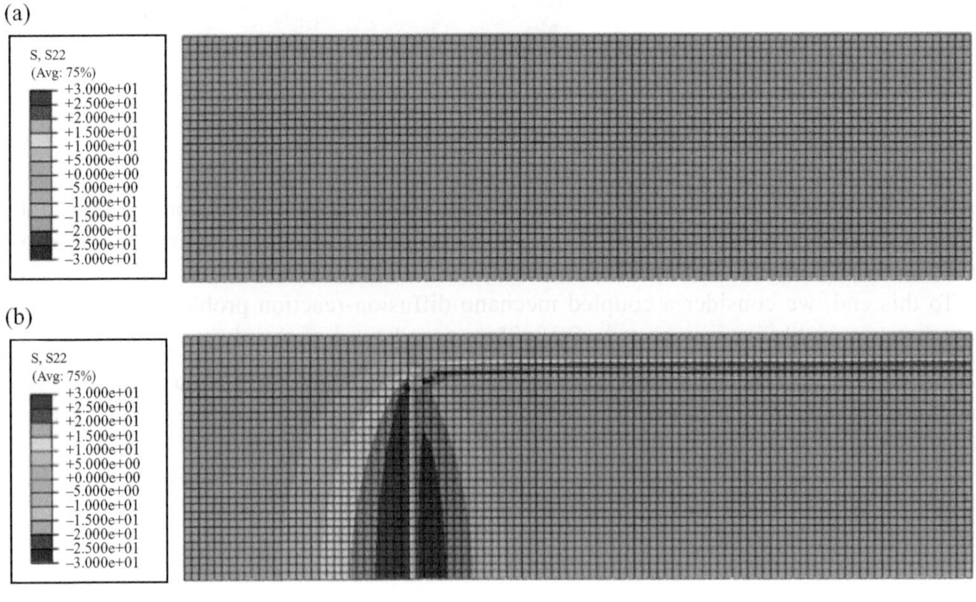

Figure 4.36 Comparison of stress in the transverse direction (a) without and (b) with coupled fields

4.10 Multiscale Characterization

4.10.1 Formulation of the Inverse Problem

Let vector $z \in R^N$ denote the unit cell parameters, describing the geometry and constitutive models of phases/interfaces. The inverse (or optimization) problem is stated as follows.

Find the vector z that minimizes the objective function

$$\Phi(z) = \frac{1}{2}\sum_{\alpha=1}^{L}\theta^{\alpha} f^{\alpha}(z) \quad l_i \le z_i \le u_i \quad (i = 1,\ldots,N) \tag{4.343}$$

where L is the total number of specific response quantities (denoted by α) that can be both measured experimentally and computed numerically; $f^{\alpha}(z)$ is the dimensionless function

$$f^{\alpha}(z) = \frac{\sum_{s=1}^{S_\alpha}\left[F_s^{\alpha} - F^{\alpha}\left(z,\tau_s^{\alpha}\right)\right]^2}{\sum_{s=1}^{S_\alpha}\left[F_s^{\alpha}\right]^2} \tag{4.344}$$

that measures the difference between the computed and experimentally measured response quantity α; τ^{α} is a parameter that defines the history of the process over the course of the experiment (e.g., a time or loading parameter), while the values $\tau_s^{\alpha}(\alpha=1,\ldots,L;\ s=1,\ldots,S_\alpha)$ define the discrete set of S_α data points; F_s^{α} is the value of the αth measured response quantity at the observation time τ_s^{α}; $F^{\alpha}\left(z,\tau_s^{\alpha}\right)$ is the value of the same response quantity obtained from the numerical simulation; θ^{α} is the weight coefficient that determines the relative contribution of information yielded by the α th set of experimental data; and l_i and u_i are side constraints stipulated by additional physical considerations, such as manufacturing, that define the search region in the space R^N of optimization parameters.

The weight coefficient θ^{α} may be chosen by various criteria: (i) an identity matrix to provide equal weight to different experiments; or (ii) a diagonal of covariance matrix (for uncorrelated quantities).

It is convenient to recast the constrained minimization problem (4.343) and (4.344) in the matrix form as

$$\Phi(z) = \frac{1}{2}\boldsymbol{F}(z)^T \boldsymbol{W}\boldsymbol{F}(z) \qquad l_i \le z_i \le u_i \quad (i = 1,\ldots,N) \tag{4.345}$$

where $\boldsymbol{F}(z) \in R^n$ and diagonal matrix $\boldsymbol{W} \in R^{n \times n}$ $\left(n = \sum_{i=1}^{L} S_i \right)$ are given by

$$\boldsymbol{F}(z) = \left[\left(F_1^1 - F^1\left(z, \tau_1^1\right)\right), \ldots, \left(F_{S_1}^1 - F^1\left(z, \tau_{S_1}^1\right)\right), \ldots, \left(F_1^N - F^N\left(z, \tau_1^N\right)\right), \ldots, \left(F_{S_N}^N - F^N\left(z, \tau_{S_N}^N\right)\right)\right]$$
$$\text{diag}(\boldsymbol{W}) = \left[\theta^1 \Big/ \sum_{s=1}^{S_1} \left[F_s^1\right]^2, \ldots, \theta^1 \Big/ \sum_{s=1}^{S_1} \left[F_s^1\right]^2, \ldots, \theta^N \Big/ \sum_{s=1}^{S_N} \left[F_s^N\right]^2, \ldots, \theta^N \Big/ \sum_{s=1}^{S_N} \left[F_s^N\right]^2 \right] \tag{4.346}$$

The well-known Gauss–Newton algorithm consists of minimizing the objective function, that is, finding z such that $\dfrac{\partial \Phi}{\partial z} = \mathbf{0}$ and using a linear approximation of $\boldsymbol{F}(z)$, which yields

$$^{k+1}z = {}^k z - \left({}^k\boldsymbol{J}^T \ \boldsymbol{W} \ {}^k\boldsymbol{J} \right)^{-1} {}^k\boldsymbol{J}^T \ \boldsymbol{W} \ {}^k\boldsymbol{F} \tag{4.347}$$

where the left superscript indicates the iteration count. ${}^k\boldsymbol{J} = \left.\dfrac{\partial \boldsymbol{F}}{\partial z}\right|_{{}^k z}$ is a Jacobian matrix that is a function of the sensitivity matrix ${}^kJ_{\alpha i} = \left.\dfrac{\partial F^\alpha}{\partial z_i}\right|_{{}^k z}$. The Gauss–Newton method makes use of the Hessian matrix approximation $\boldsymbol{H} \approx {}^k\boldsymbol{J}^T \ \boldsymbol{W} \ {}^k\boldsymbol{J}$ instead of either evaluating the exact second derivatives of the objective function $H_{ij} = \dfrac{\partial^2 \Phi}{\partial z_i \partial z_j}$ as required by the Newton method or approximating it in the quasi-Newton method.

The sensitivity matrix ${}^kJ_{\alpha i}$ can be determined either analytically or by using finite difference, which requires computation of the observations for small variations of each of the parameter values. While the latter is time consuming since the number of necessary multiscale simulations increase linearly with the number of unknown parameters, it provides a nonintrusive strategy that makes use of the existing multiscale solver as a black box.

To overcome the potential singularity or near singularity of the Hessian matrix approximation ${}^k\boldsymbol{J}^T \ \boldsymbol{W} \ {}^k\boldsymbol{J}$, the Levenberg–Marquardt algorithm [130] perturbs the Hessian matrix approximation by a positive definite matrix $\mu \boldsymbol{I}$ and guides the iterative process

$$^{k+1}\hat{z} = {}^k\hat{z} - \left({}^k\hat{\boldsymbol{J}}^T \ \hat{\boldsymbol{W}} \ {}^k\hat{\boldsymbol{J}} + \mu \boldsymbol{I} \right)^{-1} {}^k\hat{\boldsymbol{J}}^T \ \hat{\boldsymbol{W}} \ {}^k\hat{\boldsymbol{F}} \tag{4.348}$$

by an active parameter strategy, which imposes simple bounds on the Levenberg–Marquardt parameter μ and on the trust region [131] radius Δ. In (4.348), $\hat{\boldsymbol{F}}$ contains components of $\boldsymbol{F}$ for model parameters that lie inside bounds (that is, $i = 1, 2, \ldots, N_a$ such that $l_i \leq z_i \leq u_i$), $\hat{\boldsymbol{J}}$ is the Jacobian of $\hat{\boldsymbol{F}}$, and μ is selected to ensure that iterates lie within the trust region radius Δ

(that is, $\mu=0$ if $\left\| {}^k\hat{\boldsymbol{J}}^T \hat{\boldsymbol{W}} \, {}^k\hat{\boldsymbol{F}} \right\|_2 \leq \Delta$ and $\mu>0$ otherwise) where μ and Δ are computed based on a locally constrained hook step algorithm ([132], also see [133] for other variants of the method).

4.10.2 Characterization of Model Parameters in ROH

The unit cell model parameters include elastic constitutive model parameters of phases (such as Young's modulus and Poisson's ratio), inelastic constitutive model parameters of phases and interfaces (such as eigenstrain and eigenseparation evolution law parameters), and geometric unit cell parameters (such as volume fraction, parametric geometry definition of inclusions, etc.).

The sequence in which the various model parameters are characterized is as follows:

1. *Elastic constitutive model parameters of phases*
   The influence functions comprising the material database depend on elastic properties and therefore have to be characterized (identified) first. In this process, we have to consider experiments for which the unit cell geometry is known either from micrographs or from manufacturers' specifications.
2. *Inelastic constitutive model parameters of phases and interfaces*
   Once the elastic properties have been characterized (identified), the inelastic constitutive parameters of phases and interfaces can be identified. As in Step 1, only experiments for which the precise microstructural geometry is known should be considered. In many situations, however, the inverse problem (4.345) for characterization of inelastic model parameters may lead to ill-posed problems, where small variations in observable response F_s^α may lead to large variations in inelastic model parameters. One possible way to address this issue is to employ a multistep characterization process by which only a subset of parameters is characterized at a time against experiments that mostly affect these parameters.
3. *Fine-scale geometric parameters*
   When the unit cell microstructure is unknown, this step has to be performed simultaneously with the previous two steps. It involves a considerable computational cost that stems from continuous generation of the unit cell geometry and meshing. This step can be also used for unit architecture (geometry) optimization rather than characterization. The input to such a material optimization is a desired overall response, which can be stated by modifying the existing experimental response.

Problems

Problem 4.1: Derive reduced order homogenization equations for a heat transfer problem.

Problem 4.2: Derive reduced order homogenization equations for a coupled thermo-mechanical problem.

Problem 4.3: Consider a 1D model problem with inelastic behavior of phases and interfaces. The unit cell consists of two homogenous materials, each occupying one-half of the unit cell domain schematically depicted in Figure 4.37. The Young's modulus of the two materials is denoted by $E^{(1)}$ and $E^{(2)}$, respectively.

Construct the reduced order unit cell equations.

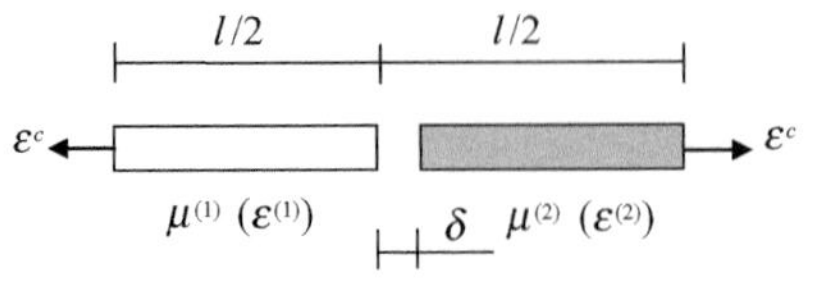

Figure 4.37 The two-scale 1D model problems with inelastic phases and imperfect interface

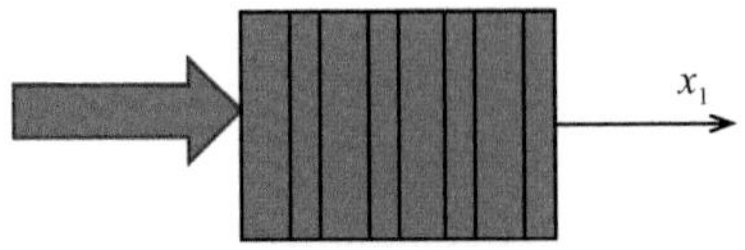

Figure 4.38 Wave propagation in a layered medium

Problem 4.4: Consider wave propagation in a layered medium. The wave is propagating in the x_1 direction, which is normal to the direction of the layering as shown in Figure 4.38. Assume the unit cell size in the physical domain to be equal to l_x and volume fractions of phases $\phi^{(\alpha)}$. Assume a plane strain condition in the x_2 and x_3 directions, so that the only nonzero coarse-scale strain component is $\varepsilon_{11}^c \neq 0$. Assume that material properties, Young's moduli, shear moduli, and densities $E^{(\alpha)}, G^{(\alpha)}, \rho^{(\alpha)}$ are given.
Calculate:

(a) Displacement influence function H_i^{mn}.
(b) Normalized acoustic impedance variation λ.
(c) Dispersion coefficient D^c.

Problem 4.5: Consider damage evolution law

$$\dot{\omega} = \left(c\ e^{\alpha\omega}\overline{\varepsilon}^{\beta}\right)\dot{\overline{\varepsilon}}$$

where $\overline{\varepsilon}$ is effective strain and α, β, c are material constants. Assume that $\overline{\varepsilon}$ is equal to constant strain amplitude and $\dot{\omega} \approx \dfrac{d\omega}{dN}$, where N is the number of cycles.

(a) Show that

$$\omega = -\frac{1}{\alpha}\text{In}\left(1 - \frac{2\alpha c}{\beta+1}\varepsilon_{\alpha}^{\beta+1}N\right)$$

(b) Find the number of cycles to failure by equating.
(c) Compare qualitatively the above approach [134] to the block cycle technique.

Problem 4.6: Show that for 1D problems, the compatible and incompatible eigenstrain influence functions are identical, that is, $\tilde{P}^{(\alpha\beta)} = P^{(\alpha\beta)}$.

References

[1] Ghosh, S. and Moorthy, S. Elastic-Plastic analysis of heterogeneous microstructures using the Voronoi cell finite element method. Computer Methods in Applied Mechanics and Engineering 1995, 121(1–4), 373–409.
[2] Ghosh, S. Micromechanical Analysis and Multi-Scale Modeling Using the Voronoi Cell Finite Element Method. CRC Series in Computational Mechanics and Applied Analysis. Taylor & Francis US, 2010.
[3] Aboudi, J. A continuum theory for fiber-reinforced elastic-viscoplastic composites. Journal of Engineering Science 1982, 20(55), 605–621.
[4] Berlyand, L.V. and Kolpakov, A.G. Network approximation in the limit of small inter-particle dispersed composite. Archive for Rational Mechanics and Analysis 2001, 159, 179–227.
[5] Mouli.nec, H. and Suquet, P. A fast numerical method for computing the linear and nonlinear properties of composites. Comptes Rendus de l'Académie de Sciences Paris II 1994, 318, 1417–1423.
[6] Moulinec, H. and Suquet, P. A numerical method for computing the overall response of nonlinear composites with complex microstructure. Computer Methods in Applied Mechanics and Engineering 1998, 157, 69–94.
[7] Cusatis, G., Mencarelli, A., Pelessone, D. and Baylot, J.T. Dynamic pull-out test simulations using the lattice discrete particle model (LDPM). In Proceedings of the 2008 ASCE Structures Congress, 2008.
[8] Cusatis, G., Pelessone, D., Mencarelli, A. and Baylot, J.T. Simulation of reinforced concrete structures under blast and penetration through lattice discrete particle modeling. In Proceedings of IMECE 2007, 2007.
[9] Chen, J.S., Pan, C., Wu, C.T. and Liu, W.K. Reproducing kernel particle methods for large deformation analysis of nonlinear structures. Computer Methods in Applied Mechanics and Engineering 1996, 139, 195–227.
[10] Chen, J.S., Wu, C.T., Yoon, S. and You, Y. A stabilized conforming nodal integration for Galerkin mesh-free methods. International Journal for Numerical Methods in Engineering 2001, 50, 435–466.
[11] Cavalcante, M.A.A., Khatam, H. and Pindera, M.J. Homogenization of elastic–plastic periodic materials by FVDAM and FEM approaches – An assessment. Composites Part B: Engineering 2011, 42(6), 1713–1730.
[12] Dvorak, G.J. Transformation field analysis of inelastic composite materials. Proceedings of the Royal Society of London, Series A 1992, 437, 311–327.
[13] Bahei-El-Din, Y.A., Rajendran, A.M. and Zikry, M.A. A micromechanical model for damage progression in woven composite systems. International Journal of Solid Structures 2004, 41, 2307–2330.
[14] Paley, M. and Aboudi, J. Micromechanical analysis of composites by the generalized cells model. Mechanics of Materials 1992, 14, 127–139.
[15] Moore, B.C. Principal component analysis in linear systems: controllability, observability, and model reduction. IEEE Transactions on Automatic Control 1981, 26, 17–32.
[16] Green, M. A relative-error bound for balanced stochastic truncation. IEEE Transactions on Automatic Control 2988, 33(10), 961–965.
[17] Glover, K. All optimal Hankel-norm approximation of linear multivariable systems and their L- error bounds. International Journal of Control 1984, 39, 1115–1193.
[18] Krysl. P., Lall, S., and Marsden, J.E. Dimensional model reduction in non-linear finite element dynamics of solids and structures. International Journal for Numerical Methods in Engineering 2001, 51(4), 479–504.
[19] Dvorak, G.J. and Benveniste, Y. On transformation strains and uniform fields in multiphase elastic media. Proceedings of the Royal Society of London, Series A 1992, 437, 291–310.
[20] Laws, N. On the thermostatics of composite materials. Journal of the Mechanics and Physics of Solids 1973, 21, 9–17.
[21] Willis, J. Variational and related methods for the overall properties of composites. In Advances in Applied Mechanics, ed. C.S. Yih. Academic Press, 1981, vol. 21, pp. 1–78.
[22] Dvorak, G.J. On uniform fields in heterogeneous media. Proceedings of the Royal Society of London, Series A 1990, 431, 89–110.
[23] Oskay, C. and Fish, J. Fatigue life prediction using 2-scale temporal asymptotic homogenization. Computational Mechanics 2008, 42(2), 181–195.
[24] Oskay, C. and Fish, J. Eigendeformation-based reduced order homogenization. Computer Methods in Applied Mechanics and Engineering 2007, 196, 1216–1243.
[25] Yuan, Z. and Fish, J. Multiple scale eigendeformation-based reduced order homogenization. Computer Methods in Applied Mechanics and Engineering 2009, 198(21–26), 2016–2038.
[26] Yuan, Z. and Fish, J. Hierarchical model reduction at multiple scales. International Journal for Numerical Methods in Engineering 2009, 79, 314–339.

[27] Fish, J. and Yuan, Z. N-scale model reduction theory. In Bridging the Scales in Science and Engineering, ed. J. Fish. Oxford University Press, 2008.
[28] Fish, J., Filonova, V. and Yuan, Z. Hybrid impotent-incompatible eigenstrain based homogenization. International Journal for Numerical Methods in Engineering 2013, 95(1), 1–32.
[29] Fish, J., Shek, K., Pandheeradi, M. and Shephard, M.S. Computational plasticity for composite structures based on mathematical homogenization: theory and practice. Computer Methods in Applied Mechanics and Engineering 1997, 148, 53–73.
[30] Fish, J., Yu, Q. and Shek, K.L. Computational damage mechanics for composite materials based on mathematical homogenization. International Journal for Numerical Methods in Engineering 1999, 45, 1657–1679.
[31] Ladeveze, P., Allix, O., Deu, J.-F. and Leveque, D. A mesomodel for localisation and damage computation in laminates. Computer Methods in Applied Mechanics and Engineering 2000, 183, 105–122.
[32] Mandel, J. Une generalisation de la theorie de la plasticite de W.T. Koiter. International Journal of Solids and Structures 1965, 1, 273–295.
[33] Levin, V.M. On the coefficients of thermal expansion of heterogeneous materials. Mechanics of Solids 1967, 2, 58–61.
[34] Mindlin, R.D. and Cheng, D.H. Nuclei of strain in the semi-infinite solid. Journal of Applied Physics 1950, 21(9), 926–930.
[35] Eshelby, J.D. The determination of the elastic field of an ellipsoidal inclusion, and related problems. Proceedings of the Royal Society of London, Series A 1957, 241(1226), 376–396.
[36] Mura, T. Micromechanics of Defects in Solids. Martinus Nijhoff, 1982.
[37] Teply, J. and Dvorak, G. Bounds on overall instantaneous properties of elastic-plastic composites. Journal of the Mechanics and Physics of Solids 1988, 36, 29–58.
[38] Suquet, P. Effective properties of nonlinear composites. In Continuum Micromechanics, ed. P. Suquet. CISM Lecture Notes, vol. 377. Springer Verlag, 1997, pp. 197–264.
[39] Chaboche, J., Kruch, S., Maire, J. and Pottier, T. Towards a micromechanics based inelastic and damage modeling of composites. International Journal of Plasticity 2001, 17, 411–439.
[40] Michel, J.C. and Suquet, P. Nonuniform transformation field analysis. International Journal of Solids and Structures 2003, 40, 6937–6955.
[41] Ueda, Y., Fukuda, K., Nakacho, K. and Endo, S.A new measuring method of residual stresses with the aid of finite element method and reliability of estimated values. Transactions of Joining and Welding Research Institute 1975, 4(2), 123–131.
[42] Reissner, H., Eigenspannungen und Eigenspannungsquellen. Schrift fuer Angewandte Mathematik und Mechanik 1931, 11(1), 1–8.
[43] Furuhashi, R. and Mura, T. On the equivalent inclusion method and impotent eigenstrains. Journal of Elasticity 1979, 9, 263–270.
[44] MDS User Manual, http://multiscale.biz.
[45] Simo, J.C. and Hughes, T.J.R. Computational Inelasticity. Interdisciplinary Applied Mathematics, Springer, 1997.
[46] Williams, K.V. and Vaziri, R. Application of a damage mechanics model for predicting the impact response of composite materials. Computers and Structures 2001, 79, 997–1011.
[47] Ladeveze, P., Allix, O., Deu, J.F. and Leveque, D. A mesomodel for localisation and damage computation in laminates. Computer Methods in Applied Mechanics and Engineering 2000, 183, 105–122.
[48] Chow, C.L. and Wang, J. An anisotropic theory of elasticity for continuum damage mechanics. International Journal of Fracture1987, 33, 3–16.
[49] Hofstetter, G., Simo, J.C. and Taylor, R.L. A modified cap model: closest point solution algorithms. Computers and Structures 1993, 46, 203–214.
[50] J. Fish, Y. Liu, V. Filonova, N. Hu, Zi. Yuan, Z. Yuan. A Phenomenological Regularized Multiscale Model International Journal for Multiscale Computational Engineering, 2013 (in press).
[51] Kouzeli, M. and Mortensen, A. Size dependent strengthening in particle reinforced aluminium, Acta Materialia 2002, 50, 39–51.
[52] Haque, M.A. and Saif, M.T.A. Strain gradient effect in nanoscale thin films. Acta Materialia 2003, 51, 3053–3061.
[53] Bazant, Z.P. Scaling of Structural Strength. Taylor & Francis, 2002.
[54] Pijaudier-Cabot, G. and Bazant, Z.P. Nonlocal damage theory. Journal of Engineering Mechanics 1987, 113, 1512–1533.

[55] Bazant, Z.P. Why continuum damage is nonlocal: justification by quasiperiodic microcrack array. Mechanics Research Communications 1987, 14, 407–419.
[56] Aero, E.L. and Kuvshinskii, E.V. 'Fundamental equations of the theory of elastic materials with rotationally interacting particles. Fizika Tverdogo Tela (St Peterburg) 1960, 2, 1399–1409.
[57] Grioli, G. Elasticita` asimmetrica. Annali di matematica pura ed applicata, Series IV 1960, 50, 389–417.
[58] Truesdell, C. and Toupin, R.A. Classical field theories of mechanics.' Handbuch der Physik, Springer, 1960, vol. III, p. 1.
[59] Mindlin, R. Micro-structure in linear elasticity. Archive for Rational Mechanics and Analysis 1964, 16, 51–78.
[60] Germain, P. The method of virtual power in continuum mechanics. Part 2: Microstructure. SIAM Journal on Applied Mathematics 1973, 25, 556–575.
[61] Maugin, G. Nonlocal theories or gradient-type theories: A matter of convenience? Archives of Mechanics 1979. 31, 15–26.
[62] Aifantis, E. The physics of plastic deformation. International Journal of Plasticity 1987, 3, 211–248.
[63] Kroner, E. Elasticity theory of materials with long range cohesive forces. International Journal of Solids and Structures 1967, 3, 731–742.
[64] Peerlings, R.H.J., Geers, M.G.D., de Borst, R. and Brekelmans, W.A.M. A critical comparison of nonlocal and gradient-enhanced softening continua, International Journal of Solids and Structures 2001, 38, 7723–7746.
[65] Fish, J., Jiang, T. and Yuan, Z. A Staggered nonlocal multiscale model for heterogeneous medium. International Journal for Numerical Methods in Engineering 2012, 91(2), 142–157.
[66] Jiang, H., Valdez, J.A., Zhu, Y.T., Beyerlein, I.J. and Lowe, T.C. The strength and toughness of cement reinforced with bone-shaped steel wires. Composites Science and Technology 2000, 60, 1753–1761.
[67] Bažant, Z. P. Instability, ductility and size effect in strain-softening concrete," Journal of the Engineering Mechanics Division, ASCE, 1976, 102(EM2), 331–344; discussions: 103, 357–358, 775–777; 104, 501–502.
[68] Song, J.H., Wang, H. and Belytschko, T. A comparative study on finite element methods for dynamic fracture., Computational Mechanics 2008, 42, 239–250.
[69] Bedford, A., Drumheller, D.S. and Sutherland, H.J. On modeling the dynamics of composite materials. In Mechanics Today, ed. S. Nemat-Nasser. Pergamon Press, 1976, vol. 3.
[70] Erofeev, V.I. Wave Processes in Solids with Microstructure. World Scientific, 2003.
[71] Sun, C.T., Achenbach, J.D. and Herrmann, G. Continuum theory for a laminated media. Journal of Applied Mechanics 1968, 35, 467–475.
[72] Murakami, H. and Hegemier, G.A. A mixture model for unidirectionally fiber-reinforced composites. Journal of Applied Mechanics 1986, 53, 765–773.
[73] Achenbach, J.D. and Herrmann, G. Dispersion of free harmonic waves in fiber-reinforced composites. American Institute of Aeronautics and Astronautics 1968, 6, 1832–1836.
[74] Bedford, A. and Stern, M. Toward a diffusing continuum theory of composite materials. Journal of Applied Mechanics 1971, 38, 8–14.
[75] Hegemier, G.A., Gurtman, G.A. and Nayfeh, A.H. A continuum mixture theory of wave propagation in laminated and fiber reinforced composites. International Journal of Solids and Structures. 1973, 9, 395–414.
[76] Boutin, C. and Auriault, J.L. Rayleigh scattering in elastic composite materials. International Journal of Engineering Science 1993, 31(12), 1669–1689.
[77] Andrianov, I.V., Bolshakov, V.I., Danishevs'kyy, V.V. and Weichert, D. Higher order asymptotic homogenization and wave propagation in periodic composite materials. Proceedings of the Royal Society A 2008, 464, 1181–1201.
[78] Sabina, F.J. and Willis, J.R. A simple self consistent analysis of wave propagation in particulate composites. Wave Motion 1988, 10, 127–142.
[79] Kanaun, S.K. and Levin, V.M. Self-consistent methods in the problem of axial elastic shear wave propagation through fiber composites. Archive of Applied Mechanics 2003, 73, 105–130.
[80] Kanaun, S.K., Levin, V.M. and Sabina, F.J. Propagation of elastic waves in composites with random set of spherical inclusions (effective medium approach). Wave Motion 2004, 40, 69–88.
[81] Santosa, F. and Symes, W.W. A dispersive effective medium for wave propagation in periodic composites. SIAM Journal of Applied Mathematics 1991, 51, 984–1005.
[82] Ting, T.C.T. Dynamic response of composites. Applied Mechanics Reviews 1980, 33, 1629–1635.
[83] Fish, J. and Chen, W. Higher-order homogenization of initial/boundary-value problem. Journal of Engineering Mechanics 2001, 127(12), 1223–1230.

[84] Chen, W. and Fish, J. A dispersive model for wave propagation in periodic heterogeneous media based on homogenization with multiple spatial and temporal scales. Journal of Applied Mechanics 2001, 68(2), 153–161.
[85] Fish, J. and Chen, W. Uniformly valid multiple spatial-temporal scale modeling for wave propagation in heterogeneous media. Mechanics of Composite Materials and Structures 2001, 8, 81–99.
[86] Fish, J., Chen, W. and Nagai, G. Non-local dispersive model for wave propagation in heterogeneous media. Part 1: One-dimensional case. International Journal for Numerical Methods in Engineering 2002, 54, 331–346.
[87] Fish, J., Chen, W. and Nagai, G. Non-local dispersive model for wave propagation in heterogeneous media. Part 2: Multi-dimensional case. International Journal for Numerical Methods in Engineering 2002, 54, 347–363.
[88] Molinari, A. and Mercier, S. Micromechanical modelling of porous materials under dynamic loading. Journal of the Mechanics and Physics of Solids 2001, 49, 1497–1516.
[89] Wang, Z.P. and Jiang, Q. A yield criterion for porous ductile media at high strain rate. Journal of Applied Mechanics 1997, 64, 503–509.
[90] Leveque, R.J. and Yong, D.H. Solitary waves in Layered Nonlinear Media. SIAM Journal of Applied Mathematics 2003, 63, 1539–1560.
[91] Wang, Z.-P. and Sun, C.T. Modeling micro-inertia in heterogeneous materials under dynamics loading. Wave Motion 2002, 36, 473–485.
[92] Fish, J., Filonova, V. and Kuznetsov, S. Micro-inertia effects in nonlinear heterogeneous media. International Journal of Numerical Methods in Engineering 2012, 91(13), 1406–1426.
[93] Yuan, Z. and Fish, J. Towards realization of computational homogenization in practice. Journal for Numerical Methods in Engineering 2008, 73(3), 361–380.
[94] Hellmich, C., Fritsch, A. and Dormieux, L. Multiscale homogenization theory: an analysis tool for revealing mechanical design principles in bone and bone replacement materials. In Biomimetics – Materials, Structures and Processes, Biological and Medical Physics, Biomedical Engineering, eds P. Gruber *et al.* Springer-Verlag, 2011.
[95] Fish, J. and Shek, K.L. Finite deformation plasticity based on the additive split of the rate of deformation and hyperelasticity. Computer Methods in Applied Mechanics and Engineering 2000, 190, 75–93.
[96] Argyris, J.H. An excursion into large rotations., Computer Methods in Applied Mechanics and Engineering 1982, 32, 85–155.
[97] Belytschko, T. and Hsieh, B.J. Nonlinear transient finite element analysis with convected coordinates. International Journal of Numerical Methods in Engineering 1973, 7, 255–271.
[98] Bergan, P.G. and Horrigmoe, G. Incremental variational principles and finite element models for nonlinear problems. Computer Methods in Applied Mechanics and Engineering. 1976, 7, 201–217.
[99] Rankin, C.C. Consistent linearization of the element-independent corotational formulation for the structural analysis of general shells. NASA Contractor Report 278428, Lockheed Palo Alto Research Laboratory, Palo Alto, CA, 1988.
[100] Paris, P.C. and Erdogan, F. A critical analysis of crack propagation laws. Journal of Basic Engineering 1963, 85, 528–534.
[101] Elber, W. Fatigue crack closure under cyclic tension. Engineering Fracture Mechanics 1970, 2, 37–45.
[102] Foreman, R.G., Keary, V.E. and Engle, R.M. Numerical analysis of crack propagation in cyclic-loaded structures. Journal of Basic Engineering 1967, 89, 459–464.
[103] Klesnil, M. and Lukas, P. Influence of strength and stress history on growth and stabilization of fatigue cracks. Engineering. Fracture Mechanics 1972, 4, 77–92.
[104] Wheeler, O.E. Spectrum loading and crack growth. Journal of Basic Engineering 1972, 94, 181–186.
[105] Nguyen, O., Repetto, E.A., Ortiz, M. and Radovitzky, R.A. A cohesive model of fatigue crack growth. International Journal of. Fracture 2001, 110(4), 351–369.
[106] Billardon, R. Etude de la rupture par la mecanique de l'endommagement. Thesis, University of Paris, 1989.
[107] Lemaitre, J. and Doghri, I. A post processor for crack initiation. Computer Methods in Applied Mechanics and Engineering. 1994, 115(3–4), 197–232.
[108] Fish, J. and Yu, Q. Computational mechanics of fatigue and life predictions for composite materials and structures. Computer Methods in Applied Mechanics and Engineering 2002, 191, 4827–4849.
[109] Paas, M.H.J.W., Schreurs, P.J.G. and Brekelmans, W.A.M. A continuum approach to brittle and fatigue damage: theory and numerical procedures. Journal of Solid Structures 1993. 30(4), 579–599.
[110] Yu, Q. and Fish, J. Temporal homogenization of viscoelastic and viscoplastic solids subjected to locally periodic loading. Computational Mechanics 2002, 29, 199–211.

[111] Oskay, C. and J. Fish, J. Fatigue life prediction using 2-scale temporal asymptotic homogenization. International Journal for Numerical Methods in Engineering 2004, 61(3), 329–359.
[112] Oskay, C. and Fish, J. Multiscale modeling of fatigue for ductile materials. International Journal of Multiscale Computational Engineering 2004, 2(3), 1–25.
[113] Fish, J. and Oskay, C. Nonlocal multiscale fatigue model. Mechanics of Advanced Materials and Structures 2005, 12(6), 485–500.
[114] Devulder, A., Aubry, D. and Puel, G. Two-time scale fatigue modeling: application to damage. Computational Mechanics 2010, 45(6), 637–646.
[115] Fish, J., Bailakanavar, M., Powers, L. and Cook, T. Multiscale fatigue life prediction model for heterogeneous materials. International Journal for Numerical Methods in Engineering 2012, 91(10), 1087–1104.
[116] Terada, K. and Kurumatani, M. Two-scale diffusion–deformation coupling model for material deterioration involving micro-crack propagation. International Journal for Numerical Methods in Engineering 2010, 83(4), 426–451.
[117] Ozdemir, I., Brekelmans, W.A.M. and Geers, M.G.D. FE2 computational homogenization for the thermo-mechanical analysis of heterogeneous solids. Computer Methods in Applied Mechanics and Engineering 2008, 198, 602–613.
[118] Yu, Q. and Fish, J. Multiscale asymptotic homogenization for multiphysics problems with multiple spatial and temporal scales: a coupled thermo-viscoelastic example problem. International Journal of Solids and Structures 1998, 39, 6429–6452.
[119] Kuznetsov, S. and Fish, J. Mathematical homogenization theory for electroactive continuum. International Journal for Numerical Methods in Engineering 2012, 91(11), 1199–1226.
[120] Fish, J. and Chen, W. Modeling and simulation of piezocomposites. Computer Methods in Applied Mechanics and Engineering 2003, 192, 3211–3232.
[121] Michopoulos, J.G., Farhat, C. and Fish, J.Survey on modeling and simulation of multiphysics systems. Journal of Computing and Information Science in Engineering 2005, 5(3), 198–213.
[122] Ruggles-Wrenn, M.B. and Broeckert, J.L. Effects of prior aging at 288°C in air and in argon environments on creep response of PMR-15 neat resin. Journal of Applied Polymer Science 2009, 111(1), 228–236.
[123] Schoeppner, G.A., Tandon, G.P. and Pochiraju, K.V. Predicting thermooxidative degradation and performance of high-temperature polymer matrix composites. In Multiscale Modeling and Simulation of Composite Materials and Structures, eds Y. Kwon, D. Allen and R. Talreja. Springer, 2007, Chapter 9.
[124] Yuan, Z., Jiang, T., Fish, J. and Morscher, G. Reduced order multiscale-multiphysics model for heterogeneous materials. International Journal for Multiscale Computational Engineering 2013 (in press).
[125] Cernocky, E.P. and Krempl, E. A nonlinear uniaxial integral constitutive equation incorporating rate effects, creep and relaxation. International Journal of Non-Linear Mechanics 1979, 14, 183–203.
[126] Yuan, Z., Ruggles-Wrenn, M. and Fish, J. Computational viscoplasticity based on overstress (CVBO) model. Journal of Computational Engineering Science 2013 (in press).
[127] Odegard, G. and Kumosa, M. Composites Science and Technology 2000, 60, 2979–2988.
[128] Rupnowski, P., Gentz, M., Sutter, J.K. and Kumosa, M. Composites Part A 2004, 33, 327–338.
[129] Schoeppner, G.A., Tandon, G.P. and Ripberger, E.R. Anisotropic oxidation and weight loss in PMR-15 composites. Composites Part A 2007, 38, 890–904.
[130] Gorke, U.-J., Bucher, A. and Kreißig, R. Ein Beitrag zur Material parameter identifikation bei finiten elastisch-plastischen Verzerrungen durch Analyse inhomogener Verschiebungsfelder mit Hilfe der FEM. Preprint SFB393 01(03), 2001.
[131] Conn, A.R., Gould, N.I.M. and Toint, P.L. Trust-region methods. MPS/SIAM Series on Optimization. SIAM, 2000.
[132] More, J.J. The Levenberg–Marquardt algorithm: implementation and theory. In Numerical Analysis, ed. G.A. Watson. Lecture Notes in Mathematics, vol. 630. Springer, pp. 105–116.
[133] Walmag, J.M.B. and Delhez, E.J.M. A trust-region method applied to parameter identification of a simple prey–predator model, Applied Mathematical Modelling 2005, 29, 289–307.
[134] Peerlings, R.H.J., Brekelmans, W.A.M., de Borst, R. and Geers, M.G.D. Gradient-enhanced damage modeling of high-cycle fatigue. International Journal of Numerical Methods in Engineering 2000, 49, 1547–1569.

5

Scale-separation-free Upscaling/ Downscaling of Continua

5.1 Introduction

Despite their noteworthy performance in some localization problems, higher order homogenization theories ([1,2], see also Chapter 2, section 2.5) are not without their shortcomings. From a theoretical point of view, they hinge on a two-scale integration scheme (or alternatively, on the Hill–Mandel macrohomogeneity condition), which assumes infinitesimality of the unit cell even though the coefficients of the enriched coarse-scale continua depend on the size of the unit cell. Moreover, these theories require consideration of higher order boundary conditions. While higher order homogenization and generalized continua theories are equipped with enriched kinematics, which approximate the fine-scale deformation, they remain local in nature. Trostel [3] and Forest and Sievert [2] referred to these types of continuum models as local, whereas Bazant and Jirasek [4] classified both the generalized continua and nonlocal gradient models as weakly nonlocal. From a computational point of view, generalized continua models require consideration of additional degrees of freedom in combination with hybrid formulations, or alternatively, they impose a C^1 continuity requirement.

Various nonlocal theories of either the integral or gradient type include a nonlocal kernel function whose support provides an internal length scale. The integral formulation reduces to the gradient type by truncating the series expansion of the nonlocality kernel [5]. By virtue of this truncation, the nonlocal gradient models assume that nonlocal interactions are limited to a close neighborhood. In the earlier works, nonlocality was introduced in the nonlocal approximations of fields and balance equations [6,7], with later works focusing on the nonlocality of internal variables [8,9], which is closely related to gradient plasticity theories [10]. The nonlocal theories reduce to those of the generalized continua or higher order homogenization theory when the coarse-scale problem size is significantly larger than the scale of heterogeneity. Selection of the nonlocal kernels and

Practical Multiscaling, First Edition. Jacob Fish.

selection of the magnitude of the internal length scale are still controversial issues. It is also unclear how to construct nonlocal kernels that would have a sufficient degree of generality for a wide range of problems in heterogeneous media. For problems in which nonlocal interactions are well understood, computational difficulties related to higher order continuity and boundary conditions can be partially alleviated by reproducing the kernel strain regularization of implicit gradient models [11]. Also noteworthy are more recent variations of higher order and nonlocal continuum theories that bring certain microstructural information to the macroscopic equation of motion [12,13,14,15,16,17].

In this chapter, the focus is on a coarse-scale continuum description that is free of scale separation and is consistent with an underlying fine-scale description for heterogeneities of *finite size*. The so-called computational continua, or simply C^2 [18,19], presented in section 5.2 possesses fine-scale features but without introducing scale separation, which can be mathematically justified provided that the fine-scale structural details are infinitesimally small. From a computational point of view, the computational continua do not require higher order continuity, introduce no new degrees of freedom, and are free of higher order boundary conditions. The C^2 description features a nonlocal quadrature scheme defined over a computational continua domain consisting of a disjoint union of computational unit cells; the positions of these unit cells are determined so as to reproduce the weak form of the governing equations on the fine scale. The nonlocal quadrature scheme for various element types is described in section 5.4.

Section 5.3 focuses on a computationally efficient framework that blends the generality and rigor of the C^2 continuum formulation with the practicality of the reduced order homogenization outlined in Chapter 4, which eliminates the bottleneck of satisfying fine-scale equilibrium equations. Blending these two methods into a single cohesive computational framework, hereafter referred to as *reduced order computational continua* or simply RC^2 [19,20], inherits the underlying characteristics of its two ingredients. The RC^2 formulation includes the development of residual-free influence functions incorporating the gradient effects and formulation of the reduced order size-dependent unit cell problem, the coarse-scale discrete equilibrium equations, and the consistent tangent operator. In section 5.5, we provide a verification of the RC^2 formulation against the direct numerical simulation (DNS) and compare the RC^2 formulation to $O(1)$ computational homogenization.

In deriving the C^2 formulation in section 5.2, we consider a strong form of the boundary value problem on a composite domain Ω_X^ζ with boundary $\partial\Omega_X^\zeta$, given as

$$\frac{\partial P_{ij}^\zeta}{\partial X_j} + B_i = 0 \quad \text{on} \quad \Omega_X^\zeta \tag{5.1}$$

$$F_{ik}^\zeta = \delta_{ik} + \frac{\partial u_i^\zeta}{\partial X_k} \tag{5.2}$$

$$P_{ij}^\zeta N_j^\zeta = \overline{T}_i \quad \text{on} \quad \partial\Omega_X^t \tag{5.3}$$

$$u_i^\zeta = \overline{u}_i \quad \text{on} \quad \partial\Omega_X^u \tag{5.4}$$

$$\partial\Omega_X^t \cup \partial\Omega_X^u = \partial\Omega_X \quad \text{and} \quad \partial\Omega_X^t \cap \partial\Omega_X^u = 0 \tag{5.5}$$

where, for simplicity, the body forces B_i and prescribed boundary conditions $\overline{T}_i, \overline{u}_i$ are assumed to be constant over a unit cell domain.

5.2 Computational Continua (C^2)

In this section, we present a computational continuum formulation that is free of the theoretical and computational limitations discussed in Chapter 2, section 2.5. The focus is on second-order computational continua that make no assumption about the *infinitesimality of the unit cell*, require C^0 continuity only, and involve no additional degrees of freedom. Furthermore, we will make no assumption about scale decomposition but will instead introduce a unit cell local coordinate system, $\boldsymbol{\chi} = \boldsymbol{X} - \hat{\boldsymbol{X}}$, into the physical unit cell domain. Both the test and trial functions will be decomposed into oscillatory weakly periodic functions and smooth coarse-scale functions. In addition to its use in various mathematical homogenization theories, such a decomposition has been used in various local enrichment methods, including enriched elements [21,22], the variational multiscale method [23], the s-version of the finite element method [24] with application to strong [25] and weak [26] discontinuities, the multigrid-like methods [27,28], and the extended finite element method [29,30].

5.2.1 Nonlocal Quadrature

To construct the coarse-scale weak form it is necessary to integrate functions on a composite domain Ω_X^ζ with *finite size* fine-scale details for which the two-scale integration scheme

$$\lim_{\zeta^\alpha \to 0^+} \int_{\Omega_X^\zeta} \Psi(\boldsymbol{X}, \boldsymbol{Y})\, d\Omega = \lim_{\zeta^\alpha \to 0^+} \int_{\Omega_X} \left(\frac{1}{|\Theta_Y|} \int_{\Theta_Y} \Psi(\boldsymbol{X}, \boldsymbol{Y})\, d\Theta \right) d\Omega \tag{5.6}$$

considered in Chapter 2 is no longer valid. The trivial solution is to replace (5.6) by a sum of integrals over unit cell domains. However, such an integration scheme is not practical for problems involving numerous unit cells. Furthermore, it would give rise to cumbersome integration over coarse-scale finite element domains whose boundaries do not coincide with unit cell boundaries. To circumvent these difficulties, the so-called *nonlocal quadrature* scheme is introduced. The integration over composite domain Ω_X^ζ is replaced by an integration over the so-called *computational continua domain* Ω_X^C, consisting of a *disjoint union* [31] (sometimes called the direct sum or free union) of *computational unit cell* domains $\Theta_{\hat{X}_I}$ and denoted as

$$\Omega_X^C = \coprod_{I=1}^{\hat{N}} \Theta_{\hat{X}_I} \tag{5.7}$$

where $\hat{\boldsymbol{X}}_I$ denotes the coordinates of centroid of the computational unit cell domain $\Theta_{\hat{X}_I}$. Note that if $\Theta_{\hat{X}_I} \cap \Theta_{\hat{X}_J} = 0,\ \forall I \neq J$, then the disjoint union reduces to a regular union.

The nonlocal quadrature scheme is then defined as

$$\int_{\Omega_X^\zeta} \Psi(\boldsymbol{X}) d\Omega = \sum_{I=1}^{\hat{N}} W_I J_I \frac{1}{\left|\Theta_{\hat{X}_I}\right|} \int_{\Theta_{\hat{X}_I}} \Psi(\hat{\boldsymbol{X}}_I, \boldsymbol{\chi})\, d\Theta \tag{5.8}$$

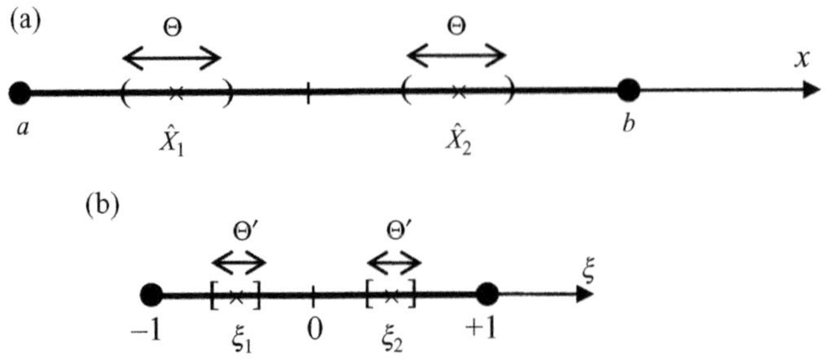

Figure 5.1 Nonlocal quadrature: physical (a) and parent element (b) domains. Unit cell domains are shown in brackets; $\hat{X}_I, \xi_I$ denote the positions of quadrature points in the physical domain and parent element domain, respectively; Θ and $\Theta' = \Theta/J$ denote the size of the unit cell in the physical domain and the parent element domain, respectively; and $J=(b-a)/2$ is the element Jacobian. Reproduced with permission from [18], © 2010 John Wiley & Sons

where W_I denotes the nonlocal quadrature weight; $\left|\Theta_{\hat{X}_I}\right|$ is the volume of the computational unit cell domain $\Theta_{\hat{X}_I}$; J_I is the Jacobian computed at the unit cell centroid mapping the coarse-scale element into a bi-unit cube (square, interval). W_I and $\hat{X}_I$ are chosen to exactly evaluate integrals (5.8) on the composite domain with integrand $\Psi(X)$ approximated by a polynomial of the order m. The pair $(W_I, \hat{X}_I)$ depends on the computational unit cell size relative to the coarse-scale finite element size, as will be subsequently discussed.

We first consider the nonlocal quadrature scheme for integrating smooth functions in one dimension. In section 5.4, we will consider generalization to multidimensions.

Consider a 1D domain $[a,b]$ mapped into a parent element domain $[-1,1]$ as shown in Figure 5.1.

The goal is to find quadrature weights and sampling points that exactly integrate polynomials of a given order. Applying the nonlocal quadrature scheme (5.8) to the 1D element domain yields

$$\mathrm{I} = \int_{-1}^{1} J\Psi(\xi)d\xi = \sum_{I=1}^{\hat{N}} \frac{W_I}{\Theta} \int_{-\Theta/2}^{\Theta/2} J\Psi(\hat{X}_I + \chi)d\chi \tag{5.9}$$

where $\hat{N}$ is the number of quadrature points in the element domain. Applying element mapping to the unit cell domain yields

$$\mathrm{I} = \sum_{I=1}^{\hat{N}} \frac{W_I}{(\Theta/J)} \int_{-(\Theta/J)/2}^{(\Theta/J)/2} J\Psi(\xi_I + \eta)d\eta \tag{5.10}$$

Further denoting $\Psi'(\xi) = J\Psi(\xi)$ and $\Theta' = \Theta/J$, equation (5.9) can be rewritten as

$$\mathrm{I} = \int_{-1}^{1} \Psi'(\xi)d\xi = \sum_{I=1}^{\hat{N}} \frac{W_I}{\Theta'} \int_{-\Theta'/2}^{\Theta'/2} \Psi'(\xi_I + \eta)d\eta \tag{5.11}$$

Let us now approximate the integrand $\Psi'(\xi)$ by a polynomial function

$$\Psi'(\xi) = \sum_{J=1}^{m} \alpha_J \xi^{J-1} \tag{5.12}$$

Inserting the polynomial expansion (5.12) into the quadrature scheme (5.9) yields

$$\mathrm{I} = \sum_{I=1}^{\hat{N}} \frac{W_I}{\Theta'} \int_{-\Theta'/2}^{\Theta'/2} \sum_{J=1}^{m} \alpha_J (\xi_I + \eta)^{J-1} d\eta \tag{5.13}$$

A closed-form integration of the above yields

$$\begin{aligned} \mathrm{I} &= \sum_{I=1}^{\hat{N}} W_I \left(\alpha_1 + \alpha_2 \xi_I + \alpha_3 \left(\xi_I^2 + \frac{\Theta'^2}{12} \right) + \alpha_4 \left(\xi_I^3 + \frac{\xi_I \Theta'^2}{4} \right) + \cdots \right) \\ &= \alpha_1 \sum_{I=1}^{\hat{N}} W_I + \alpha_2 \sum_{I=1}^{\hat{N}} W_I \xi_I + \alpha_3 \left(\frac{\Theta'^2}{12} + \sum_{I=1}^{\hat{N}} W_I \xi_I^2 \right) + \alpha_4 \left(\sum_{I}^{\hat{N}} W_I \left(\xi_I^3 + \frac{\xi_I \Theta'^2}{4} \right) \right) \cdots \end{aligned} \tag{5.14}$$

On the other hand, exact integration of (5.13) gives

$$\mathrm{I} = \int_{-1}^{1} \Psi'(\xi) d\xi = \int_{-1}^{1} \sum_{I=1}^{m} \alpha_I \xi^{I-1} d\xi = 2\alpha_1 + 0\alpha_2 + \frac{2}{3}\alpha_3 + 0\alpha_4 + \cdots \tag{5.15}$$

Requiring the nonlocal quadrature to exactly integrate polynomials for arbitrary α_I yields a nonlinear system of equations from which the quadrature weights and positions of the computational unit cells can be determined. For instance, for a two-point nonlocal quadrature we get

$$\begin{bmatrix} 1 & 1 \\ \xi_1 & \xi_2 \\ \xi_1^2 + \frac{\Theta'^2}{12} & \xi_2^2 + \frac{\Theta'^2}{12} \\ \xi_1^3 + \xi_1 \frac{\Theta'^2}{4} & \xi_2^3 + \xi_2 \frac{\Theta'^2}{4} \end{bmatrix} \begin{bmatrix} W_1 \\ W_2 \end{bmatrix} = \begin{bmatrix} 2 \\ 0 \\ 2/3 \\ 0 \end{bmatrix} \tag{5.16}$$

which yields

$$\xi_{1,2} = \pm \sqrt{\frac{1}{3} - \frac{\Theta'^2}{12}}; \quad W_{1,2} = 1 \tag{5.17}$$

Remark 5.1 As expected, in the limit as $\Theta' \to 0$ the above quadrature reduces to the usual Gauss quadrature, hereafter referred to as *local quadrature*. When the unit cell size is equal to one-half of the element size, $\Theta' = 1$, we get $\xi_{1,2} = \pm 0.5$. As the unit cell size further increases, the two sampling points move toward the origin, and when the size of the unit cell coincides with that of the element, we get $\xi_{1,2} = 0$, which is equivalent to having one quadrature point in the middle of the interval with quadrature weight equal to 2. Note that while the nonlocal one-point quadrature element maintains full rank (provided that the unit cell is fully integrated), it should be used only if the unit cell size is close to that of the element. Also, it can be seen that $\Theta' \le 2$, that is, the unit cell must be smaller than the element.

Remark 5.2 For a three-point nonlocal quadrature, the weights and quadrature points are

$$W_{1,3} = \frac{5(4-\Theta'^2)}{3(12-7\Theta'^2)}; \quad \xi_{1,3} = \pm\frac{\sqrt{60-35\Theta'^2}}{10}$$
$$W_2 = 2-2W_1; \qquad \xi_2 = 0$$

Note that when $\Theta'=0$, we have $W_{1,3}=5/9$ and $\xi_{1,3}=\pm 0.774596692$, which reduces to the local three-point Gauss quadrature. For $\Theta'=1$, the three-point nonlocal quadrature reduces to a two-point nonlocal quadrature with $W_{1,3}=1$, $W_2=0$, and $\xi_{1,3}=\pm 0.5$. The three-point nonlocal quadrature should be limited to $\Theta' \le 1$ to avoid negative values of weights.

Remark 5.3 The extension of nonlocal quadrature to multidimensions for elements whose edges (faces in 3D) are parallel to those of the unit cell is trivial. The positions of quadrature points and weights in each space direction a of the parent element (W_{Ia}, ξ_{Ia}) can be determined by $\Theta'_a=\Theta_a/J_a$, where Θ_a is the length of the unit cell and J_a is one-half of the element length along the axis ξ_a. For general quadrilateral, hexahedral, triangular, and tetrahedral elements, see section 5.4.

Remark 5.4 In principle, the nonlocal quadrature can be extended to unit cells that are larger than coarse-scale elements. However, if the above approach is used, it is necessary to carry out the integration over the unit cell domain rather than over the elements. This, of course, complicates the integration as it requires either additional triangulation or an increased number of quadrature points [24].

5.2.2 Coarse-Scale Problem

Consider the coarse-scale weak form

$$\int_{\Omega_X^\zeta} w_i^c \left(\frac{\partial}{\partial X_k} P_{ik}^\zeta + B_i \right) d\Omega = 0 \quad \forall \mathbf{w}^c \in W_{\Omega_X} \tag{5.18}$$

where $\mathbf{w}^c$ is the C^0 continuous test function on Ω_X, satisfying homogeneous boundary conditions on $\partial\Omega_X^u$

$$W_{\Omega_X} = \left\{ \mathbf{w}^c \text{ defined in } \Omega_X,\ C^0(\Omega_X),\ \mathbf{w}^c = \mathbf{0} \quad \text{on} \quad \partial\Omega_X^u \right\} \tag{5.19}$$

Integrating by parts the divergence terms and applying the nonlocal quadrature scheme yields

$$\sum_{I=1}^{\hat{N}} W_I J_I \frac{1}{\left|\Theta_{\hat{X}_I}\right|} \int_{\Theta_{\hat{X}_I}} \frac{\partial w_i^c}{\partial X_j} P_{ij}^\zeta (\hat{X}_I, \chi) d\Theta = \int_{\partial\Omega_X^t} w_i^c \bar{T}_i d\Gamma + \int_{\Omega_X} w_i^c B_i d\Omega \tag{5.20}$$

Consider a second-order computational continua approach by which $\partial w_i^c/\partial X_j$ is approximated by a linear approximation defined as

$$\frac{\partial w_i^c}{\partial X_j}(\hat{\boldsymbol{X}}_I,\boldsymbol{\chi}) \approx \left\{\overline{\frac{\partial w_i^c}{\partial X_j}}\right\}_{\hat{X}_I} + \left\{\overline{D_k \frac{\partial w_i^c}{\partial X_j}}\right\}_{\hat{X}_I} \chi_k \tag{5.21}$$

where the averaging operators are defined as

$$\left\{\overline{\frac{\partial w_i^c}{\partial X_j}}\right\}_{\hat{X}_I} = \frac{1}{\left|\Theta_{\hat{X}_I}\right|}\int_{\Theta_{\hat{X}_I}} \frac{\partial w_i^c}{\partial X_j}(\hat{\boldsymbol{X}}_I,\boldsymbol{\chi})d\Theta = \frac{1}{\left|\Theta_{\hat{X}_I}\right|}\int_{\partial\Theta_{\hat{X}_I}} w_i^c(\hat{\boldsymbol{X}}_I,\boldsymbol{\chi})N_j^{\Theta}d\gamma \tag{5.22}$$

$$\left\{\overline{D_k \frac{\partial w_i^c}{\partial X_j}}\right\}_{\hat{X}_I} = \frac{1}{\left|\Theta_{\hat{X}_I}\right|}\int_{\Theta_{\hat{X}_I}} \frac{\partial}{\partial \chi_k}\left(\frac{\partial w_i^c}{\partial X_j}(\hat{\boldsymbol{X}}_I,\boldsymbol{\chi})\right)d\Theta = \frac{1}{\left|\Theta_{\hat{X}_I}\right|}\int_{\partial\Theta_{\hat{X}_I}} \frac{\partial w_i^c}{\partial X_j}(\hat{\boldsymbol{X}}_I,\boldsymbol{\chi})N_k^{\Theta}d\gamma \tag{5.23}$$

Inserting (5.21) into (5.20) yields

$$\sum_{I=1}^{\hat{N}} W_I J_I \left(\left\{\overline{\frac{\partial w_i^c}{\partial X_j}}\right\}_{\hat{X}_I} \overline{P}_{ij}(\hat{\boldsymbol{X}}_I) + \left\{\overline{D_k \frac{\partial w_i^c}{\partial X_j}}\right\}_{\hat{X}_I} \overline{Q}_{ijk}(\hat{\boldsymbol{X}}_I)\right) = \int_{\partial\Omega_X^t} w_i^c \overline{T}_i \, d\Gamma + \int_{\Omega_X} w_i^c B_i \, d\Omega \tag{5.24}$$

where the overall stress $\overline{P}_{ij}$ and the stress couple $\overline{Q}_{ijk}$ are defined as

$$\overline{P}_{ij}(\hat{\boldsymbol{X}}_I) = \frac{1}{\left|\Theta_{\hat{X}_I}\right|}\int_{\Theta_{\hat{X}_I}} P_{ij}^{\zeta}(\hat{\boldsymbol{X}}_I,\boldsymbol{\chi})d\Theta \tag{5.25}$$

and

$$\overline{Q}_{ijk}(\hat{\boldsymbol{X}}_I) = \frac{1}{\left|\Theta_{\hat{X}_I}\right|}\int_{\Theta_{\hat{X}_I}} P_{ij}^{\zeta}(\hat{\boldsymbol{X}}_I,\boldsymbol{\chi})\chi_k d\Theta \tag{5.26}$$

Similar to the coarse-scale test function (5.21), the coarse-scale deformation gradient F_{ij}^c is approximated as a linear function as

$$F_{ij}^c(\hat{\boldsymbol{X}}_I,\boldsymbol{\chi}) \approx \overline{F}_{ij}(\hat{\boldsymbol{X}}_I) + \overline{F}_{ij,k}(\hat{\boldsymbol{X}}_I)\chi_k \tag{5.27a}$$

$$\begin{aligned}\overline{F}_{ij}(\hat{\boldsymbol{X}}_I) &= \delta_{ij} + \frac{1}{\left|\Theta_{\hat{X}_I}\right|}\int_{\Theta_{\hat{X}_I}} \frac{\partial u_i^c(\hat{\boldsymbol{X}}_I,\boldsymbol{\chi})}{\partial X_j}d\Theta = \delta_{ij} + \frac{1}{\left|\Theta_{\hat{X}_I}\right|}\int_{\partial\Theta_{\hat{X}_I}} u_i^c N_j^{\Theta} d\gamma \\ &= \delta_{ij} + \frac{1}{\left|\Theta_{\hat{X}_I}\right|}\int_{\partial\Theta_{\hat{X}_I}} u_i^{\zeta} N_j^{\Theta} d\gamma\end{aligned} \tag{5.27b}$$

$$\begin{aligned}\overline{F}_{ij,k}(\hat{\boldsymbol{X}}_I) &= \frac{1}{\left|\Theta_{\hat{X}_I}\right|}\int_{\Theta_{\hat{X}_I}} \frac{\partial}{\partial \chi_k}\left(\frac{\partial u_i^c(\hat{\boldsymbol{X}}_I,\boldsymbol{\chi})}{\partial X_j}\right)d\Theta = \frac{1}{\left|\Theta_{\hat{X}_I}\right|}\int_{\partial\Theta_{\hat{X}_I}} \frac{\partial u_i^c}{\partial X_j} N_k^{\Theta} d\gamma \\ &= \frac{1}{\left|\Theta_{\hat{X}_I}\right|}\int_{\partial\Theta_{\hat{X}_I}} \frac{\partial u_i^{\zeta}}{\partial X_j} N_k^{\Theta} d\gamma\end{aligned} \tag{5.27c}$$

where the substitution of the course-scale displacement u_i^c by the total displacement u_i^ζ follows from the periodicity assumption of $u_i^{(1)} = u_i^\zeta - u_i^c$ and its spatial derivative.

We now proceed to the Galerkin approximation of the coarse-scale problem (5.24). The coarse-scale trial u_i^c and test w_i^c functions are discretized, using the same $C^0(\Omega_X)$ continuous coarse-scale shape functions $N_{i\alpha}^c$, as

$$\begin{aligned} u_i^c &= N_{i\alpha}^c d_\alpha^c \\ w_i^c &= N_{i\alpha}^c c_\alpha^c \end{aligned} \tag{5.28}$$

where d_α^c and c_α^c denote the nodal values of the trial and test functions. As in the previous chapters, Greek subscripts are reserved for tensor components, and summation convention over repeated indices is employed.

Inserting (5.28) into the coarse-scale weak form (5.24) yields the discrete coarse-scale problem, which states

Given $\overline{T}_i$, $\overline{B}_i$, and $\overline{\boldsymbol{d}}$, find Δd_α^c such that

$$\begin{aligned} r_\alpha^c(\Delta \boldsymbol{d}^c) \equiv f_\alpha^{\text{int}} - f_\alpha^{\text{ext}} = 0 \\ \boldsymbol{d}^c = \overline{\boldsymbol{d}} \text{ on } \partial\Omega_X^u \end{aligned} \tag{5.29}$$

where, for simplicity, we omit the indices referring to load increment and iteration counts. r_α^c and Δd_α^c are the coarse-scale residual and displacement increments, respectively, and

$$f_\alpha^{\text{int}} = \sum_{I=1}^{\hat{N}} W_I J_I \left(\left\{ \overline{\frac{\partial N_{i\alpha}^c}{\partial X_j}} \right\}_{\hat{X}_I} \overline{P}_{ij}(\hat{\boldsymbol{X}}_I) + \left\{ D_k \overline{\frac{\partial N_{i\alpha}^c}{\partial X_j}} \right\}_{\hat{X}_I} \overline{Q}_{ijk}(\hat{\boldsymbol{X}}_I) \right) \tag{5.30}$$

$$f_\alpha^{\text{ext}} = \int_{\Omega_X} N_{i\alpha}^c \overline{B}_i d\Omega + \int_{\partial\Omega_X^t} N_{i\alpha}^c \overline{T}_i \, d\Gamma \tag{5.31}$$

where f_α^{int} and f_α^{ext} are the internal and external forces, respectively. The average operators in (5.30) are defined in (5.22) and (5.23) with $N_{i\alpha}^c$ replacing w_i^c.

Remark 5.5 An alternative approximation to the averaging operators in (5.21) is a Taylor series expansion of $\partial w_i^c(\hat{\boldsymbol{X}}_I, \chi)/\partial X_j$ around the element centroid. However, (5.21) is advantageous over the Taylor series expansion because (i) it is more accurate for large unit cells and (ii) it avoids calculating second-order derivatives by utilization of Green's theorem in equation (5.22) and equation (5.23).

Remark 5.6 The formulation of the *B-matrix* in the internal force (5.30) falls into the category of an assumed strain formulation in the form of a B-bar approach [32]. In the limit, as the number of terms in the expansion (5.21) is increased, the assumed strain approximation will approach the displacement-based formulation. For higher than quadratic coarse-scale elements, it is necessary to include higher than second-order derivatives and higher order stresses along the lines of equation (5.21), equation (5.25), and equation (5.26).

Remark 5.7 In [18] it was shown that there exists a linear coarse-scale stress function $P_{ij}^{c}(\hat{\boldsymbol{X}}_I,\boldsymbol{\chi})$, defined as

$$\begin{aligned} P_{ij}^{c}(\hat{\boldsymbol{X}}_I,\boldsymbol{\chi}) &= \overline{P}_{ij}(\hat{\boldsymbol{X}}_I)+Q_{ijk}(\hat{\boldsymbol{X}}_I)\chi_k \\ Q_{ijk}(\hat{\boldsymbol{X}}_I) &= \frac{12}{l_{\underline{k}}^2}\overline{Q}_{ij\underline{k}}(\hat{\boldsymbol{X}}_I) \end{aligned} \tag{5.32}$$

that replaces the fine-scale stress $P_{ij}^{\zeta}(\hat{\boldsymbol{X}}_I,\boldsymbol{\chi})$ in (5.20). The proof follows from replacing $P_{ij}^{\zeta}(\hat{\boldsymbol{X}}_I,\boldsymbol{\chi})$ with $P_{ij}^{c}(\hat{\boldsymbol{X}}_I,\boldsymbol{\chi})$ as defined in (5.32), which yields (5.24). In (5.32), the bar under the subscript k denotes no summation over the subscript k.

5.2.3 Computational Unit Cell Problem

The computational unit cell problem is constructed by defining the space of test functions over the computational unit cell domains as

$$W_{\Theta_{\hat{X}_I}} = \left\{ \boldsymbol{w}^{(1)}(\hat{\boldsymbol{X}}_I,\boldsymbol{\chi}) \text{ defined in } \Theta_{\hat{X}_I},\ C^0(\Theta_{\hat{X}_I}),\ \boldsymbol{\chi}-\text{periodic},\ \int_{\Theta_{\hat{X}_I}} w_i^{(1)} d\Theta = 0 \right\} \tag{5.33}$$

and requiring the equilibrium equation (5.1) to be satisfied in the weak sense

$$\int_{\Theta_{\hat{X}_I}} w_i^{(1)} \left(\frac{\partial}{\partial \chi_k} P_{ik}^{\zeta} + B_i \right) d\Theta = 0 \quad \forall \boldsymbol{w}^{(1)} \in W_{\Theta_{\hat{X}_I}} \tag{5.34}$$

Integration by parts of the above yields

$$\int_{\Theta_{\hat{X}_I}} \frac{\partial w_i^{(1)}}{\partial \chi_k} P_{ik}^{\zeta}\, d\Theta = \int_{\partial\Theta_{\hat{X}_I}} w_i^{(1)} P_{ik}^{\zeta}\, N_k^{\Theta}\, d\gamma \quad \forall \boldsymbol{w}^{(1)} \in W_{\Theta_{\hat{X}_I}} \tag{5.35}$$

subjected to periodicity and the normalization constraints [see equation (2.30b) in Chapter 2] and defined as

$$\int_{\Theta_{\hat{X}_I}} w_i^{(1)} d\Theta = 0 \tag{5.36}$$

Note that since B_i has been assumed to be constant over the unit cell domain, the body force term does not appear in the unit cell problem due to the normalization condition (5.36).

If the coarse-scale deformation gradient is constant in the unit cell domain, that is, $F_{ij}^{c}(\hat{\boldsymbol{X}}_I,\boldsymbol{\chi}) = \overline{F}_{ij}(\hat{\boldsymbol{X}}_I)$, then stress P_{ik}^{ζ} will be a periodic function and, consequently, the right-hand side of (5.35) will vanish due to periodicity. On the other hand, when the coarse-scale deformation gradient is linear and its constant part is zero, that is, $F_{ij}^{c}(\hat{\boldsymbol{X}}_I,\boldsymbol{\chi}) = \overline{F}_{ij,k}(\hat{\boldsymbol{X}}_I)\chi_k$, then the stress is no longer periodic. It is common in practice to assume that stress has a linear component (see Remark 5.7) in addition to the periodic field

$$P_{ik}^{\zeta} = \alpha_{ikj}\chi_j + P_{ik}^{\text{periodic}} \tag{5.37}$$

Inserting (5.37) into (5.35) and requiring it to be satisfied for arbitrary α_{ikj} yields the usual weak form of the unit cell equilibrium equation

$$\int_{\Theta_{\hat{X}_I}} \frac{\partial w_i^{(1)}}{\partial \chi_k} P_{ik}^{\zeta}\, d\Theta = 0 \quad \forall \boldsymbol{w}^{(1)} \in W_{\Theta_{\hat{X}_I}} \tag{5.38}$$

and an additional constraint equation

$$\int_{\partial\Theta_{\hat{X}_I}} w_i^{(1)} \chi_j N_k^{\Theta}\, d\gamma = 0 \tag{5.39}$$

where we take advantage of the $\boldsymbol{\chi}$-periodicity of P_{ik}^{periodic}.

Combining the definition of the coarse-scale deformation gradient in (5.27b) and (5.39) for $k=j$ yields the homogeneous constraints on each of the unit cell bounding surfaces $\partial\Theta_{\hat{X}_I}^j$ (see Problem 5.6)

$$\int_{\partial\Theta_{\hat{X}_I}^j} w_i^{(1)} d\gamma = 0 \quad \forall j = 1, 2, \ldots, 6 \tag{5.40}$$

where $\bigcup_{j=1}^{6} \partial\Theta_{\hat{X}_I}^j = \partial\Theta_{\hat{X}_I}$ and $\partial\Theta_{\hat{X}_I}^i \cap \partial\Theta_{\hat{X}_I}^j = 0 \quad \forall i \neq j$. Equation (5.40) can be interpreted as a weak compatibility condition between adjacent unit cells. In [33], it was shown that equation (5.40) is necessary to pass the patch test in a mesh consisting of finite elements enriched with a fine-scale kinematics interfacing standard finite elements with homogenized material properties. Equation (5.40) is also employed in [1] in conjunction with periodic boundary conditions.

It is instructive to point out that if (5.40) were imposed on $u_i^{(1)}$ arising from the constant deformation gradient, that is, $\bar{F}_{ij,k}(\hat{\boldsymbol{X}}_I) = 0$, the resulting solution would not converge to the classical $O(1)$ homogenization theory due to the existence of the additional constraint (5.40). In fact, the results illustrated by the model problem (section 5.2.4) show that if $u_i^{(1)}$ arising from the combined effect of $\bar{F}_{ij}(\hat{\boldsymbol{X}}_I)$ and $\bar{F}_{ij,k}(\hat{\boldsymbol{X}}_I)$ is constrained based on (5.40), the computational continua performance is even worse than that of the order $O(1)$ homogenization it is supposed to improve upon.

To preserve consistency with the $O(1)$ homogenization theory, on the one hand, and to satisfy periodicity, normalization (5.36), and weak compatibility (5.40) conditions when the unit cell is subjected to $\bar{F}_{ij,k}(\hat{\boldsymbol{X}}_I)$ only, on the other hand, we define a new space $W_{\Theta_{\hat{X}_I}}^{\nabla}$

$$W_{\Theta_{\hat{X}_I}}^{\nabla} = \left\{ \begin{matrix} w_i^{(1)}(\hat{\boldsymbol{X}}_I, \boldsymbol{\chi}) \text{ defined in } \Theta_{\hat{X}_I}, C^0(\Theta_{\hat{X}_I}), \boldsymbol{\chi}-\text{periodic}, \\ \int_{\Theta_{\hat{X}_I}} w_i^{(1)} d\Theta = 0, \quad \int_{\partial\Theta_{\hat{X}_I}^j} w_i^{(1)} d\gamma = 0 \quad \forall j = 1,2\ldots,6 \end{matrix} \right\} \tag{5.41}$$

There are a number of possible ways to proceed. One possible variant is to employ a two-step stress update approach. In Step 1, the unit cell is subjected to the increment $\Delta\bar{F}_{ij}(\hat{\boldsymbol{X}}_I)$ and the test and trial functions are selected from $W_{\Theta_{\hat{X}_I}}$. In Step 2, the unit cell is subjected to the increment $\Delta\bar{F}_{ij,k}(\hat{\boldsymbol{X}}_I)$ and the test and trial functions are selected from $W^{\nabla}_{\Theta_{\hat{X}_I}}$. An alternative one-step approach, illustrated for a linear model problem in section 5.2.4 and generalized to 3D nonlinear problems in section 5.3, consists of additively decomposing $u_i^{(1)}$ into residual-free components that either depend on $\Delta\bar{F}_{ij}(\hat{\boldsymbol{X}}_I)$ or $\Delta\bar{F}_{ij,k}(\hat{\boldsymbol{X}}_I)$ and then selecting appropriate spaces for each component.

We now proceed to the Galerkin discretization of the computational unit cell test and trial functions

$$
\begin{aligned}
u_i^{\zeta}(\hat{\boldsymbol{X}}_I,\boldsymbol{\chi}) &= u_i^c(\hat{\boldsymbol{X}}_I)+N_{i\beta}^f(\boldsymbol{\chi})d_\beta^{(1)}(\hat{\boldsymbol{X}}_I) \\
w_i^{(1)}(\hat{\boldsymbol{X}}_I,\boldsymbol{\chi}) &= N_{i\beta}^f(\boldsymbol{\chi})c_\beta^{(1)}(\hat{\boldsymbol{X}}_I)
\end{aligned}
\tag{5.42}
$$

where $N_{i\beta}^f(\boldsymbol{\chi})$ is a unit cell shape function, $u_i^c(\hat{\boldsymbol{X}}_I)$ denotes the coarse-scale displacement, and $d_\beta^{(1)}(\hat{\boldsymbol{X}}_I)$ is a discrete fine-scale perturbation.

Let $U_{\Theta_{\hat{X}_I}}$ and $U^{\nabla}_{\Theta_{\hat{X}_I}}$ be the discrete spaces corresponding to $W_{\Theta_{\hat{X}_I}}$ and $W^{\nabla}_{\Theta_{\hat{X}_I}}$, respectively, which impose homogeneous linear constraints on $d_\beta^{(1)}(\hat{\boldsymbol{X}}_I)$ and $c_\beta^{(1)}(\hat{\boldsymbol{X}}_I)$. The discrete constraint equation (5.40) and equation (5.36) can be enforced using the penalty method, the Lagrange multiplier method, the augmented Lagrangian method [34], or, when possible, by the method of eliminating dependent degrees of freedom. For simplicity of exposition of ideas, the latter method is employed here. Let $\tilde{d}_\eta^{(1)}(\hat{\boldsymbol{X}}_I)$ and $\tilde{c}_\eta^{(1)}(\hat{\boldsymbol{X}}_I)$ be the independent degrees of freedom, and then the linear transformation operator $T^*_{\eta\beta}$ can be defined as

$$
d_\eta^{(1)}(\hat{\boldsymbol{X}}_I) = T^*_{\eta\beta}\,\tilde{d}_\beta^{(1)}(\hat{\boldsymbol{X}}_I); \quad c_\eta^{(1)}(\hat{\boldsymbol{X}}_I) = T^*_{\eta\beta}\,\tilde{c}_\beta^{(1)}(\hat{\boldsymbol{X}}_I)
\tag{5.43}
$$

Note that when subjecting the unit cell to $\Delta\bar{F}_{ij}$, $T^*_{\eta\beta}=T_{\eta\beta}$ is based on the discrete space $U_{\Theta_{\hat{X}_I}}$, whereas when subjecting the unit cell to $\Delta\bar{F}_{ij,k}$, $T^*_{\eta\beta}=T^{\nabla}_{\eta\beta}$ is chosen to satisfy $U^{\nabla}_{\Theta_{\hat{X}_I}}$.

In summary, the discrete unit cell problem is formulated as follows:

Given the coarse-scale displacements $\boldsymbol{u}^c(\hat{\boldsymbol{X}}_I)$ and prior converged solution $\boldsymbol{d}^{(1)}(\hat{\boldsymbol{X}}_I)$, find $\Delta\boldsymbol{d}^{(1)}(\hat{\boldsymbol{X}}_I)$ such that

$$
\int_{\Theta_{\hat{X}_I}} T^*_{\eta\beta}\frac{\partial N_{i\eta}^f(\boldsymbol{\chi})}{\partial \chi_j}P_{ij}^{\zeta}(\hat{\boldsymbol{X}}_I,\boldsymbol{\chi})d\Theta = 0 \quad \forall \hat{\boldsymbol{X}}_I
\tag{5.44}
$$

The computational algorithm consists of an evaluation of nonlocal quadrature points in the preprocessing stage, solution of the computational unit cell problems, evaluation of the coarse-scale stress function, and solution of the coarse-scale problem. The information flow is summarized in Figure 5.2.

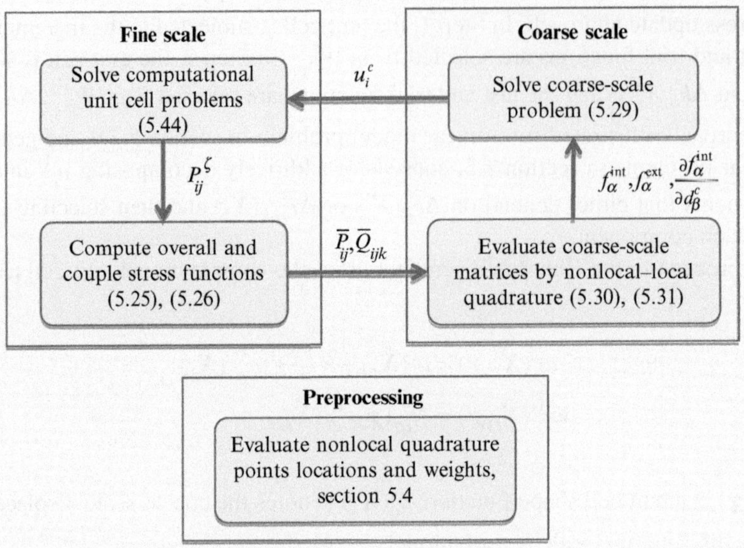

Figure 5.2 Computational Continua: the information flow

Remark 5.8 Adding the fine-scale weak form (5.38) and

$$\int_{\Theta_{\hat{X}_I}} \frac{\partial w_i^c}{\partial X_j} P_{ij}^\zeta\left(\hat{X}_I,\chi\right) d\Theta = \left(\left\{\overline{\frac{\partial w_i^c}{\partial X_j}}\right\}_{\hat{X}_I} \bar{P}_{ij}\left(\hat{X}_I\right) + \left\{\overline{D_k \frac{\partial w_i^c}{\partial X_j}}\right\}_{\hat{X}_I} \bar{Q}_{ijk}\left(\hat{X}_I\right)\right) \left|\Theta_{\hat{X}_I}\right|$$

which arises from (5.20) and (5.24), yields the Hill–Mandel-like macrohomogeneity condition

$$\frac{1}{\left|\Theta_{\hat{X}_I}\right|} \int_{\Theta_{\hat{X}_I}} \frac{\partial w_i^\zeta}{\partial X_j} P_{ij}^\zeta\left(\hat{X}_I,\chi\right) d\Theta = \left\{\overline{\frac{\partial w_i^c}{\partial X_j}}\right\}_{\hat{X}_I} \bar{P}_{ij}\left(\hat{X}_I\right) + \left\{\overline{D_k \frac{\partial w_i^c}{\partial X_j}}\right\}_{\hat{X}_I} \bar{Q}_{ijk}\left(\hat{X}_I\right) \tag{5.45}$$

where we account for the test function additive decomposition $w_i^\zeta = w_i^c + w_i^{(1)}$.

5.2.4 One-Dimensional Model Problem

To illustrate the theoretical framework and performance of the C^2 formulation, we consider a 1D bar consisting of a periodic arrangement of two linear elastic material phases. The bar is fixed at both ends and is subjected to a constant body force, $b=1$. Θ denotes the unit cell length and $\chi \in [-\Theta/2,+\Theta/2]$ is a local coordinate of the unit cell. The unit cell configuration is depicted in Figure 4.4 in Chapter 4.

The total displacement u^ζ is additively decomposed into the coarse-scale displacement u^c and the periodic fine-scale perturbation $u^{(1)}$ as

$$u^{\zeta}(\hat{X}_I,\chi)=u^c(\hat{X}_I,\chi)+u^{(1)}(\hat{X}_I,\chi)$$

The fine-scale perturbation is assumed to be periodic, that is,

$$u^{(1)}(\partial\Theta_L)=u^{(1)}(\partial\Theta_R) \tag{5.46}$$

and satisfy normalization condition (5.36)

$$\int_{\Theta_{\hat{X}_I}} u^{(1)}(\hat{X}_I,\chi)\,d\Theta=0 \tag{5.47}$$

where $\partial\Theta_L$, $\partial\Theta_R$ are the left and right boundaries of the unit cell, respectively. The total strain is given as

$$\varepsilon^{\zeta}(\hat{X}_I,\chi)=\varepsilon^c(\hat{X}_I,\chi)+\varepsilon^*(\hat{X}_I,\chi)=\frac{du^c(\hat{X}_I,\chi)}{d\chi}+\frac{du^{(1)}(\hat{X}_I,\chi)}{d\chi} \tag{5.48}$$

Following (5.27), the coarse-scale strain $\varepsilon^c(\hat{X}_I,\chi)$ is approximated as a linear function

$$\begin{aligned}
\varepsilon^c(\hat{X}_I,\chi)&=\bar{\varepsilon}(\hat{X}_I)+\nabla\bar{\varepsilon}(\hat{X}_I)\chi\\
\bar{\varepsilon}(\hat{X}_I)&=\frac{1}{\Theta}\int_{\Theta_{\hat{X}_I}}\varepsilon^{\zeta}(\hat{X}_I,\chi)\,d\Theta=\frac{1}{\Theta}\int_{\Theta_{\hat{X}_I}}\varepsilon^c(\hat{X}_I,\chi)\,d\Theta\\
\nabla\bar{\varepsilon}(\hat{X}_I)&=\frac{1}{\Theta}\int_{\Theta_{\hat{X}_I}}\frac{\partial\varepsilon^{\zeta}(\hat{X}_I,\chi)}{\partial\chi}\,d\Theta=\frac{1}{\Theta}\int_{\Theta_{\hat{X}_I}}\frac{\partial\varepsilon^c(\hat{X}_I,\chi)}{\partial\chi}\,d\Theta
\end{aligned} \tag{5.49}$$

where we exploit the periodicity of $u^{(1)}$ and its derivative.

Consider now the unit cell weak form defined in (5.38)

$$\int_{\Theta_{\hat{X}_I}}\frac{\partial w^{(1)}}{\partial\chi}L(\chi)\,\varepsilon^{\zeta}(\hat{X}_I,\chi)\,d\Theta=0\quad\forall w^{(1)}\in W_{\Theta_{\hat{X}_I}},W^{\nabla}_{\Theta_{\hat{X}_I}} \tag{5.50}$$

where $L^{\zeta}(X)=L(\chi)$ is a periodic piecewise constant Young's modulus.

For 1D problems, the weak compatibility constraint (5.40), which must be satisfied when the unit cell is subjected to $\nabla\bar{\varepsilon}(\hat{X}_I)$, is given by

$$u^{(1)}(\partial\Theta_L)=u^{(1)}(\partial\Theta_R)=0 \tag{5.51}$$

The trial solution $u^{(1)}$ and the test function $w^{(1)}$ are discretized using the fine-scale shape functions $\boldsymbol{N}^f(\chi)$ as

$$u^{(1)}(\hat{X}_I,\chi)=\boldsymbol{N}^f(\chi)\boldsymbol{d}^{(1)}(\hat{X}_I);\quad w^{(1)}(\hat{X}_I,\chi)=\boldsymbol{N}^f(\chi)\boldsymbol{c}^{(1)}(\hat{X}_I) \tag{5.52}$$

where $\boldsymbol{d}^{(1)}$ and $\boldsymbol{c}^{(1)}$ are the nodal values of the trial solution and the test function, respectively. When the unit cell is subjected to $\bar{\varepsilon}(\hat{X}_I)$, $\boldsymbol{d}^{(1)}(\hat{X}_I)$ and $\boldsymbol{c}^{(1)}(\hat{X}_I)$ should satisfy (5.46) and (5.47). On the other hand, when the unit cell is subjected to $\nabla\bar{\varepsilon}(\hat{X}_I)$, $\boldsymbol{d}^{(1)}(\hat{X}_I)$ and $\boldsymbol{c}^{(1)}(\hat{X}_I)$ should satisfy (5.47) and (5.51) instead. Following (5.43), the independent degrees of freedom are constructed as

$$\boldsymbol{d}^{(1)}(\hat{X}_I) = \boldsymbol{T}^*\tilde{\boldsymbol{d}}^{(1)}(\hat{X}_I); \quad \boldsymbol{c}^{(1)}(\hat{X}_I) = \boldsymbol{T}^*\tilde{\boldsymbol{c}}^{(1)}(\hat{X}_I) \tag{5.53}$$

where the transformation matrix $\boldsymbol{T}^*$ is equal to $\boldsymbol{T}$ when the unit cell is subjected to $\bar{\varepsilon}(\hat{X}_I)$; it will be equal to $\boldsymbol{T}^\nabla$ when the unit cell is subjected to $\nabla\bar{\varepsilon}(\hat{X}_I)$. Inserting (5.49), (5.52), and (5.53) into (5.50) and requiring arbitrariness of $\tilde{\boldsymbol{c}}^{(1)}(\hat{X}_I)$ yields

$$\left((\boldsymbol{T}^*)^T \boldsymbol{K}^f \boldsymbol{T}^*\right)\tilde{\boldsymbol{d}}^{(1)}(\hat{X}_I) = (\boldsymbol{T}^*)^T \boldsymbol{f}^f \bar{\varepsilon}(\hat{X}_I) + (\boldsymbol{T}^*)^T \boldsymbol{f}^\nabla \nabla\bar{\varepsilon}(\hat{X}_I) \tag{5.54}$$

where

$$\begin{aligned}
\boldsymbol{K}^f &= \int_{\Theta_{\hat{X}_I}} \left(\frac{\partial \boldsymbol{N}^f(\chi)}{\partial \chi}\right)^T L(\chi)\left(\frac{\partial \boldsymbol{N}^f(\chi)}{\partial \chi}\right) d\Theta \\
\boldsymbol{f}^f &= -\int_{\Theta_{\hat{X}_I}} \left(\frac{\partial \boldsymbol{N}^f(\chi)}{\partial \chi}\right)^T L(\chi)\, d\Theta \\
\boldsymbol{f}^\nabla &= -\int_{\Theta_{\hat{X}_I}} \left(\frac{\partial \boldsymbol{N}^f(\chi)}{\partial \chi}\right)^T L(\chi)\,\chi\, d\Theta
\end{aligned} \tag{5.55}$$

We require the linear system (5.54) to be valid for arbitrary $\bar{\varepsilon}(\hat{X}_I)$ and $\nabla\bar{\varepsilon}(\hat{X}_I)$ with appropriate constraint equations, which yields

$$\boldsymbol{d}^{(1)} = \boldsymbol{H}_1\bar{\varepsilon} + \boldsymbol{H}_2\nabla\bar{\varepsilon} \tag{5.56}$$

where

$$\begin{aligned}
\boldsymbol{H}_1 &= \boldsymbol{T}(\boldsymbol{T}^T\boldsymbol{K}^f\boldsymbol{T})^{-1}\boldsymbol{T}^T\boldsymbol{f}^f \\
\boldsymbol{H}_2 &= \boldsymbol{T}^\nabla\left((\boldsymbol{T}^\nabla)^T \boldsymbol{K}^f \boldsymbol{T}^\nabla\right)^{-1}(\boldsymbol{T}^\nabla)^T \boldsymbol{f}^\nabla
\end{aligned} \tag{5.57}$$

We now proceed to the weak form of the coarse-scale problem, which is given by

$$\sum_{I=1}^{\hat{N}} \frac{W_I J_I}{\Theta} \int_{\Theta_{\hat{X}_I}} \frac{\partial w^c}{\partial X} L(\chi)\left(\varepsilon^c(\hat{X}_I,\chi) + \frac{du^{(1)}(\hat{X}_I,\chi)}{d\chi}\right) d\Theta = \int_{\Omega_X} w^c b\, d\Omega \tag{5.58}$$

The coarse-scale trial solution and test function are discretized as

$$u^c(X) \equiv u^c(\hat{X}_I,\chi) = \boldsymbol{N}^c(\hat{X}_I,\chi)\boldsymbol{d}^c; \quad w^c(X) \equiv w^c(\hat{X}_I,\chi) = \boldsymbol{N}^c(\hat{X}_I,\chi)\boldsymbol{c}^c \tag{5.59}$$

where $\boldsymbol{d}^c$ and $\boldsymbol{c}^c$ are the nodal values of the trial solution and test function, respectively, and $\boldsymbol{N}^c(\hat{X}_I,\chi)$ are the coarse-scale shape functions expressed in the unit cell physical domain $\chi \in [-\Theta/2, +\Theta/2]$, using the mapping $X = \chi + \hat{X}_I$.

Following the approximation of the coarse-scale trial solution and test function gradients in (5.21), we have

$$\begin{aligned} \varepsilon^c(\hat{X}_I,\chi) &= \left(\boldsymbol{B}_I^c + \boldsymbol{B}_I^\nabla \chi\right)\boldsymbol{d}^c \\ \frac{\partial w^c(\hat{X}_I,\chi)}{\partial \chi} &= \left(\boldsymbol{B}_I^c + \boldsymbol{B}_I^\nabla \chi\right)\boldsymbol{c}^c \end{aligned} \tag{5.60}$$

where

$$\boldsymbol{B}_I^c = \frac{1}{\Theta}\int_{\Theta_{\hat{X}_I}} \frac{\partial \boldsymbol{N}^c(\hat{X}_I,\chi)}{\partial \chi} d\chi; \quad \boldsymbol{B}_I^\nabla = \frac{1}{\Theta}\int_{\Theta_{\hat{X}_I}} \frac{\partial^2 \boldsymbol{N}^c(\hat{X}_I,\chi)}{\partial \chi^2} d\chi \tag{5.61}$$

Inserting (5.56) and (5.60) into (5.58) yields the discrete linear system of equations for the coarse-scale problem

$$\begin{aligned} \boldsymbol{K}^c \boldsymbol{d}^c &= \int_{\Omega_X} (\boldsymbol{N}^c)^T b\, d\Omega \\ \boldsymbol{K}^c &= \sum_{I=1}^{\hat{N}} \frac{W_I J_I}{\Theta} \int_{\Theta_{\hat{X}_I}} \left(\boldsymbol{B}_I^c + \boldsymbol{B}_I^\nabla \chi\right)^T L(\chi)\left[\left(\boldsymbol{B}_I^c + \boldsymbol{B}_I^\nabla \chi\right) + \frac{\partial N^f(\chi)}{\partial \chi}\left(\boldsymbol{H}_1 \boldsymbol{B}_I^c + \boldsymbol{H}_2 \boldsymbol{B}_I^\nabla\right)\right] d\Theta \end{aligned} \tag{5.62}$$

The model problem is solved for one, two, four, and eight quadratic coarse-scale elements and various unit cell sizes, $\Theta = \alpha L_e$, where $\alpha = \{0.10, 0.20, 0.25, 0.33, 0.50, 1\}$ and L_e is a coarse-scale element size. For all cases considered, the fiber volume fraction is $\phi^f = 0.5$, the ratio between the Young's moduli of the two phases is 10, and the dimensionless body force is $b = 1$. Results of the C^2 formulation are compared with $O(1)$ homogenization, and the reference solution is obtained by DNS. To compare the C^2 to DNS, the number of unit cells n_{uc} considered in DNS is taken as $n_{uc} = 1/\Theta = N_e/\alpha$. Figure 5.3 depicts the relative error in strain energy for C^2 and $O(1)$ formulations with respect to the DNS. It is apparent that as the size of the unit cell is approaching zero, C^2 and $O(1)$ approximations converge to the DNS. On the other hand, for larger unit cells, the strain-gradient effects prevail and C^2 clearly outperforms $O(1)$ homogenization. When the number of coarse-scale elements is increased, the advantage of the C^2 over the $O(1)$ homogenization becomes less pronounced as both converge to the DNS even for relatively large unit cells.

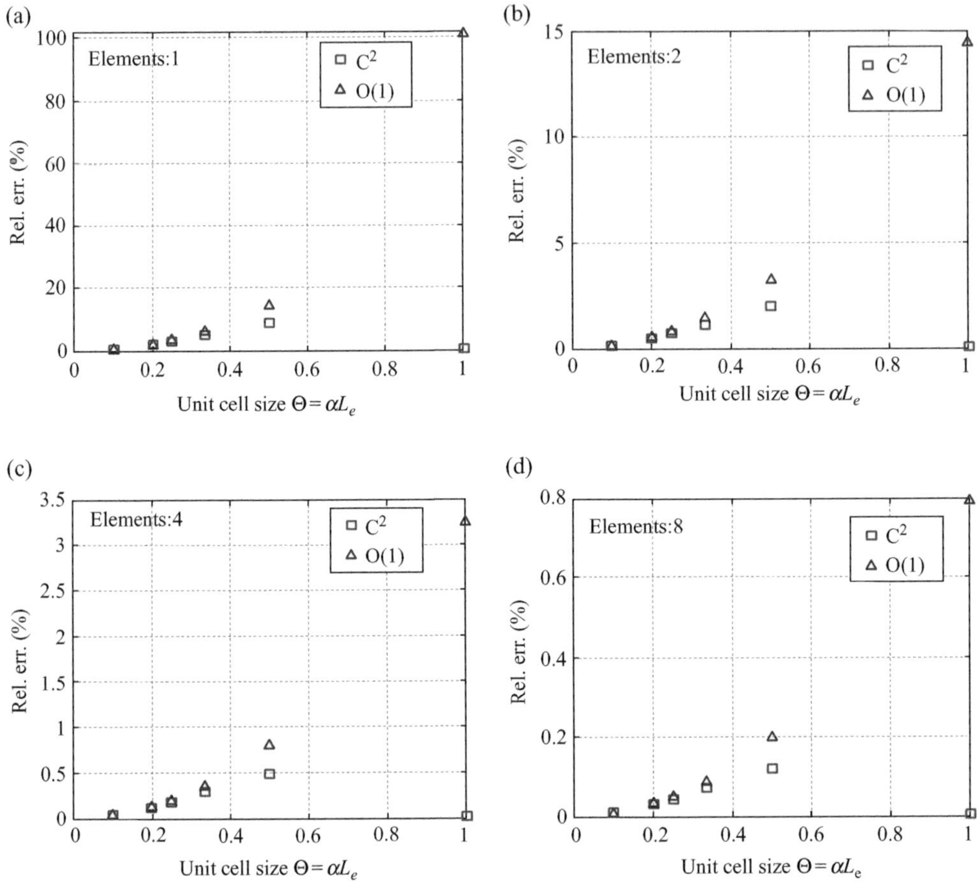

Figure 5.3 Relative error in the strain energy of C^2 and $O(1)$ formulations as a function of the unit cell size for (a) one element, (b) two elements, (c) four elements, and (d) eight elements

To this end, we study the performance of alternative C^2 formulations. We consider two coarse-scale quadratic elements and the same range of unit cell sizes as in the previous example. The following two alternatives are considered:

(1) C^2 case-1: This approach is the same as the aforementioned C^2 formulation, except that instead of constraint (5.51), the boundary term (5.35) is included in the unit cell problem subjected to the coarse-scale strain gradient.
(2) C^2 case-2: This approach is the same as the aforementioned C^2 formulation, except that $u^{(1)} \in W^{\nabla}_{\Theta_{\hat{x}_I}}$ and $w^{(1)} \in W^{\nabla}_{\Theta_{\hat{x}_I}}$, that is, the constraint (5.51) is imposed for the entire perturbation field.

Figure 5.4 suggests that the performance of the aforementioned C^2 formulation is clearly superior to the above two alternative formulations. In fact, the alternative formulations are

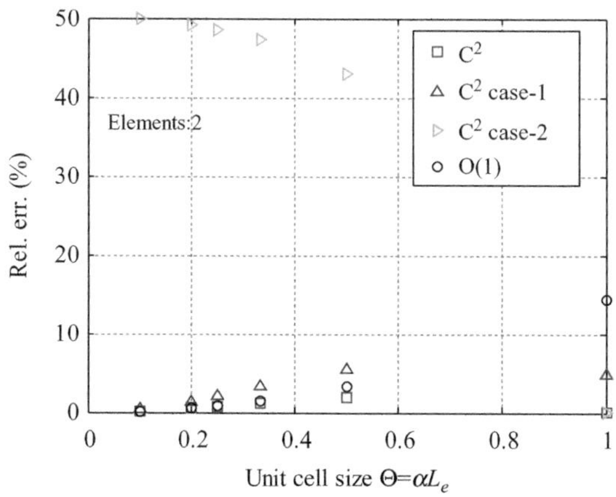

Figure 5.4 Performance of alternative C² formulations for two elements

even less accurate than the $O(1)$ homogenization except when the unit cell coincides with the coarse-scale element.

5.3 Reduced Order Computational Continua (RC²)

In the remainder of this chapter, we consider a unit cell undergoing small distortions. Large deformations can be accounted for via the corotational formulation outlined in Chapter 4.

Consider the governing equations stated at the fine scale of interest [equation (4.1), equation (4.2), equation (4.3), equation (4.4), equation (4.5), and equation (4.6) in Chapter 4]. The body force $b_i^\zeta(\boldsymbol{x})$ is assumed to be constant over the unit cell domain. We further assume an additive decomposition of displacements $\boldsymbol{u}^\zeta$ into a smooth coarse-scale function $\boldsymbol{u}^c$ and an oscillatory function $\boldsymbol{u}^{(1)}$

$$\boldsymbol{u}^\zeta = \boldsymbol{u}^c + \boldsymbol{u}^{(1)} \tag{5.63}$$

The total strain $\boldsymbol{\varepsilon}^\zeta(\hat{\boldsymbol{x}}_I, \boldsymbol{\chi})$ is additively decomposed into coarse-scale strain $\boldsymbol{\varepsilon}^c$ and the fine-scale perturbation $\boldsymbol{\varepsilon}^*$ as

$$\varepsilon_{ij}^\zeta(\hat{\boldsymbol{x}}_I, \boldsymbol{\chi}) = \varepsilon_{ij}^c(\hat{\boldsymbol{x}}_I, \boldsymbol{\chi}) + \varepsilon_{ij}^*(\hat{\boldsymbol{x}}_I, \boldsymbol{\chi}) \tag{5.64}$$

defined as a symmetric spatial derivative of displacements

$$\varepsilon_{ij}^c(\hat{\boldsymbol{x}}_I, \boldsymbol{\chi}) = u_{(i,x_j)}^c(\hat{\boldsymbol{x}}_I, \boldsymbol{\chi}); \quad \varepsilon_{ij}^*(\hat{\boldsymbol{x}}_I, \boldsymbol{\chi}) = u_{(i,\chi_j)}^{(1)}(\hat{\boldsymbol{x}}_I, \boldsymbol{\chi}) \tag{5.65}$$

Following the linear approximation of the deformation gradient (5.27), the coarse-scale strain ε_{ij}^c is approximated by a linear function over the computational unit cell domain consisting of an average strain $\bar{\varepsilon}_{ij}$ and an average strain gradient $\bar{\varepsilon}_{ij,m}$ defined as

$$\begin{gathered}\varepsilon_{ij}^c(\hat{\boldsymbol{x}}_I, \boldsymbol{\chi}) = \bar{\varepsilon}_{ij}(\hat{\boldsymbol{x}}_I) + \bar{\varepsilon}_{ij,m}(\hat{\boldsymbol{x}}_I)\chi_m \\ \bar{\varepsilon}_{ij}(\hat{\boldsymbol{x}}_I) = \frac{1}{\left|\Theta_{\hat{x}_I}\right|}\int_{\Theta_{\hat{x}_I}} \varepsilon_{ij}^c(\hat{\boldsymbol{x}}_I, \boldsymbol{\chi})\,d\Theta; \quad \bar{\varepsilon}_{ij,m}(\hat{\boldsymbol{x}}_I) = \frac{1}{\left|\Theta_{\hat{x}_I}\right|}\int_{\Theta_{\hat{x}_I}} \frac{\partial \varepsilon_{ij}^c}{\partial \chi_m}(\hat{\boldsymbol{x}}_I, \boldsymbol{\chi})\,d\Theta\end{gathered} \tag{5.66}$$

Recall from section 5.2.4 that $\boldsymbol{u}^{(1)}$ satisfies periodicity and normalization (5.36) conditions when the unit cell is subjected to $\bar{\varepsilon}_{ij}(\hat{\boldsymbol{x}}_I)$ and that $\boldsymbol{u}^{(1)}$ satisfies periodicity, normalization (5.36), and weak compatibility (5.40) conditions when the unit cell is subjected to $\bar{\varepsilon}_{ij,m}(\hat{\boldsymbol{x}}_I)$.

Employing (5.66) and the constitutive relation in (4.2), the equilibrium equation can be expressed as

$$\left\{L_{ijkl}(\boldsymbol{\chi})\left[\bar{\varepsilon}_{kl}(\hat{\boldsymbol{x}}_I)+\bar{\varepsilon}_{kl,m}(\hat{\boldsymbol{x}}_I)\chi_m+u^{(1)}_{(k,\chi_l)}(\hat{\boldsymbol{x}}_I,\boldsymbol{\chi})-\mu^{\zeta}_{kl}(\hat{\boldsymbol{x}}_I,\boldsymbol{\chi})\right]\right\}_{,\chi_j}+b^{\zeta}_i(\hat{\boldsymbol{x}}_I,\boldsymbol{\chi})=0 \tag{5.67}$$

where, for simplicity, we assume that elastic properties depend on the fine-scale coordinates only, $L^{\zeta}_{ijkl}(\boldsymbol{x})=L_{ijkl}(\boldsymbol{\chi})$, that is, they are the same for all computational unit cells $\Theta_{\hat{x}_I}$.

5.3.1 Residual-Free Computational Unit Cell Problem

For large unit cells over which the coarse-scale fields are no longer constant, the residual-free expansion considered in Chapter 4 has to be modified to include higher order residual-free coarse-scale fields. This is accomplished by introducing the following decomposition of fine-scale displacements

$$\begin{aligned} u^{(1)}_i(\hat{\boldsymbol{x}}_I,\boldsymbol{\chi}) &= H^{mn}_i(\hat{\boldsymbol{x}}_I,\boldsymbol{\chi})\bar{\varepsilon}_{mn}(\hat{\boldsymbol{x}}_I)+H^{mnp}_i(\hat{\boldsymbol{x}}_I,\boldsymbol{\chi})\bar{\varepsilon}_{mn,p}(\hat{\boldsymbol{x}}_I) \\ &+\int_{\Theta_{\hat{x}_I}} \tilde{h}^{mn}_i(\hat{\boldsymbol{x}}_I,\boldsymbol{\chi},\tilde{\boldsymbol{\chi}})\mu^{\zeta}_{mn}(\hat{\boldsymbol{x}}_I,\tilde{\boldsymbol{\chi}})\,d\tilde{\Theta}+\int_{S_{\hat{x}_I}} \breve{h}^{\breve{m}}_i(\hat{\boldsymbol{x}}_I,\boldsymbol{\chi},\breve{\boldsymbol{\chi}})\delta^{\zeta}_{\breve{m}}(\hat{\boldsymbol{x}}_I,\breve{\boldsymbol{\chi}})\,d\breve{S} \end{aligned} \tag{5.68}$$

where H^{mn}_i, H^{mnp}_i, $\tilde{h}^{mn}_i$, and $\breve{h}^{\breve{m}}_i$ are the influence functions; $\delta^{\zeta}_{\breve{m}}$ is an eigenseparation; and the subscript $\breve{m}$ denotes components in the local Cartesian coordinate system of the unit cell interface, denoted by $S_{\hat{x}_I}$.

Inserting (5.66) and (5.68) into the constitutive relation (4.2) yields the residual-free field

$$\sigma^{\zeta}_{ij}(\hat{\boldsymbol{x}}_I,\boldsymbol{\chi})=L_{ijkl}(\boldsymbol{\chi})\left(\begin{aligned} &E^{mn}_{kl}(\hat{\boldsymbol{x}}_I,\boldsymbol{\chi})\bar{\varepsilon}_{mn}(\hat{\boldsymbol{x}}_I)+E^{mnp}_{kl}(\hat{\boldsymbol{x}}_I,\boldsymbol{\chi})\bar{\varepsilon}_{mn,p}(\hat{\boldsymbol{x}}_I) \\ &+\int_{S_{\hat{x}_I}} \breve{h}^{\breve{m}}_{(k,\chi_l)}(\hat{\boldsymbol{x}}_I,\boldsymbol{\chi},\breve{\boldsymbol{\chi}})\delta^{\zeta}_{\breve{m}}(\hat{\boldsymbol{x}}_I,\breve{\boldsymbol{\chi}})\,d\breve{S} \\ &+\int_{\Theta_{\hat{x}_I}} \tilde{h}^{mn}_{(k,\chi_l)}(\hat{\boldsymbol{x}}_I,\boldsymbol{\chi},\tilde{\boldsymbol{\chi}})\,\mu^{\zeta}_{mn}(\hat{\boldsymbol{x}}_I,\tilde{\boldsymbol{\chi}})\,d\tilde{\Theta}-\mu^{\zeta}_{kl}(\hat{\boldsymbol{x}}_I,\boldsymbol{\chi}) \end{aligned}\right) \tag{5.69}$$

where

$$E^{mn}_{kl}(\hat{\boldsymbol{x}}_I,\boldsymbol{\chi})=I_{klmn}+H^{mn}_{(k,\chi_l)}(\hat{\boldsymbol{x}}_I,\boldsymbol{\chi});\quad E^{mnp}_{kl}(\hat{\boldsymbol{x}}_I,\boldsymbol{\chi})=I_{klmn}\chi_p+H^{mnp}_{(k,\chi_l)}(\hat{\boldsymbol{x}}_I,\boldsymbol{\chi}) \tag{5.70}$$

The resulting equilibrium equation (5.67) takes the following form:

$$\frac{\partial}{\partial \chi_j}\left\{ L_{ijkl}(\boldsymbol{\chi}) \left(\begin{array}{l} E_{kl}^{mn}(\hat{\boldsymbol{x}}_I,\boldsymbol{\chi})\overline{\varepsilon}_{mn}(\hat{\boldsymbol{x}}_I) + E_{kl}^{mnp}(\hat{\boldsymbol{x}}_I,\boldsymbol{\chi})\overline{\varepsilon}_{mn,p}(\hat{\boldsymbol{x}}_I) \\ + \int_{S_{\hat{x}_I}} \breve{h}_{(k,\chi_l)}^{\breve{m}}(\hat{\boldsymbol{x}}_I,\boldsymbol{\chi},\breve{\boldsymbol{\chi}})\delta_{\breve{m}}^{\zeta}(\hat{\boldsymbol{x}}_I,\breve{\boldsymbol{\chi}})\,d\breve{S} \\ + \int_{\Theta_{\hat{x}_I}} \tilde{h}_{(k,\chi_l)}^{mn}(\hat{\boldsymbol{x}}_I,\boldsymbol{\chi},\tilde{\boldsymbol{\chi}})\mu_{mn}^{\zeta}(\hat{\boldsymbol{x}}_I,\tilde{\boldsymbol{\chi}})\,d\tilde{\Theta} - \mu_{kl}^{\zeta}(\hat{\boldsymbol{x}}_I,\boldsymbol{\chi}) \end{array} \right) \right\} + b_i(\hat{\boldsymbol{x}}_I) = 0 \tag{5.71}$$

The weak form of the unit cell problem is constructed by multiplying (5.71) by a test function $\boldsymbol{w}^{(1)}$, the space of which will be subsequently defined, and then integrating over the unit cell domain, which gives

$$\int_{\Theta_{\hat{x}_I}} w_i^{(1)}(\hat{\boldsymbol{x}}_I,\boldsymbol{\chi})\frac{\partial}{\partial \chi_j}\left\{ L_{ijkl}(\boldsymbol{\chi}) \left(\begin{array}{l} E_{kl}^{mn}(\hat{\boldsymbol{x}}_I,\boldsymbol{\chi})\overline{\varepsilon}_{mn}(\hat{\boldsymbol{x}}_I) + E_{kl}^{mnp}(\hat{\boldsymbol{x}}_I,\boldsymbol{\chi})\overline{\varepsilon}_{mn,p}(\hat{\boldsymbol{x}}_I) \\ + \int_{S_{\hat{x}_I}} \breve{h}_{(k,l)}^{\breve{m}}(\hat{\boldsymbol{x}}_I,\boldsymbol{\chi},\breve{\boldsymbol{\chi}})\delta_{\breve{m}}^{\zeta}(\hat{\boldsymbol{x}}_I,\breve{\boldsymbol{\chi}})\,d\breve{S} \\ + \int_{\Theta_{\hat{x}_I}} \tilde{h}_{(k,l)}^{mn}(\hat{\boldsymbol{x}}_I,\boldsymbol{\chi},\tilde{\boldsymbol{\chi}})\mu_{mn}^{\zeta}(\hat{\boldsymbol{x}}_I,\tilde{\boldsymbol{\chi}})\,d\tilde{\Theta} - \mu_{kl}^{\zeta}(\hat{\boldsymbol{x}}_I,\boldsymbol{\chi}) \end{array} \right) \right\} d\Theta = 0 \quad \forall \boldsymbol{w}^{(1)} \tag{5.72}$$

where we exploit the fact that the body forces are constant over the unit cell domain.

5.3.1.1 Model Reduction

The primary objective of model reduction is to reduce the computational complexity of solving a sequence of unit cell problems. As in Chapter 4, model reduction is introduced by discretizing eigenstrains and eigenseparations. Following second-order computational continua, eigenstrains are approximated by a piecewise constant approximation over each phase partition

$$\begin{array}{l} \mu_{kl}^{\zeta}(\hat{\boldsymbol{x}}_I,\boldsymbol{\chi}) = \sum\limits_{\alpha=1}^{\tilde{M}} \tilde{N}^{(\alpha)}\left(\hat{\boldsymbol{x}}_I,\boldsymbol{\chi}^{(\alpha)}\right)\mu_{kl}^{(\alpha)}(\hat{\boldsymbol{x}}_I) \\ \mu_{ij}^{(\alpha)}\left(\hat{\boldsymbol{x}}_I,\boldsymbol{\chi}^{(\alpha)}\right) \equiv \overline{\mu}_{ij}^{(\alpha)}(\hat{\boldsymbol{x}}_I) + \overline{\mu}_{ij,m}^{(\alpha)}(\hat{\boldsymbol{x}}_I)\chi_m^{(\alpha)} \\ \tilde{N}^{(\alpha)}\left(\hat{\boldsymbol{x}}_I,\boldsymbol{\chi}^{(\alpha)}\right) \equiv \begin{cases} 1 & \boldsymbol{\chi}^{(\alpha)} \in \Theta_{\hat{x}_I}^{(\alpha)} \\ 0 & \boldsymbol{\chi}^{(\alpha)} \notin \Theta_{\hat{x}_I}^{(\alpha)} \end{cases}; \quad \Theta_{\hat{x}_I} = \bigcup\limits_{\alpha=1}^{\tilde{M}} \Theta_{\hat{x}_I}^{(\alpha)}; \quad \varnothing = \bigcap\limits_{\alpha=1}^{\tilde{M}} \Theta_{\hat{x}_I}^{(\alpha)} \end{array} \tag{5.73}$$

and the eigenseparations are approximated by a $C^0(\Theta_{\hat{x}_I})$ piecewise linear function (see Chapter 4, section 4.2.2)

$$\delta_{\breve{q}}^{\zeta}(\hat{x}_I,\chi)=\sum_{\xi=1}^{\breve{M}}\breve{N}^{(\xi)}(\hat{x}_I,\chi)\delta_{\breve{q}}^{(\xi)}(\hat{x}_I)$$
$$\breve{N}^{(\xi)}(\hat{x}_I,\chi)\equiv\begin{cases}\sum_{r\in S_{\hat{x}_I}^{(\xi)}}N_r(\hat{x}_I,\chi) & \chi\in S_{\hat{x}_I}^{(\xi)}\\ 0 & \chi\notin S_{\hat{x}_I}^{(\xi)}\end{cases} \tag{5.74}$$

where $\overline{\mu}_{ij}^{(\alpha)}$ and $\overline{\mu}_{ij,m}^{(\alpha)}$ are the average and linear variations of the eigenstrain, respectively; $\chi^{(\alpha)}=\chi-\hat{\chi}^{(\alpha)}$ denotes the local coordinate system positioned at the centroid $\hat{\chi}^{(\alpha)}$ of phase partition α; and $\Theta_{\hat{x}_I}^{(\alpha)}$, $S_{\hat{x}_I}^{(\xi)}$, $\tilde{M}$, and $\breve{M}$ are defined as in section 4.2.2 of Chapter 4.

Inserting discretized eigenstrain (5.73) into (5.72) yields

$$\int_{\Theta_{\hat{x}_I}} w_i^{(1)}(\hat{x}_I,\chi)\frac{\partial}{\partial\chi_j}\left\{L_{ijkl}(\chi)\left(\begin{array}{l}E_{kl}^{mn}(\hat{x}_I,\chi)\overline{\varepsilon}_{mn}(\hat{x}_I)+E_{kl}^{mnp}(\hat{x}_I,\chi)\overline{\varepsilon}_{mn,p}(\hat{x}_I)\\ +\Lambda_{kl}^{mn(\alpha)}(\hat{x}_I,\chi)\overline{\mu}_{mn}^{(\alpha)}(\hat{x}_I)+\Lambda_{kl}^{mnr(\alpha)}(\hat{x},\chi)\overline{\mu}_{mn,r}^{(\alpha)}(\hat{x}_I)\\ +Q_{kl}^{\breve{m}(\xi)}(\hat{x}_I,\chi)\delta_{\breve{m}}^{(\xi)}(\hat{x}_I)\end{array}\right)\right\}d\Theta=0\quad\forall w^{(1)} \tag{5.75}$$

where

$$\begin{aligned}
&\Lambda_{kl}^{mn(\alpha)}(\hat{x}_I,\chi)=P_{kl}^{mn(\alpha)}(\hat{x}_I,\chi)-I_{klmn}^{(\alpha)}(\hat{x}_I,\chi)\\
&\Lambda_{kl}^{mnr(\alpha)}(\hat{x}_I,\chi)=P_{kl}^{mnr(\alpha)}(\hat{x}_I,\chi)-I_{klmn}^{(\alpha)}(\hat{x}_I,\chi)\chi_r^{(\alpha)}\\
&\tilde{h}_k^{mn(\alpha)}(\hat{x}_I,\chi)=\int_{\Theta_{\hat{x}_I}}\tilde{N}^{(\alpha)}(\hat{x}_I,\tilde{\chi})\,\tilde{h}_k^{mn}(\hat{x}_I,\chi,\tilde{\chi})\,d\tilde{\Theta}\\
&\tilde{h}_k^{mnr(\alpha)}(\hat{x}_I,\chi)=\int_{\Theta_{\hat{x}_I}}\tilde{N}^{(\alpha)}(\hat{x}_I,\tilde{\chi})\tilde{h}_k^{mn}(\hat{x}_I,\chi,\tilde{\chi})\tilde{\chi}_r^{(\alpha)}d\tilde{\Theta}\\
&\breve{h}_i^{\breve{n}(\xi)}(\hat{x}_I,\chi)=\int_{S_{\hat{x}_I}}\breve{N}^{(\xi)}(\hat{x}_I,\breve{\chi})\breve{h}_i^{\breve{n}}(\hat{x}_I,\chi,\breve{\chi})\,d\breve{S}\\
&Q_{ij}^{\breve{n}(\xi)}(\hat{x}_I,\chi)=\breve{h}_{(i,\chi_j)}^{\breve{n}(\xi)}(\hat{x}_I,\chi)\\
&P_{kl}^{mn(\alpha)}(\hat{x}_I,\chi)=\tilde{h}_{(k,\chi_l)}^{mn(\alpha)}(\hat{x}_I,\chi)\\
&P_{kl}^{mnr(\alpha)}(\hat{x}_I,\chi)=\tilde{h}_{(k,\chi_l)}^{mnr(\alpha)}(\hat{x}_I,\chi)
\end{aligned} \tag{5.76}$$

and

$$I_{klmn}^{(\alpha)}(\hat{x}_I,\chi^{(\alpha)})=I_{klmn}\tilde{N}^{(\alpha)}(\hat{x}_I,\chi^{(\alpha)})=\begin{cases}I_{klmn} & \chi^{(\alpha)}\in\Theta_{\hat{x}_I}^{(\alpha)}\\ 0 & \chi^{(\alpha)}\notin\Theta_{\hat{x}_I}^{(\alpha)}\end{cases} \tag{5.77}$$

The influence functions H_i^{mn}, H_i^{mnp}, $\tilde{h}_k^{mn(\alpha)}$, $\tilde{h}_k^{mnr(\alpha)}$, and $\breve{h}_i^{\breve{n}(\xi)}$ are constructed to satisfy (5.72) for arbitrary $\overline{\varepsilon}_{mn}$, $\overline{\varepsilon}_{mn,p}$, $\overline{\mu}_{mn}^{(\alpha)}$, $\overline{\mu}_{mn,r}^{(\alpha)}$, and $\delta_{\breve{m}}^{(\xi)}$. We will employ a Galerkin weak form where the trial solution and test function belong to the same space. In the spirit of the formulation in the previous section, we will assume that the influence functions $H_i^{mn} \in W_{\Theta_{\hat{x}_I}}$, $\tilde{h}_k^{mn(\alpha)} \in W_{\Theta_{\hat{x}_I}}$, and $\breve{h}_i^{\breve{m}} \in W_{\Theta_{\hat{x}_I}}$ satisfy periodicity and normalization (5.36) conditions, whereas $H_i^{mnp} \in W_{\Theta_{\hat{x}_I}}^{\nabla}$ and $\tilde{h}_k^{mnr(\alpha)} \in W_{\Theta_{\hat{x}_I}}^{\nabla}$ satisfy an additional weak compatibility (5.40) condition.

Thus, in the weak forms resulting from the arbitrariness of the gradients $\overline{\varepsilon}_{mn,p}$ and $\overline{\mu}_{mn,r}^{(\alpha)}$, the test functions are selected from the space $W_{\Theta_{\hat{x}_I}}^{\nabla}$, whereas for the remaining three problems, the trial solution and test function will be selected from the space $W_{\Theta_{\hat{x}_I}}$, which yields

$$\begin{aligned}
&\int_{\Theta_{\hat{x}_I}} w_i^{(1)}(\hat{\boldsymbol{x}}_I,\boldsymbol{\chi})\{L_{ijkl}(\boldsymbol{\chi})E_{kl}^{mn}(\hat{\boldsymbol{x}}_I,\boldsymbol{\chi})\}_{,\chi_j}\,d\Theta = 0 && \forall \boldsymbol{w}^{(1)} \in W_{\Theta_{\hat{x}_I}}\\
&\int_{\Theta_{\hat{x}_I}} w_i^{(1)}(\hat{\boldsymbol{x}}_I,\boldsymbol{\chi})\{L_{ijkl}(\boldsymbol{\chi})\Lambda_{kl}^{mn(\alpha)}(\hat{\boldsymbol{x}}_I,\boldsymbol{\chi})\}_{,\chi_j}\,d\Theta = 0 && \forall \boldsymbol{w}^{(1)} \in W_{\Theta_{\hat{x}_I}}\\
&\int_{\Theta_{\hat{x}_I}} w_i^{(1)}(\hat{\boldsymbol{x}}_I,\boldsymbol{\chi})\{L_{ijkl}(\boldsymbol{\chi})Q_{kl}^{\breve{m}(\xi)}(\hat{\boldsymbol{x}}_I,\boldsymbol{\chi})\}_{,\chi_j}\,d\Theta = 0 && \forall \boldsymbol{w}^{(1)} \in W_{\Theta_{\hat{x}_I}}\\
&\int_{\Theta_{\hat{x}_I}} w_i^{(1)}(\hat{\boldsymbol{x}}_I,\boldsymbol{\chi})\{L_{ijkl}(\boldsymbol{\chi})E_{kl}^{mnp}(\hat{\boldsymbol{x}}_I,\boldsymbol{\chi})\}_{,\chi_j}\,d\Theta = 0 && \forall \boldsymbol{w}^{(1)} \in W_{\Theta_{\hat{x}_I}}^{\nabla}\\
&\int_{\Theta_{\hat{x}_I}} w_i^{(1)}(\hat{\boldsymbol{x}}_I,\boldsymbol{\chi})\{L_{ijkl}(\boldsymbol{\chi})\Lambda_{kl}^{mnr(\alpha)}(\hat{\boldsymbol{x}}_I,\boldsymbol{\chi})\}_{,\chi_j}\,d\Theta = 0 && \forall \boldsymbol{w}^{(1)} \in W_{\Theta_{\hat{x}_I}}^{\nabla}
\end{aligned} \tag{5.78}$$

Integrating (5.78) by parts and exploiting $\boldsymbol{\chi}$-periodicity yields the weak form of the elastic influence function problem

$$\begin{aligned}
&\int_{\Theta_{\hat{x}_I}} w_{(i,\chi_j)}^{(1)}(\hat{\boldsymbol{x}}_I,\boldsymbol{\chi})L_{ijkl}(\boldsymbol{\chi})E_{kl}^{mn}(\hat{\boldsymbol{x}}_I,\boldsymbol{\chi})d\Theta = 0 && \forall \boldsymbol{w}^{(1)} \in W_{\Theta_{\hat{x}_I}}\\
&\int_{\Theta_{\hat{x}_I}} w_{(i,\chi_j)}^{(1)}(\hat{\boldsymbol{x}}_I,\boldsymbol{\chi})L_{ijkl}(\boldsymbol{\chi})\Lambda_{kl}^{mn(\alpha)}(\hat{\boldsymbol{x}}_I,\boldsymbol{\chi})d\Theta = 0 && \forall \boldsymbol{w}^{(1)} \in W_{\Theta_{\hat{x}_I}}\\
&\int_{\Theta_{\hat{x}_I}} w_{(i,\chi_j)}^{(1)}(\hat{\boldsymbol{x}}_I,\boldsymbol{\chi})L_{ijkl}(\boldsymbol{\chi})Q_{kl}^{\breve{m}(\xi)}(\hat{\boldsymbol{x}}_I,\boldsymbol{\chi})d\Theta = 0 && \forall \boldsymbol{w}^{(1)} \in W_{\Theta_{\hat{x}_I}}\\
&\int_{\Theta_{\hat{x}_I}} w_{(i,\chi_j)}^{(1)}(\hat{\boldsymbol{x}}_I,\boldsymbol{\chi})L_{ijkl}(\boldsymbol{\chi})E_{kl}^{mnp}(\hat{\boldsymbol{x}}_I,\boldsymbol{\chi})d\Theta = 0 && \forall \boldsymbol{w}^{(1)} \in W_{\Theta_{\hat{x}_I}}^{\nabla}\\
&\int_{\Theta_{\hat{x}_I}} w_{(i,\chi_j)}^{(1)}(\hat{\boldsymbol{x}}_I,\boldsymbol{\chi})L_{ijkl}(\boldsymbol{\chi})\Lambda_{kl}^{mnr(\alpha)}(\hat{\boldsymbol{x}}_I,\boldsymbol{\chi})d\Theta = 0 && \forall \boldsymbol{w}^{(1)} \in W_{\Theta_{\hat{x}_I}}^{\nabla}
\end{aligned} \tag{5.79}$$

The test function $w_i^{(1)}$ and the trial solutions H_i^{mn}, H_i^{mnp}, $\tilde{h}_k^{mn(\alpha)}$, $\tilde{h}_k^{mnr(\alpha)}$, and $\breve{h}_i^{\breve{n}(\xi)}$ are discretized using the Galerkin finite element method. The resulting discrete influence function problems, which are an algebraic linear system of equations, are solved prior to nonlinear analysis.

The residual-free reduced order strain field [see (5.68)] is given as

$$\begin{aligned}\varepsilon_{kl}^{\zeta}(\hat{\boldsymbol{x}}_I,\boldsymbol{\chi}) &= E_{kl}^{mn}(\hat{\boldsymbol{x}}_I,\boldsymbol{\chi})\bar{\varepsilon}_{mn}(\hat{\boldsymbol{x}}_I)+E_{kl}^{mnp}(\hat{\boldsymbol{x}}_I,\boldsymbol{\chi})\bar{\varepsilon}_{mn,p}(\hat{\boldsymbol{x}}_I)+\sum_{\alpha=1}^{\tilde{M}}P_{kl}^{mn(\alpha)}(\hat{\boldsymbol{x}}_I,\boldsymbol{\chi})\bar{\mu}_{mn}^{(\alpha)}(\hat{\boldsymbol{x}}_I)\\ &+\sum_{\alpha=1}^{\tilde{M}}P_{kl}^{mnr(\alpha)}(\hat{\boldsymbol{x}}_I,\boldsymbol{\chi})\bar{\mu}_{mn,r}^{(\alpha)}(\hat{\boldsymbol{x}}_I)+\sum_{\xi=1}^{\breve{M}}Q_{kl}^{\breve{m}(\xi)}(\hat{\boldsymbol{x}}_I,\boldsymbol{\chi})\delta_{\breve{m}}^{(\xi)}(\hat{\boldsymbol{x}}_I)\end{aligned} \tag{5.80}$$

and the residual-free stress follows from the constitutive relation in (4.2) and (5.80)

$$\sigma_{ij}^{\zeta}(\hat{\boldsymbol{x}}_I,\boldsymbol{\chi}) = L_{ijkl}(\boldsymbol{\chi})\left(\begin{aligned}&E_{kl}^{mn}(\hat{\boldsymbol{x}}_I,\boldsymbol{\chi})\bar{\varepsilon}_{mn}(\hat{\boldsymbol{x}}_I)+E_{kl}^{mnp}(\hat{\boldsymbol{x}}_I,\boldsymbol{\chi})\bar{\varepsilon}_{mn,p}(\hat{\boldsymbol{x}}_I)\\ &+\sum_{\alpha=1}^{\tilde{M}}\Lambda_{kl}^{mn(\alpha)}(\hat{\boldsymbol{x}}_I,\boldsymbol{\chi})\bar{\mu}_{mn}^{(\alpha)}(\hat{\boldsymbol{x}}_I)+\sum_{\alpha=1}^{\tilde{M}}\Lambda_{kl}^{mnr(\alpha)}(\hat{\boldsymbol{x}}_I,\boldsymbol{\chi})\bar{\mu}_{mn,r}^{(\alpha)}(\hat{\boldsymbol{x}}_I)\\ &+\sum_{\xi=1}^{\breve{M}}Q_{kl}^{\breve{m}(\xi)}(\hat{\boldsymbol{x}}_I,\boldsymbol{\chi})\delta_{\breve{m}}^{(\xi)}(\hat{\boldsymbol{x}}_I)\end{aligned}\right) \tag{5.81}$$

The traction along the unit cell interfaces is given by

$$t_{\breve{q}}(\hat{\boldsymbol{x}}_I,\breve{\boldsymbol{\chi}}) = a_{\breve{q}i}(\hat{\boldsymbol{x}}_I,\breve{\boldsymbol{\chi}})\sigma_{ij}^{\zeta}(\hat{\boldsymbol{x}}_I,\breve{\boldsymbol{\chi}})\breve{n}_j^{S}(\hat{\boldsymbol{x}}_I,\breve{\boldsymbol{\chi}}) \tag{5.82}$$

where $a_{\breve{q}i}(\hat{\boldsymbol{x}}_I,\breve{\boldsymbol{\chi}})$ is the transformation matrix from the global coordinates system to the local interface coordinate system and $\breve{\boldsymbol{n}}^{S}(\hat{\boldsymbol{x}}_I,\breve{\boldsymbol{\chi}})$ is a unit normal to the interface $S_{\hat{x}_I}$.

5.3.1.2 Reduced Order Computational Unit Cell Problem

Following the eigenstrain approximation, the strain at each partition of the unit cell domain is assumed to have a linear variation

$$\begin{aligned}\varepsilon_{ij}^{\zeta}(\hat{\boldsymbol{x}}_I,\boldsymbol{\chi}^{(\beta)}) &= \sum_{\beta=1}^{\tilde{M}}\tilde{N}^{(\beta)}(\hat{\boldsymbol{x}}_I,\boldsymbol{\chi}^{(\beta)})\varepsilon_{ij}^{(\beta)}(\hat{\boldsymbol{x}}_I,\boldsymbol{\chi})\\ \varepsilon_{ij}^{(\beta)}(\hat{\boldsymbol{x}}_I,\boldsymbol{\chi}^{(\beta)}) &= \bar{\varepsilon}_{ij}^{(\beta)}(\hat{\boldsymbol{x}}_I)+\bar{\varepsilon}_{ij,m}^{(\beta)}(\hat{\boldsymbol{x}}_I)\chi_m^{(\beta)}\end{aligned} \tag{5.83}$$

Integrating (5.80) over the partition of the unit cell yields

$$\bar{\varepsilon}_{kl}^{(\beta)}(\hat{\boldsymbol{x}}_I)=E_{kl}^{mn(\beta)}(\hat{\boldsymbol{x}}_I)\bar{\varepsilon}_{mn}(\hat{\boldsymbol{x}}_I)+E_{kl}^{mnp(\beta)}(\hat{\boldsymbol{x}}_I)\bar{\varepsilon}_{mn,p}(\hat{\boldsymbol{x}}_I)+\sum_{\alpha=1}^{\tilde{M}}P_{kl}^{mn(\beta\alpha)}(\hat{\boldsymbol{x}}_I)\bar{\mu}_{mn}^{(\alpha)}(\hat{\boldsymbol{x}}_I)$$
$$+\sum_{\alpha=1}^{\tilde{M}}P_{kl}^{mnr(\beta\alpha)}(\hat{\boldsymbol{x}}_I)\bar{\mu}_{mn,r}^{(\alpha)}(\hat{\boldsymbol{x}}_I)+\sum_{\xi=1}^{\breve{M}}Q_{kl}^{\breve{m}(\beta\xi)}(\hat{\boldsymbol{x}}_I)\delta_{\breve{m}}^{(\xi)}(\hat{\boldsymbol{x}}_I) \tag{5.84}$$

where $E_{kl}^{mn(\beta)}(\hat{\boldsymbol{x}}_I)$, $P_{kl}^{mn(\beta\alpha)}(\hat{\boldsymbol{x}}_I)$, and $Q_{kl}^{\breve{m}(\beta\xi)}(\hat{\boldsymbol{x}}_I)$ are defined as in (4.42) and

$$E_{kl}^{mnp(\beta)}(\hat{\boldsymbol{x}}_I)=\frac{1}{\left|\Theta_{\hat{x}_I}^{(\beta)}\right|}\int_{\Theta_{\hat{x}_I}^{(\beta)}}E_{kl}^{mnp}(\hat{\boldsymbol{x}}_I,\boldsymbol{\chi})\,d\Theta \tag{5.85a}$$

$$P_{kl}^{mnr(\beta\alpha)}(\hat{\boldsymbol{x}}_I)=\frac{1}{\left|\Theta_{\hat{x}_I}^{(\beta)}\right|}\int_{\Theta_{\hat{x}_I}^{(\beta)}}P_{kl}^{mnr(\alpha)}(\hat{\boldsymbol{x}}_I,\boldsymbol{\chi})\,d\Theta \tag{5.85b}$$

Remark 5.9 To alleviate locking for lower order approximations of eigenstrains, $P_{kl}^{mn(\beta\alpha)}(\hat{\boldsymbol{x}}_I)$ and $P_{kl}^{mnr(\beta\alpha)}$ in (5.84) has to be redefined according to section 4.3 in Chapter 4 for the matrix-dominated mode of deformation. For RC2 it gives (see [19] for details)

$$\begin{aligned}
\tilde{P}_{kl}^{mn(\alpha)}(\hat{\boldsymbol{x}}_I,\boldsymbol{\chi})&=\tilde{N}^{(\alpha)}(\hat{\boldsymbol{x}}_I,\boldsymbol{\chi}^{(\alpha)})I_{klmn}-\phi^{(\alpha)}E_{kl}^{mn}(\hat{\boldsymbol{x}}_I,\boldsymbol{\chi})-\phi^{(\alpha)}\hat{\chi}_r^{(\alpha)}E_{kl}^{mnr}(\hat{\boldsymbol{x}}_I,\boldsymbol{\chi})\\
\tilde{P}_{kl}^{mnr(\alpha)}(\hat{\boldsymbol{x}}_I,\boldsymbol{\chi})&=\tilde{N}^{(\alpha)}(\hat{\boldsymbol{x}}_I,\boldsymbol{\chi}^{(\alpha)})I_{klmn}\chi_r^{(\alpha)}-\phi^{(\alpha)}\lambda_{pr}^{(\alpha)}(\hat{\boldsymbol{x}}_I)E_{kl}^{mnp}(\hat{\boldsymbol{x}}_I,\boldsymbol{\chi})\\
\tilde{P}_{kl}^{mn(\beta\alpha)}(\hat{\boldsymbol{x}}_I)&=\delta_{\alpha\beta}I_{klmn}-\phi^{(\alpha)}E_{kl}^{mn(\beta)}(\hat{\boldsymbol{x}}_I)-\phi^{(\alpha)}\hat{\chi}_r^{(\alpha)}E_{kl}^{mnr(\beta)}(\hat{\boldsymbol{x}}_I)\\
\tilde{P}_{kl}^{mnr(\beta\alpha)}(\hat{\boldsymbol{x}}_I)&=-\phi^{(\alpha)}\lambda_{pr}^{(\alpha)}(\hat{\boldsymbol{x}}_I)E_{kl}^{mnp(\beta)}(\hat{\boldsymbol{x}}_I)
\end{aligned} \tag{5.86}$$

Multiplying (5.80) by $\chi_p^{(\beta)}\in\Theta_{\hat{x}_I}^{(\beta)}$ and integrating over the volume partition gives

$$\bar{\varepsilon}_{kl,m}^{(\beta)}(\hat{\boldsymbol{x}}_I)\lambda_{mp}^{(\beta)}(\hat{\boldsymbol{x}}_I)=E_{klp}^{\nabla mn(\beta)}(\hat{\boldsymbol{x}}_I)\bar{\varepsilon}_{mn}(\hat{\boldsymbol{x}}_I)+E_{klp}^{\nabla mnr(\beta)}(\hat{\boldsymbol{x}}_I)\bar{\varepsilon}_{mn,r}(\hat{\boldsymbol{x}}_I)+\sum_{\alpha=1}^{\tilde{M}}P_{klp}^{\nabla mn(\beta\alpha)}(\hat{\boldsymbol{x}}_I)\bar{\mu}_{mn}^{(\alpha)}(\hat{\boldsymbol{x}}_I)$$
$$+\sum_{\alpha=1}^{\tilde{M}}P_{klp}^{\nabla mnr(\beta\alpha)}(\hat{\boldsymbol{x}}_I)\bar{\mu}_{mn,r}^{(\alpha)}(\hat{\boldsymbol{x}}_I)+\sum_{\xi=1}^{\breve{M}}Q_{klp}^{\nabla\breve{m}(\beta\xi)}(\hat{\boldsymbol{x}}_I)\delta_{\breve{m}}^{(\xi)}(\hat{\boldsymbol{x}}_I) \tag{5.87}$$

where

$$E_{klp}^{\nabla mn(\beta)}(\hat{\boldsymbol{x}}_I)=\frac{1}{\left|\Theta_{\hat{x}_I}^{(\beta)}\right|}\int_{\Theta_{\hat{x}_I}^{(\beta)}}E_{kl}^{mn}\chi_p^{(\beta)}d\Theta;\quad E_{klp}^{\nabla mnr(\beta)}(\hat{\boldsymbol{x}}_I)=\frac{1}{\left|\Theta_{\hat{x}_I}^{(\beta)}\right|}\int_{\Theta_{\hat{x}_I}^{(\beta)}}E_{kl}^{mnr}\chi_p^{(\beta)}d\Theta$$
$$P_{klp}^{\nabla mn(\beta\alpha)}(\hat{\boldsymbol{x}}_I)=\frac{1}{\left|\Theta_{\hat{x}_I}^{(\beta)}\right|}\int_{\Theta_{\hat{x}_I}^{(\beta)}}P_{kl}^{mn(\alpha)}\chi_p^{(\beta)}d\Theta;\quad P_{klp}^{\nabla mnr(\beta\alpha)}(\hat{\boldsymbol{x}}_I)=\frac{1}{\left|\Theta_{\hat{x}_I}^{(\beta)}\right|}\int_{\Theta_{\hat{x}_I}^{(\beta)}}P_{kl}^{mnr(\alpha)}\chi_p^{(\beta)}d\Theta \tag{5.88}$$
$$Q_{klp}^{\nabla\breve{m}(\beta\xi)}(\hat{\boldsymbol{x}}_I)=\frac{1}{\left|\Theta_{\hat{x}_I}^{(\beta)}\right|}\int_{\Theta_{\hat{x}_I}^{(\beta)}}Q_{kl}^{\breve{m}(\xi)}\chi_p^{(\beta)}d\Theta;\quad \lambda_{mp}^{(\beta)}(\hat{\boldsymbol{x}}_I)=\frac{1}{\left|\Theta_{\hat{x}_I}^{(\beta)}\right|}\int_{\Theta_{\hat{x}_I}^{(\beta)}}\chi_m^{(\beta)}\chi_p^{(\beta)}d\Theta$$

Similarly, we define the partitioned traction as an average traction over the interface partition $S_{\hat{x}_I}^{(\eta)}$

$$t_{\breve{q}}^{(\eta)}(\hat{\boldsymbol{x}}_I)=\frac{1}{\left|S_{\hat{x}_I}^{(\eta)}\right|}\int_{S_{\hat{x}_I}^{(\eta)}} t_{\breve{q}}(\hat{\boldsymbol{x}}_I,\breve{\boldsymbol{\chi}})\,d\breve{S} \tag{5.89}$$

which from (5.81) and (5.82) yields

$$\begin{aligned} t_{\breve{q}}^{(\eta)}(\hat{\boldsymbol{x}}_I)&=T_{\breve{q}}^{mn(\eta)}(\hat{\boldsymbol{x}}_I)\bar{\varepsilon}_{mn}(\hat{\boldsymbol{x}}_I)+T_{\breve{q}}^{mnp(\eta)}(\hat{\boldsymbol{x}}_I)\bar{\varepsilon}_{mn,p}(\hat{\boldsymbol{x}}_I)+\sum_{\alpha=1}^{\tilde{M}}C_{\breve{q}}^{mn(\eta\alpha)}(\hat{\boldsymbol{x}}_I)\bar{\mu}_{mn}^{(\alpha)}(\hat{\boldsymbol{x}}_I)\\ &+\sum_{\alpha=1}^{\tilde{M}}C_{\breve{q}}^{mnr(\eta\alpha)}(\hat{\boldsymbol{x}}_I)\bar{\mu}_{mn,r}^{(\alpha)}(\hat{\boldsymbol{x}}_I)+\sum_{\xi=1}^{\breve{M}}D_{\breve{q}}^{\breve{m}(\eta\xi)}(\hat{\boldsymbol{x}}_I)\delta_{\breve{m}}^{(\xi)}(\hat{\boldsymbol{x}}_I) \end{aligned} \tag{5.90}$$

where $C_{\breve{q}}^{mn(\eta\alpha)}(\hat{\boldsymbol{x}}_I)$, $D_{\breve{q}}^{\breve{m}(\eta\xi)}(\hat{\boldsymbol{x}}_I)$, and $T_{\breve{q}}^{mn(\eta)}(\hat{\boldsymbol{x}}_I)$ are defined as in (4.45) and

$$\begin{aligned} T_{\breve{q}}^{mnp(\eta)}(\hat{\boldsymbol{x}}_I)&=\frac{1}{\left|S_{\hat{x}_I}^{(\eta)}\right|}\int_{S_{\hat{x}_I}^{(\eta)}} a_{\breve{q}i}(\hat{\boldsymbol{x}}_I,\breve{\boldsymbol{\chi}})L_{ijkl}(\breve{\boldsymbol{\chi}})E_{kl}^{mnp}(\hat{\boldsymbol{x}}_I,\breve{\boldsymbol{\chi}})\breve{n}_j^S(\hat{\boldsymbol{x}}_I,\breve{\boldsymbol{\chi}})\,d\breve{S}^{(\eta)}\\ C_{\breve{q}}^{mnr(\eta\alpha)}(\hat{\boldsymbol{x}}_I)&=\frac{1}{\left|S_{\hat{x}_I}^{(\eta)}\right|}\int_{S_{\hat{x}_I}^{(\eta)}} a_{\breve{q}i}(\hat{\boldsymbol{x}}_I,\breve{\boldsymbol{\chi}})L_{ijkl}(\breve{\boldsymbol{\chi}})\Lambda_{kl}^{mnr(\alpha)}(\hat{\boldsymbol{x}}_I,\boldsymbol{\chi})\breve{n}_j^S(\hat{\boldsymbol{x}}_I,\breve{\boldsymbol{\chi}})\,d\breve{S}^{(\eta)} \end{aligned} \tag{5.91}$$

We further assume that cohesive law is specified, that is, partition traction $t_{\breve{q}}^{(\eta)}(\hat{\boldsymbol{x}}_I)$ along the interface partition is prescribed as a function of the eigenseparation. Also assume that the partitioned average eigenstrain and its gradient $\bar{\mu}_{ij}^{(\alpha)}(\hat{\boldsymbol{x}}_I)$, $\bar{\mu}_{ij,k}^{(\alpha)}(\hat{\boldsymbol{x}}_I)$ can be obtained from the constitutive law defined in the volume partition as described in section 5.3.1.3. Schematically, this is denoted as

$$\begin{aligned} &t_{\breve{p}}^{(\xi)}(\hat{\boldsymbol{x}}_I)=g\left(\delta_{\breve{q}}^{(\xi)}(\hat{\boldsymbol{x}}_I)\right);\ \bar{\mu}_{mn}^{(\alpha)}(\hat{\boldsymbol{x}}_I)=f_1\left(\bar{\varepsilon}_{kl}^{(\alpha)}(\hat{\boldsymbol{x}}_I),\bar{\varepsilon}_{kl,p}^{(\alpha)}(\hat{\boldsymbol{x}}_I)\right)\\ &\bar{\mu}_{mn,r}^{(\alpha)}(\hat{\boldsymbol{x}}_I)=f_2\left(\bar{\varepsilon}_{kl}^{(\alpha)}(\hat{\boldsymbol{x}}_I),\bar{\varepsilon}_{kl,p}^{(\alpha)}(\hat{\boldsymbol{x}}_I)\right) \end{aligned} \tag{5.92}$$

Equation (5.84), equation (5.87), and equation (5.90) comprise the reduced order nonlinear system of equations for the unknown vector $\boldsymbol{\theta}^{(\alpha,\xi)}\equiv\left[\bar{\varepsilon}_{kl}^{(\alpha)}\quad \bar{\varepsilon}_{kl,p}^{(\alpha)}\quad \delta_{\breve{q}}^{(\xi)}\right]^{\mathrm{T}}$. The reduced order residual system of equations is given by

$$\mathbf{0}=\boldsymbol{r}^{(\beta,\eta)}(\hat{\boldsymbol{x}}_I)=\begin{bmatrix} E_{kl}^{mn(\beta)}\,\overline{\varepsilon}_{mn}+E_{kl}^{mnp(\beta)}\,\overline{\varepsilon}_{mn,p} \\ E_{klp}^{\nabla mn(\beta)}\,\overline{\varepsilon}_{mn}+E_{klp}^{\nabla mnr(\beta)}\,\overline{\varepsilon}_{mn,r} \\ T_{\breve{q}}^{mn(\eta)}\,\overline{\varepsilon}_{mn}+T_{\breve{q}}^{mnp(\eta)}\,\overline{\varepsilon}_{mn,p} \end{bmatrix} - \begin{bmatrix} \overline{\varepsilon}_{kl}^{(\beta)}-\sum_{\alpha=1}^{\tilde{M}}P_{kl}^{mn(\beta\alpha)}\overline{\mu}_{mn}^{(\alpha)}-\sum_{\alpha=1}^{\tilde{M}}P_{kl}^{mnr(\beta\alpha)}\overline{\mu}_{mn,r}^{(\alpha)}-\sum_{\xi=1}^{\breve{M}}Q_{kl}^{\breve{m}(\beta\xi)}\delta_{\breve{m}}^{(\xi)} \\ \overline{\varepsilon}_{kl,m}^{(\beta)}\lambda_{mp}^{(\beta)}-\sum_{\alpha=1}^{\tilde{M}}P_{klp}^{\nabla mn(\beta\alpha)}\overline{\mu}_{mn}^{(\alpha)}-\sum_{\alpha=1}^{\tilde{M}}P_{klp}^{\nabla mnr(\beta\alpha)}\overline{\mu}_{mn,r}^{(\alpha)}-\sum_{\xi=1}^{\breve{M}}Q_{klp}^{\nabla\breve{m}(\beta\xi)}\delta_{\breve{m}}^{(\xi)} \\ t_{\breve{q}}^{(\eta)}-\sum_{\alpha=1}^{\tilde{M}}C_{\breve{q}}^{mn(\eta\alpha)}\overline{\mu}_{mn}^{(\alpha)}-\sum_{\alpha=1}^{\tilde{M}}C_{\breve{q}}^{mnr(\eta\alpha)}\overline{\mu}_{mn,r}^{(\alpha)}-\sum_{\xi=1}^{\breve{M}}D_{\breve{q}}^{\breve{m}(\eta\xi)}\delta_{\breve{m}}^{(\xi)} \end{bmatrix} \tag{5.93}$$

The nonlinear system of equations (5.93) is solved by the Newton method, which is nested with the coarse-scale Newton iteration. Note that the second term of (5.93) is independent of $\theta^{(\beta,\eta)}$ and is a forcing term governed by the coarse-scale iteration.

5.3.1.3 Constitutive Equations of Phases

Consider the constitutive relation (4.2) for phase partitions

$$\mu_{kl}^{(\alpha)}\left(\hat{\boldsymbol{x}}_I,\boldsymbol{\chi}^{(\alpha)}\right)=f\left(\varepsilon_{kl}^{(\alpha)}\left(\hat{\boldsymbol{x}}_I,\boldsymbol{\chi}^{(\alpha)}\right)\right)\equiv\varepsilon_{kl}^{(\alpha)}\left(\hat{\boldsymbol{x}}_I,\boldsymbol{\chi}^{(\alpha)}\right)-M_{klmn}^{(\alpha)}\sigma_{mn}^{(\alpha)}\left(\varepsilon_{mn}^{(\alpha)}\left(\hat{\boldsymbol{x}}_I,\boldsymbol{\chi}^{(\alpha)}\right)\right) \tag{5.94}$$

where elastic compliance modulus $M_{klij}^{(\alpha)}$ is assumed to be constant for each phase partition. The constitutive relation defines the function $\sigma_{ij}^{(\alpha)}\left(\varepsilon_{ij}^{(\alpha)}\right)$, from which the tangent $\partial\Delta\sigma_{ij}^{(\alpha)}/\partial\Delta\varepsilon_{kl}^{(\alpha)}$ can be determined, and thus provides $f\left(\varepsilon_{kl}^{(\alpha)}\right)$ and $\partial f\left(\varepsilon_{kl}^{(\alpha)}\right)/\partial\varepsilon_{ij}^{(\alpha)}$.

Consider the following expansion of $f\left(\varepsilon_{kl}^{(\alpha)}\left(\hat{\boldsymbol{x}}_I,\boldsymbol{\chi}^{(\alpha)}\right)\right)$ around the average value $\varepsilon_{kl}^{(\alpha)}(\hat{\boldsymbol{x}}_I)=\overline{\varepsilon}_{kl}^{(\alpha)}(\hat{\boldsymbol{x}}_I)$ at $\boldsymbol{\chi}^{(\alpha)}=\mathbf{0}$, which is the position of the phase centroid $\hat{\boldsymbol{\chi}}^{(\alpha)}$.

$$f\left(\varepsilon_{kl}^{(\alpha)}\left(\hat{\boldsymbol{x}}_I,\boldsymbol{\chi}^{(\alpha)}\right)\right)\approx f\left(\varepsilon_{kl}^{(\alpha)}\left(\hat{\boldsymbol{x}}_I,\boldsymbol{\chi}^{(\alpha)}\right)\right)\Big|_{\boldsymbol{\chi}^{(\alpha)}=\mathbf{0}}+\left.\frac{\partial f\left(\varepsilon_{kl}^{(\alpha)}\left(\hat{\boldsymbol{x}}_I,\boldsymbol{\chi}^{(\alpha)}\right)\right)}{\partial\varepsilon_{ij}^{(\alpha)}\left(\hat{\boldsymbol{x}}_I,\boldsymbol{\chi}^{(\alpha)}\right)}\right|_{\boldsymbol{\chi}^{(\alpha)}=\mathbf{0}}\overline{\varepsilon}_{ij,m}^{(\alpha)}(\hat{\boldsymbol{x}}_I)\chi_m^{(\alpha)} \tag{5.95}$$

The average phase eigenstrain and the eigenstrain gradient are defined as

$$\overline{\mu}_{kl}^{(\alpha)}(\hat{\boldsymbol{x}}_I)=f\left(\varepsilon_{kl}^{(\alpha)}\left(\hat{\boldsymbol{x}}_I,\boldsymbol{\chi}^{(\alpha)}\right)\right)\Big|_{\boldsymbol{\chi}^{(\alpha)}=\mathbf{0}};\quad \overline{\mu}_{kl,m}^{(\alpha)}(\hat{\boldsymbol{x}}_I)=\left.\frac{\partial f\left(\varepsilon_{kl}^{(\alpha)}\left(\hat{\boldsymbol{x}}_I,\boldsymbol{\chi}^{(\alpha)}\right)\right)}{\partial\varepsilon_{ij}^{(\alpha)}\left(\hat{\boldsymbol{x}}_I,\boldsymbol{\chi}^{(\alpha)}\right)}\right|_{\boldsymbol{\chi}^{(\alpha)}=\mathbf{0}}\overline{\varepsilon}_{ij,m}^{(\alpha)}(\hat{\boldsymbol{x}}_I) \tag{5.96}$$

where

$$\frac{\partial f\left(\varepsilon_{kl}^{(\alpha)}\right)}{\partial \varepsilon_{ij}^{(\alpha)}} = I_{klij} - M_{klmn}^{(\alpha)} \frac{\partial \sigma_{mn}^{(\alpha)}}{\partial \varepsilon_{ij}^{(\alpha)}} \tag{5.97}$$

The derivatives of strain and strain gradient are evaluated as

$$\frac{\partial \bar{\mu}_{kl}^{(\alpha)}(\hat{\boldsymbol{x}}_I)}{\partial \bar{\varepsilon}_{ij}^{(\alpha)}(\hat{\boldsymbol{x}}_I)} = \left.\frac{\partial f\left(\varepsilon_{kl}^{(\alpha)}\left(\hat{\boldsymbol{x}}_I,\boldsymbol{\chi}^{(\alpha)}\right)\right)}{\partial \varepsilon_{ij}^{(\alpha)}\left(\hat{\boldsymbol{x}}_I,\boldsymbol{\chi}^{(\alpha)}\right)}\right|_{\boldsymbol{\chi}^{(\alpha)}=\boldsymbol{0}} ; \quad \frac{\partial \bar{\mu}_{kl,p}^{(\alpha)}(\hat{\boldsymbol{x}}_I)}{\partial \bar{\varepsilon}_{ij,q}^{(\alpha)}(\hat{\boldsymbol{x}}_I)} = \left.\frac{\partial f\left(\varepsilon_{kl}^{(\alpha)}\left(\hat{\boldsymbol{x}}_I,\boldsymbol{\chi}^{(\alpha)}\right)\right)}{\partial \varepsilon_{ij}^{(\alpha)}\left(\hat{\boldsymbol{x}}_I,\boldsymbol{\chi}^{(\alpha)}\right)}\right|_{\boldsymbol{\chi}^{(\alpha)}=\boldsymbol{0}} \delta_{pq} \tag{5.98}$$

Given $\mu_{kl}^{(\alpha)}\left(\hat{\boldsymbol{x}}_I,\boldsymbol{\chi}^{(\alpha)}\right)$ in (5.94) and $\partial\mu_{ij}^{(\alpha)}\left(\hat{\boldsymbol{x}}_I,\boldsymbol{\chi}^{(\alpha)}\right)/\partial\varepsilon_{kl}^{(\alpha)}\left(\hat{\boldsymbol{x}}_I,\boldsymbol{\chi}^{(\alpha)}\right)$ in (5.97), functions $\bar{\mu}_{kl}^{(\alpha)}(\hat{\boldsymbol{x}}_I) = f_1(\hat{\boldsymbol{x}}_I)$, $\bar{\mu}_{kl,m}^{(\alpha)}(\hat{\boldsymbol{x}}_I) = f_2(\hat{\boldsymbol{x}}_I)$ in (5.92) are defined by (5.96). Furthermore, (5.98) provides the Jacobian of the residual (5.93).

5.3.2 The Coarse-Scale Weak Form

Following (5.24), the coarse-scale weak form is given as

$$\sum_{I=1}^{\hat{N}} W_I J_I \left(\left\{ \frac{\partial w_i^c}{\partial x_j} \right\}_{\hat{x}_I} \bar{\sigma}_{ij}(\hat{\boldsymbol{x}}_I) + \left\{ D_k \frac{\partial w_i^c}{\partial x_j} \right\}_{\hat{x}_I} \bar{q}_{ijk}(\hat{\boldsymbol{x}}_I) \right) = \int_{\partial\Omega^t} w_i^c \bar{t}_i d\Gamma + \int_{\Omega} w_i^c b_i d\Omega \quad \forall w_i^c \in W_\Omega \tag{5.99}$$

where

$$\begin{aligned} \bar{\sigma}_{ij}(\hat{\boldsymbol{x}}_I) &= \frac{1}{\left|\Theta_{\hat{x}_I}\right|} \int_{\Theta_{\hat{x}_I}} \sigma_{ij}(\hat{\boldsymbol{x}}_I,\boldsymbol{\chi})\, d\Theta \\ \bar{q}_{ijk}(\hat{\boldsymbol{x}}_I) &= \frac{1}{\left|\Theta_{\hat{x}_I}\right|} \int_{\Theta_{\hat{x}_I}} \sigma_{ij}(\hat{\boldsymbol{x}}_I,\boldsymbol{\chi})\, \chi_k\, d\Theta \end{aligned} \tag{5.100}$$

The coefficients $\bar{\sigma}_{ij}$ and $\bar{q}_{ijk}$ can be directly calculated by inserting (5.81) into (5.100), which yields

$$\begin{aligned} \bar{\sigma}_{ij}(\hat{\boldsymbol{x}}_I) = {} & L_{ijmn}^c(\hat{\boldsymbol{x}}_I)\bar{\varepsilon}_{mn}(\hat{\boldsymbol{x}}_I) + L_{ijmnp}^c(\hat{\boldsymbol{x}}_I)\bar{\varepsilon}_{mn,p}(\hat{\boldsymbol{x}}_I) \\ & + \sum_{\alpha=1}^{\tilde{M}} A_{ijmn}^{c(\alpha)}(\hat{\boldsymbol{x}}_I)\bar{\mu}_{mn}^{(\alpha)}(\hat{\boldsymbol{x}}_I) + \sum_{\alpha=1}^{\tilde{M}} A_{ijmnp}^{c(\alpha)}(\hat{\boldsymbol{x}}_I)\bar{\mu}_{mn,p}^{(\alpha)}(\hat{\boldsymbol{x}}_I) \\ & + \sum_{\xi=1}^{\breve{M}} B_{ij\breve{m}}^{c(\xi)}(\hat{\boldsymbol{x}}_I)\delta_{\breve{m}}^{(\xi)}(\hat{\boldsymbol{x}}_I) \end{aligned} \tag{5.101a}$$

$$\overline{q}_{ijr}(\hat{\boldsymbol{x}}_I) = L^{\nabla}_{ijmnr}(\hat{\boldsymbol{x}}_I)\overline{\varepsilon}_{mn}(\hat{\boldsymbol{x}}_I) + L^{\nabla}_{ijmnpr}(\hat{\boldsymbol{x}}_I)\overline{\varepsilon}_{mn,p}(\hat{\boldsymbol{x}}_I) + \sum_{\alpha=1}^{\tilde{M}} A^{\nabla(\alpha)}_{ijmnr}(\hat{\boldsymbol{x}}_I)\overline{\mu}^{(\alpha)}_{mn}(\hat{\boldsymbol{x}}_I)$$
$$+ \sum_{\alpha=1}^{\tilde{M}} A^{\nabla(\alpha)}_{ijmnpr}(\hat{\boldsymbol{x}}_I)\overline{\mu}^{(\alpha)}_{mn,p}(\hat{\boldsymbol{x}}_I) + \sum_{\xi=1}^{\breve{M}} B^{\nabla(\xi)}_{ij\breve{m}r}(\hat{\boldsymbol{x}}_I)\delta^{(\xi)}_{\breve{m}}(\hat{\boldsymbol{x}}_I) \tag{5.101b}$$

where $L^{c}_{ijmn}(\hat{\boldsymbol{x}}_I)$, $A^{c(\alpha)}_{ijmn}(\hat{\boldsymbol{x}}_I)$, $B^{c(\xi)}_{ij\breve{m}}(\hat{\boldsymbol{x}}_I)$ are defined as in (4.48) and

$$\begin{aligned}
&L^{c}_{ijmnp}(\hat{\boldsymbol{x}}_I) \equiv \frac{1}{|\Theta_{\hat{x}_I}|}\int_{\Theta_{\hat{x}_I}} L_{ijkl}E^{mnp}_{kl}\,d\Theta; \qquad A^{c(\alpha)}_{ijmnp}(\hat{\boldsymbol{x}}_I) \equiv \frac{1}{|\Theta_{\hat{x}_I}|}\int_{\Theta_{\hat{x}_I}} L_{ijkl}\Lambda^{mnp(\alpha)}_{kl}\,d\Theta\\
&L^{\nabla}_{ijmnr}(\hat{\boldsymbol{x}}_I) \equiv \frac{1}{|\Theta_{\hat{x}_I}|}\int_{\Theta_{\hat{x}_I}} L_{ijkl}E^{mn}_{kl}\chi_r\,d\Theta; \qquad L^{\nabla}_{ijmnpr}(\hat{\boldsymbol{x}}_I) \equiv \frac{1}{|\Theta_{\hat{x}_I}|}\int_{\Theta_{\hat{x}_I}} L_{ijkl}E^{mnp}_{kl}\chi_r\,d\Theta\\
&A^{\nabla(\alpha)}_{ijmnr}(\hat{\boldsymbol{x}}_I) \equiv \frac{1}{|\Theta_{\hat{x}_I}|}\int_{\Theta_{\hat{x}_I}} L_{ijkl}\Lambda^{mn(\alpha)}_{kl}\chi_r\,d\Theta; \quad A^{\nabla(\alpha)}_{ijmnpr}(\hat{\boldsymbol{x}}_I) \equiv \frac{1}{|\Theta_{\hat{x}_I}|}\int_{\Theta_{\hat{x}_I}} L_{ijkl}\Lambda^{mnp(\alpha)}_{kl}\chi_r\,d\Theta\\
&B^{\nabla(\xi)}_{ij\breve{m}r}(\hat{\boldsymbol{x}}_I) \equiv \frac{1}{|\Theta_{\hat{x}_I}|}\int_{\Theta_{\hat{x}_I}} L_{ijkl}Q^{\breve{m}(\xi)}_{kl}\chi_r\,d\Theta
\end{aligned} \tag{5.102}$$

Note that all the influence functions and overall coefficients in (5.102) are computed in the preprocessing stage prior to nonlinear analysis.

5.3.3 Coarse-Scale Consistent Tangent Stiffness Matrix

The coarse-scale trial solution and test function approximation is shown in (5.28). Following the B-bar averaging scheme in (5.21), (5.22), and (5.23), the coarse-scale strain and the symmetric gradient of the test function are given as

$$\begin{aligned}
&\varepsilon^{c}_{ij}(\hat{\boldsymbol{x}}_I,\boldsymbol{\chi}) \approx \left(\left\{\overline{N^{c}_{(i,x_j)\alpha}}\right\}_{\hat{x}_I} + \left\{\overline{D_m N^{c}_{(i,x_j)\alpha}}\right\}_{\hat{x}_I}\chi_m\right)d^{c}_{\alpha}\\
&\frac{\partial w^{c}_{i}}{\partial x_j}(\hat{\boldsymbol{x}}_I,\boldsymbol{\chi}) \approx \left(\left\{\overline{N^{c}_{(i,x_j)\alpha}}\right\}_{\hat{x}_I} + \left\{\overline{D_m N^{c}_{(i,x_j)\alpha}}\right\}_{\hat{x}_I}\chi_m\right)c^{c}_{\alpha}\\
&\overline{\varepsilon}_{ij}(\hat{\boldsymbol{x}}_I) = \left\{\overline{N^{c}_{(i,x_j)\alpha}}\right\}_{\hat{x}_I} d^{c}_{\alpha}; \quad \overline{\varepsilon}_{ij,m}(\hat{\boldsymbol{x}}_I) = \left\{\overline{D_m N^{c}_{(i,x_j)\alpha}}\right\}_{\hat{x}_I} d^{c}_{\alpha}
\end{aligned} \tag{5.103}$$

Following (5.30), the coarse-scale internal force vector is given by

$$f^{\mathrm{int}}_{\alpha} = \sum_{I=1}^{\hat{N}} W_I J_I \left(\left\{\overline{N^{c}_{(i,x_j)\alpha}}\right\}_{\hat{x}_I}\overline{\sigma}_{ij}(\hat{\boldsymbol{x}}_I) + \left\{\overline{D_k N^{c}_{(i,x_j)\alpha}}\right\}_{\hat{x}_I}\overline{q}_{ijk}(\hat{\boldsymbol{x}}_I)\right) \tag{5.104}$$

The consistent tangent stiffness matrix is derived by consistent linearization of the internal force

$$K_{\alpha\beta}=\sum_{I=1}^{\hat{N}}W_I J_I\left(\left\{\overline{N^c_{(i,x_j)\alpha}}\right\}_{\hat{x}_I}\frac{\partial\Delta\bar{\sigma}_{ij}(\hat{\boldsymbol{x}}_I)}{\partial\Delta d^c_\beta}+\left\{\overline{D_r N^c_{(i,x_j)\alpha}}\right\}_{\hat{x}_I}\frac{\partial\Delta\bar{q}_{ijr}(\hat{\boldsymbol{x}}_I)}{\partial\Delta d^c_\beta}\right) \tag{5.105}$$

where Δ denotes the incremental value.

We now focus on calculation of $\partial\Delta\bar{\sigma}_{ij}(\hat{\boldsymbol{x}}_I)/\partial\Delta d^c_\beta$ and $\partial\Delta\bar{q}_{ijr}(\hat{\boldsymbol{x}}_I)/\partial\Delta d^c_\beta$. Using (5.101) and (5.103) yields

$$\begin{aligned}
\frac{\partial\Delta\bar{\sigma}_{ij}(\hat{\boldsymbol{x}}_I)}{\partial\Delta d^c_\beta}&=L^c_{ijkl}(\hat{\boldsymbol{x}}_I)\left\{\overline{N^c_{(k,x_l)\beta}}\right\}_{\hat{x}_I}+L^c_{ijmnp}(\hat{\boldsymbol{x}}_I)\left\{\overline{D_p N^c_{(m,x_n)\beta}}\right\}_{\hat{x}_I}\\
&+\sum_{\alpha=1}^{\tilde{M}}\left(A^{c(\alpha)}_{ijmn}(\hat{\boldsymbol{x}}_I)\frac{\partial\Delta\bar{\mu}^{(\alpha)}_{mn}(\hat{\boldsymbol{x}}_I)}{\partial\Delta d^c_\beta}+A^{c(\alpha)}_{ijmnp}(\hat{\boldsymbol{x}}_I)\frac{\partial\Delta\bar{\mu}^{(\alpha)}_{mn,p}(\hat{\boldsymbol{x}}_I)}{\partial\Delta d^c_\beta}\right)\\
&+\sum_{\xi=1}^{\breve{M}}B^{c(\xi)}_{ij\breve{m}}(\hat{\boldsymbol{x}}_I)\frac{\partial\Delta\delta^{(\xi)}_{\breve{m}}(\hat{\boldsymbol{x}}_I)}{\partial\Delta d^c_\beta}\\
\frac{\partial\Delta\bar{q}_{ijr}(\hat{\boldsymbol{x}}_I)}{\partial\Delta d^c_\beta}&=L^\nabla_{ijklr}(\hat{\boldsymbol{x}}_I)\left\{\overline{N^c_{(k,x_l)\beta}}\right\}_{\hat{x}_I}+L^\nabla_{ijmnpr}(\hat{\boldsymbol{x}}_I)\left\{\overline{D_p N^c_{(m,x_n)\beta}}\right\}_{\hat{x}_I}\\
&+\sum_{\alpha=1}^{\tilde{M}}\left(A^{\nabla(\alpha)}_{ijmnr}(\hat{\boldsymbol{x}}_I)\frac{\partial\Delta\bar{\mu}^{(\alpha)}_{mn}}{\partial\Delta d^c_\beta}(\hat{\boldsymbol{x}}_I)+A^{\nabla(\alpha)}_{ijmnpr}(\hat{\boldsymbol{x}}_I)\frac{\partial\Delta\bar{\mu}^{(\alpha)}_{mn,p}(\hat{\boldsymbol{x}}_I)}{\partial\Delta d^c_\beta}\right)\\
&+\sum_{\xi=1}^{\breve{M}}B^{\nabla(\xi)}_{ij\breve{m}r}(\hat{\boldsymbol{x}}_I)\frac{\partial\Delta\delta^{(\xi)}_{\breve{m}}(\hat{\boldsymbol{x}}_I)}{\partial\Delta d^c_\beta}
\end{aligned} \tag{5.106}$$

For elastic problems, the consistent tangent stiffness matrix denoted by $K^{\text{el}}_{\alpha\beta}$ is given by

$$K^{\text{el}}_{\alpha\beta}=\sum_{I=1}^{\hat{N}}W_I J_I\left(\begin{aligned}&\left\{\overline{N^c_{(i,x_j)\alpha}}\right\}_{\hat{x}_I}\left(L^c_{ijkl}(\hat{\boldsymbol{x}}_I)\left\{\overline{N^c_{(k,x_l)\beta}}\right\}_{\hat{x}_I}+L^c_{ijmnp}(\hat{\boldsymbol{x}}_I)\left\{\overline{D_p N^c_{(m,x_n)\beta}}\right\}_{\hat{x}_I}\right)\\
&+\left\{\overline{D_r N^c_{(i,x_j)\alpha}}\right\}_{\hat{x}_I}\left(L^\nabla_{ijklr}(\hat{\boldsymbol{x}}_I)\left\{\overline{N^c_{(k,x_l)\beta}}\right\}_{\hat{x}_I}+L^\nabla_{ijmnpr}(\hat{\boldsymbol{x}}_I)\left\{\overline{D_p N^c_{(m,x_n)\beta}}\right\}_{\hat{x}_I}\right)\end{aligned}\right) \tag{5.107}$$

To this end, we focus on evaluating the unknown derivatives in (5.106), $\partial\Delta\bar{\mu}^{(\alpha)}_{mn}/\partial\Delta d^c_\beta$, $\partial\Delta\bar{\mu}^{(\alpha)}_{mn,p}/\partial\Delta d^c_\beta$, and $\partial\Delta\delta^{(\xi)}_{\breve{m}}/\partial\Delta d^c_\beta$. Differentiating the residual system of equations (5.93) with respect to Δd^c_γ yields

$$\frac{\partial \boldsymbol{r}\left(\Delta\boldsymbol{\theta}^{(\beta,\eta)}\right)}{\partial \Delta\boldsymbol{\theta}^{(\alpha,\xi)}}\frac{\partial \Delta\boldsymbol{\theta}^{(\alpha,\xi)}}{\partial \Delta d^c_\gamma}=\begin{bmatrix} E^{mn(\beta)}_{kl}\overline{\left\{N^c_{(m,x_n)\gamma}\right\}}_{\hat{x}_I}+E^{mnp(\beta)}_{kl}\overline{\left\{D_pN^c_{(m,x_n)\gamma}\right\}}_{\hat{x}_I} \\ E^{\nabla mn(\beta)}_{klp}\overline{\left\{N^c_{(m,x_n)\gamma}\right\}}_{\hat{x}_I}+E^{\nabla mnr(\beta)}_{klp}\overline{\left\{D_pN^c_{(m,x_n)\gamma}\right\}}_{\hat{x}_I} \\ T^{mn(\eta)}_{\breve{q}}\overline{\left\{N^c_{(m,x_n)\gamma}\right\}}_{\hat{x}_I}+T^{mnp(\eta)}_{\breve{q}}\overline{\left\{D_pN^c_{(m,x_n)\gamma}\right\}}_{\hat{x}_I} \end{bmatrix} \tag{5.108}$$

where

$$\frac{\partial \Delta\boldsymbol{\theta}^{(\alpha,\xi)}}{\partial \Delta d^c_\gamma}=\left[\frac{\partial \Delta\bar{\varepsilon}^{(\alpha)}_{ij}}{\partial \Delta d^c_\gamma}\quad\frac{\partial \Delta\bar{\varepsilon}^{(\alpha)}_{ij,p}}{\partial \Delta d^c_\gamma}\quad\frac{\partial \Delta\delta^{(\xi)}_{\breve{q}}}{\partial \Delta d^c_\gamma}\right]^{\mathrm{T}} \tag{5.109}$$

and the Jacobian of the residual can be approximated as

$$\frac{\partial \boldsymbol{r}\left(\Delta\boldsymbol{\theta}^{(\beta,\eta)}\right)}{\partial \Delta\boldsymbol{\theta}^{(\alpha,\xi)}}$$

$$=\begin{bmatrix} \delta_{\alpha\beta}I_{klij}-P^{mn(\beta\alpha)}_{kl}\dfrac{\partial\Delta\bar{\mu}^{(\alpha)}_{mn}}{\partial\Delta\bar{\varepsilon}^{(\alpha)}_{ij}} & -P^{mnr(\beta\alpha)}_{kl}\dfrac{\partial\Delta\bar{\mu}^{(\alpha)}_{mn,r}}{\partial\Delta\bar{\varepsilon}^{(\alpha)}_{ij,s}} & -Q^{\breve{q}(\beta\xi)}_{kl} \\ -P^{\nabla mn(\beta\alpha)}_{klp}\dfrac{\partial\Delta\bar{\mu}^{(\alpha)}_{mn}}{\partial\Delta\bar{\varepsilon}^{(\alpha)}_{ij}} & I_{klij}\lambda^{(\beta)}_{sp}\delta_{\alpha\beta}-P^{\nabla mnr(\beta\alpha)}_{klp}\dfrac{\partial\Delta\bar{\mu}^{(\alpha)}_{mn,r}}{\partial\Delta\bar{\varepsilon}^{(\alpha)}_{ij,s}} & -Q^{\nabla\breve{q}(\beta\xi)}_{klp} \\ -C^{mn(\eta\alpha)}_{\breve{q}}\dfrac{\partial\Delta\bar{\mu}^{(\alpha)}_{mn}}{\partial\Delta\bar{\varepsilon}^{(\alpha)}_{ij}} & -C^{mnr(\eta\alpha)}_{\breve{q}}\dfrac{\partial\Delta\bar{\mu}^{(\alpha)}_{mn,r}}{\partial\Delta\bar{\varepsilon}^{(\alpha)}_{ij,s}} & \delta_{\eta\xi}\dfrac{\partial\Delta t^{(\xi)}_{\breve{q}}}{\partial\Delta\delta^{(\xi)}_{\breve{r}}}-D^{\breve{r}(\eta\xi)}_{\breve{q}} \end{bmatrix} \tag{5.110}$$

Given the constitutive equations and cohesive laws of fine-scale phases and interfaces (5.92), we may determine derivatives $\partial\Delta\bar{\mu}^{(\alpha)}_{mn}/\partial\Delta\bar{\varepsilon}^{(\alpha)}_{ij}$, $\partial\Delta\bar{\mu}^{(\alpha)}_{mn,r}/\partial\Delta\bar{\varepsilon}^{(\alpha)}_{ij,s}$ (see section 5.3.1.3), and $\partial\Delta t^{(\xi)}_{\breve{q}}/\partial\Delta\delta^{(\xi)}_{\breve{r}}$, which are necessary for computing the Jacobian (5.110). For closed form expressions of $\partial\Delta\bar{\mu}^{(\alpha)}_{mn}/\partial\Delta\bar{\varepsilon}^{(\alpha)}_{ij}$ for damage, plasticity, and viscoplasticity constitutive equations, we refer to [35].

The unknown derivatives in (5.109) follow from (5.108)

$$\frac{\partial \Delta\boldsymbol{\theta}^{(\alpha,\xi)}(\hat{\boldsymbol{x}}_I)}{\partial \Delta d_\gamma}=\left[\frac{\partial \boldsymbol{r}\left(\Delta\boldsymbol{\theta}^{(\beta,\eta)}(\hat{\boldsymbol{x}}_I)\right)}{\partial \Delta\boldsymbol{\theta}^{(\alpha,\xi)}(\hat{\boldsymbol{x}}_I)}\right]^{-1}\begin{bmatrix} E^{mn(\beta)}_{kl}(\hat{\boldsymbol{x}}_I)\overline{\left\{N^c_{(m,x_n)\gamma}\right\}}_{\hat{x}_I}+E^{mnp(\beta)}_{kl}(\hat{\boldsymbol{x}}_I)\overline{\left\{D_pN^c_{(m,x_n)\gamma}\right\}}_{\hat{x}_I} \\ E^{\nabla mn(\beta)}_{klp}(\hat{\boldsymbol{x}}_I)\overline{\left\{N^c_{(m,x_n)\gamma}\right\}}_{\hat{x}_I}+E^{\nabla mnr(\beta)}_{klp}(\hat{\boldsymbol{x}}_I)\overline{\left\{D_pN^c_{(m,x_n)\gamma}\right\}}_{\hat{x}_I} \\ T^{mn(\eta)}_{\breve{q}}(\hat{\boldsymbol{x}}_I)\overline{\left\{N^c_{(m,x_n)\gamma}\right\}}_{\hat{x}_I}+T^{mnp(\eta)}_{\breve{q}}(\hat{\boldsymbol{x}}_I)\overline{\left\{D_pN^c_{(m,x_n)\gamma}\right\}}_{\hat{x}_I} \end{bmatrix} \tag{5.111}$$

Using the solution of (5.111), the derivatives $\partial\Delta\bar{\mu}^{(\alpha)}_{mn}/\partial\Delta d^c_\gamma$ and $\partial\Delta\bar{\mu}^{(\alpha)}_{mn,p}/\partial\Delta d^c_\gamma$ can be evaluated from

$$\frac{\partial \Delta \bar{\mu}_{mn}^{(\alpha)}}{\partial \Delta d_{\gamma}^{c}} = \frac{\partial \Delta \bar{\mu}_{mn}^{(\alpha)}}{\partial \Delta \bar{\varepsilon}_{ij}^{(\alpha)}} \frac{\partial \Delta \bar{\varepsilon}_{ij}^{(\alpha)}}{\partial \Delta d_{\gamma}^{c}}; \quad \frac{\partial \Delta \bar{\mu}_{mn,p}^{(\alpha)}}{\partial \Delta d_{\gamma}^{c}} = \frac{\partial \Delta \bar{\mu}_{mn,p}^{(\alpha)}}{\partial \Delta \bar{\varepsilon}_{ij,s}^{(\alpha)}} \frac{\partial \Delta \bar{\varepsilon}_{ij,s}^{(\alpha)}}{\partial \Delta d_{\gamma}^{c}} \tag{5.112}$$

Note that (5.110) is computed during the Newton iterations of the nonlinear unit cell problem (5.93). Finally, given (5.111), (5.112), and (5.98), the consistent tangent stiffness matrix follows from (5.106) and (5.105).

5.4 Nonlocal Quadrature in Multidimensions

Both the hexahedral (quadrilateral in 2D) and tetrahedral (triangular in 2D) elements are considered. For convenience the physical coordinates are denoted by x, y, z as opposed to x_i in the previous sections.

5.4.1 Tetrahedral Elements

In this section, we derive a nonlocal quadrature scheme for a quadratic tetrahedral element having four planar faces. The centroid of the unit cell is positioned at the nonlocal quadrature point. All unit cells are assumed to be the same size and parallel to each other. We consider the nonlocal quadrature of a quadratic element with respect to the volume coordinates (r, s, t, q). The integration is carried out over the physical domain of a tetrahedral element defined in the Cartesian coordinate system.

The goal is to calculate the integral over the composite domain of a tetrahedral Ω^{ζ}

$$\int_{\Omega^{\zeta}} f(r(x,y,z), s(x,y,z), t(x,y,z), q(x,y,z))\, d\Omega \tag{5.113}$$

The volume coordinates, r, s, t, q, are related to the physical Cartesian coordinates by [36] as shown in Figure 5.5

$$k(x,y,z) = \frac{a_k + b_k x + c_k y + d_k z}{6V}; \quad \{k\} = r,s,t,q \tag{5.114a}$$

$$r = \frac{V_{PBCD}}{V_{ABCD}}; \quad s = \frac{V_{PACD}}{V_{ABCD}}; \quad t = \frac{V_{PABD}}{V_{ABCD}}; \quad q = \frac{V_{PABC}}{V_{ABCD}} \tag{5.114b}$$

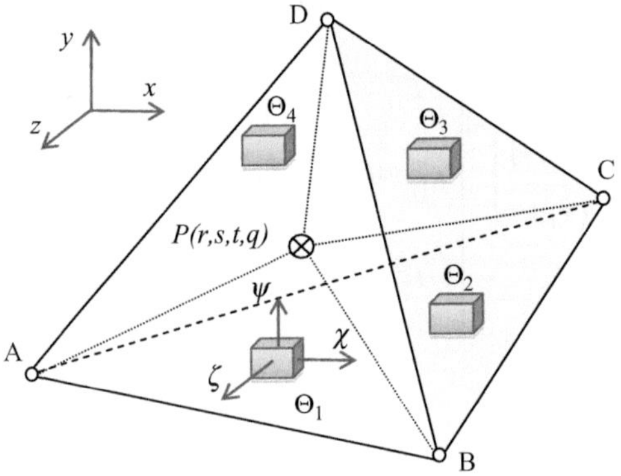

Figure 5.5 A tetrahedral element domain with unit cells positioned at the nonlocal quadrature points

and satisfy

$$r+s+t+q=1 \tag{5.115}$$

In the following, the volume of a tetrahedral element is denoted by V and is equal to

$$V=\frac{1}{6}J; \quad J\equiv\det\left(\frac{\partial(x,y,z)}{\partial(r,s,t,q)}\right)=\det\left(\begin{bmatrix}1 & x_A & y_A & z_A\\ 1 & x_B & y_B & z_B\\ 1 & x_C & y_C & z_C\\ 1 & x_D & y_D & z_D\end{bmatrix}\right) \tag{5.116}$$

where (x_A,y_A,z_A), (x_B,y_B,z_B), (x_C,y_C,z_C), and (x_D,y_D,z_D) denote the coordinates of element vertices (Figure 5.5), which in the volume coordinates are at $r=1$, $s=1$, $t=1$, and $q=1$, respectively. For any point in the interior of the tetrahedron, $0<r<1$, $0<s<1$, and $0<t<1-r-s$.

The coefficients a_k, b_k, c_k, d_k for $k=r$ are the minor determinants of the Jacobian matrix in (5.116)

$$\begin{aligned} a_r&=\det\begin{bmatrix}x_B & y_B & z_B\\ x_C & y_C & z_C\\ x_D & y_D & z_D\end{bmatrix}; \quad b_r=-\det\begin{bmatrix}1 & y_B & z_B\\ 1 & y_C & z_C\\ 1 & y_D & z_D\end{bmatrix}\\ c_r&=\det\begin{bmatrix}x_B & 1 & z_B\\ x_C & 1 & z_C\\ x_D & 1 & z_D\end{bmatrix}; \quad d_r=-\det\begin{bmatrix}x_B & y_B & 1\\ x_C & y_C & 1\\ x_D & y_D & 1\end{bmatrix}\end{aligned} \tag{5.117}$$

The remaining constants for $k=s,t,q$ are defined by a permutation of subscripts A, B, C, and D.

The Cartesian coordinates can be expressed in terms of the coordinates of the tetrahedron's vertices as

$$\begin{aligned} x&=x_Ar+x_Bs+x_Ct+x_Dq\\ y&=y_Ar+y_Bs+y_Ct+y_Dq\\ z&=z_Ar+z_Bs+z_Ct+z_Dq\end{aligned} \tag{5.118}$$

Consider a composite domain consisting of a disjoint union of finite cuboidal unit cells denoted by $\Theta_I=[l_\chi\times l_\psi\times l_\zeta]$. The local Cartesian coordinates (χ,ψ,ζ) are defined with respect to the unit cell centroid $(\hat{x}_I,\hat{y}_I,\hat{z}_I)$ as $\chi=x-\hat{x}_I$, $\psi=y-\hat{y}_I$, and $\zeta=z-\hat{z}_I$. Following the definition of the nonlocal quadrature in (5.8), the integration over the composite domain of a tetrahedral element is given by

$$\begin{aligned} &\int_{\Omega^\zeta} f(r(x,y,z),s(x,y,z),t(x,y,z),q(x,y,z))\,dxdydz\\ &=V\sum_{I=1}^{\hat{N}}W_I\frac{1}{|\Theta_I|}\int_{\Theta_I} f\begin{pmatrix} r(\hat{x}_I+\chi,\hat{y}_I+\psi,\hat{z}_I+\zeta),\\ s(\hat{x}_I+\chi,\hat{y}_I+\psi,\hat{z}_I+\zeta),\\ t(\hat{x}_I+\chi,\hat{y}_I+\psi,\hat{z}_I+\zeta),\\ q(\hat{x}_I+\chi,\hat{y}_I+\psi,\hat{z}_I+\zeta)\end{pmatrix} d\chi d\psi d\zeta\end{aligned} \tag{5.119}$$

Integration of the monomials $f(r,s,t,q)=r^\alpha s^\beta t^\gamma q^\delta$ over the tetrahedral element domain is given by [32,36]

$$\int_{\Omega^\zeta} r^\alpha s^\beta t^\gamma q^\delta \, dxdydz = \frac{\alpha!\beta!\gamma!\delta!}{(\alpha+\beta+\gamma+\delta+2)} 6V \tag{5.120}$$

where coordinate q can be excluded from consideration due to (5.115).

To evaluate (5.120) using the nonlocal quadrature, we first express $f(r,s,t)=r^\alpha s^\beta t^\gamma$ in the local Cartesian coordinates using (5.114a), then integrate it over the unit cell domain, which yields

$$\frac{1}{(6V)^{\alpha+\beta+\gamma}} \int_{\Theta_I} \begin{matrix} (r_I 6V + b_r\chi + c_r\psi + d_r\zeta)^\alpha (s_I 6V + b_s\chi + c_s\psi + d_s\zeta)^\beta \\ (t_I 6V + b_t\chi + c_t\psi + d_t\zeta)^\gamma \, d\chi d\psi d\zeta \end{matrix} \tag{5.121}$$

where the nonlocal quadrature points are expressed in volume coordinates as

$$k_I \equiv \frac{a_k + b_k\hat{x}_I + c_k\hat{y}_I + d_k\hat{z}_I}{6V}; \ \{k\} = r,s,t,q \tag{5.122}$$

Integrating monomials $1, r, s, t, rs, rt, st, r^2, s^2, t^2$ in (5.121) gives

$$\begin{aligned}
&\int_{\Theta_I} d\chi d\psi d\zeta = |\Theta_I| \\
&\frac{1}{6V}\int_{\Theta_I} (r_I 6V + b_r\chi + c_r\psi + d_r\zeta) d\chi d\psi d\zeta = r_I |\Theta_I| \\
&\frac{1}{6V}\int_{\Theta_I} (s_I 6V + b_s\chi + c_s\psi + d_s\zeta) d\chi d\psi d\zeta = s_I |\Theta_I| \\
&\frac{1}{6V}\int_{\Theta_I} (t_I 6V + b_t\chi + c_t\psi + d_t\zeta) d\chi d\psi d\zeta = t_I |\Theta_I| \\
&\frac{1}{(6V)^2}\int_{\Theta_I} (r_I 6V + b_r\chi + c_r\psi + d_r\zeta)(s_I 6V + b_s\chi + c_s\psi + d_s\zeta) d\chi d\psi d\zeta \\
&\quad = r_I s_I |\Theta_I| + \frac{|\Theta_I|}{12(6V)^2}\left(b_r b_s l_\chi^2 + c_r c_s l_\psi^2 + d_r d_s l_\zeta^2\right) \\
&\frac{1}{(6V)^2}\int_{\Theta_I} (r_I 6V + b_r\chi + c_r\psi + d_r\zeta)(t_I 6V + b_t\chi + c_t\psi + d_t\zeta) d\chi d\psi d\zeta \\
&\quad = r_I t_I |\Theta_I| + \frac{|\Theta_I|}{12(6V)^2}\left(b_r b_t l_\chi^2 + c_r c_t l_\psi^2 + d_r d_t l_\zeta^2\right) \\
&\frac{1}{(6V)^2}\int_{\Theta_I} (s_I 6V + b_s\chi + c_s\psi + d_s\zeta)(t_I 6V + b_t\chi + c_t\psi + d_t\zeta) d\chi d\psi d\zeta \\
&\quad = s_I t_I |\Theta_I| + \frac{|\Theta_I|}{12(6V)^2}\left(b_s b_t l_\chi^2 + c_s c_t l_\psi^2 + d_s d_t l_\zeta^2\right)
\end{aligned} \tag{5.123}$$

$$\frac{1}{(6V)^2}\int_{\Theta_I}\left(r_I 6V + b_r\chi + c_r\psi + d_r\zeta\right)^2 d\chi d\psi d\zeta$$
$$= r_I^2\left|\Theta_I\right| + \frac{\left|\Theta_I\right|}{12(6V)^2}\left(b_r^2 l_\chi^2 + c_r^2 l_\psi^2 + d_r^2 l_\zeta^2\right)$$
$$\frac{1}{(6V)^2}\int_{\Theta_I}\left(s_I 6V + b_s\chi + c_s\psi + d_s\zeta\right)^2 d\chi d\psi d\zeta$$
$$= s_I^2\left|\Theta_I\right| + \frac{\left|\Theta_I\right|}{12(6V)^2}\left(b_s^2 l_\chi^2 + c_s^2 l_\psi^2 + d_s^2 l_\zeta^2\right)$$
$$\frac{1}{(6V)^2}\int_{\Theta_I}\left(t_I 6V + b_t\chi + c_t\psi + d_t\zeta\right)^2 d\chi d\psi d\zeta$$
$$= t_I^2\left|\Theta_I\right| + \frac{\left|\Theta_I\right|}{12(6V)^2}\left(b_t^2 l_\chi^2 + c_t^2 l_\psi^2 + d_t^2 l_\zeta^2\right)$$

Inserting (5.123) into the right-hand side of (5.119) and exploiting the exact integration (5.120) yields the following nonlinear algebraic system of equations

$$1 = \sum_{I=1}^{4} W_I; \quad \frac{1}{4} = \sum_{I=1}^{4} W_I r_I; \quad \frac{1}{4} = \sum_{I=1}^{4} W_I s_I; \quad \frac{1}{4} = \sum_{I=1}^{4} W_I t_I \tag{5.124a}$$

$$\frac{1}{20} - \frac{1}{12(6V)^2}\left(b_r b_s l_\chi^2 + c_r c_s l_\psi^2 + d_r d_s l_\zeta^2\right) = \sum_{I=1}^{4} W_I r_I s_I \tag{5.124b}$$

$$\frac{1}{20} - \frac{1}{12(6V)^2}\left(b_r b_t l_\chi^2 + c_r c_t l_\psi^2 + d_r d_t l_\zeta^2\right) = \sum_{I=1}^{4} W_I r_I t_I \tag{5.124c}$$

$$\frac{1}{20} - \frac{1}{12(6V)^2}\left(b_s b_t l_\chi^2 + c_s c_t l_\psi^2 + d_s d_t l_\zeta^2\right) = \sum_{I=1}^{4} W_I s_I t_I \tag{5.124d}$$

$$\frac{1}{10} - \frac{1}{12(6V)^2}\left(b_r^2 l_\chi^2 + c_r^2 l_\psi^2 + d_r^2 l_\zeta^2\right) = \sum_{I=1}^{4} W_I r_I^2 \tag{5.124e}$$

$$\frac{1}{10} - \frac{1}{12(6V)^2}\left(b_s^2 l_\chi^2 + c_s^2 l_\psi^2 + d_s^2 l_\zeta^2\right) = \sum_{I=1}^{4} W_I s_I^2 \tag{5.124f}$$

$$\frac{1}{10} - \frac{1}{12(6V)^2}\left(b_t^2 l_\chi^2 + c_t^2 l_\psi^2 + d_t^2 l_\zeta^2\right) = \sum_{I=1}^{4} W_I t_I^2 \tag{5.124g}$$

where equation (5.124a) was employed to simplify the other equations.

The system (5.124) consists of 10 equations with 16 unknowns, including W_I, r_I, s_I, t_I for $I=1,2,3,4$. To solve such a system of equations, additional assumptions have to be introduced. The four-point Gauss quadrature possesses the following characteristics:

(i) equal weights

$$W_1 = W_2 = W_3 = W_4 = \frac{1}{4} \tag{5.125}$$

(ii) symmetry

$$P_1=\{\beta,\alpha,\beta,\beta\};\quad P_2=\{\beta,\beta,\alpha,\beta\};\quad P_4=\{\beta,\beta,\beta,\alpha\};\quad P_4=\{\alpha,\beta,\beta,\beta\} \tag{5.126}$$

If we were to adopt both assumptions, then the system of equations (5.124) will have no solution except for the trivial case when the unit cell size is equal to zero. Similarly, if we assume symmetry (ii) but not equal weights (i), then, once again, there is no solution.

For the nonlocal quadrature, we will assume equal weights (i), but instead of total symmetry (ii), we will assume partial symmetry as follows

$$r_1 = r_2 = r_3 = A;\quad s_2 = s_4 = B \tag{5.127}$$

where A, B are unknown parameters.

With equal weights and the partial symmetry assumption, the system (5.124) now has an equal number of unknowns and equations and, therefore, can be solved analytically. For the purpose of illustration consider $l_\chi = l_\psi = l_\varsigma = l$. Then (5.124) yields

$$\begin{aligned}
&1 = r_4 + 3A;\quad 1 = s_1 + s_3 + 2B;\quad 1 = t_1 + t_2 + t_3 + t_4\\
&\frac{1}{5} - K_{rs} l^2 - B(1-2A) = A(s_1 + s_3)\\
&\frac{1}{5} - K_{rt} l^2 = A(t_1 + t_2 + t_3) + (1-3A)t_4\\
&\frac{1}{5} - K_{st} l^2 = s_1 t_1 + s_3 t_3 + B(t_2 + t_4)\\
&\frac{2}{5} - K_r l^2 = 3A^2 + (1-3A)^2\\
&\frac{2}{5} - K_s l^2 = s_1^2 + s_3^2 + 2B^2\\
&\frac{2}{5} - K_t l^2 = t_1^2 + t_2^2 + t_3^2 + t_4^2
\end{aligned} \tag{5.128}$$

where the geometry-dependent coefficients are given by

$$\begin{aligned}
&K_{rs} = \frac{b_r b_s + c_r c_s + d_r d_s}{3(6V)^2};\quad K_{rt} = \frac{b_r b_t + c_r c_t + d_r d_t}{3(6V)^2};\quad K_{st} = \frac{b_s b_t + c_s c_t + d_s d_t}{3(6V)^2}\\
&K_r = \frac{b_r^2 + c_r^2 + d_r^2}{3(6V)^2};\quad K_s = \frac{b_s^2 + c_s^2 + d_s^2}{3(6V)^2};\quad K_t = \frac{b_t^2 + c_t^2 + d_t^2}{3(6V)^2}
\end{aligned} \tag{5.129}$$

The solution of (5.128) yields

$$A=\frac{1}{4}\left(1-\sqrt{\frac{D_A}{15}}\right);\quad B=\frac{1-5A-5K_{rs}l^2}{5(1-4A)};\quad r_4=\frac{1}{4}\left(1+3\sqrt{\frac{D_A}{15}}\right)$$
$$s_1=\frac{1}{2}\left(1-2B+\sqrt{D_s}\right);\quad s_3=\frac{1}{2}\left(1-2B-\sqrt{D_s}\right);\quad t_4=\frac{1-5A-5K_{rt}l^2}{5(1-4A)} \tag{5.130}$$

with

$$D_A=3-20K_rl^2>0 \tag{5.131}$$

$$D_s=-8B^2+4B-\frac{1}{5}-2K_sl^2\geq 0 \tag{5.132}$$

Consider first the case of $s_1\neq B$, then

$$t_3^+=\frac{\lambda\gamma-(1-\lambda-t_4)(\gamma-1)+\sqrt{D_t}}{2\left(\gamma^2-\gamma+1\right)}\quad\text{or}\quad t_3^-=\frac{\lambda\gamma-(1-\lambda-t_4)(\gamma-1)-\sqrt{D_t}}{2\left(\gamma^2-\gamma+1\right)}$$
$$t_1^\pm=\lambda-\gamma t_3^\pm;\quad t_2^\pm=1-\lambda-t_4+(\gamma-1)t_3^\pm \tag{5.133}$$

with

$$D_t=\left[(1-\lambda-t_4)(\gamma-1)-\lambda\gamma\right]^2$$
$$-2\left(\gamma^2-\gamma+1\right)\left(\lambda^2+t_4^2+(1-\lambda-t_4)^2+K_tl^2-\frac{2}{5}\right)\geq 0 \tag{5.134}$$

$$\lambda=\frac{1-5B-5K_{st}l^2}{5(s_1-B)};\quad \gamma=\frac{s_3-B}{s_1-B} \tag{5.135}$$

Otherwise, if $s_1=B$, then

$$t_3=\frac{1-5B-5K_{st}l^2}{5(s_3-B)}$$
$$t_1^+=\frac{1}{2}\left(1-t_3-t_4+\sqrt{\tilde{D}_t}\right)\ \text{or}\ t_1^-=\frac{1}{2}\left(1-t_3-t_4-\sqrt{\tilde{D}_t}\right) \tag{5.136}$$
$$t_2^\pm=1-t_3-t_4-t_1^\pm$$

with

$$\tilde{D}_t=(1-t_3-t_4)^2-2\left(t_3^2+t_4^2+(1-t_3-t_4)^2+K_tl^2-\frac{2}{5}\right)\geq 0 \tag{5.137}$$

Following (5.115), the fourth volume coordinate q_I is given by $q_I=1-s_I-r_I-t_I$.

Inserting B into the constraint (5.132) yields

$$-\sqrt{3D_A D_B} \leq 60K_{rs}l^2 + 3 \leq \sqrt{3D_A D_B} \tag{5.138a}$$

$$D_B = 3 - 20K_s l^2 > 0 \tag{5.138b}$$

Following the strong inequities, (5.131) and (5.138b), it can be seen that the singularity arising from $A = 1/4$ or $B = 1/4$ is eliminated.

In section 5.2.1 we saw that the unit cell size is not arbitrary. For 1D quadratic elements, the unit cell cannot be larger than the element, whereas for 1D cubic elements, the unit cell cannot be larger than half of the element. Likewise, the constraints (5.131), (5.138) combined with either (5.134) or (5.137) define the limitations of the unit cell size for which the solution (5.130) combined with either (5.133) or (5.136) exists.

Constraint equation (5.134) and equation (5.137) can be solved for l numerically for each tetrahedron. It is instructive to point out that replacing (5.125) and (5.127) with another set of assumptions will yield a different location of quadrature points.

We now consider specific shapes of tetrahedral elements. Satisfaction of constraints (5.131), (5.138) combined with either (5.134) or (5.137) is checked numerically for each unit cell size l, and the maximum unit cell size $l^{\max}$ is approximated for each tetrahedral element. In the case of a zero unit cell size, the nonlocal quadrature points coincide with the local Gauss points for any tetrahedral element (e.g., Table 5.1).

(1) *Regular tetrahedron* with vertices positioned at

$$\begin{aligned}(x_A, y_A, z_A) &= (1, 1, 1)\\ (x_B, y_B, z_B) &= (-1, -1, 1)\\ (x_C, y_C, z_C) &= (-1, 1, -1)\\ (x_D, y_D, z_D) &= (1, -1, -1)\end{aligned} \tag{5.139}$$

The above corresponds to a tetrahedron edge equal to $l^{\text{edge}} = \sqrt{2}$ and a volume of $V = 2.6667$. The maximum unit cell size and the volume are approximately equal to $l^{\max} \approx 0.566\ l^{\text{edge}}$ and $V_{UC}^{\max} \approx 0.192\ V$, respectively. The nonlocal quadrature points are listed in Table 5.1.

(2) *Trirectangular tetrahedron* with vertices positioned at

$$\begin{aligned}(x_A, y_A, z_A) &= (0, 0, 0)\\ (x_B, y_B, z_B) &= (1, 0, 0)\\ (x_C, y_C, z_C) &= (0, 1, 0)\\ (x_D, y_D, z_D) &= (0, 0, 1)\end{aligned} \tag{5.140}$$

The trirectangular tetrahedron volume is $V = 0.1667$. The maximum unit cell size and the volume are approximately equal to $l^{\max} \approx 0.223$ and $V_{UC}^{\max} \approx 0.0665\ V$, respectively. The nonlocal quadrature points for different unit cell sizes are listed in Table 5.2. Note that for large l, one set of points has a negative value of t_I^+, and, therefore, this solution should be excluded.

Table 5.1 Nonlocal quadrature points in the regular tetrahedron for different unit cell sizes

l	P_1				P_2				P_3				P_4			
	r_1	s_1	t_1^-	q_1^-	r_2	s_2	t_2^-	q_2^-	r_3	s_3	t_3^-	q_3^-	r_4	s_4	t_4	q_4
			t_1^+	q_1^+			t_2^+	q_2^+			t_3^+	q_3^+				
0	0.1382	0.5854	0.1382	0.1382	0.1382	0.1382	0.5854	0.1382	0.1382	0.1382	0.1382	0.5854	0.5854	0.1382	0.1382	0.1382
			0.1382	0.1382			0.1382	0.5854			0.5854	0.1382				
0.5	0.1442	0.5625	0.1442	0.1491	0.1442	0.1196	0.5675	0.1688	0.1442	0.1983	0.1442	0.5133	0.5675	0.1196	0.1442	0.1688
			0.0717	0.2216			0.2323	0.5039			0.5518	0.1057				
0.8	0.1543	0.4328	0.1543	0.2587	0.1543	0.0846	0.5372	0.2239	0.1543	0.3980	0.1543	0.2935	0.5372	0.0846	0.1543	0.2239
			0.1164	0.2966			0.5330	0.2281			0.1963	0.2514				

Table 5.2 Nonlocal quadrature points in the trirectangular tetrahedron for different unit cell sizes

l	P_1				P_2				P_3				P_4			
	r_1	s_1	t_1^-	q_1^-	r_2	s_2	t_2^-	q_2^-	r_3	s_3	t_3^-	q_3^-	r_4	s_4	t_4	q_4
			t_1^+	q_1^+			t_2^+	q_2^+			t_3^+	q_3^+				
0.1	0.1420	0.5805	0.1507	0.1268	0.1420	0.1266	0.5813	0.1501	0.1420	0.1664	0.1260	0.5656	0.5740	0.1266	0.1420	0.1574
			0.1109	0.1667			0.1673	0.5642			0.5799	0.1117				
0.2	0.1543	0.5321	0.2269	0.0866	0.1543	0.0846	0.5543	0.2069	0.1543	0.2986	0.0645	0.4826	0.5372	0.0846	0.1543	0.2239
			0.0147	0.2989			0.3225	0.4386			0.5086	0.0386				
0.223	0.1586	0.4522	0.3340	0.0552	0.1586	0.0679	0.4966	0.2769	0.1586	0.4119	0.0108	0.4186	0.5242	0.0679	0.1586	0.2493
			–0.0021	0.3913			0.4572	0.3163			0.3863	0.0432				

Table 5.3 Nonlocal quadrature points in the general tetrahedron for different unit cell sizes

l	P_1				P_2				P_3				P_4			
	r_1	s_1	t_1^-	q_1^-	r_2	s_2	t_2^-	q_2^-	r_3	s_3	t_3^-	q_3^-	r_4	s_4	t_4	q_4
			t_1^+	q_1^+			t_2^+	q_2^+			t_3^+	q_3^+				
0.3	0.1411	0.5803	0.1527	0.1258	0.1411	0.1287	0.5753	0.1548	0.1411	0.1622	0.1249	0.5718	0.5766	0.1287	0.1470	0.1476
			0.1194	0.1592			0.1587	0.5715			0.5749	0.1218				
0.6	0.1504	0.5460	0.2206	0.0831	0.1504	0.0963	0.5303	0.2230	0.1504	0.2615	0.0729	0.5153	0.5488	0.0963	0.1763	0.1786
			0.0561	0.2475			0.2471	0.5063			0.5206	0.0676				
0.729	0.1568	0.4405	0.3633	0.0395	0.1568	0.0714	0.4230	0.3489	0.1568	0.4167	0.0162	0.4104	0.5297	0.0714	0.1976	0.2013
			0.0139	0.3889			0.3989	0.3729			0.3896	0.0369				

(3) *General tetrahedron* with vertices positioned at

$$\begin{aligned}(x_A, y_A, z_A) &= (1,2,3)\\(x_B, y_B, z_B) &= (4,1,2)\\(x_C, y_C, z_C) &= (2,4,4)\\(x_D, y_D, z_D) &= (3,2,5)\end{aligned} \tag{5.141}$$

The tetrahedron edges and volume are

$$\begin{aligned}&L_{AB} = L_{DB} = \sqrt{11};\ L_{AD} = 2\sqrt{2}\\&L_{BC} = 5;\ L_{AC} = L_{DC} = \sqrt{6}\\&V = 2.6667\end{aligned} \tag{5.142}$$

The maximum unit cell size and the volume are approximately equal to $l^{\max} \approx 0.729$ and $V_{UC}^{\max} \approx 0.145\ V$, respectively. The nonlocal quadrature points for different unit cell sizes are listed in Table 5.3.

5.4.2 Triangular Elements

In this section, we will derive the nonlocal quadrature scheme for the straight-sided triangular element. We consider the nonlocal quadrature of a quadratic element with respect to the area coordinates (r, s, t). The integration is carried out over the physical domain of a triangle defined in the Cartesian coordinate system. The computational continua domain consists of a disjoint union of finite rectangular computational unit cells denoted by $\Theta_I = [l_\chi \times l_\psi]$, as shown in Figure 5.6.

The integration over the composite domain of a triangle is given by

$$\begin{aligned}&\int_{\Omega^\zeta} f(r(x,y), s(x,y))\,dxdy\\&= \Delta \sum_{I=1}^{\breve{N}} W_I \frac{1}{|\Theta_I|} \int_{\Theta_I} f(r(\hat{x}_I + \chi, \hat{y}_I + \psi), s(\hat{x}_I + \chi, \hat{y}_I + \psi))\,d\chi d\psi\end{aligned} \tag{5.143}$$

where Δ is the triangle area equal to

$$\Delta = \frac{1}{2}J; \quad J \equiv \det\left(\frac{\partial(x,y)}{\partial(r,s,t)}\right) = \det\left(\begin{bmatrix}1 & x_A & y_A\\1 & x_B & y_B\\1 & x_C & y_C\end{bmatrix}\right) \tag{5.144}$$

and the area coordinates are related to the Cartesian coordinates by [36]

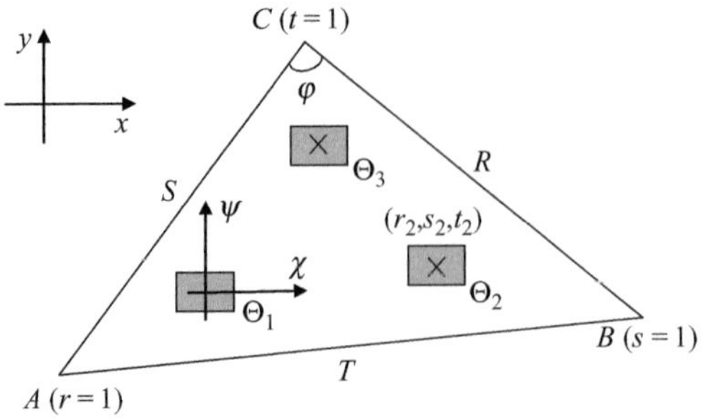

Figure 5.6 A triangular domain with unit cells placed at nonlocal quadrature points, where *R*, *S*, and *T* are the lengths of the three edges and φ is the angle between two edges

$$
\begin{aligned}
r &= \frac{a_r + b_r x + c_r y}{2\Delta}; \quad b_r = y_B - y_C; \quad c_r = x_C - x_B; \quad a_r = x_B y_C - x_C y_B \\
s &= \frac{a_s + b_s x + c_s y}{2\Delta}; \quad b_s = y_C - y_A; \quad c_s = x_A - x_C; \quad a_s = x_C y_A - x_A y_C \\
t &= \frac{a_t + b_t x + c_t y}{2\Delta}; \quad b_t = y_A - y_B; \quad c_t = x_B - x_A; \quad a_t = x_A y_B - x_B y_A \\
r &+ s + t = 1
\end{aligned}
\tag{5.145}
$$

where (x_A, y_A), (x_B, y_B), and (x_C, y_C) denote the coordinates of element vertices, which in the area coordinates are at $r=1$, $s=1$, and $t=1$, respectively. For any point in the interior of the triangle, $0<r<1$ and $0<s<1-r$. The local Cartesian coordinates (χ, Ψ) for each quadrature point are defined by $\chi = x - \hat{x}_I$ and $\psi = y - \hat{y}_I$ where $(\hat{x}_I, \hat{y}_I)$ is the unit cell centroid.

Following the procedure described above for the tetrahedral element, we derive the system of equations for the triangular element

$$
\begin{aligned}
&1 = \sum_{I=1}^{3} W_I; \quad \frac{1}{3} = \sum_{I=1}^{3} W_I r_I; \quad \frac{1}{3} = \sum_{I=1}^{3} W_I s_I \\
&\frac{1}{12}\left(1 - \frac{1}{(2\Delta)^2}\left(b_r b_s l_\chi^2 + c_r c_s l_\psi^2\right)\right) = \sum_{I=1}^{3} W_I r_I s_I \\
&\frac{1}{12}\left(2 - \frac{1}{(2\Delta)^2}\left(b_r^{\,2} l_\chi^2 + c_r^{\,2} l_\psi^2\right)\right) = \sum_{I=1}^{3} W_I r_I^{\,2} \\
&\frac{1}{12}\left(2 - \frac{1}{(2\Delta)^2}\left(b_s^{\,2} l_\chi^2 + c_s^{\,2} l_\psi^2\right)\right) = \sum_{I=1}^{3} W_I s_I^{\,2}
\end{aligned}
\tag{5.146}
$$

where the unknown position of integration points (or unit cell centroids) is expressed in the triangular coordinates as

$$r_I = \frac{a_r + b_r \hat{x}_I + c_r \hat{y}_I}{2\Delta}; \quad s_I = \frac{a_s + b_s \hat{x}_I + c_s \hat{y}_I}{2\Delta} \tag{5.147}$$

To find a unique solution of (5.146), we further assume

(i) equal weights

$$W_1 = W_2 = W_3 = \frac{1}{3} \tag{5.148}$$

(ii) partial symmetry by making the first coordinate of the two integration points equal

$$r_1 = r_2 \tag{5.149}$$

With the above two assumptions, the system (5.146) now has an equal number of unknowns and equations and, therefore, can be solved analytically

$$\begin{gathered} r_1 = r_2; \quad 1 = \sum_{I=1}^{3} r_I; \quad 1 = \sum_{I=1}^{3} s_I \\ \frac{1}{4}\left(1 - \frac{1}{(2\Delta)^2}\left(b_r b_s l_\chi^2 + c_r c_s l_\psi^2\right)\right) = \sum_{I=1}^{3} r_I s_I \\ \frac{1}{4}\left(2 - \frac{1}{(2\Delta)^2}\left(b_r{}^2 l_\chi^2 + c_r{}^2 l_\psi^2\right)\right) = \sum_{I=1}^{3} r_I{}^2 \\ \frac{1}{4}\left(2 - \frac{1}{(2\Delta)^2}\left(b_s{}^2 l_\chi^2 + c_s{}^2 l_\psi^2\right)\right) = \sum_{I=1}^{3} s_I{}^2 \end{gathered} \tag{5.150}$$

As an example, consider a square unit cell, $l_\chi = l_\psi = l$. Then the system of equations (5.150) will have two sets of solutions

$$\begin{gathered} \text{(a)}\ r_1 = r_2 = \frac{4 + \sqrt{D_A}}{12}; \quad r_3 = \frac{2 - \sqrt{D_A}}{6}; \quad s_3 = \frac{\sqrt{D_A} - c}{3\sqrt{D_A}} \\ \text{(b)}\ r_1 = r_2 = \frac{4 - \sqrt{D_A}}{12}; \quad r_3 = \frac{2 + \sqrt{D_A}}{6}; \quad s_3 = \frac{\sqrt{D_A} + c}{3\sqrt{D_A}} \\ s_1 = \frac{1 - s_3 - \sqrt{D_s}}{2}; \quad s_2 = \frac{1 - s_3 + \sqrt{D_s}}{2} \end{gathered} \tag{5.151}$$

where

$$D_A = 2\left(2-3\lambda_R^2\right); \quad D_B = 2\left(2-3\lambda_S^2\right)$$
$$c = 3\lambda_R\lambda_S\cos\varphi - 1; \quad D_s = -3s_3{}^2 + 2s_3 - \frac{\lambda_S{}^2}{2} \tag{5.152}$$
$$\lambda_R = \frac{lR}{2\Delta}; \quad \lambda_S = \frac{lS}{2\Delta}$$

In (5.152), R and S are the lengths of the two edges that are opposite to vertices $r=1$ and $s=1$, respectively, and φ is an angle between the two edges as shown in Figure 5.6. Note that the third coordinate follows from (5.151) by $t_I = 1 - s_I - r_I$.

For a unit cell size equal to zero, the integration points are symmetric and thus coincide with the Gauss quadrature points [32].

$$\text{(a)}\ r_1 = r_2 = \frac{1}{2}; \quad r_3 = 0; \quad s_3 = \frac{1}{2}; \quad s_1 = 0; \quad s_2 = \frac{1}{2}$$
$$\text{(b)}\ r_1 = r_2 = \frac{1}{6}; \quad r_3 = \frac{2}{3}; \quad s_3 = \frac{1}{6}; \quad s_1 = \frac{2}{3}; \quad s_2 = \frac{1}{6} \tag{5.153}$$

The solution of (5.151) exists only when the following constraints are satisfied

$$|2c| \le \sqrt{D_A D_B}$$
$$\lambda_R < \sqrt{\frac{2}{3}}; \quad \lambda_S \le \sqrt{\frac{2}{3}} \tag{5.154}$$

The above implies the following constraints on the unit cell size

$$\text{if} \quad l_0 \le l_2 \quad \text{then} \quad 0 \le l < \min\{l_0, l_1\} \tag{5.155a}$$

$$\text{if} \quad l_0 > l_2 \quad \text{then} \quad 0 \le l \le l_1 \ \text{or} \ l_2 \le l \le l_0 \tag{5.155b}$$

with

$$l_1 = \sqrt{\frac{1}{6}\left(R^2 + S^2 + T^2 - \sqrt{d}\right)}; \quad l_2 = \sqrt{\frac{1}{6}\left(R^2 + S^2 + T^2 + \sqrt{d}\right)}$$
$$d = \left(R^2 + S^2 + T^2\right)^2 - 48\Delta^2; \quad l_0 = \min\left\{2\sqrt{\frac{2}{3}\frac{\Delta}{R}}, 2\sqrt{\frac{2}{3}\frac{\Delta}{S}}\right\} \tag{5.156}$$

where T is the length of the edge opposite to vertex $t=1$, as shown in Figure 5.6. When $l=l_1$, then two integration points (5.151) coincide, that is, $s_1 = s_2$, $D_s = 0$. To avoid a unit-cell-size gap in (5.155b), we choose $0 \le l < \min\{l_0, l_1\}$.

Examples of unit-cell-size constraints and nonlocal quadrature points [based on solution (a) in (5.151)] are shown in Table 5.4 and Table 5.5 for various triangle shapes.

Table 5.5 Nonlocal quadrature points for various triangles

Triangle			UC size	$P_1^{(a)}$			$P_2^{(a)}$			$P_3^{(a)}$		
R	S	φ	l	r_1	s_1	t_1	r_2	s_2	t_2	r_3	s_3	t_3
1	1	60°	0.5	0.4512	0.0976	0.4512	0.4512	0.4512	0.0976	0.0976	0.4512	0.4512
3	4	90°	1.9	0.4689	0.1896	0.3415	0.4689	0.2721	0.2590	0.0622	0.5382	0.3995
1	$\sqrt{2}$	45°	0.5	0.4651	0.1701	0.3649	0.4651	0.4439	0.0910	0.0698	0.3860	0.5442
1	2	60°	0.67	0.4801	0.2623	0.2576	0.4801	0.3001	0.2198	0.0398	0.4376	0.5226

Table 5.4 Upper bound of the unit cell size for various triangle shapes

R	S	φ	Triangle area	l_0	l_1	l_2	Upper bound for l	Upper UC area (l^2)
1	1	60°	0.433	0.7071	0.7071	0.7071	$l_0=l_1=l_2$	0.5
3	4	90°	6	2.4495	1.9242	3.6006	l_1	3.7025
1	$\sqrt{2}$	45°	0.5	0.5774	0.5774	1	$l_0=l_1$	0.3333
1	2	60°	0.866	0.7071	0.6719	1.4884	l_1	0.4514

Remark 5.10 The nonlocal quadrature for an equilateral triangle with edges $R=S=T=h$ is symmetric

$$\alpha=\frac{1\mp\sqrt{1-2\zeta^2}}{3};\quad \beta=\frac{2\pm\sqrt{1-2\zeta^2}}{6};\quad \zeta=\frac{l}{h}$$
$$r_1=t_1=r_2=s_2=s_3=t_3=\beta;\quad s_1=t_2=r_3=\alpha \tag{5.157}$$

with the following constraint

$$l\le l_0=l_1=l_2=\frac{h}{\sqrt{2}} \tag{5.158}$$

If $l=h/\sqrt{2}$, then $\alpha=\beta$ and all three integration points coincide.

5.4.3 Quadrilateral and Hexahedral Elements

For elements whose edges (faces in 3D) are parallel to those of the unit cell, the nonlocal quadrature scheme has been discussed in Remark 5.3 of Section 5.2.1. In general, however, a rectangular unit cell in the physical domain becomes distorted following the element mapping shown for a quadrilateral element in Figure 5.7.

In this section we derive the nonlocal quadrature scheme for a general quadrilateral element by matching 2D monomials. For more details and extension to hexahedral elements, see [19].

Consider the nonlocal quadrature rule (5.8) expressed in the parent element domain.

$$\begin{aligned}\int_{\Omega_X^\zeta}\Psi(X)d\Omega&=\sum_{I=1}^{\hat N}W_IJ_I\frac{1}{\left|\Theta_{\hat X_I}\right|}\int_{\Theta_{\hat X_I}}\Psi(\hat X_I,\chi)d\Theta\\&=\sum_{I=1}^{\hat N}W_I\frac{1}{\left|\Theta_{\hat X_I}\right|}\int_{\Theta_{\hat X_I}}J'(\hat X_I,\chi)\Psi'(\hat X_I,\chi)d\Theta\\&=\int_\square J(\square)\Psi(\square)d\square=\int_\square\Psi'(\square)d\square\end{aligned} \tag{5.159}$$

where $\Psi'(\hat X_I,\chi)=J(\hat X_I,\chi)\Psi(\hat X_I,\chi)$, $J(\square)\Psi(\square)=\Psi'(\square)$, and $J'(\hat X_I,\chi)=J_I/J(\hat X_I,\chi)$.

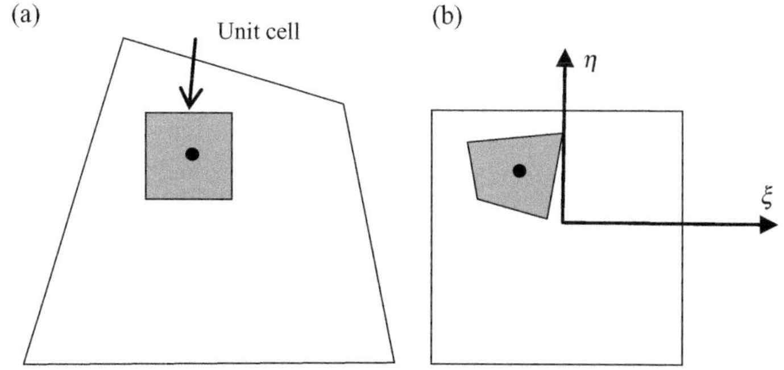

Figure 5.7 Definition of the effective unit cell size in the parent element domain: (a) physical element and unit cell domains; and (b) parent element on a bi-unit square; shaded area denotes the mapped unit cell

We focus on the four-point nonlocal quadrature scheme ($\hat{N} = 4$) for a general quadrilateral element, $d\square = d\xi d\eta$. There are twelve unknowns including the positions $(\hat{\xi}_I, \hat{\eta}_I)$ and weights W_I of the four nonlocal quadrature points. The twelve unknowns are computed using the Newton method applied to the system of twelve nonlinear equations

$$\int_{-1}^{1}\int_{-1}^{1} \xi^{\alpha}\eta^{\beta} d\xi d\eta = \sum_{I=1}^{4} W_I \frac{1}{\left|\Theta_{\hat{X}_I}\right|}\int_{\Theta_{\hat{X}_I}} J'(\hat{X}_I, \chi)\xi^{\alpha}\eta^{\beta} d\Theta \tag{5.160}$$

which correspond to the twelve monomials depicted in Figure 5.8

The left-hand side of (5.160) is calculated analytically. The right-hand side of (5.160) is evaluated numerically by integrating over the local Gauss quadrature points (r_i, s_i) in the parent unit cell domain as shown in Figure 5.9. Note that an evaluation of $\xi(\hat{X}_I, \hat{Y}_I, \chi(r_i), \psi(s_i))$ and $\eta(\hat{X}_I, \hat{Y}_I, \chi(r_i), \psi(s_i))$ in (5.160) requires inverse mapping for every quadrature point $\chi(r_i)$, $\psi(s_i)$ in the physical unit cell domain positioned at $(\hat{X}_I, \hat{Y}_I)$.

An efficient numerical solution of (5.160) for the unknowns $(\hat{\xi}_I, \hat{\eta}_I, W_I)$ requires a good initial guess $({}^0\hat{\xi}_I, {}^0\hat{\eta}_I, {}^0W_I)$. If the element edges were parallel to the unit cell edges, then the solution $(\hat{\xi}_I, \hat{\eta}_I, W_I)$ can be obtained by the tensor product rule discussed in Remark 5.3. The nonlocal quadrature scheme in one dimension depends on the ratio between the unit cell length and the element size. The extension of the tensor product rule to a general quadrilateral element requires definition of the equivalent rectangular element whose edges are parallel to those of the unit cell. The equivalent rectangular element is defined to have the same area and centroid as the original quadrilateral element, and its dimensions L_x, L_y are determined by a least square minimization of the second moments of area I_{xx}, I_{yy} of the equivalent rectangular element and the original quadrilateral. Consequently, the initial guess of the nonlocal quadrature points in a general quadrilateral element is defined as

$${}^0\hat{\xi}_I = \pm\sqrt{\frac{1}{3} - \frac{\theta_{\xi}'^2}{12}}; \quad {}^0\hat{\eta}_I = \pm\sqrt{\frac{1}{3} - \frac{\theta_{\eta}'^2}{12}}; \quad {}^0W_I = 1 \tag{5.161}$$

where $\theta'_{\xi}, \theta'_{\eta}$ are the normalized unit cell dimensions defined as

$$1$$
$$\xi \quad \eta$$
$$\xi^2 \quad \xi\eta \quad \eta^2$$
$$\xi^3 \quad \xi^2\eta \quad \xi\eta^2 \quad \eta^3$$
$$\xi^3\eta \qquad \xi\eta^3$$

Figure 5.8 Monomials matched in the four-point nonlocal quadrature scheme

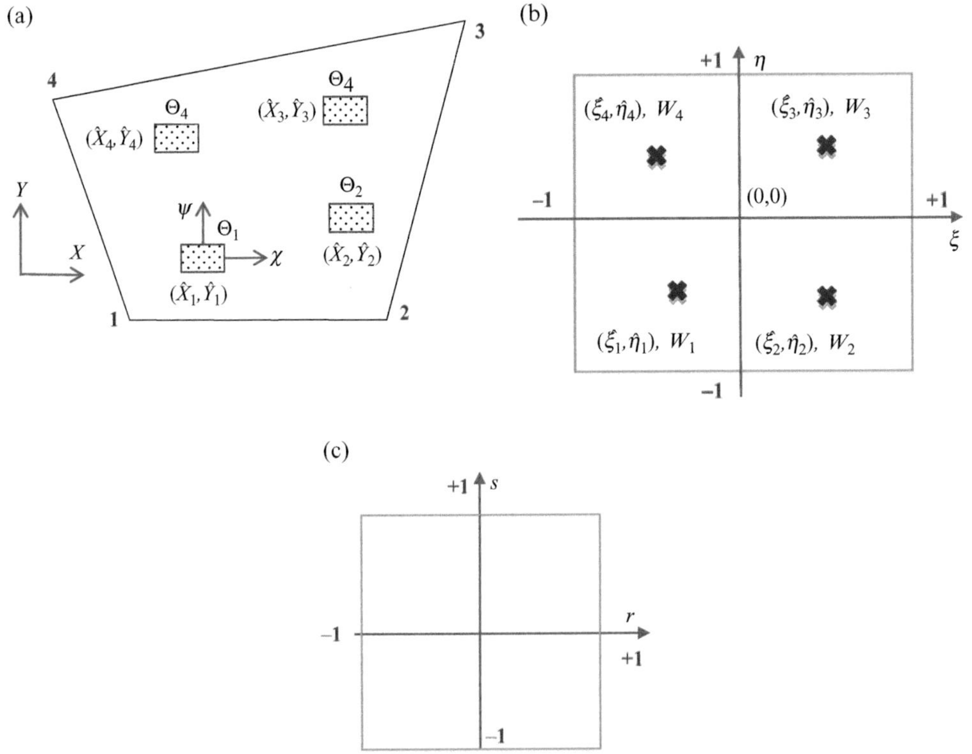

Figure 5.9 Mapping of a quadrilateral element and its unit cells: (a) an element and four unit cells in the physical domain; (b) an element in the parent element domain with unit cell centroids denoted by a cross; and (c) a unit cell in the parent element domain

$$\theta'_\xi = 2\frac{l_\chi}{L_x}; \quad \theta'_\eta = 2\frac{l_\psi}{L_y} \tag{5.162}$$

and l_χ, l_ψ are the unit cell dimensions in the physical domain.

To illustrate the nonlocal quadrature scheme, we consider a quadrilateral element with vertices at (1, 1), (3, 0.8), (3.2, 3), and (0.8, 2.8) resulting in the element area equal to $|\Omega| = 4.4$. The dimensions of the equivalent rectangular element are $L_x = 2.1902$ and $L_y = 2.0089$. Positions and weights of the nonlocal quadrature points for the quadrilateral element are given in Figure 5.10 and Table 5.6. We consider several unit cell sizes ranging from 0 to 1.6. It can be

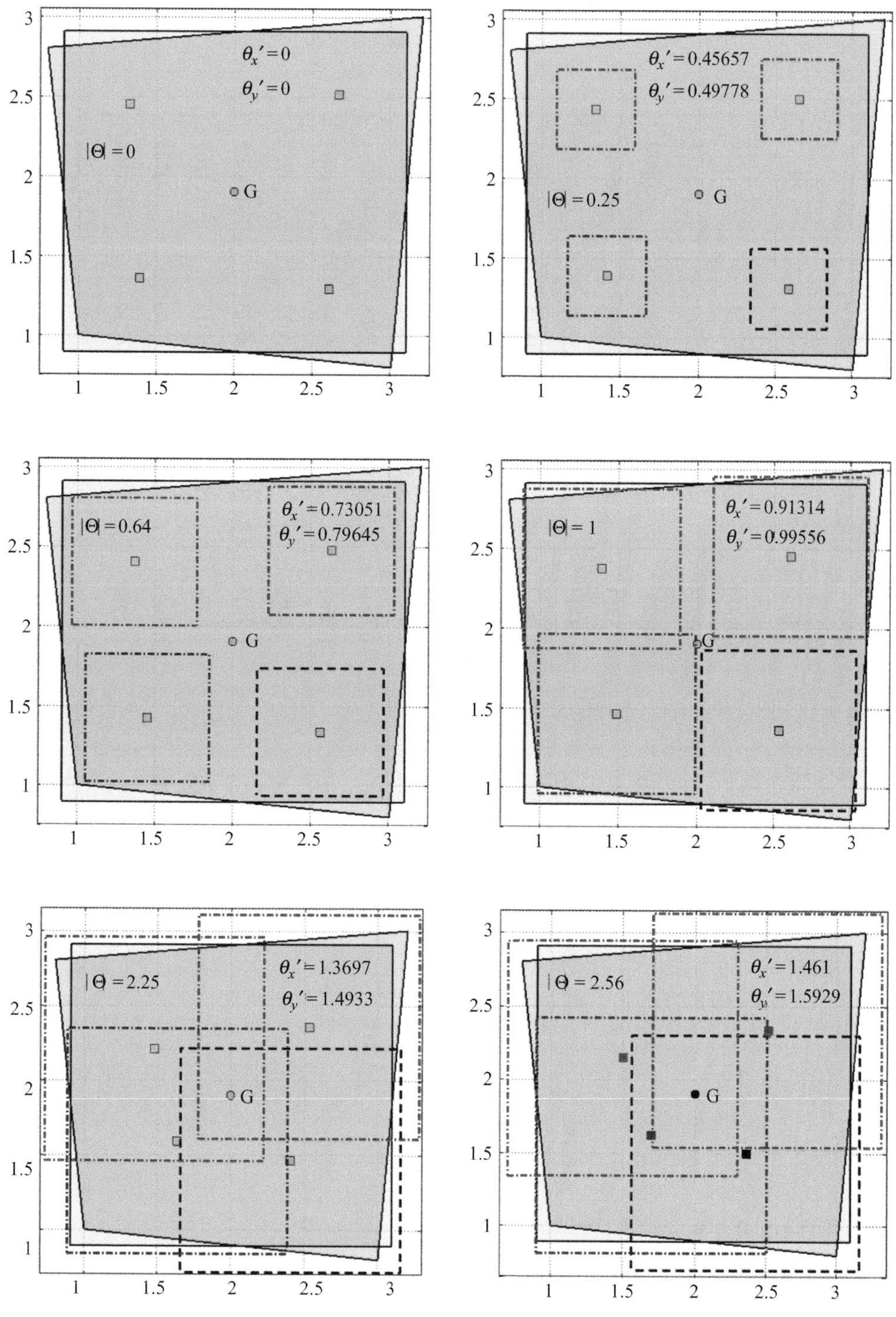

Figure 5.10 Positions of four nonlocal quadrature points for various unit cell sizes $\theta'_{\xi}, \theta'_{\eta}$ for $|\Omega| = 4.4$. The quadrilateral element is denoted in cyan; the equivalent rectangular element is in yellow; the dotted lines depict the boundaries of the unit cells

Table 5.6 Nonlocal quadrature points for a quadrilateral element

l_χ	θ'_ξ	θ'_η	Quadrilateral vertex coordinates: (1,1) ; (3,0.8) ; (3.2,3) ; (0.8,2.8)				
			Unknowns	Θ_1	Θ_2	Θ_3	Θ_4
			$\hat{\xi}_I$	−0.5774	0.5774	0.5774	−0.5774
0	0	0	$\hat{\eta}_I$	−0.5774	−0.5774	0.5774	0.5774
			W_I	1.0000	1.0000	1.0000	1.0000
			$\hat{\xi}_I$	−0.5551	0.5610	0.5661	−0.5648
0.5	0.4566	0.4978	$\hat{\eta}_I$	−0.5511	−0.5618	0.5633	0.5577
			W_I	1.0011	0.9979	1.0015	0.9980
			$\hat{\xi}_I$	−0.5193	0.5338	0.5481	−0.5449
0.8	0.7305	0.7965	$\hat{\eta}_I$	−0.5090	−0.5369	0.5404	0.5245
			W_I	1.0014	0.9950	1.0043	0.9955
			$\hat{\xi}_I$	−0.4848	0.5065	0.5308	−0.5264
1	0.9131	0.9956	$\hat{\eta}_I$	−0.4686	−0.5131	0.5182	0.4905
			W_I	0.9986	0.9931	1.0073	0.9951
			$\hat{\xi}_I$	−0.3410	0.3832	0.4673	−0.4596
1.5	1.3697	1.4933	$\hat{\eta}_I$	−0.3205	−0.4208	0.4317	0.3207
			W_I	0.8892	1.0105	1.0149	1.0723
			$\hat{\xi}_I$	−0.2760	0.3422	0.4517	−0.4394
1.6	1.4610	1.5929	$\hat{\eta}_I$	−0.2975	−0.3926	0.4077	0.2483
			W_I	0.7549	1.0254	1.0105	1.1943

seen that the positions and weights of the nonlocal quadrature points are close but not equal to those obtained by the tensor product rule (5.17). Furthermore, the solution of the nonlinear system of equations with twelve unknowns is found in just 3 to 6 iterations when the initial guess is taken from the product rule (5.161).

5.5 Model Verification

We consider a 20-node serendipity quadratic hexahedral element as a coarse-scale element where the unit cell faces are parallel to the coarse-scale element faces. A general case is studied in [19].

We define a nondimensional parameter ζ_k that denotes the ratio of the unit cell size l_k along the kth direction to the size of the coarse-scale domain L_k^{Ω} along the kth direction

$$\zeta_k = \frac{l_k}{L_k^{\Omega}} \tag{5.163}$$

The size effect parameter ζ is defined as $\zeta = \max\limits_{k=1...3} \zeta_k$. For $\zeta \to 0$, the nonlocal integration points coincide with local element Gauss quadrature points. For unit cell size approaching the coarse-scale element size, that is, $\zeta \to 1$, the nonlocal quadrature points approach the element centroid.

For the nonlocal quadrature, we use eight integration points. When the unit cell size is less than the coarse-scale element size, that is, $\zeta < 1$, we employ eight local Gauss quadrature points. Consequently, the total number of quadrature points in the coarse-scale element is 64, and therefore, the stiffness matrix full rank is preserved. For $\zeta = 1$ a single nonlocal quadrature point is used resulting in rank deficiency of the stiffness matrix. The rank deficiency can be circumvented in several ways: (i) using $\zeta = 0.99$; (ii) increasing the number of local quadrature points; or (iii) employing stabilization.

For model verification, we will consider a beam consisting of eight coarse-scale elements. For simplicity, both the coarse-scale element and the computational unit cell are in the form of a cube. The unit cell contains a spherical or cylindrical inclusion placed at the unit cell centroid. The diameter of the inclusion cross section is chosen to be 0.6 of the unit cell length. For the reference solution, we will consider DNS, which employs a sufficiently fine finite element mesh where the element size is considerably smaller than the size of the inclusion. The results of the RC^2 are also compared with the $O(1)$ reduced order formulation.

We will study the dependence of the solution accuracy on the unit cell size. For the RC^2 formulation, any value of ζ in [0,1] can be considered. However, the reference solution cannot be constructed for some choices of ζ, such as, for instance, $1 > \zeta > 0.5$, if the coarse-scale problem is built from spatial repetition of complete unit cells. Thus, the reference solution is constructed for $\zeta = 1$ and $\zeta = 0.5$, which corresponds to a coarse-scale element consisting of one and eight unit cells, respectively. Figure 5.11 depicts two examples: (i) a coarse-scale element constructed from eight unit cells, $\zeta = 0.5$; and (ii) a beam with the coarse-scale element equal to the unit cell, $\zeta = 1$. Note that the results obtained by the $O(1)$ homogenization method do not depend on the value of ζ.

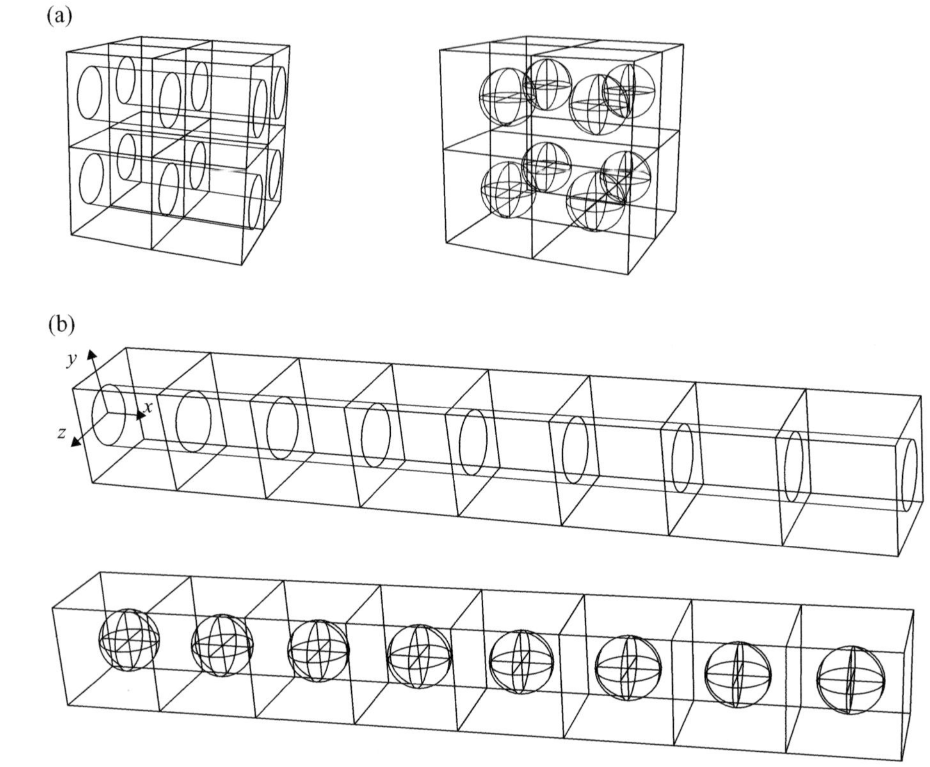

Figure 5.11 (a) a single coarse-scale finite element with spherical inclusion (top left) and cylindrical inclusion (top right) for $\zeta = 0.5$; and (b) the beam with eight unit cells for $\zeta = 1$

We consider three types of boundary conditions, hereafter labeled as BC (a), BC (b), and BC (c), with a prescribed linear distribution of displacements in the x direction (U_x) on the right face of the beam

$$\mathrm{BC(a)}\; U_x = A(y+1) + x_R; \quad U_y = 0; \quad U_z = 0; \quad A = 0.0195$$

$$\mathrm{BC(b)}\; U_x = Ay + x_R; \quad U_y = 0; \quad U_z = 0; \quad A = 0.01$$

$$\mathrm{BC(c)}\; U_x = \begin{cases} A_1 t + x_R, & t \in (0, t_1) \\ A_2 y(t - t_1) + A_1 t_1 + x_R, & t \in [t_1, t_2] \end{cases} \quad U_y = 0; \quad U_z = 0$$

Here, x_R denotes the coordinate of the right face of the beam. The left side in all problems is clamped. The boundary condition BC (c) is considered for nonlinear problems only where t denotes the quasi-time that controls the magnitude of loading. Constants A_1, A_2 and t_1, t_2 will be specified later. For $t_1 = 0$ and fixed t, the boundary condition BC (c) coincides with BC (b).

First, we consider a linear isotropic elastic material with different elastic properties for each phase and each spherical inclusion. The nondimensional Young's modulus and Poisson ratio for the two phases are $E_{\mathrm{matrix}} = 4 \times 10^{10}$, $\nu_{\mathrm{matrix}} = 0.3$ and $E_{\mathrm{inclusion}} = 4$, $\nu_{\mathrm{inclusion}} = 0.3$. For the inelastic case, we consider classical isotropic hardening plasticity for the matrix phase, with the inclusion being elastic. The hardening modulus H is assumed to be constant. The

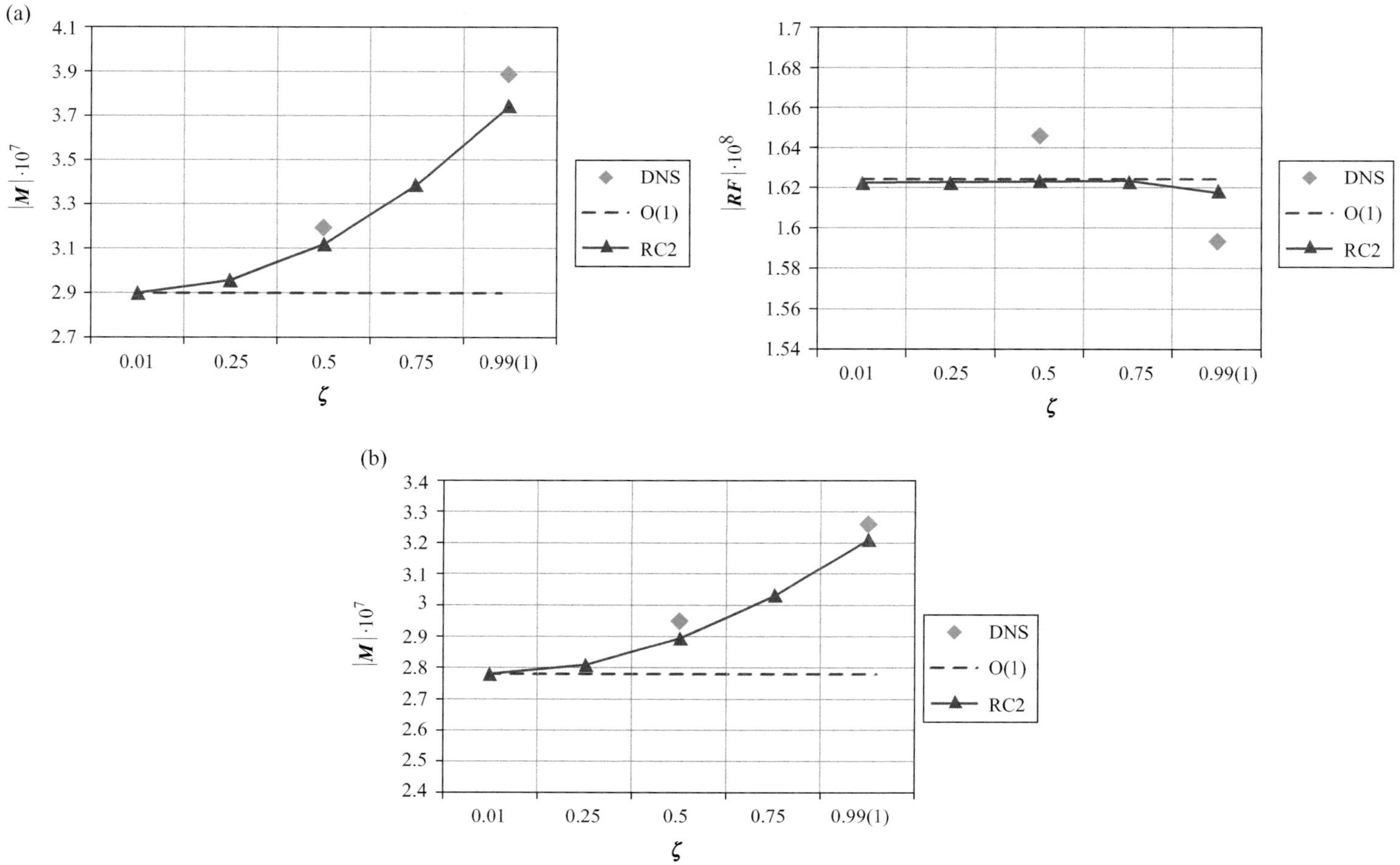

Figure 5.12 Comparison of reaction forces and moments for various unit cell sizes: (a) BC (a); and (b) BC (b)

nondimensional inelastic parameters (the yield stress and the hardening modulus for the matrix phase) are $\sigma_Y=380.e+10^6$, $H=200.e+10^8$, respectively, while the elastic moduli for the two phases are $E_{matrix}=200\times10^9$, $\nu_{matrix}=0.3$ and $E_{inclusion}=1000\times10^9$, $\nu_{inclusion}=0.3$. For simplicity, we consider one partition for each material phase.

The unit cell finite mesh considered in both RC^2 and $O(1)$ formulations is made of linear tetrahedral elements totaling 1807 (for the unit cell with a sphere) and 1523 (for the unit cell with a cylindrical inclusion).

5.5.1 The Beam Problem

Elastic case. The beam problem with eight coarse-scale elements and a spherical inclusion is depicted in Figure 5.11. The beam is first subjected to boundary conditions of type BC (a) and BC (b). It can be seen that the RC^2 solution agrees well with the DNS solution. For $\zeta=1$,

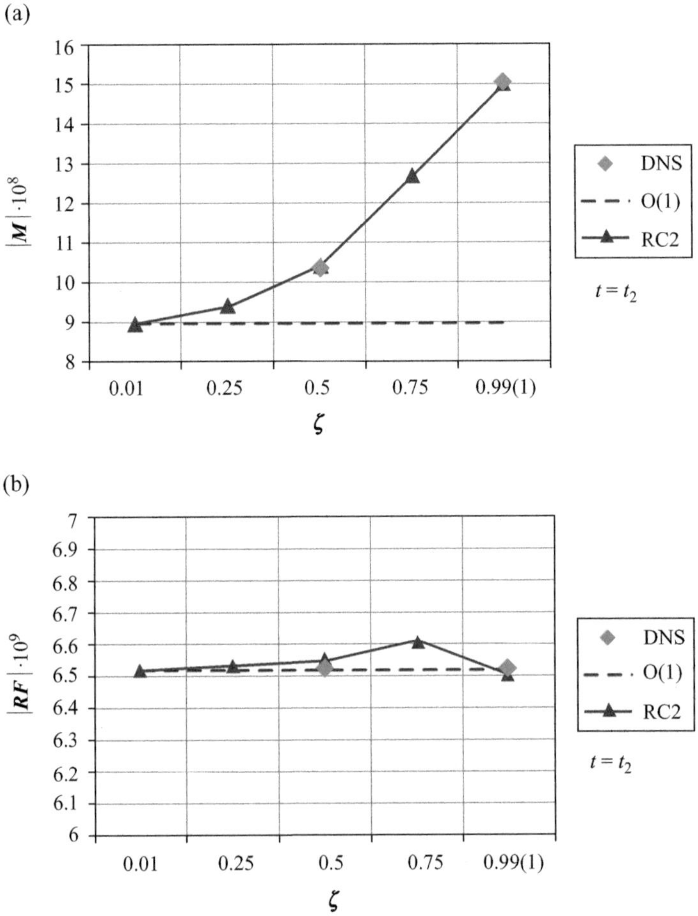

Figure 5.13 (a) Moment and (b) reaction force evolution for various unit cell sizes. The axis of the cylindrical inclusion is parallel to the axis of the beam. The BC (c) is applied along the beam

the $O(1)$ formulation results in a 25% error in moment for boundary conditions of type BC (a) [Figure 5.12(a), left] and a 15% error for boundary conditions of type BC (b) [Figure 5.12(b)]. The RC^2 and $O(1)$ formulations coincide at ζ=0.01.The error in reaction forces is considerably smaller for both boundary conditions [Figure 5. 12(a), right].

Plasticity with isotropic hardening. Consider the beam problem with a cylindrical inclusion where the axis of the cylinder is parallel to the axis of the beam. The boundary conditions BC (c) is considered with parameters A_1=0.08 , A_2=0.04, t_1=1, and t_1=2. The results for bending moment and reaction force are given in Figure 5.13 at t=t_2. It can be seen that the RC^2 formulation is in excellent agreement with DNS (error is 1%), while the $O(1)$ solution results in a 40% error in the moment for ζ=1. Here the reaction forces [Figure 5.13(b)] are almost the identical for all formulations.

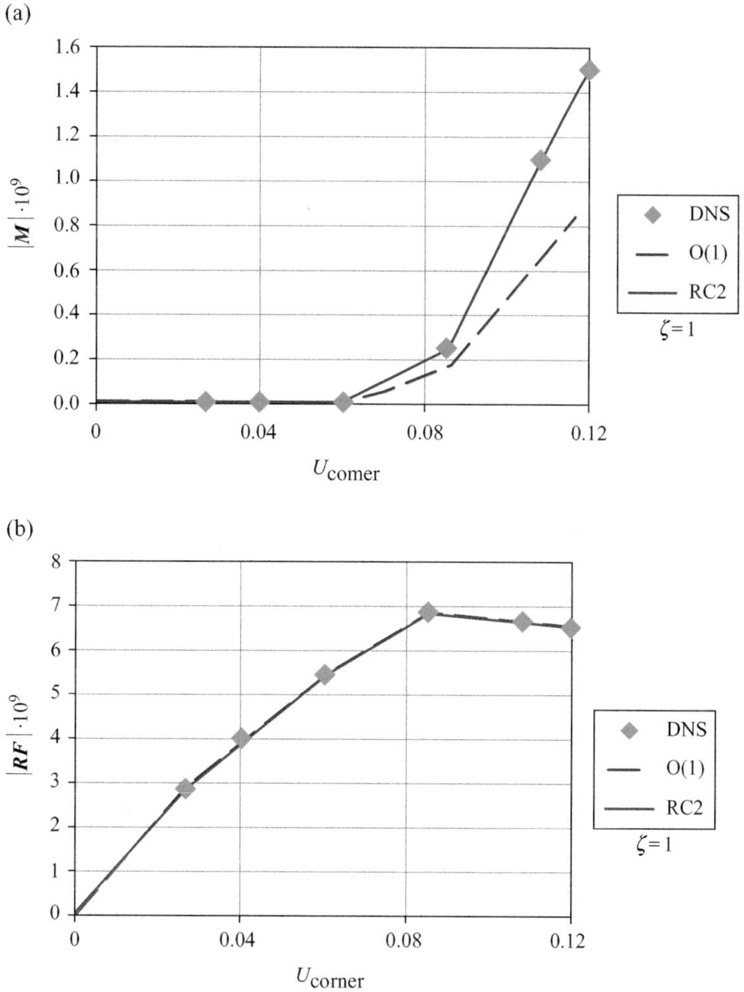

Figure 5.14 (a) Moment and (b) reaction force evolution versus displacement at the right upper corner of the beam

Figure 5.14 depicts the evolution of moment as a function of displacement at the upper corner U_{corner} of the beam. The coarse-scale element is chosen to be equal to the unit cell size, $\zeta=1$. The results show good agreement between RC² and DNS during the entire loading process. When only tension exists (the bending moment is absent), the reaction force evolution shows that plasticity takes place; note the change in slope near $U_{\text{corner}}=0.04$ in Figure 5.14(b). When the bending initiates, the $O(1)$ formulation continues to correctly predict the value of the reaction force but errs badly in predicting the moment.

Problems

Problem 5.1: Prove that

$$\int_{\Theta_{\hat{X}_I}} \frac{\partial w_i^c}{\partial X_j}\left(P_{ij}^{\zeta}(\hat{X}_I,\chi)-\bar{P}_{ij}(\hat{X}_I)-Q_{ijm}(\hat{X}_I)\chi_m\right)d\Theta=0$$

where χ_k is a local coordinate with respect to the centroid of the unit cell, $\partial w_i^c/\partial X_j$ is approximated by linear function, and Q_{ijm} is defined as in Remark 5.7.

Problem 5.2: Show that the coarse-scale stiffness matrix (5.62) for a linear model problem in section 5.2.4 is symmetric.

Problem 5.3: For the 1D model considered in section 5.2.4, compare the RC² and C² formulations.

Problem 5.4: For the overall stress (5.101a) in an RC² formulation, indicate the tensor coefficients that are not size dependent and that have linear dependency on the unit cell size.

Problem 5.5: Consider constraints for a triangular element

$$-\sqrt{D_A D_B}\le 6\lambda_R\lambda_S\cos\varphi-2\le\sqrt{D_A D_B}$$

$$\lambda_R<\sqrt{\frac{2}{3}};\quad \lambda_S\le\sqrt{\frac{2}{3}}$$

Derive

$$\begin{aligned}&\text{if}\quad l_0\le l_2\quad\text{then}\quad 0\le l<\min\{l_0,l_1\}\\&\text{if}\quad l_0>l_2\quad\text{then}\quad 0\le l\le l_1\ \text{ or }\ l_2\le l\le l_0\end{aligned}$$

by utilizing

$$D_A=2\left(2-3\lambda_R^2\right);\quad D_B=2\left(2-3\lambda_S^2\right);\quad \lambda_R=\frac{lR}{2\Delta};\quad \lambda_S=\frac{lS}{2\Delta}$$

$$d=(R^2+S^2+T^2)^2-48\Delta^2;\quad l_0=\min\left\{2\sqrt{\frac{2}{3}}\frac{\Delta}{R},2\sqrt{\frac{2}{3}}\frac{\Delta}{S}\right\}$$

$$l_1=\sqrt{\frac{1}{6}(R^2+S^2+T^2-\sqrt{d})};\quad l_2=\sqrt{\frac{1}{6}(R^2+S^2+T^2+\sqrt{d})}$$

where R, S, T are the lengths of the triangle edges and Δ is an area of the triangle.

Problem 5.6: Consider a rectangular unit cell whose local coordinate system axes are parallel to the unit cell sides. The displacement $u_i^{(1)}$ is $\boldsymbol{\chi}$-periodic. Show that combining

(i) periodicity condition $\int_{\partial\Theta_{\hat{X}_I}} u_i^{(1)} N_k^{\Theta} d\gamma = 0$

and

(ii) $\int_{\partial\Theta_{\hat{X}_I}} u_i^{(1)} \chi_j N_k^{\Theta} d\gamma = 0$ for $k{=}j$

yields homogeneous constraints on each of the four bounding edges $\partial\Theta_{\hat{X}_I}^j$ of the unit cell

$$\int_{\partial\Theta_{\hat{X}_I}^j} u_i^{(1)} d\gamma = 0 \quad \forall j = 1,...,4$$

where $\bigcup_{j=1}^{4} \partial\Theta_{\hat{X}_I}^j = \partial\Theta_{\hat{X}_I}$ and $\partial\Theta_{\hat{X}_I}^i \cap \partial\Theta_{\hat{X}_I}^j = 0 \quad \forall i \neq j.$

References

[1] Geers, M.G.D. Kouznetsova, V. and Brekelmans, W.A.M. Gradient-enhanced computational homogenization for the micro–macro scale transition. Journal de Physique IV 2001, 11, 145–152.

[2] Forest, S. and Sievert, R. Nonlinear microstrain theories. International Journal of Solids and Structures 2006, 43, 7224–7245.

[3] Trostel, R. Gedanken zur Konstruktion mechanischer Theorien II. Forschungsbericht, No. 7, 2. Technical University of Berlin, 1988.

[4] Bazant, Z.P. and Jirasek, M. Nonlocal integral formulations of plasticity and damage: survey of progress. Journal of Engineering Mechanics 2002, 128(11), 1119–1149.

[5] Eringen, A.C. Theory of non-local elasticity and some applications. Res Mechanica Letters 1987, 21, 313–342.

[6] Eringen, A.C. and Edelen, D.G.B. On non-local elasticity. International Journal of Enginering Science 1972, 10, 233–248.

[7] Kroner, E. Elasticity theory of materials with long range cohesive forces. International Journal of Solids and Structures 1967, 3, 731–742.

[8] Belytschko, T., Bazant, Z.P., Hyun, Y.-W. and Chang, T.-P. Strain softening materials and finite element solutions. Computers and Structures 1986, 23(2), 163–180.

[9] Bazant, Z.P. and Pijaudier-Cabot, G. Nonlocal continuum damage, localization instability and convergence. Journal of Applied Mechanics 1988, 55, 287–293.

[10] Fleck, N.A. and Hutchinson, J.W. A reformulation of strain gradient plasticity. Journal of the Mechanics and Physics of Solids 2001, 49, 2245–2271.

[11] Chen, J.S., Zhang, X. and Belytschko, T. An implicit gradient model by a reproducing kernel strain regularization in strain localization problems. Computer Methods in Applied Mechanics and Engineering 2004, 193, 2827–2844.

[12] Zohdi, T.I., Oden, J.T. and Rodin, G.J. Hierarchical modeling of heterogeneous bodies. Computer Methods in Applied Mechanics and Engineering 1996, 138, 273–298.

[13] E, W., Engquist, B. and Huang, Z. Heterogeneous multiscale method: A general methodology for multiscale modeling. Physical Review B 2003, 67(9), 1–4.

[14] Loehnert, S. and Belytschko, T. A multiscale projection method for macro/microcrack simulations. International Journal for Numerical Methods in Engineering 2007, 71(12), 1466–1482.

[15] Belytschko, T., Loehnert, S. and. Song, J. Multiscale aggregating discontinuities: A method for circumventing loss of material stability. International Journal for Numerical Methods in Engineering 2008, 73(6), 869–894.

[16] Ghosh, S., Bai, J. and Raghavan, P. Concurrent multi-level model for damage evolution in microstructurally debonding composites. Mechanics of Materials 2007, 39(3), 241–266.

[17] Fish, J., Jiang, T. and Yuan, Z. A staggered nonlocal multiscale model for heterogeneous medium. International Journal for Numerical Methods in Engineering 2012, 91(2), 142–157.

[18] Fish, J. and Kuznetsov, S. Computational continua. International Journal for Numerical Methods in Engineering 2010, 84, 774–802.

[19] Fish, J., Filonova, V. and Fafalis, D. Computational continua revisited. International Journal for Numerical Methods in Engineering 2013 (in press).

[20] Fish, J., Filonova, V. and Yuan, Z. Reduced order computational continua. Computer Methods in Applied Mechanics and Engineering 2012, 221–222, 104–116.

[21] Belytschko, T., Fish, J. and Engelmann, B.E. A finite element with embedded localization zones. Computer Methods in Applied Mechanics and Engineering 1988, 70, 59–89.

[22] Simo, J.C., Oliver, J. and Armero, F. An analysis of strong discontinuities induced by strain-softening in rate-independent inelastic solids. Computational Mechanics 1993, 12, 277–296.

[23] Hughes, T.J.R. Multiscale phenomena: Green's functions, the Dirichlet to Neumann formulation, subgrid scale models, bubbles and the origin of stabilized methods. Computer Methods in Applied Mechanics and Engineering 1995, 127, 387–401.

[24] Fish, J.The s-version of the finite element method. Computers and Structures 1992, 43(3), 539–547.

[25] Fish, J. Hierarchical modeling of discontinuous fields. Communications in Applied Numerical Methods 1992, 8, 443–453.

[26] Fish and, J. Wagiman, A. Multiscale finite element method for heterogeneous medium. Computational Mechanics 1993, 12, 1–17.

[27] Fish, J. and Belsky, V. Multigrid method for a periodic heterogeneous medium. Part I: Convergence studies for one-dimensional case. Computer Methods in Applied Mechanics and Engineering 1995, 126, 1–16.

[28] Fish, J. and Chen, W. Discrete-to-continuum bridging based on multigrid principles. Computer Methods in Applied Mechanics and Engineering 2004, 193, 1693–1711.

[29] Belytschko, T. and Black, T. Elastic crack growth in finite elements with minimal remeshing. International Journal for Numerical Methods in Engineering 1999, 45(5), 601–620.

[30] Belytschko, T., Moës, N., Usui, S. and Parimi, C. Arbitrary discontinuities in finite element. International Journal for Numerical Methods in Engineering 2001, 50, 993–1013.

[31] Armstrong, M.A. Basic Topology, revised edition. Springer-Verlag, 1997.

[32] Hughes, T.J.R. The Finite Element Method. Dover Publications, Inc., 2000.

[33] Fish, J. and Yuan, Z. Multiscale enrichment based on the partition of unity. International Journal for Numerical Methods in Engineering 2005, 62(10), 1341–1359.

[34] Belytschko, T., Liu, W.K. and Moran, B. Nonlinear Finite Elements for Continua and Structures. John Wiley & Sons, Ltd, 2000.

[35] Wu, W., Yuan, Z., Fish, J. and Aitharaju, V. On the canonical structure of the eigendeformation-based reduced order homogenization. International Journal of Multiscale Computational Engineering 2010, 8(6), 615–629.

[36] Zienkiewicz, O.C., Taylor, R.L. and Zhu, J.Z. The Finite Element Method: Its Basis and Fundamentals, 6th edition. Elsevier, 2005.

6

Multiscale Design Software

6.1 Introduction

Multiscale science and engineering is a relatively new discipline, but its impact is being felt not only in academia but also in industry. This impact is evidenced by a growing number of commercial multiscale software codes, such as Helius [1] from Firehole, MAC/GMC [2] from NASA, CZone [3] from Engenuity, DIGIMAT [4] from eXstream Engineering, Genoa [5] from AlfaSTAR, and MDS [6] from a newcomer, Multiscale Design Systems. The focus of this chapter is on MDS code not only because its theoretical framework is detailed in this book, but also because MDS-Lite, a student (lite) version of multiscale software, is included with this book.

MDS [6] is a complete environment for the analysis and design of structural components made of composites (CMCs, PMCs, or MMCs), polycrystals, concrete, soil, or any other material system involving microstructure. The MDS framework explicitly accounts for fine-scale details in component design. It is based on the reduced order scale-separation-free unit cell solution (Chapter 4 and Chapter 5), and is fully integrated with a macro solver of choice (commercial or in-house) and an optimization engine for model characterization (or parameter identification) and validation. The characteristic material length scale in MDS is identified based on user-specified experimental data at a coupon level, which ultimately provides mesh insensitive results at a component level. MDS-C is a feature module of MDS aimed at linking continuum scales; it is equipped with an extensible library of parametric unit cell models generated automatically using MDS built-in CAD and meshing tools. The current library of parametric unit cell models includes fibrous, particulate, woven, fabric, and several random inclusion microstructures. The parametric library of unit cell models eliminates the overhead involved in generating complex unit cells and linking them to the coarse-scale finite element engine of choice. In addition to the parametric library, unit cell models (CAD and/or meshes) can be imported or generated entirely in MDS. MDS is equipped with forward and

Practical Multiscaling, First Edition. Jacob Fish.

inverse stochastic multiscale capabilities that reverse engineer experimental uncertainties into unit cell geometry and constitutive model uncertainties and consequently predict component level uncertainties, such as mean, standard deviation, and A- and B-basis design allowables (MDS-UQ). MDS also provides the microstructural optimization (MDS-OP), multiscale-multiphysics (MDS-MP), and multiscale fatigue (MDS-FT) capabilities outlined in Chapter 4.

The focus of this chapter is on the student (lite) version of MDS, hereafter referred to as MDS-Lite. MDS-Lite is a two-scale solver integrated with an optimization engine for material model parameter identification. The built-in macro solver (MDS-Macro) in MDS-Lite is a 3D large deformation solver with an extensive library of nonlinear single-scale material models. MDS-Lite has a built-in CAD module, mesh generator, optimization engine, preprocessor, and postprocessor. The unit cell solver in MDS-Lite is based on the linear direct homogenization method and is equipped with a complete library of parametric unit cell models. The coarse-scale (macro) finite elements can be assigned either single-scale or multiscale material properties. MDS-Lite provides one-way multiscale coupling capablities only, in the sense that the unit cell problem is linear and solved prior to the coarse-scale analysis. The coarse-scale analysis can be either linear or nonlinear, where the homogenized linear properties provide the input to the local or nonlocal nonlinear phenomenological anisotropic material model [7].

Advanced multiscale capabilities are offered by MDS-C, including an extensive library of nonlinear microscale models, scale-separation-free multiscale capablities, and reduced order unit cell models fully coupled with several commercial macro solvers. The complete suite of MDS programs includes microstructure optmization (MDS-OP), fatigue life prediction (MDS-FT), multiphysics at multiple scales (MDS-MP), linking discrete and continuum scales (MDS-D), and stochastic multiscale solutions (MDS-UQ). For more details and a complete software download, visit the MDS website [6].

The nonlinear macro solver in MDS-Lite (MDS-Macro) is a 3D nonlinear multiphysics finite element engine integrated with a complete MDS-Macro graphical user interface (GUI), Gmsh geometic modeler, mesher, and pre- and postprocessor [8]. The MDS-Macro solver is integrated with a sensitivity-based optimization engine for parameter identification and a linear direct homogenization engine with a complete MDS-Micro GUI and a Gmsh-based pre- and postprocessor.

The multiscale design process utilizing MDS-Lite consists of a sequence of analyses taking place at the micro- or macroscales (as shown in Figure 6.1). The flowchart in Figure 6.1 depicts the major analysis functions with numbers indicating their sequence in the multiscale process. Assuming that experimental data are available at a coupon level, the multiscale process will consist of a sequence of 10 stages. First, the unit cell geometry and mesh are constructed. The user provides the initial (or tentative) linear microscale properties from which homogenized linear properties are computed using the direct homogenization approach outlined in Chapter 2. A constrained optimization module is then invoked to identify microscale properties that minimize the error between experimental observations and simulations at a coupon level[1] (see Chapter 4). The macroanalysis stage involves specification of macro geometry, load, initial (optional) conditions, and boundary conditions, followed by the finite element mesh generation. The homogenized linear properties computed at the microanalysis stage are inputted into the MDS-Macro.[2] If macroanalysis is linear, then Step 7 in Figure 6.1

[1] If experimental data are not available, then the homogenized properties computed from the user-defined microscale properties are saved for the macroanalysis stage.

[2] Some macro elements can be assigned to a nonlinear single-scale material model.

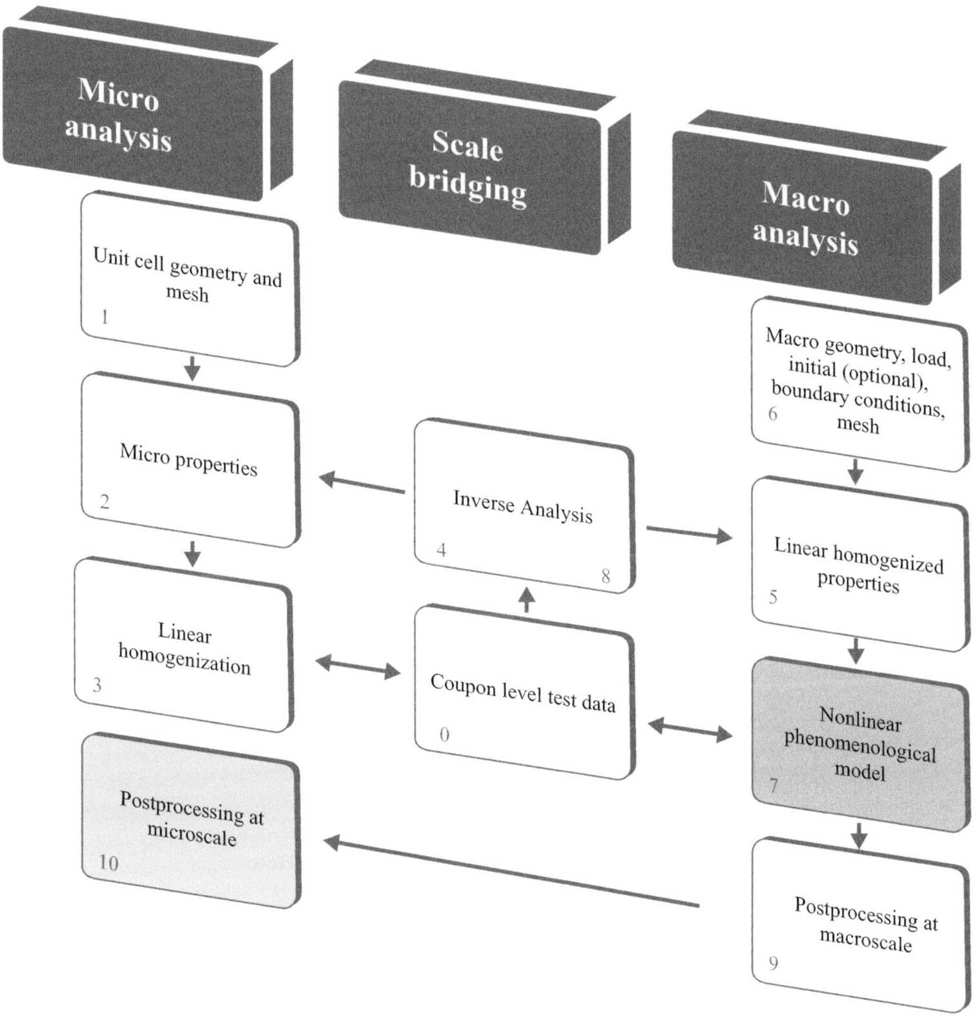

Figure 6.1 Major analysis steps and information flow in MDS-Lite

is bypassed. For the nonlinear phenomenological multiscale analysis in MDS-Lite, homogenized linear properties serve as input to the nonlinear anisotropic phenomenological material model. If experimental data in the nonlinear regime are available, the phenomenological material constants are calibrated to the user-specified experimental data (or reference solution). Finally, the solution and output control parameters are specified and the macro problem is analyzed. For nonlinear problems, only phenomenological data can be postprocessed, whereas for linear problems, various fine-scale fields at any macro quadrature point can be postprocessed (Step 10 in Figure 6.1).

A complete GUI is designed to assist the user at each analysis step by collecting input data, invoking backend programs, and reviewing the resulting data or graphs. MDS-Lite is invoked by double-clicking on the "MDS_Lite.exe" file at the root folder. The main

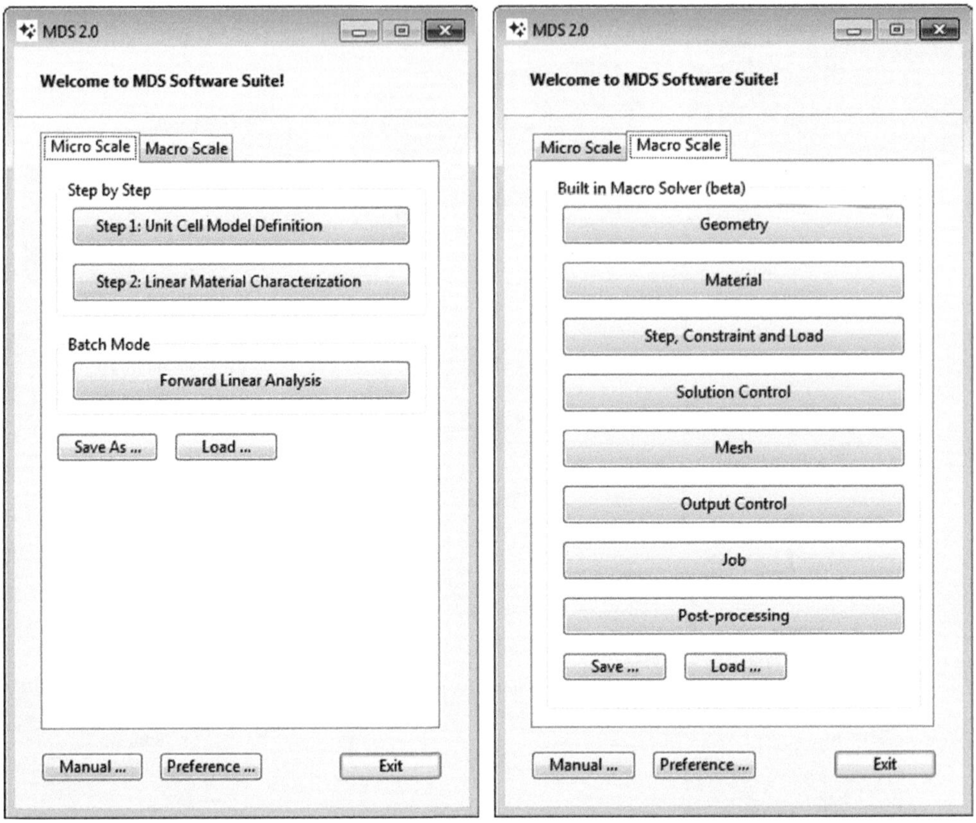

Figure 6.2 The main control window in MDS-Lite

control window pops up as shown in Figure 6.2.[3] Users can access each step through this window.[4] The microanalysis process in MDS-Lite is described in section 6.2, whereas section 6.3 focuses on macroanalysis and postprocessing at the macro- and microscales.

6.2 Microanalysis with MDS-Lite

The microanalysis module in MDS-Lite consists of two steps: (i) geometry, mesh, and attributes definition; and (ii) direct homogenization and linear material parameter characterization. If macroanalysis is linear, the unit cell (micro)analysis module is again invoked to postprocess fine-scale fields.

[3] The tutorial is based on MDS-Lite Version 2.0. For updates, visit the MDS website [6].

[4] Note that various attributes of unit cell and macro models are stored in different data files. It is important to define a *Working Folder* prior to the analysis where various unit cell and macro data files will be stored. Several examples are given in the "example" folder. The location of the *Working Folder* is arbitrary, and the default location is ".\temp\" where dot "." represents your MDS root folder. For different unit cells, the creation of different subfolders is recommended.

- In Step 1, the unit cell geometry and mesh can be defined in one of three ways: (1) using the parametric unit cell model definition; (2) importing the unit cell model; or (3) creating a unit cell model in MDS-Lite. The first and most efficient option is to utilize the MDS-Lite library of unit cells defined parametrically. Alternatively (option two), unit cell mesh data generated by a third-party preprocessor can be imported into MDS-Lite. Finally (option three), the unit cell model (geometry and mesh) can be constructed directly in MDS-Lite using the Gmsh modeling tool that is integrated into MDS-Lite. The parametric model definition option generates unit cell attributes (microphases, orientations, and their association with element sets). For the last two options, the MDS-Lite GUI is used to define the unit cell attributes. Note that Step 1 is completely independent of the material characterization steps.
- In Step 2, linear (transport or mechanical) material properties are characterized (identified) by solving either a forward homogenization problem or an inverse optimization problem. For the forward problem, the micro properties are user-defined and the corresponding macro properties are evaluated. For the inverse problem, selected macro properties are user-defined (based on either experimental data or target design values), and subsequently, unknown micro properties are characterized. MDS-Lite provides a laminate functionality by which the overall properties of the laminate, which consists of a stack of unit cell layers with different orientations, can be predicted. Micro properties can also be identified directly from the overall laminate properties by solving an inverse problem. With the completion of Step 2, a detailed stress field in the unit cell subjected to any unit overall strain can be visualized.

Remark 6.1 In addition to a step-by-step analysis, MDS-Lite GUI also provides two batch modes for running the forward analysis when Step 1 and Step 2 are combined. There is a universal "Save …" button to store the input data in a data file, which can be reloaded using the "Load …" button to import the previous input. This comes in handy when you want to reload your previous work or if multiple similar analyses are conducted with only a few input data modifications. You can save input data at any time when a step is finished and then load the previous input data when needed. You can also set up a default working environment through the "Preference" window shown in Figure 6.3.

6.2.1 *Familiarity with the GUI*

In each step, there is a single input window for collecting user input, executing backend programs, and accessing the resulting review window. Figure 6.4 depicts the input collection window in Step 1.

As depicted in Figure 6.4, the window's title indicates the step title. The window is resizable and may include several tab pages for collecting input data for different analysis options. At the bottom of the window there are four buttons, with one on the left and three on the right. The "Help …" button will access the help documentation specific to the current step. The "Run …" button is used to invoke the backend programs aimed at performing a specific analysis or task. The Step 1 backend program window is shown in Figure 6.5. The processing, output data, and potential error information are displayed in the backend program window. Press any key to close this window. The "Result …" button is used to invoke the result-viewing window. For Step 1, the unit cell mesh with all its attributes can be viewed as shown in Figure 6.6. Click on the "Close" button to close the result-viewing and input windows, leaving the main control window open.

Each step has a similar workflow.

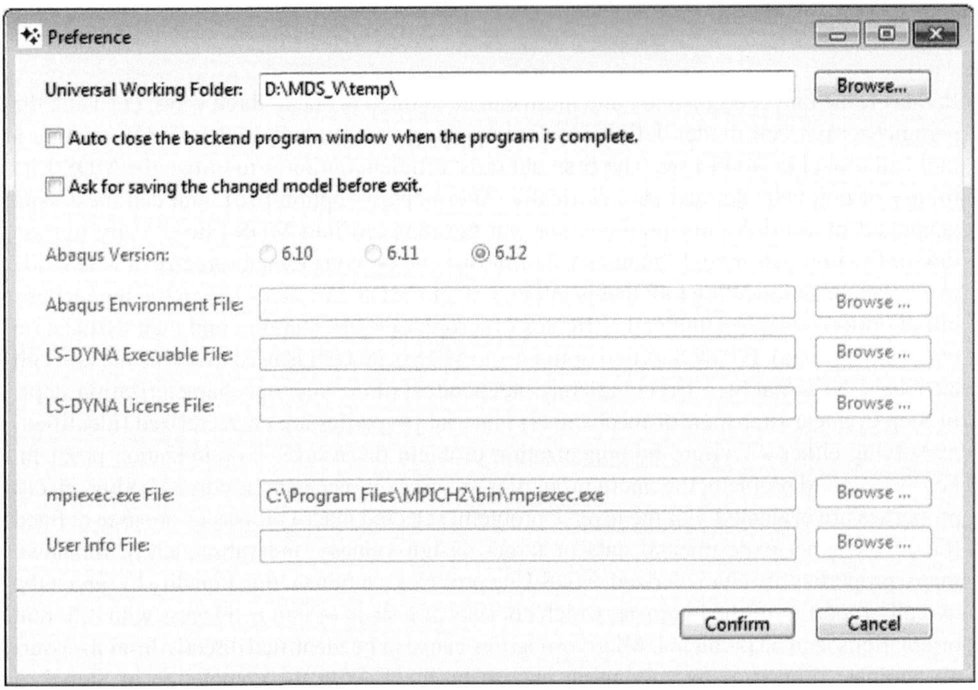

Figure 6.3 The "Preference" window

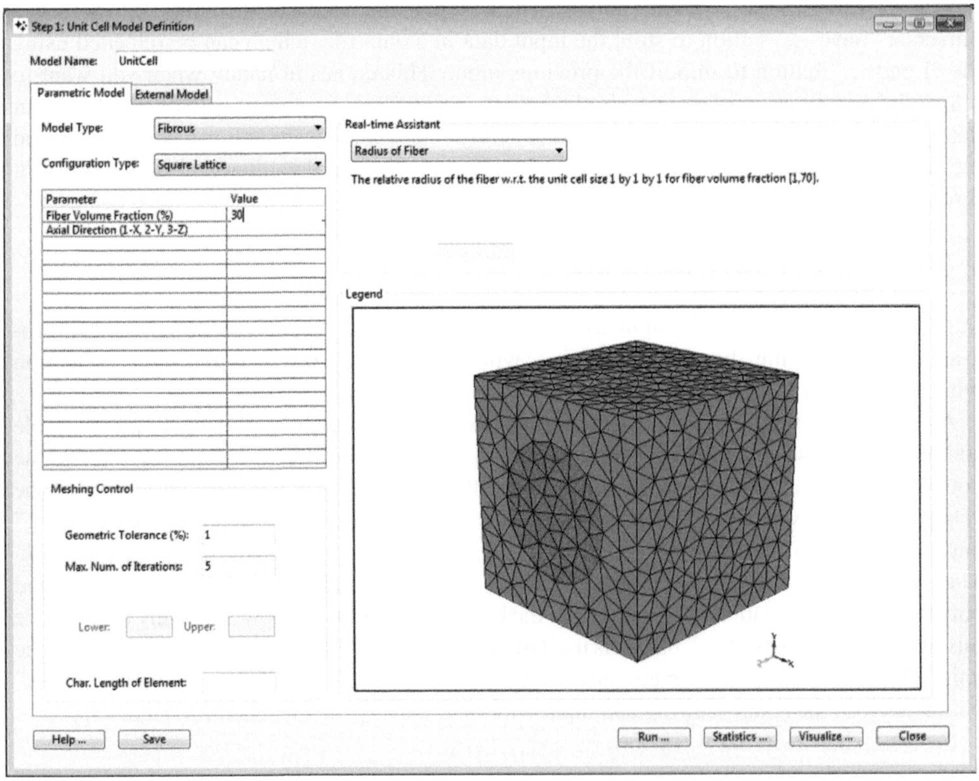

Figure 6.4 The Step 1 input window of MDS-Lite

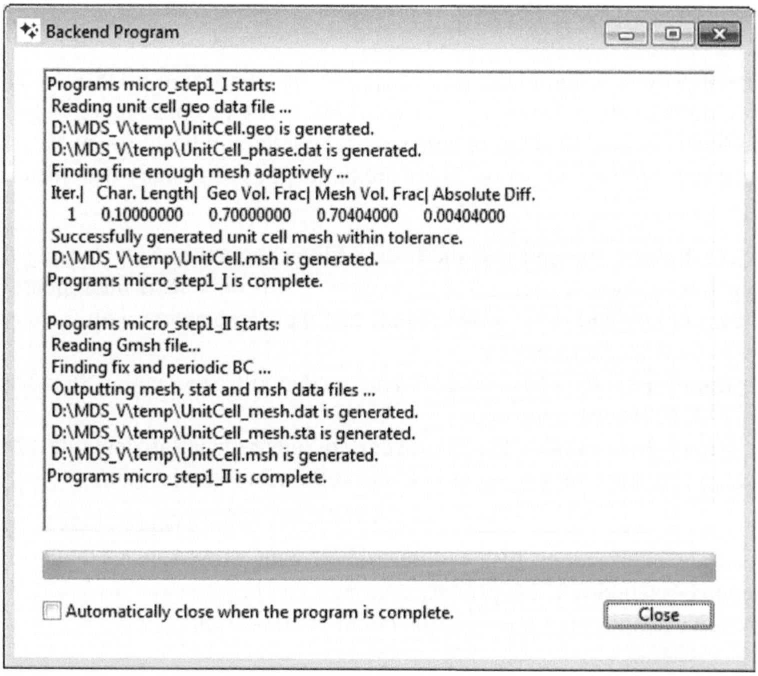

Figure 6.5 The Step 1 backend program window of MDS-Lite

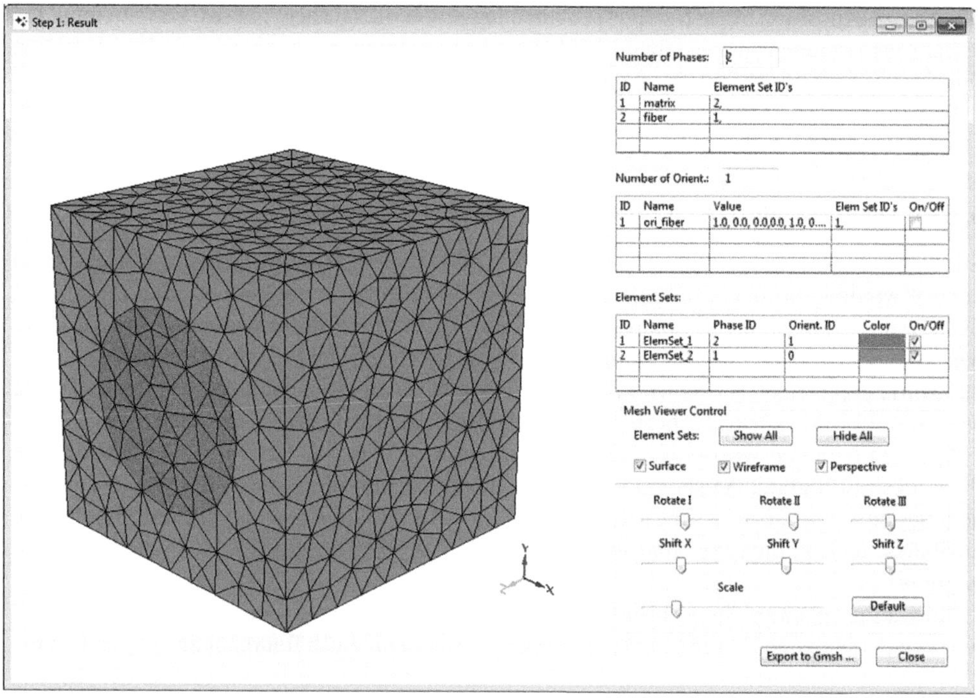

Figure 6.6 The Step 1 result-viewing window of MDS-Lite

6.2.2 Labeling Data Files

The unit cell database is stored in the individual data files (*.dat) in MDS-Lite. The specific name suffixes are defined to distinguish between different types of unit cell data. In the following, **bold** letters denote mandatory suffixes, *italic* letters indicate the names that can be changed, and # indicates the numerals. There are three types of primary data files:

1. *UCname*_**mesh.dat** – the unit cell mesh data file generated in Step 1 using the unified MDS data format, which includes the standard mesh data with microphase attributes (microphase number and name, orientations, and their association with the element sets) and periodic boundary information.
2. *UCname*_**Lmatl.dat** – the unit cell linear material data file generated in Step 2 using either the forward or the inverse approach.
3. *UCname*_**Lmatl_homo.dat** – the homogenized linear material data file generated in Step 2, which is consistent with identified unit cell linear material data.

All other files are intermediate files for interaction with MDS-Lite GUI and backend or built-in programs. Examples of the primary data files can be generated by going through the first walkthrough example, which is explained in the next section.

6.2.3 The First Walkthrough MDS-Micro Example

In this section, a complete two-step analysis of a fibrous unit cell is illustrated. Note that this example is intended to illustrate the use of the most basic options in MDS-Lite. A more comprehensive exposition of each step is provided in section 6.2.4. There are two ways to input the various parameters. You can either directly type in everything, as shown in each demonstration figure, or you can load the data using the "Load …" button.

Double-click on "MDS-Lite.exe" to start the analysis.

6.2.3.1 Step 1: Unit Cell Geometry Definition

Start Step 1 by clicking on the "Step 1: Unit Cell Geometry Definition" button in the main control window. In fact, you have already seen all the Step 1 windows for this example in Section 6.2.1. At the input window, we will use the first option—a parametric model to let MDS automatically generate mesh data and related microphase attributes based on the geometric parameters. As shown in Figure 6.4, we leave all the default settings, that is, name as "UnitCell" and type as "Fibrous." We set the fiber volume fraction as 30 (%).

We are now done with inputting unit cell data. Click on the "Run …" button to invoke the backend program. As shown in Figure 6.5, the backend program automatically generates the unit cell mesh, defines the periodic boundary conditions and related microphase attributes, and outputs the final mesh data. Press any key to close the backend program window.

Now let us examine the resulting data. Click on the "Result …" button to view the result window. As shown in Figure 6.6, we have the finite element mesh figure on the left, and on the

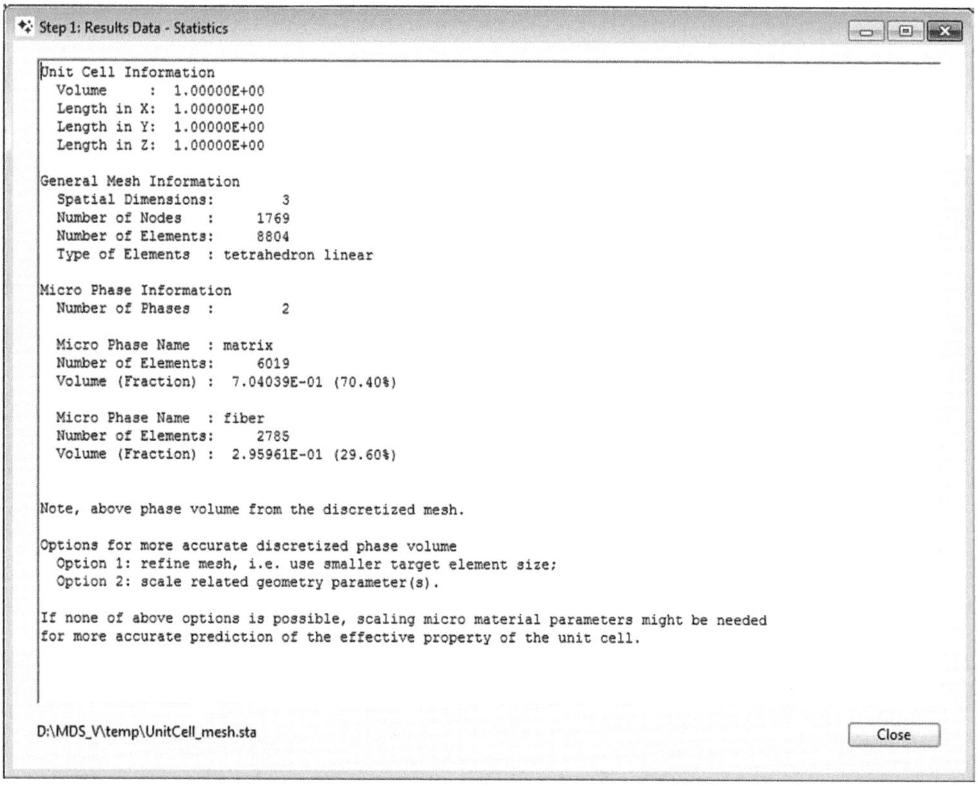

Figure 6.7 The Step 1 result statistics window of MDS-Lite

right we have the microphase information, the material orientation information, and its association with element sets. There are a number of viewing control options to view the mesh and material orientation(s).

Click on the "Statistics ..." button to see more information about the unit cell, as shown in Figure 6.7.

Close both the result window and the input window to complete Step 1.

6.2.3.2 Step 2: Linear Material Characterization

Invoke Step 2 by clicking on the "Step 2: Linear Material Characterization" button in the main control window. The Step 2 input window will pop up. Note that the UnitCell_mesh.dat file generated in Step 1 is automatically loaded with the information on the number of microphases and their names as shown in the "Micro Material Property" frame.

First, we choose the "Direct Homogenization" option where we provide the linear micro material properties while MDS computes the homogenized linear properties. Designate the material type as "Elasticity" and the symmetry type for both fiber and matrix as "Isotropic."

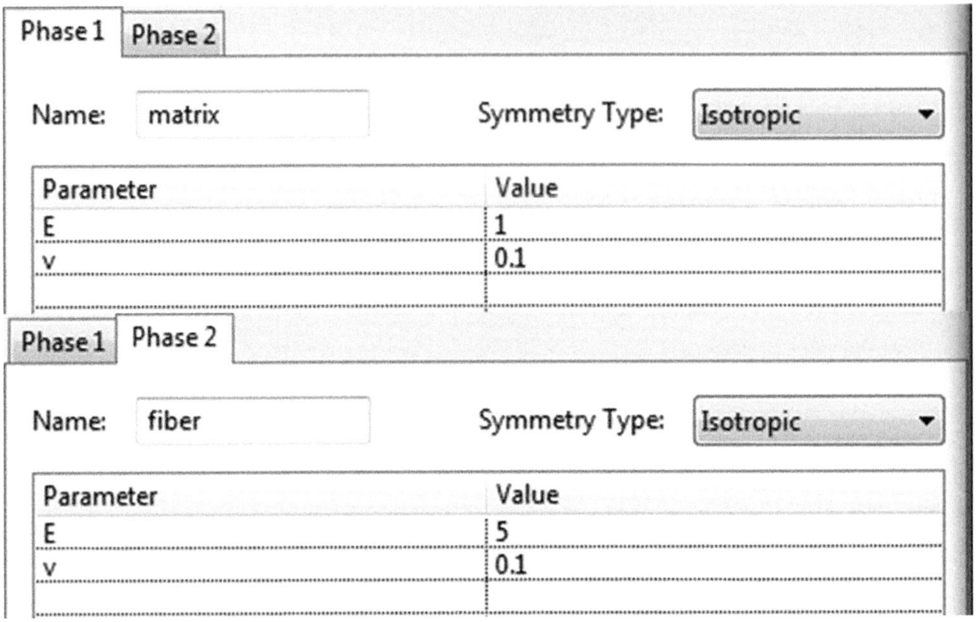

Figure 6.8 The first walkthrough MDS-Micro example: Step 2 – direct homogenization option, micro material property window

Now let us input Young's modulus $E = 1$ and Poisson's ratio $v = 0.1$ for Phase 1 (matrix) and $E = 5$ and $v = 0.1$ for Phase 2 (fiber) as shown in Figure 6.8. The list of homogenized properties to be computed is shown on the right. You have the option to view the stresses by checking the box "Output Unit Cell Stress Data for Visualization." The ready-to-run input window is shown in Figure 6.9. Note that you can save these data and then reload them if needed.

We are now done with inputting data for Step 2. Click on the "Run ..." button to invoke the backend program for direct homogenization. As shown in Figure 6.10, the backend program performs linear homogenization and creates linear micro and macro material data files. Close this window.

Now let us examine the resulting data. Click on the "Result ..." button to view the results window. The homogenized properties tab page is shown in Figure 6.11. The homogenized properties are provided in two forms: (i) engineering material constants, such as Young's modulus, shear modulus, and Poisson's ratio; and (ii) a full 6×6 homogenized elastic tensor which can be used for linear macroanalysis.

Now let us view the unit cell stress data. Click on the "View Stress Field ..." button to pop up the Gmsh postprocessing window, which shows the stress fields corresponding to six overall strain modes. The Mode XX (tension in the X direction) result is shown in Figure 6.12. Gmsh provides numerous options for viewing and postprocessing the stress data. Some useful options will be discussed in the Step 2 section.

Close the Gmsh and result windows to complete the direct homogenization analysis.

Next let us examine the inverse characterization option. In the inverse approach, the backend program solves a constrained optimization problem where some of the macro properties

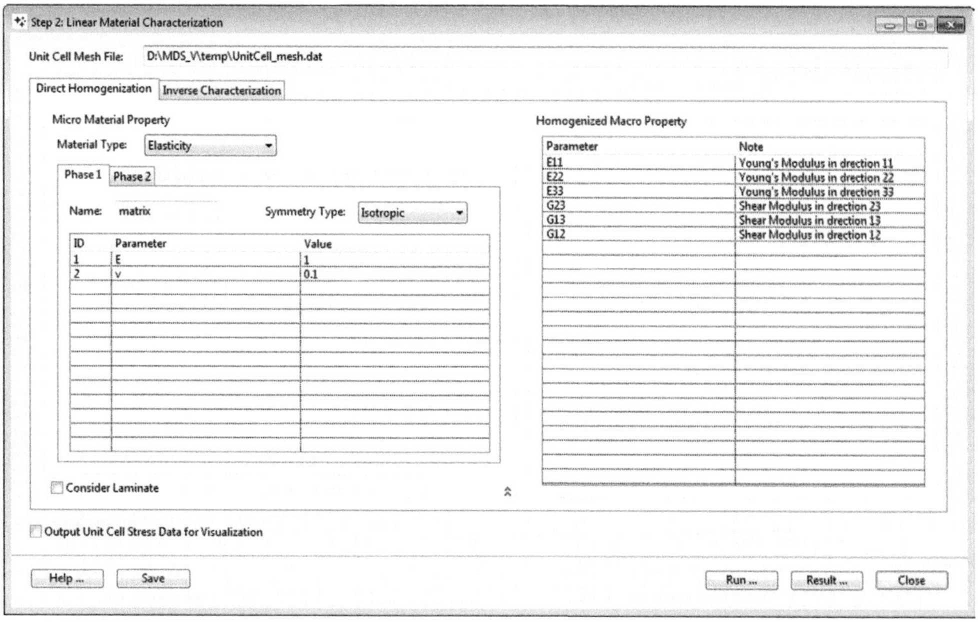

Figure 6.9 The first walkthrough MDS-Micro example: Step 2 input window for direct linear homogenization

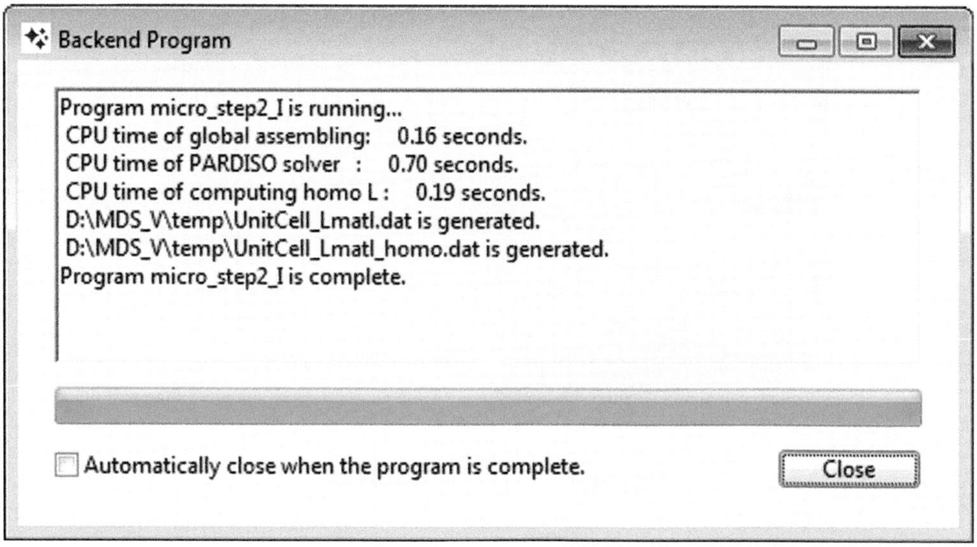

Figure 6.10 The first walkthrough MDS-Micro example: Step 2 backend program window for direct homogenization

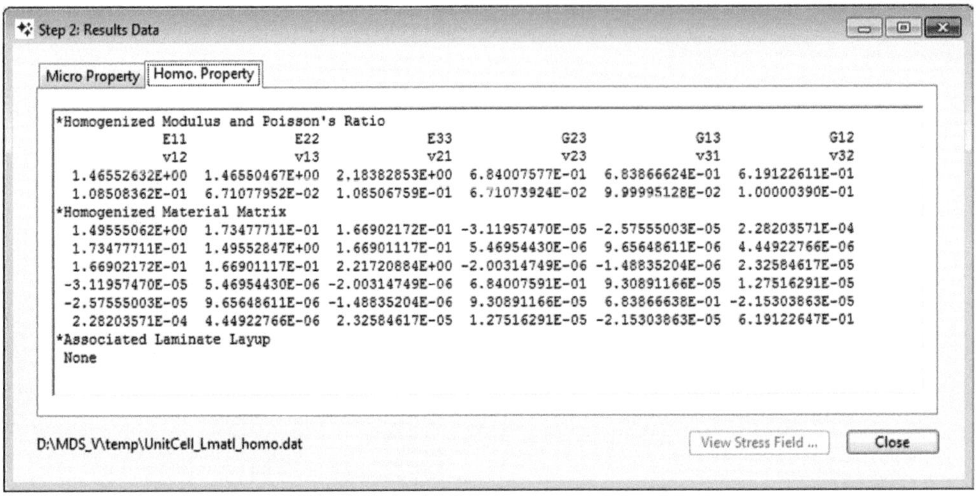

Figure 6.11 The first walkthrough MDS-Micro example: Step 2 result window – homogenized property

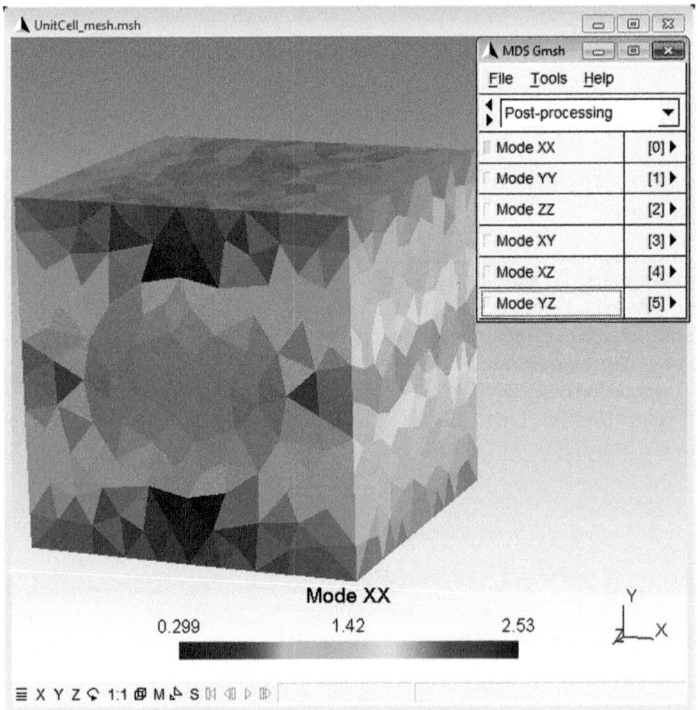

Figure 6.12 The first walkthrough MDS-Micro example: Step 2 Gmsh postprocessing – unit cell stress field

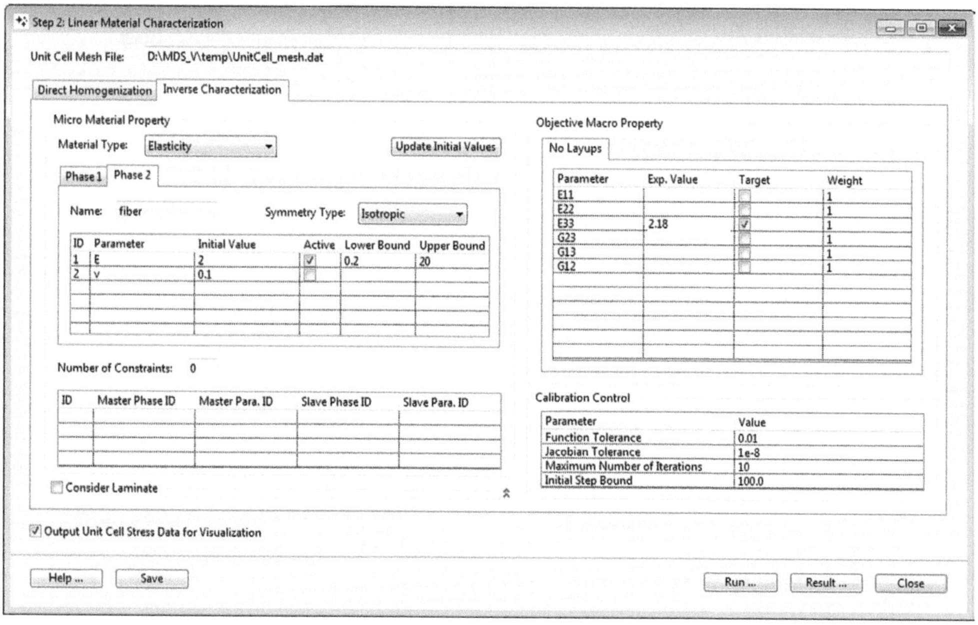

Figure 6.13 The first walkthrough MDS-Micro example: Step 2 input window – inverse characterization

are prescribed by the experiment (or are design target values) and some of the micro material properties are active parameters to be identified. The input window collects all the necessary information for this optimization problem. There are two options for inputting data into the inverse characterization problem: basic and advanced. For the basic option, most of the default parameters for the optimization problem are preset. For the advanced option, which is typically chosen by an experienced user, there is more control over the optimization process. In the present example, we will demonstrate the basic input option.

Let us set the Young's modulus of the fiber phase as a single active parameter equal to 2 and the observed (or target) macro modulus E_{33} to be 2.18 as shown in Figure 6.13. Note that the overall modulus selected coincides with the overall property obtained from the direct homogenization in the previous example, that is, the Young's modulus of the fiber phase should converge to 5 in the inverse problem.

Click on the "Run …" button to invoke the backend program for the inverse characterization. As shown in Figure 6.14, the backend program solves the constrained optimization problem and outputs linear micro and macro material data files. Since we provided an initial guess that was close to optimal in this example, the calibrated value was found in a single iteration. (The first two residual calculations are for the Jacobian calculation obtained by the finite difference method.) Close the "Backend Program" window.

Now let us examine the resulting data. Click on the "Result …" button to view the result window (Figure 6.15). Note the resulting homogenized E_{33} modulus and fiber Young's modulus values. Both of these are within 1% of the prescribed value. To get a higher accuracy, use the "advanced" option to set a tighter tolerance.

Close the result and input windows to complete the inverse characterization analysis.

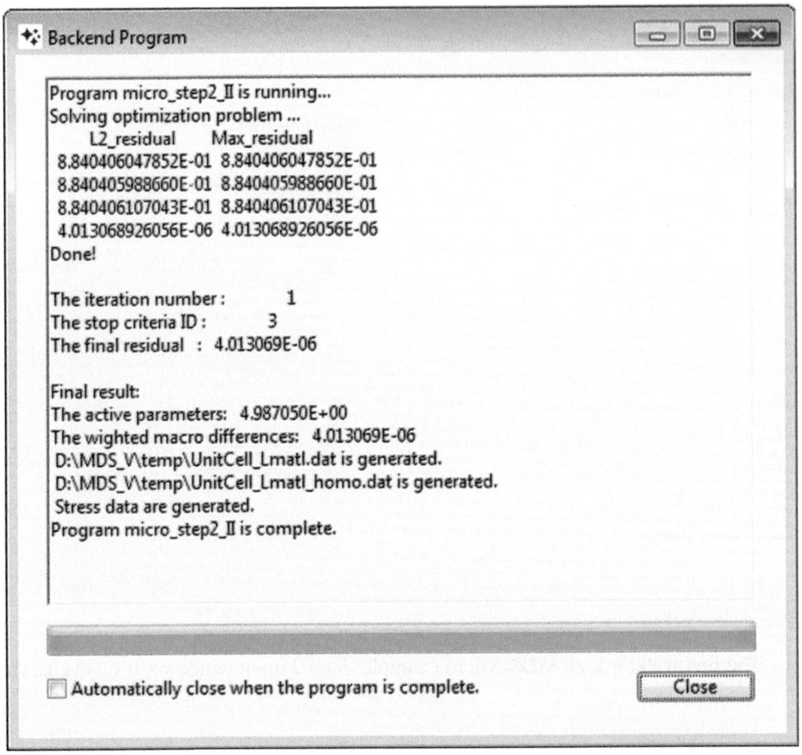
Backend Program
Program micro_step2_II is running...
Solving optimization problem ...
L2_residual Max_residual
8.840406047852E-01 8.840406047852E-01
8.840405988660E-01 8.840405988660E-01
8.840406107043E-01 8.840406107043E-01
4.013068926056E-06 4.013068926056E-06
Done!
The iteration number : 1
The stop criteria ID : 3
The final residual : 4.013069E-06
Final result:
The active parameters: 4.987050E+00
The wighted macro differences: 4.013069E-06
D:\MDS_V\temp\UnitCell_Lmatl.dat is generated.
D:\MDS_V\temp\UnitCell_Lmatl_homo.dat is generated.
Stress data are generated.
Program micro_step2_II is complete.
Automatically close when the program is complete.
Close

Figure 6.14 The first walkthrough MDS-Micro example: Step 2 backend program window – inverse characterization

```
*Homogenized Modulus and Poisson's Ratio
            E11             E22             E33             G23             G13             G12
            v21             v31             v12             v32             v13             v23
 1.46484154E+00  1.46482010E+00  2.17999526E+00  6.83399674E-01  6.83487046E-01  6.18761511E-01
 1.08440083E-01  6.71949969E-02  1.08438495E-01  6.71939862E-02  1.00000424E-01  1.00000383E-01
                                 fiber
                                 ISOTROPIC
                                 ENGINEERING
                                         2,  4.98702727E+00,  1.00000000E-01,
```

Figure 6.15 The first walkthrough MDS-Micro example: Step 2 result window (top, homogenized property; bottom, calibrated property)

6.2.4 The Second Walkthrough MDS-Micro Example

In this section, the two microanalysis steps are described in more detail.

6.2.4.1 Step 1: Geometry, Mesh, and Attributes Definition

The architecture of Step 1 is shown in Figure 6.16. A complete unit cell geometry, mesh, and model attributes definition consists of the following three steps:

1. Feature specification and mesh generation
2. Microphase attributes specification (phase number and names, material orientations, and their associations with element sets)
3. Periodic boundary information

MDS-Lite provides the user three options for generating the Step 1 information: a parametric model, an import model, and a build model. In the following, each option is described in detail.

Parametric Model

With this option selected, the user defines the geometric attribute(s) of one of the unit cells built in MDS-Lite, and all Step 1 information is automatically generated by MDS-Lite. In the first walkthrough example, you already saw one such example where a unidirectional fibrous unit cell was generated just by defining the volume fraction of the fiber.

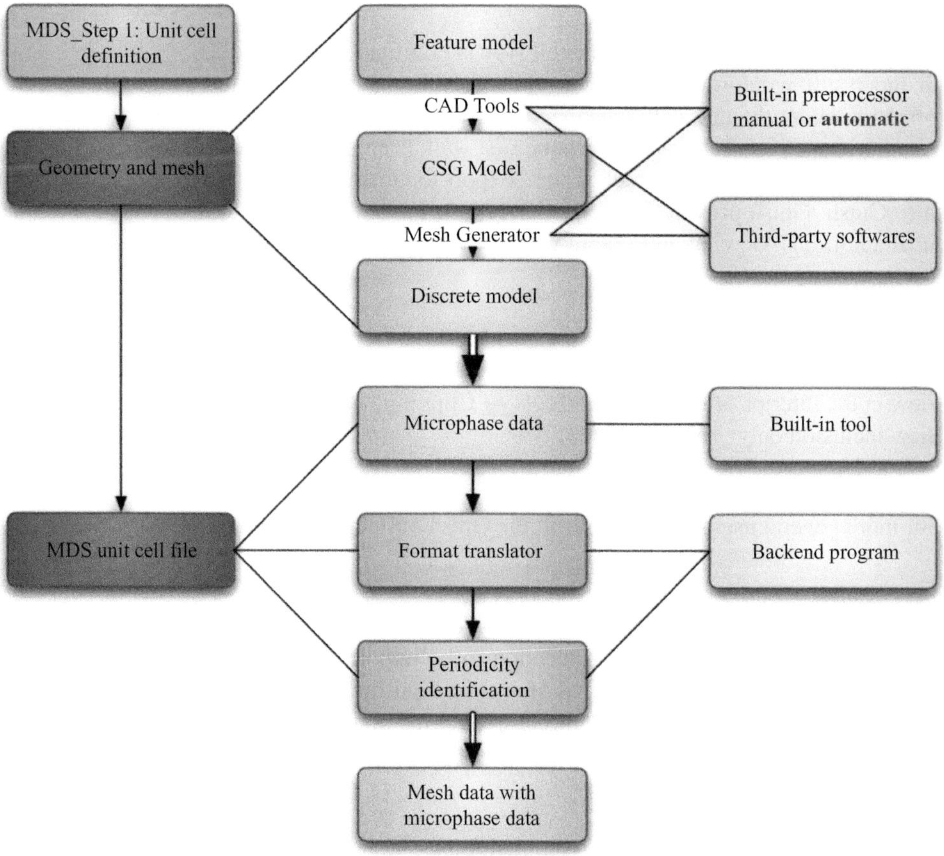

Figure 6.16 The architecture for Step 1: unit cell geometry, mesh, and model attributes definition

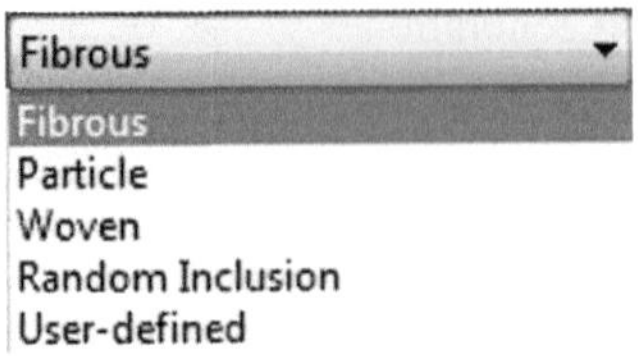

Figure 6.17 Step 1 parametric model types[5]

The following features are available to control the model parameters:

1. Define the unit cell name (default name "UnitCell").
2. Select the type of unit cell you want to generate. In version 1.4, MDS-Lite provides several types, as shown in Figure 6.17.
3. Define the corresponding geometric attribute(s).
4. Consult the "Real-time Assistant" to obtain allowable values for model attributes.
5. Specify the target element size for mesh density.

Complete information on each unit cell type is provided in section 6.2.5.

In the result window, the user can click the "Statistics …" button to get the general information about the unit cell, mesh, and each microphase. Figure 6.18 shows the statistics for the fibrous unit cell consistent with the first walkthrough example.

In the result window, you can click the "Export to Gmsh …" button to export all the mesh data to Gmsh. Gmsh provides more elaborate information about the finite element mesh, as well as additional viewing and output options. A number of useful options are summarized below:

- View element surface: Tools → Options → Mesh → General → check "Volume faces"
- Find element number: Tools → Statistics
- View in the interior of the unit cell: Tools → Clipping
- View inclusion only: Tools → Visibility
- Save as a picture: File → Save as → select save type (PNG, GIF, JPEG, etc.)

For more general use of Gmsh, consult the Gmsh online documentation [8].

Import Model

If this option is selected, the Step 1 information can be imported in the following three steps. First, identify the original mesh file location to be imported. Currently, three types of data format are supported by MDS-Lite, as shown in Figure 6.19. Secondly, specify the microphase attribute information. It can be directly specified in the MDS-Lite built-in GUI tool if it has not been defined in the imported model.[6] Lastly, complete Step 1 by clicking the "Run …"

[5] The library of parametric unit cell models is continuously expanding.

[6] For mesh data using an Abaqus input format with material, orientation, and solid section definitions, MDS-Lite supports directly loading this information in the "Step 1: Microphase Data Generation" window.

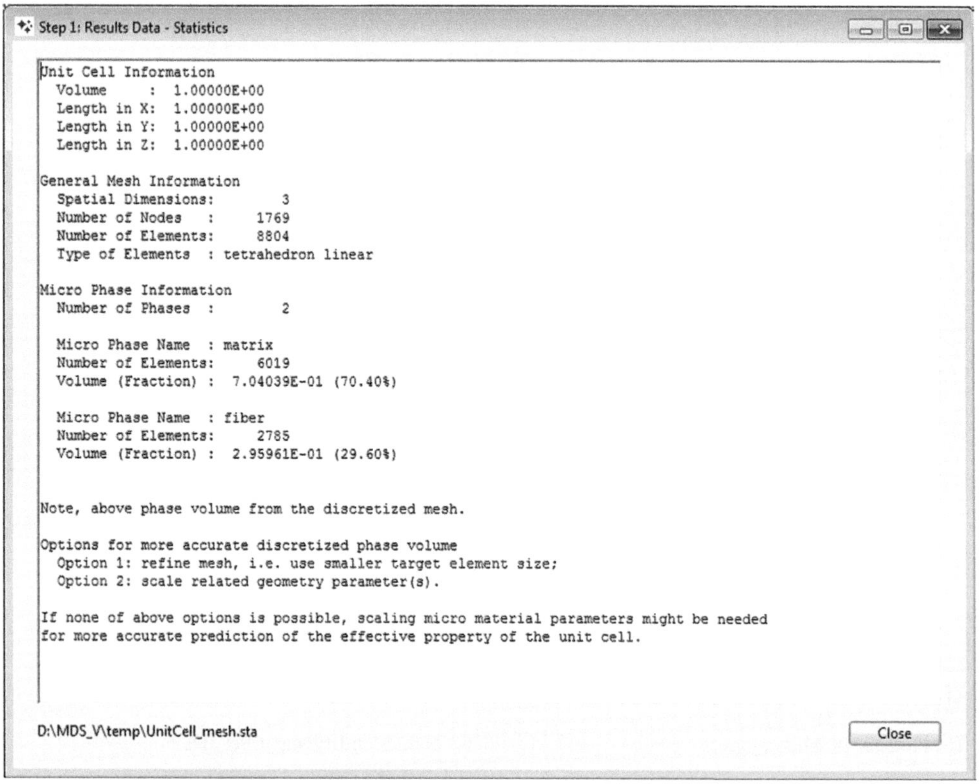

Figure 6.18 Step 1 statistics window

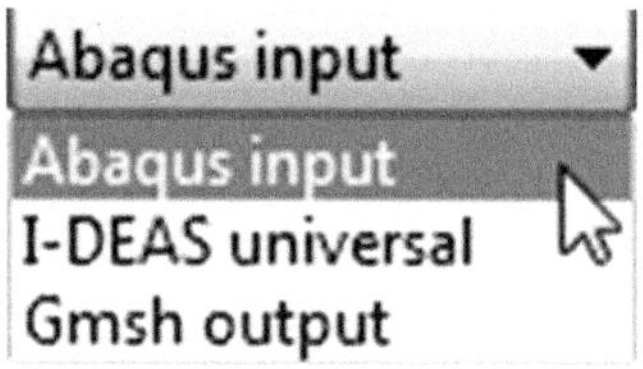

Figure 6.19 Step 1 import mesh format types

button. The backend program processes the mesh data, collects the microphase information, and identifies the periodic boundary conditions.

In the remainder of this section, we will focus on the specification of the unit cell attributes in the MDS-Lite built-in GUI. Click the "Open …" button to pop up the "Step 1: Microphase Data Generation" window as shown in Figure 6.20.

Various options for viewing and controlling the microphase attributes definition are set forth in Figure 6.20. Labels ① and ② show the unit cell mesh and global coordinates,

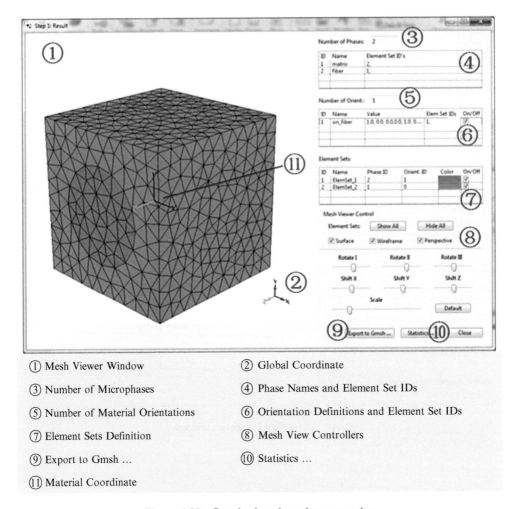

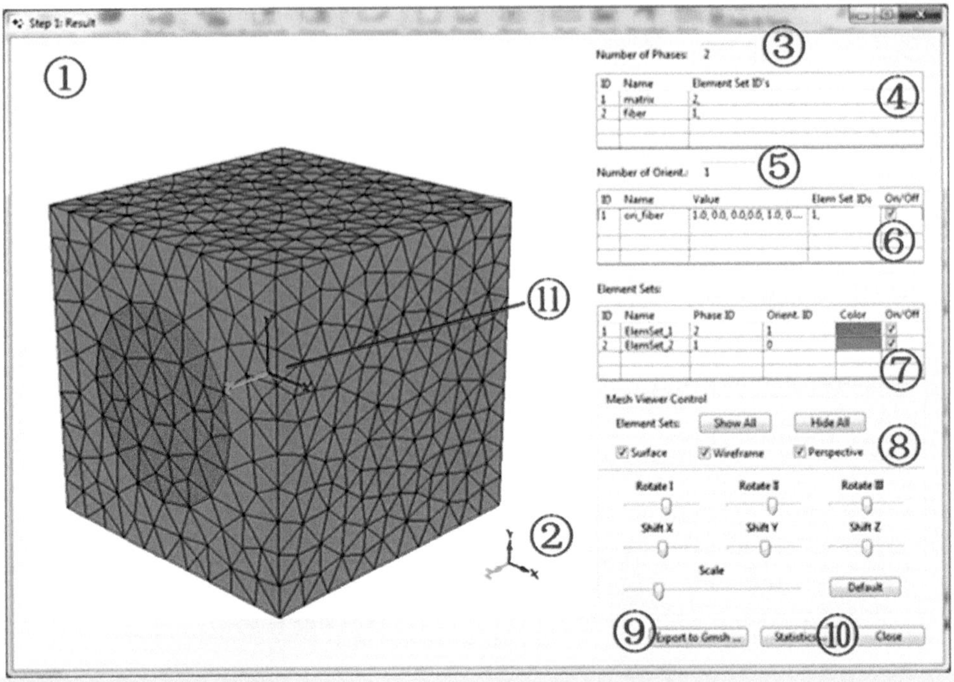

① Mesh Viewer Window ② Global Coordinate

③ Number of Microphases ④ Phase Names and Element Set IDs

⑤ Number of Material Orientations ⑥ Orientation Definitions and Element Set IDs

⑦ Element Sets Definition ⑧ Mesh View Controllers

⑨ Export to Gmsh ... ⑩ Statistics ...

⑪ Material Coordinate

Figure 6.20 Step 1 microphase data generation

respectively. The number of microphases is defined in Box ③. Table ④ assigns the element sets to different microphases. By clicking in the second and third columns of Table ④, you can input the names of phases and associate element sets with each phase. The number of material orientation directions is defined in Box ⑤, with 0 denoting that all phases are isotropic. Table ⑥ assigns element sets to different material orientations. By clicking in the second, third, fourth, and fifth columns of Table ⑥, you can change the names of the material orientations, set the (cos(**X′**,**X**)) values for the transformation matrix, associate element sets, and display or turn off the material coordinates using the "On/Off" button. Table ⑦ relates phases and material orientation information for each element set. By clicking in the second, third, fourth, fifth, and sixth columns of Table ⑦, you can associate phase and material orientation information for each element set, modify element set color, and control the visibility of each element set (using the “On/Off” button). Various controllers in ⑧ provide additional options for mesh viewing, including rotation, scaling, shifting, hiding/displaying mesh surfaces, mesh edges, and the possibility of perspective viewing of the model. The “Export to Gmsh” button ⑨ allows MDS to visualize microphase information in Gmsh. The “Statistics” button ⑩ provides information on the number of elements and nodes, as well as the discretized volume fraction of phases, which may be different from the geometric volume fraction. The coordinate system ① is a local material coordinate denoted as X′, Y′, Z′.

Note that all the attributes association information can be defined in Table ④ and Table ⑥, or alternatively, in Table ⑦. For one phase, or if material orientation is associated with multiple element sets, the first possibility is preferable. Note also the similarity between the “Step 1: Microphase Data Generation” window and the “Step 1: Result” window. The difference between the two windows is that in the result window, you can only review the inputted microphase data; you will not be able to make any changes.

Build Model

If this option is selected, the entire unit cell model, including geometry, meshing, and attribute definition, is defined in MDS. The corresponding Step 1 input window is shown in Figure 6.21. Model geometry and mesh are all defined in Gmsh, which is invoked by clicking the “Gmsh …” button. There are three methods for creating mesh data:

1. Construct both the CAD and mesh data in Gmsh. Note that Gmsh is equipped with limited CAD capabilities (Geometry module) but excellent meshing tools (Mesh module). This option should be pursued if specialized CAD software is not available.
2. Import the CAD model to Gmsh and use the Gmsh meshing tool to generate the mesh. Gmsh supports BREP, STEP, and IGES formats.

Figure 6.21 Step 1 input window build model

3. Use Gmsh to translate mesh data formats. Import the mesh data in a format that is not currently supported by MDS but is supported by Gmsh. Then export the model in the Gmsh format.

The last two options can be accessed through File → Open ⋯ → select your file type in the Gmsh environment.

6.2.4.2 Step 2: Linear Material Parameter Characterization

The overall architecture of Step 2 is shown in Figure 6.22. Three types of fields – scalar field (such as density), differential scalar field (such as thermal or concentration fields), and differential vector field (such as mechanical displacement field) – are supported. For each type of field, either a forward or an inverse approach can be used to identify linear material properties. In the forward approach, all the micro properties are user-prescribed and the corresponding overall properties are computed by solving the direct homogenization problem.

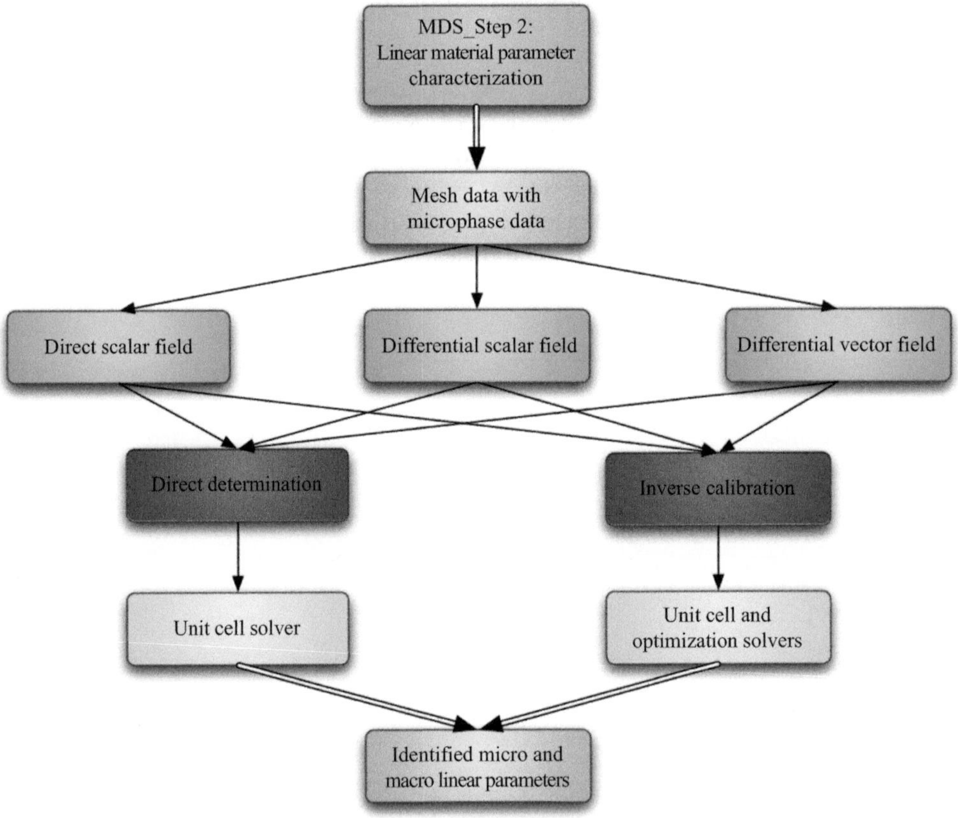

Figure 6.22 The architecture for Step 2: linear material parameter characterization

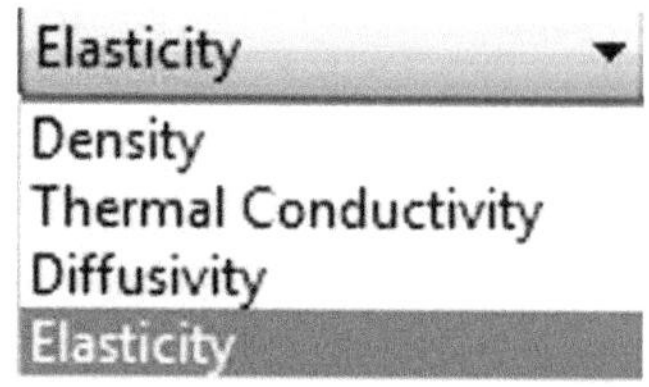

Figure 6.23 Step 2 material type options

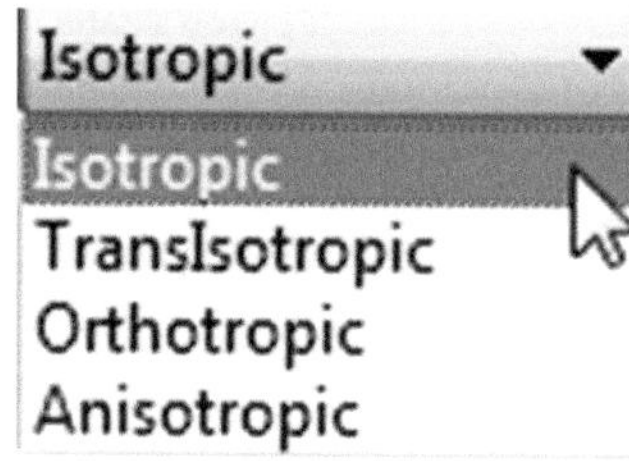

Figure 6.24 Step 2 material symmetry type options

In the inverse approach, some of the overall properties are prescribed (either from experimental data or from target values) and the unknown (active) micro properties are characterized by solving the inverse problem. The latter approach is often exercised in practice since some of the in situ properties of micro constituents are difficult to measure, while most of the experimental data are available at the macroscale.

As shown in the first walkthrough example, once you open the Step 2 input window, the mesh and microphase attributes information from the previous step is automatically loaded into Step 2.

Consider the "Micro Material Property" frame. There are several options for controlling the input micro material data for both direct homogenization and inverse characterization, including:

1. Material type. In MDS-Lite, you can perform a multiscale analysis of density calculation, steady-state heat transfer, or elasticity, as shown in Figure 6.23.Note that the micro and macro property list (homogenized for the forward approach or targeted for the inverse approach) are automatically updated according to the user's selection.
2. Material symmetry type. For the two differential fields (thermal conductivity and elasticity), four symmetry types are available, including isotropic, transversely isotropic, orthotropic, and anisotropic (Figure 6.24).

Note that this option is available for each microphase, that is, different symmetry types can be assigned to different phases. Notice that the material parameter list is automatically updated once the selection is made.

For direct homogenization, the values of the material parameters for each microphase have to be specified a priori. If you click the "Run …" button, MDS-Lite will compute the homogenized linear properties. For example, the homogenized linear property list for elasticity is shown in Figure 6.25.

Homogenized Macro Property

Parameter	Note
E11	Young's Modulus in direction 11
E22	Young's Modulus in direction 22
E33	Young's Modulus in direction 33
G23	Shear Modulus in direction 23
G13	Shear Modulus in direction 13
G12	Shear Modulus in direction 12
v21	Poisson's Ratio in direction 21
v31	Poisson's Ratio in direction 31
v12	Poisson's Ratio in direction 12
v32	Poisson's Ratio in direction 32
v13	Poisson's Ratio in direction 13
v23	Poisson's Ratio in direction 23

Figure 6.25 Step 2 homogenized macro property list for elasticity

For inverse characterization, initial values for each parameter have to be specified. The microphase properties, which are set as active parameters (Figure 6.26), are selected by the optimization program to fit the user-specified (Figure 6.27) observed (target) macro properties. The optimization control parameters are shown in Figure 28.

Note that the total number of active parameters should be less than or equal to the number of observed (target) macro parameters. MDS-Lite GUI checks this condition once the inverse characterization option is invoked.

The advanced inverse characterization option provides additional flexibility for controlling the convergence of the constrained optimization problem. Several additional input parameters can be specified, including:

- the lower and upper bounds of active micro properties, as shown in Figure 6.28;
- the weight indicating the importance (or confidence level) of the observed macro parameters, as shown in Figure 6.29 (when the inverse solver constructs the objective function, the error between the observed and computed values is scaled by the corresponding weight);
- the optimization control parameters, as shown in Figure 6.30.

MDS-Lite provides laminate functionality with support for multiple layups, as shown in Figure 6.29. In this case, the macro property for direct homogenization or inverse characterization is defined with respect to the overall laminate rather than for a single unit cell (lamina). Each layup consists of a stack of unit cells (layers) (generated in Step 1) with different material orientation angles defined with respect to the layup configuration (X, Y, or Z) and a different thickness for each layer. The rotation angle equaling zero denotes that the orientation of the unit cell is same as defined at Step 1. The positive angle denotes the counterclockwise rotation (viewed from the top of the layup direction) and the negative angle denotes the clockwise rotation. Note that the orientation angle feature can be used for a single layer of the unit cell to directly obtain the off-axis properties of the unit cell.

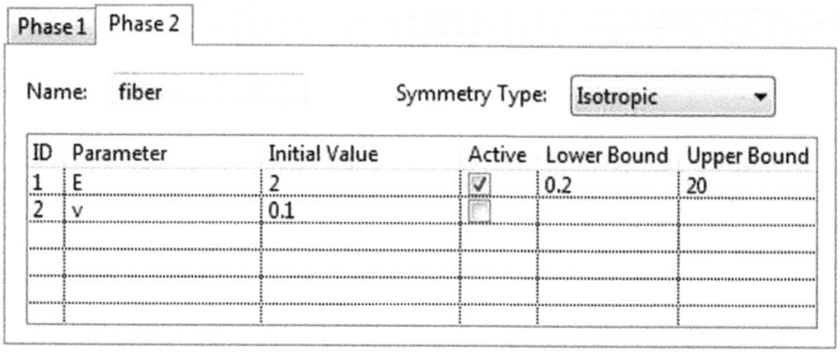

ID	Parameter	Initial Value	Active	Lower Bound	Upper Bound
1	E	2	☑	0.2	20
2	v	0.1	☐		

Figure 6.26 Step 2 active parameter setting for inverse characterization

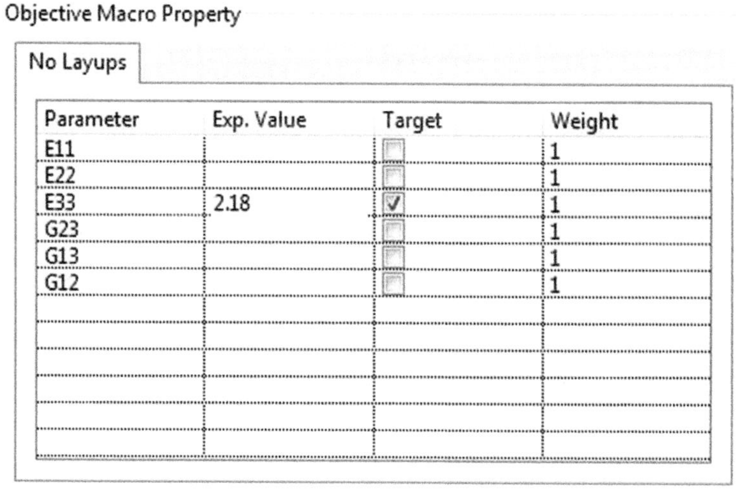

Parameter	Exp. Value	Target	Weight
E11		☐	1
E22		☐	1
E33	2.18	☑	1
G23		☐	1
G13		☐	1
G12		☐	1

Figure 6.27 Step 2 target macro parameter setting for inverse characterization

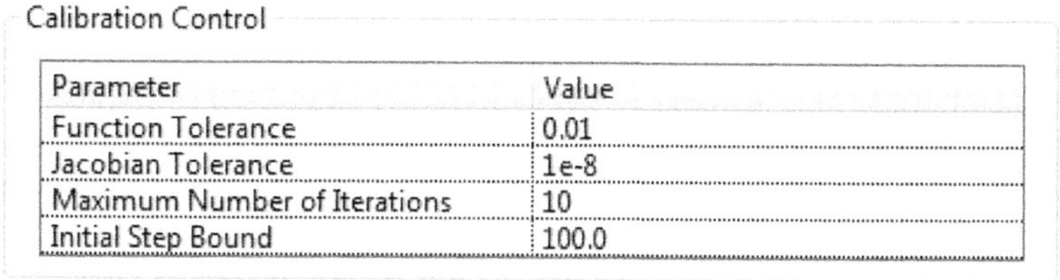

Parameter	Value
Function Tolerance	0.01
Jacobian Tolerance	1e-8
Maximum Number of Iterations	10
Initial Step Bound	100.0

Figure 6.28 Step 2 calibration control parameter setting for inverse characterization

As shown in Figure 6.29, multiple layups with different layers can be defined. Accordingly, in the result window, MDS-Lite provides the homogenized properties for single unit cells and for each layup (laminate), as shown in Figure 6.30. Note that the laminate layup information is also given in the resulting data files.

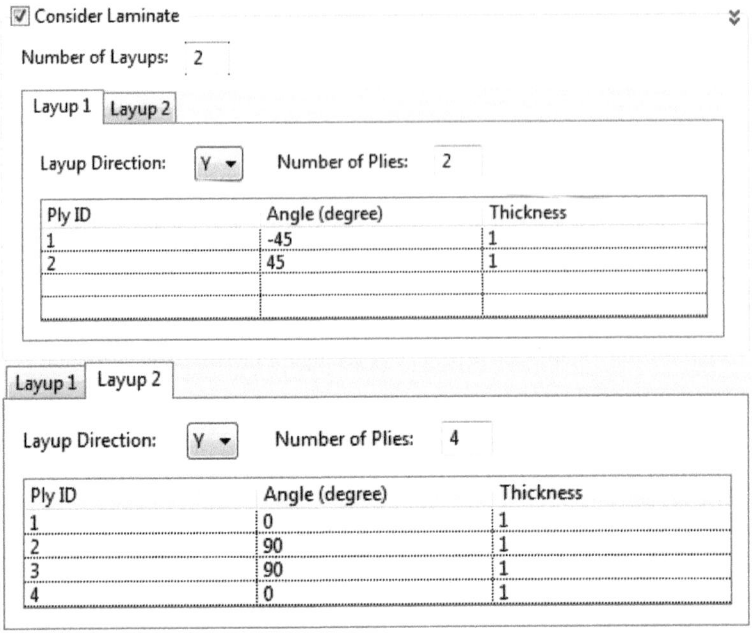

Figure 6.29 Step 2 calibration control parameter setting for inverse characterization – advanced option

For the inverse calibration process, the objective macro properties come from each layup considered, as shown in Figure 6.31. This provides the possibility of calibrating the micro properties based on macro experimental data from different layups or loading directions.

Another useful feature provided in Step 2 is the ability to output unit cell stress data to Gmsh for visualization. Once this option is checked, as shown in Figure 6.32, the stress distribution in the unit cell domain for six overall strain modes is exported to Gmsh for postprocessing. In the case of inverse characterization, the stress data are generated for the calibrated micro material properties.

Click the "Run …" button to execute the direct homogenization or inverse characterization problems. Once the forward or inverse problems have been executed by the MDS-Lite backend programs, the resulting data files *UCname*_Lmatl.dat and *UCname*_Lmatl_homo.dat are generated. The resulting micro and macro linear material parameters can be then viewed in the result window.

Note that in the case of inverse characterization, it is a good practice to check whether the homogenized modulus is close to the observed (target) value, the accuracy of which can be controlled by the user-defined convergence tolerance. If the two differ significantly, check the backend program screen shown in Figure 6.33 to see if the optimization problem indeed converged. MDS-Lite has several convergence criteria. Convergence criterion with ID=3 corresponds to the L_2 norm of the objective function.

If you have selected to output stress data for visualization in the input window, the "View Stress Field …" button in the bottom right corner of the result window will be enabled. By

Micro Property | Homo. Property | Layup 1 Property | Layup 2 Property

```
*Homogenized Modulus and Poisson's Ratio
           E11              E22              E33              G23              G13              G12
           v21              v31              v12              v32              v13              v23
  1.40093251E+00  1.40197724E+00  1.45007268E+00  6.34680311E-01  6.36229792E-01  6.26859066E-01
  1.03629269E-01  1.01687545E-01  1.03706550E-01  1.00544274E-01  1.05254414E-01  1.03993489E-01
*Homogenized Material Matrix
  1.43542744E+00  1.65918570E-01  1.68368275E-01 -5.80404109E-05 -3.14859240E-03  1.08403203E-02
  1.65918570E-01  1.43609123E+00  1.66829440E-01 -3.50670051E-03  8.34001199E-04  9.17741783E-03
  1.68368275E-01  1.66829440E-01  1.48515697E+00 -1.50291333E-03 -1.85461610E-03  3.30138753E-03
 -5.80404109E-05 -3.50670051E-03 -1.50291333E-03  6.34749312E-01  6.11998612E-03 -6.39304770E-04
 -3.14859240E-03  8.34001199E-04 -1.85461610E-03  6.11998612E-03  6.36303712E-01 -1.82565249E-03
  1.08403203E-02  9.17741783E-03  3.30138753E-03 -6.39304770E-04 -1.82565249E-03  6.26991956E-01
*Associated Laminate Layup
 None
```

Micro Property | Homo. Property | Layup 1 Property | Layup 2 Property

```
*Homogenized Modulus and Poisson's Ratio
           E11              E22              E33              G23              G13              G12
           v21              v31              v12              v32              v13              v23
  1.41363217E+00  1.40204345E+00  1.41374518E+00  6.30852658E-01  6.45949635E-01  6.30788770E-01
  1.02944248E-01  1.10913125E-01  1.02100328E-01  1.02151690E-01  1.10921992E-01  1.03004269E-01
*Homogenized Material Matrix
  1.45063395E+00  1.66374005E-01  1.78026527E-01  7.39050752E-04  6.46988084E-04  9.32733253E-03
  1.66374005E-01  1.43609123E+00  1.66374005E-01 -2.47961175E-03  2.23469715E-11  6.48941447E-03
  1.78026527E-01  1.66374005E-01  1.45063395E+00 -1.84281175E-03 -6.46988218E-04  6.72365126E-04
  7.39050752E-04 -2.47961175E-03 -1.84281175E-03  6.30870634E-01  2.66541533E-03 -1.71301306E-11
  6.46988084E-04  2.23469715E-11 -6.46988218E-04  2.66541533E-03  6.45961964E-01 -5.10839749E-04
  9.32733253E-03  6.48941447E-03  6.72365126E-04 -1.71301306E-11 -5.10839749E-04  6.30870634E-01
*Associated Laminate Layup
 Layup Direction:        2
 Number of Plies:        2
  Rotation Angle,      Thickness
          -45.00, 1.00000000E+00
           45.00, 1.00000000E+00
```

Micro Property | Homo. Property | Layup 1 Property | Layup 2 Property

```
*Homogenized Modulus and Poisson's Ratio
           E11              E22              E33              G23              G13              G12
           v21              v31              v12              v32              v13              v23
  1.42556261E+00  1.40204345E+00  1.42554445E+00  6.30814748E-01  6.36270740E-01  6.30826676E-01
  1.03850546E-01  1.03456294E-01  1.02137203E-01  1.02114816E-01  1.03454976E-01  1.03826461E-01
*Homogenized Material Matrix
  1.46029220E+00  1.66374005E-01  1.68368275E-01 -1.67971394E-03 -6.46987559E-04  4.66870370E-03
  1.66374005E-01  1.43609123E+00  1.66374005E-01 -6.34205922E-03  1.22033285E-11  2.83535878E-03
  1.68368275E-01  1.66374005E-01  1.46029220E+00 -6.17161688E-03  6.46988225E-04  1.62167354E-03
 -1.67971394E-03 -6.34205922E-03 -6.17161688E-03  6.30870634E-01  2.14716684E-03  1.03928797E-10
 -6.46987559E-04  1.22033285E-11  6.46988225E-04  2.14716684E-03  6.36303712E-01 -3.97281930E-03
  4.66870370E-03  2.83535878E-03  1.62167354E-03  1.03928766E-10 -3.97281930E-03  6.30870634E-01
*Associated Laminate Layup
 Layup Direction:        2
 Number of Plies:        4
  Rotation Angle,      Thickness
            0.00, 1.00000000E+00
           90.00, 1.00000000E+00
           90.00, 1.00000000E+00
            0.00, 1.00000000E+00
```

Figure 6.30 Step 2 result window for multiple layups

clicking it, you can use Gmsh to visualize stress data in the unit cell domain subjected to six overall strain modes (tensile in X – XX mode, Y – YY mode, Z – ZZ mode, shear in YZ mode, XZ mode, and XY mode). For example, the stress distribution (von Mises) in a triaxial braided unit cell subjected to the overall XX mode is shown in Figure 6.34.

Objective Macro Property

Layup 1 | **Layup 2**

Parameter	Experimental Value	Target
E11	1.5	✓
E22		☐
E33		☐
G23		☐
G13		☐
G12		☐

Objective Macro Property

Layup 1 | **Layup 2**

Parameter	Experimental Value	Target
E11	1.7	✓
E22		☐
E33		☐
G23		☐
G13		☐
G12		☐

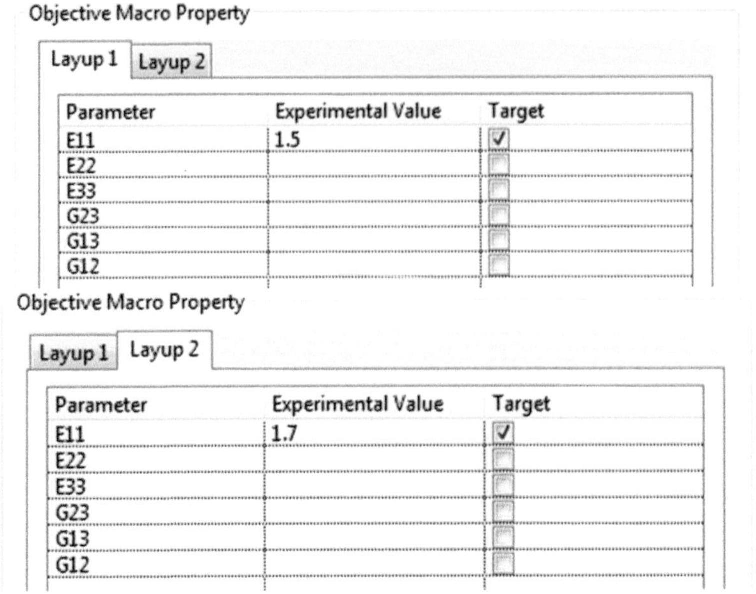

Figure 6.31   Step 2 calibration with consideration of multiple layups

☑ Output Unit Cell Stress Data for Visualization

Figure 6.32   Step 2 input window output stress option

```
❖ Backend Program                                    ⬓ ▢ ✕

Program MDS_step2_IC is running...
Solving optimization problem ...
    L2_residual      Max_residual
 8.840375333907E-01  8.840375333907E-01
 8.840375274715E-01  8.840375274715E-01
 8.840375393100E-01  8.840375393100E-01
 4.984949641962E-06  4.984949641962E-06
Done!

The iteration number :       1
The stop criteria ID :       3
The final residual  :  4.984950E-06

Final result:
The active parameters:  4.987006E+00
The wighted macro differences:  4.984950E-06
C:\MDS\develop\v1.3_beta\MDS_v1.3\temp\UnitCell_Lmatl.dat is generated.
C:\MDS\develop\v1.3_beta\MDS_v1.3\temp\UnitCell_Lmatl_homo.dat is generated.
Stress data are generated.
Program MDS_step2_IC is complete.



☐ Automatically close when the program is complete.        Close
```

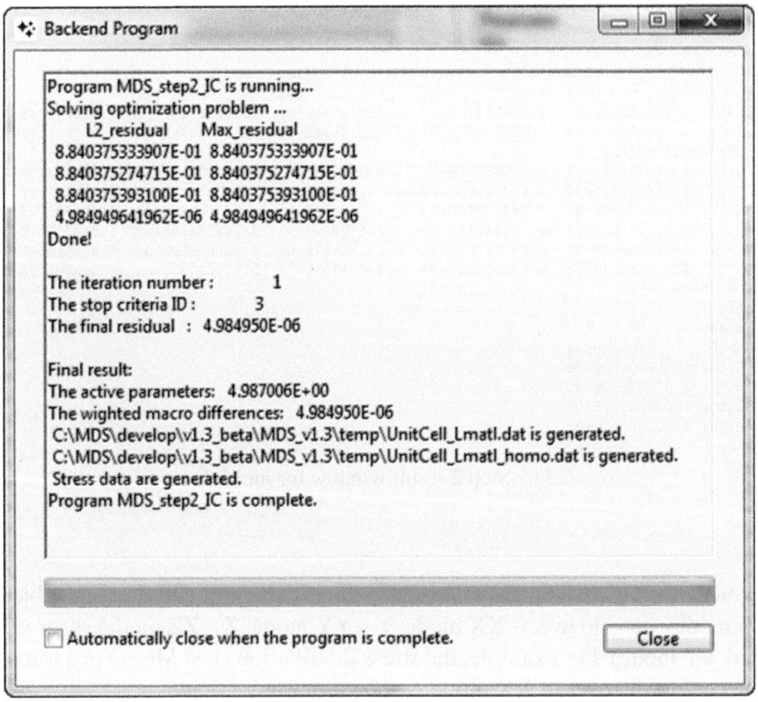

Figure 6.33   Step 2 backend program output for inverse characterization

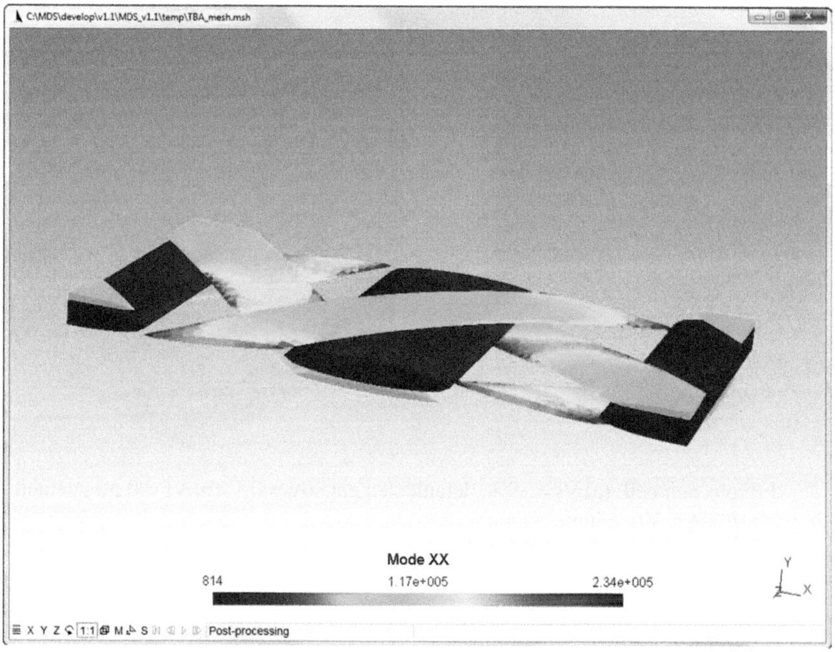

Figure 6.34 Step 2 stress distribution in the tows for a triaxial braided unit cell

Note that Gmsh has a number of useful features for data postprocessing and viewing, including:

- View inclusion(s) or matrix only: Tools → Visibility → select physical group
- View in any cross section: Tools → Clipping
- Compute minimum and maximum for current view: Tools → Options → Custom Range Mode: Min; Max
- Compute principle direction: Tools → Plugins → Eigenvectors

6.2.5 Parametric Library of Unit Cell Models

Various unit cell types are shown in Figure 6.35–6.50. The real-time assistant provides the range of possible parameters.

6.2.5.1 Unidirectional Fibrous Unit Cell

For a fibrous unit cell, the fiber volume fraction (%) is the only parameter that needs to be defined. The allowable values for the volume fraction are in the range of [0.05, 85]. For the volume fraction in the range of [0.05, 70], a unit cell with a fiber phase in the Z direction is generated; for the range of (70, 85], a unit cell in a hexagonal array in the Z direction is generated to achieve the desired volume fraction.

Figure 6.35 and Figure 6.36 illustrate various fiber volume fractions and mesh densities.

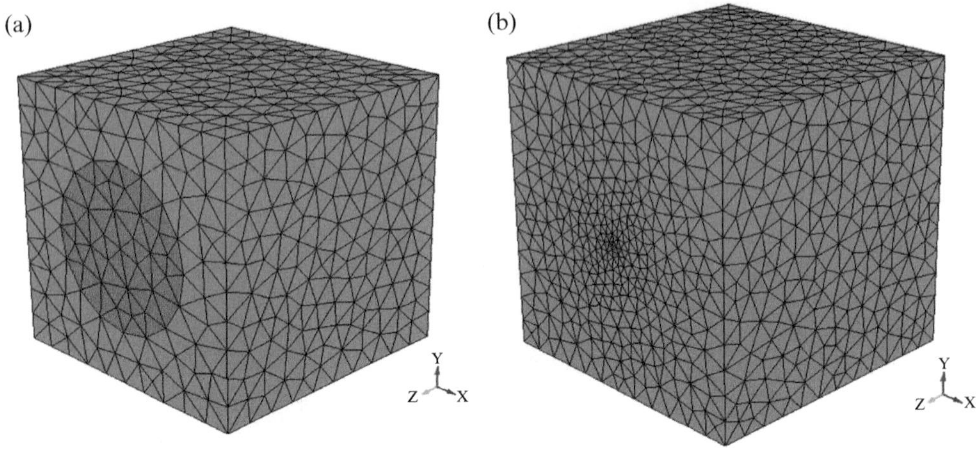

Figure 6.35 Fibrous unit cell. (a) Vf = 30%, default element size = 0.1. (b) Vf = 0.5%, default element size = 0.0797885. Vf, volume fraction

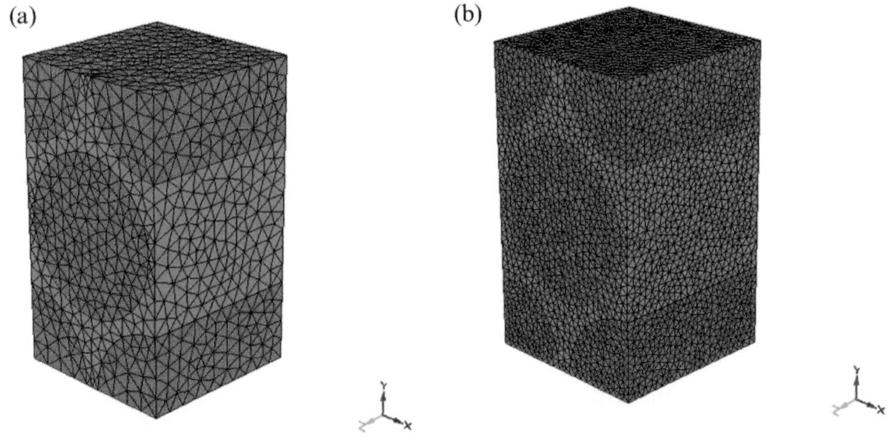

Figure 6.36 Fibrous unit cell with Vf = 75%. (a) Finite element mesh with default element size = 0.1. (b) Finite element with specified element size = 0.05

6.2.5.2 Unidirectional Fibrous Unit Cell with Interphase

For a fibrous unit cell with interphase, the two parameters that need to be defined are the fiber volume fraction (%) and the ratio of the interphase thickness to the fiber radius (t/r). The allowable values for fiber plus interphase volume fraction are in the range of (0, 85]. For the range of (0, 70], a $1\times1\times1$ cubic unit cell with a fiber phase in the Z direction is generated; for the range of (70, 85], a $1\times1.732\times1$ unit cell in a hexagonal pattern of fibers in the Z direction is generated to achieve the desired volume fraction.

Figure 6.37 and Figure 6.38 illustrate various fiber volume fractions and mesh densities.

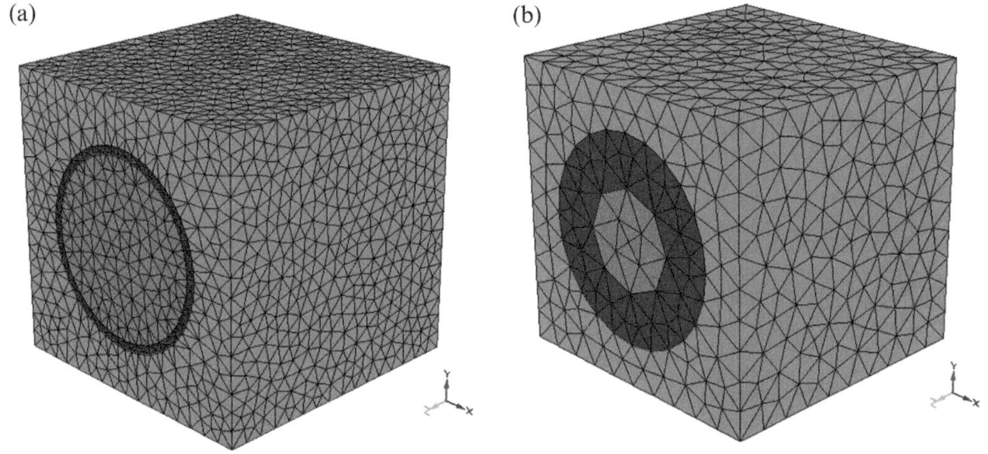

Figure 6.37 Fibrous unit cell with interphase. (a) Vf = 30%, $t/r = 0.1$, default element size = 0.0618039. (b) Vf = 10%, $t/r = 1$, default element size = 0.1

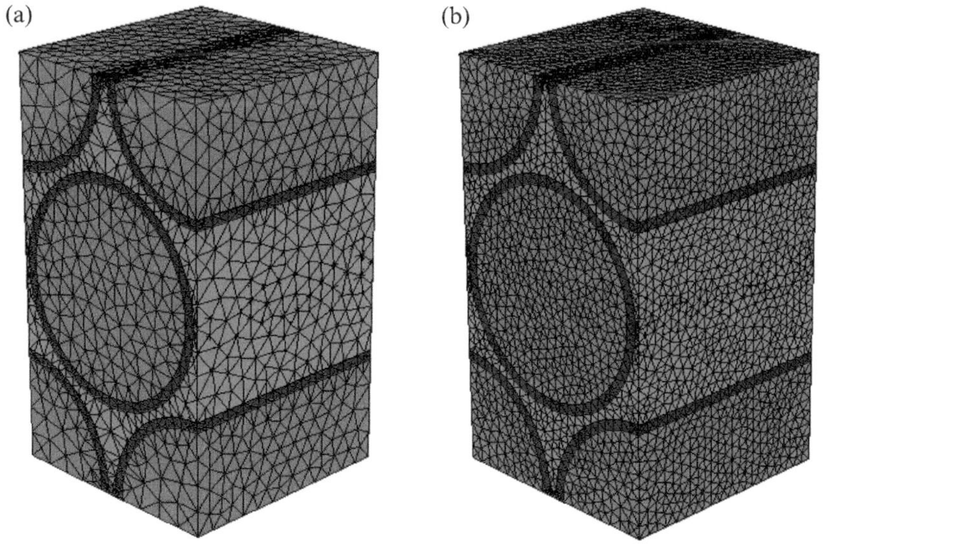

Figure 6.38 Fibrous unit cell with interphase. Vf = 70%, $t/r = 0.1$. (a) Default element size = 0.0878556. (b) Element size = 0.05

6.2.5.3 Spherical Particle Unit Cell

For a spherical inclusion unit cell, the only parameter that needs to be specified is the particle volume fraction (%). The allowable values are in the range of [0.001, 65]. For the range of [0.001, 45], a $1 \times 1 \times 1$ unit cell with a spherical particle at its center is generated; for the range of (45, 65], a $1 \times 1 \times 1$ unit cell with a body-centered cubic (BCC) pattern of particles is generated to achieve the desired volume fraction.

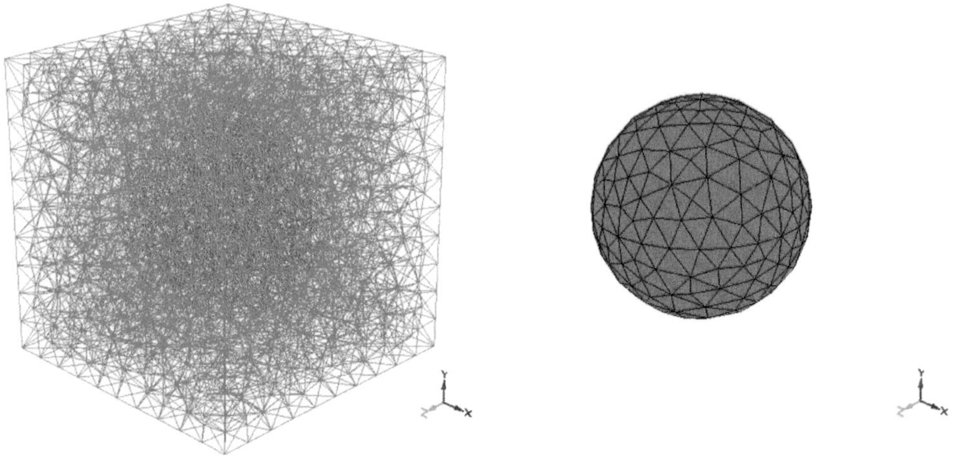

Figure 6.39 Spherical particle unit cell. Vf = 20%, default element size = 0.1

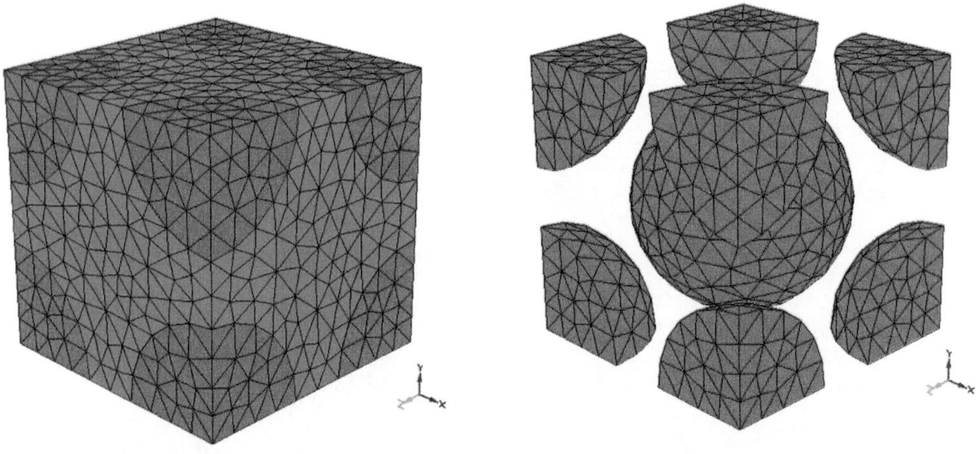

Figure 6.40 Spherical particle unit cell. Vf = 50%, default element size = 0.1

Figure 6.39, Figure 6.40, and Figure 6.41 illustrate various fiber volume fractions and mesh densities for a spherical particle unit cell.

6.2.5.4 Spherical Particle Unit Cell with Interphase

For a spherical inclusion unit cell with interphase, the two parameters that need to be defined are the particle volume fraction (%) and the ratio of the interphase thickness to the particle radius. The allowable values for the particle plus interphase volume fraction are in the range of [0.001, 65]. For the range of [0.001, 45], a $1 \times 1 \times 1$ cubic unit cell with a particle at its center is generated; for the range of (45, 65], a $1 \times 1 \times 1$ unit cell in a BCC array of particles is generated to achieve the desired volume fraction.

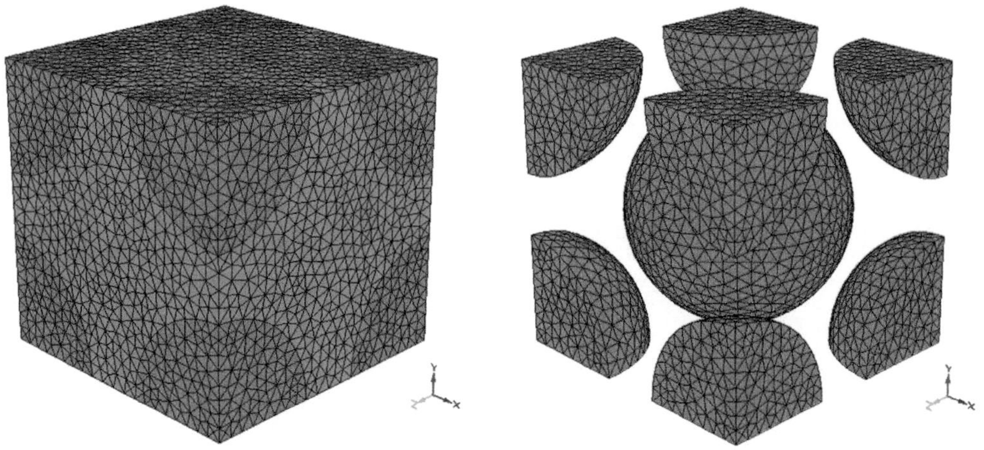

Figure 6.41 Spherical particle unit cell. Vf = 50%, element size = 0.05

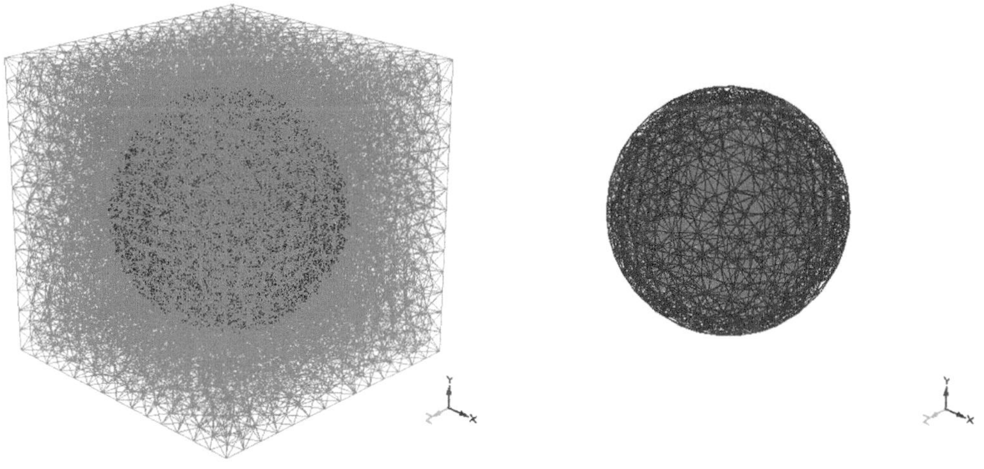

Figure 6.42 Spherical particle unit cell with interphase. Vf = 20%, $t/r = 0.1$, default element size selected by MDS-Lite

Figure 6.42 and Figure 6.43 illustrate various fiber volume fractions and mesh densities for a spherical particle unit cell with interphase.

6.2.5.5 Plain Weave Unit Cell

For a plain weave unit cell, five parameters have to be defined. The cross section of tows is assumed to be an ellipse. The weave plane is in the X and Z directions. You can define different tow spacing values in the two in-plane directions. The input parameters are as

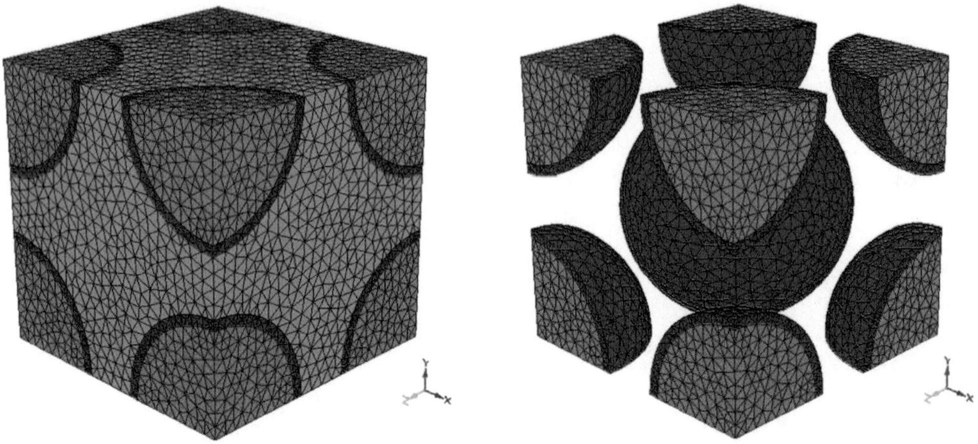

Figure 6.43 Spherical particle unit cell with interphase. Vf = 40%, $t/r = 0.1$, user-specified element size = 0.05

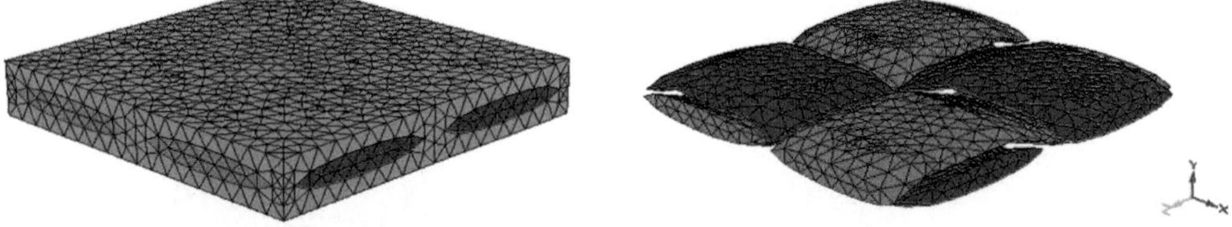

Figure 6.44 Plain weave unit cell. Parameter list {6,1,14,14,60}, default element size = 1.536131

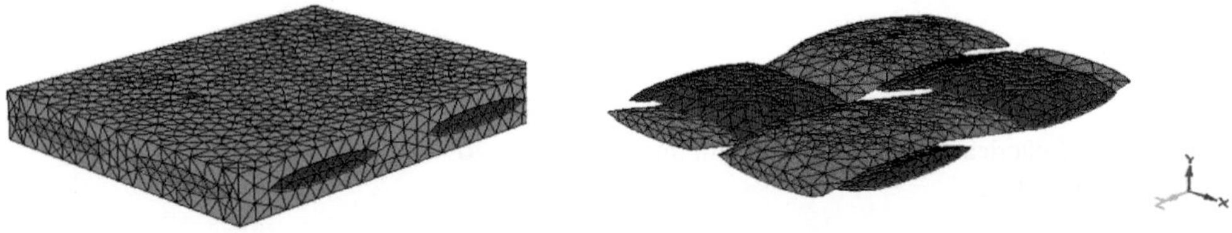

Figure 6.45 Plain weave unit cell. Parameter list {6,1,14,18,50}, default element size = 1.701622

follows: major radius of a tow; minor radius of a tow; spacing in the X direction, that is, the distance between two adjacent tows oriented in the Z direction; spacing in the Z direction, that is, the distance between two adjacent tows oriented in the X direction; and the tow volume fraction (%).

Figure 6.44 and Figure 6.45 illustrate various fiber volume fractions and mesh densities for a plain weave unit cell.

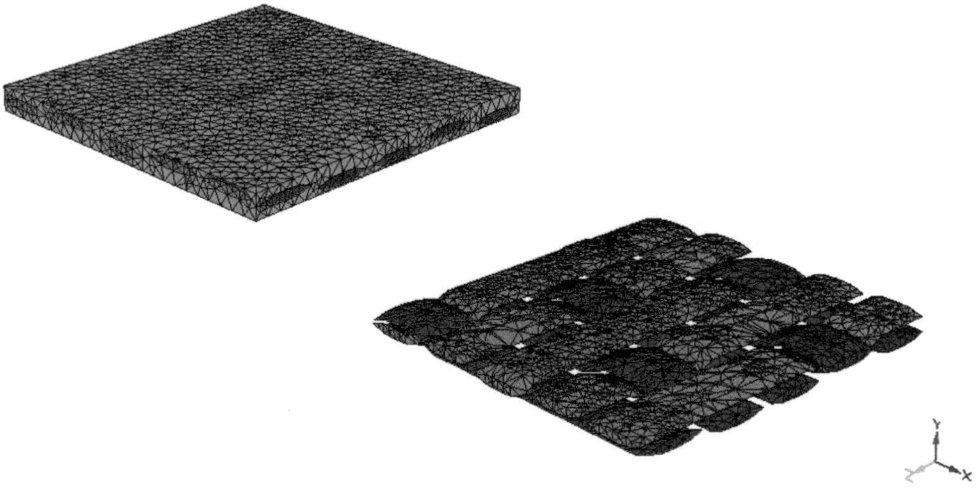

Figure 6.46 Five harness weave unit cell. Parameter list {6,1,14,14,55}, default element size = 2.895675

6.2.5.6 Five Harness Weave Unit Cell

For a five harness weave unit cell, five parameters need to be defined. As in the plain weave, the cross section of tows is assumed to be an ellipse. The in-plane is in the X and Z axes. You can define different spacing values in two in-plane directions. The input parameters are as follows: major radius of the ellipse, minor radius of the ellipse, spacing in the X direction, spacing in the Z direction, and the tow volume fraction (%).

Figure 6.46 and Figure 6.47 illustrate various fiber volume fractions and mesh densities for a five harness weave unit cell.

6.2.5.7 Random Chopped Fiber Unit Cell in Two Dimensions

For a random chopped fiber unit cell, six parameters need to be defined. The cross section of fibers is again assumed to be an ellipse. The in-plane is defined in the X and Z axes. You can define different unit cell sizes (length and width) in two in-plane directions. The six parameters are as follows: fiber volume fraction (%), major radius of the ellipse, minor radius of the ellipse, fiber length projection to the X–Z plane, unit cell length in the X direction, and unit cell width in the Z direction.

Figure 6.48 depicts a random chopped fiber unit cell in two dimensions.

6.2.5.8 Random Chopped Fiber Unit Cell in Three Dimensions

For the random chopped fiber unit cell in three dimensions, fourteen parameters should be defined. The cross section of the fiber is assumed to be an ellipse. Fibers are originally generated in the global coordinate system with their cross section lying in-plane (X–Z) and their

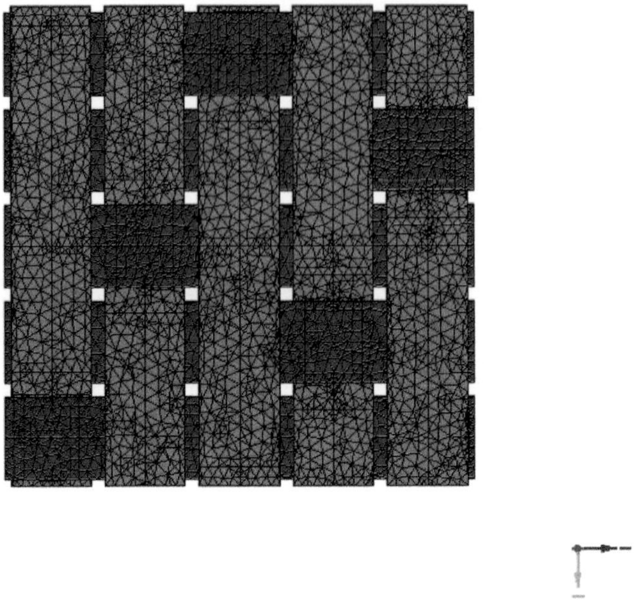

Figure 6.47 Five harness weave unit cell. Parameter list {6,1,14,14,55}, user-specified element size = 2.0

Figure 6.48 Random chopped fiber unit cell in two dimensions. Parameter list {30,1,0.5,20,40,40}, default element size = 0.5 (view fibers only)

spine along the out-of-plane (Y) direction, and subsequently rotated. You can choose either to generate fibers randomly or to specify upper and lower bounds for the major and minor axes of the fiber cross section, as well as the rotation angles. The in-plane rotation angle and the out-of-plane rotation angle are user-defined parameters. The input parameters are as follows:

1. fiber volume fraction (%)
2. major radius of the fiber (lower bound)
3. major radius of the fiber (upper bound)
4. minor radius of the fiber (lower bound)
5. minor radius of the fiber (upper bound)
6. fiber length (lower bound)
7. fiber length (upper bound)
8. out-of-plane rotation angle – rotation around the global Y axis (lower bound)
9. out-of-plane rotation angle – rotation around the global Y axis (upper bound)
10. in-plane rotation angle – rotation around the global Z axis (lower bound)
11. in-plane rotation angle – rotation around the global Z axis (upper bound)
12. unit cell length in the X direction
13. unit cell length in the Y direction
14. unit cell width in the Z direction

Figure 6.49 depicts a random chopped fiber unit cell in three dimensions.

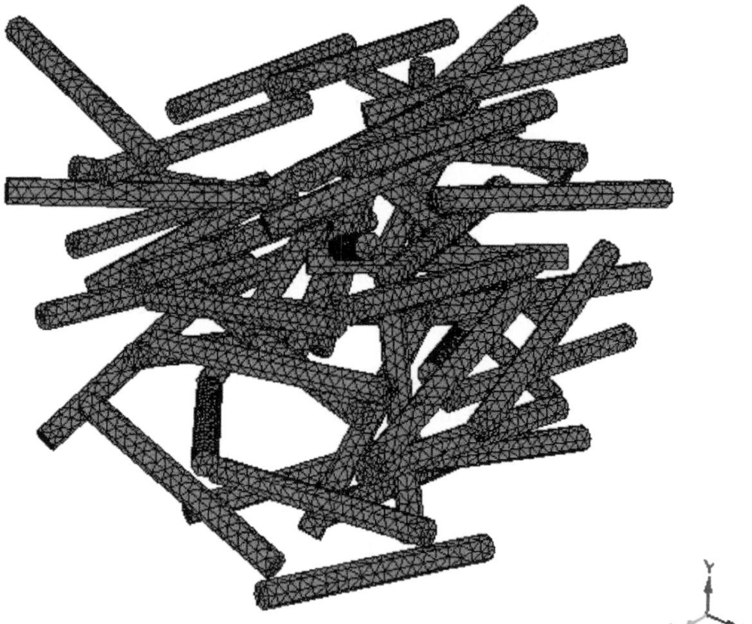

Figure 6.49 Random chopped fiber unit cell in three dimensions. Parameter list {5, 0.5, 0.5, 0.5, 0.5, 10, 10, 0, 180, 0, 360, 20, 20, 20}, default element size = 0.5 (view fibers only)

6.2.5.9 Random Ellipsoids Unit Cell in Three Dimensions

For a random pack of ellipsoids unit cell, fourteen parameters have to be defined. The ellipsoids are originally generated in the global coordinate system with semi-major axis a in the X direction, semi-minor axis b in the Y direction, and semi-minor axis c in the Z direction. Subsequently, the ellipsoids are rotated. You can either choose to generate the ellipsoids completely randomly or to specify upper and lower bounds for the major and minor axes of the ellipsoids, as well as the rotation angles. The in-plane angle and the out-of-plane angle are user-defined parameters. The in-plane is defined in the X and Z axes. The input parameters are as follows:

1. particle volume fraction (%)
2. major radius a of the ellipsoid (lower bound)
3. major radius a of the ellipsoid (upper bound)
4. first minor radius b of the ellipse for the cross section (lower bound)
5. first minor radius b of the ellipse for the cross section (upper bound)
6. second minor radius c of the ellipse for the cross section (lower bound)
7. second minor radius c of the ellipse for the cross section (upper bound)
8. out-of-plane rotation angle – rotation around the global Y axis (lower bound)
9. out-of-plane rotation angle – rotation around the global Y axis (upper bound)
10. in-plane rotation angle – rotation around the global Z axis (lower bound)
11. in-plane rotation angle – rotation around the global Z axis (upper bound)
12. unit cell length in the X direction
13. unit cell length in the Y direction
14. unit cell width in the Z direction

Figure 6.50 depicts a random ellipsoids unit cell.

6.3 Macroanalysis with MDS-Lite

The MDS-Macro solver is a complete finite element analysis environment for solving general static and dynamic solid mechanics problems. It consists of three major components:

1. Gmsh [8] as the geometry, mesh, and postprocessing engine.
2. The MDS finite element analysis engine.
3. The MDS-Macro GUI as the main driver to define the macro problem and to control the Gmsh and the finite element engine.

The major components of the MDS-Macro GUI are shown in Figure 6.51. The geometry of the model is defined using Gmsh, followed by the five steps from "Material" to "Output Control" that control every aspect of the macro problem definition. After the job is submitted and the finite element analysis is completed, the resulting data and/or graphs can be viewed using "Postprocessing."

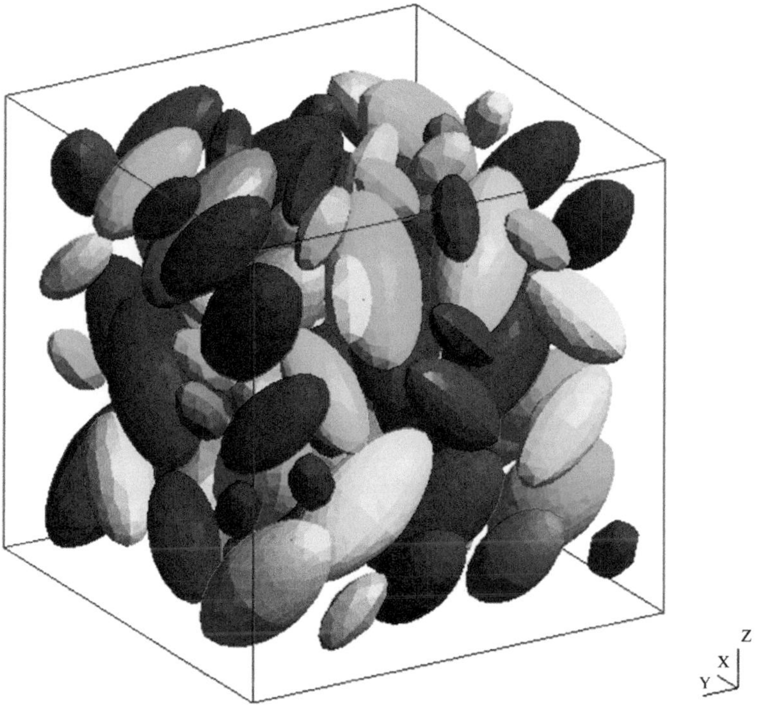

Figure 6.50 Random ellipsoids unit cell. Parameter list {28, 0.1, 0.2, 0.05, 0.1, 0.05, 0.1, 45, 45, 0, 360, 1, 1, 1}, default element size = 0.05 (view ellipsoids only)

In the four tutorial examples set out below, we will consider the classic plate with a hole problem depicted in Figure 6.52. The analytical stress concentration factor $K=\sigma^{\max}/\sigma$ is 3 for a homogeneous isotropic elastic infinite plate with a hole. The numerical solution should be close to 3 for $w/d \gg 1$ and sufficiently fine discretization.

Due to symmetry, only the upper right quarter of a 3D plate with a hole is modeled. The plate dimensions are $w/2=20$, $d/2=2.5$, $t=5$ and loading is $\sigma=0.05$. In the first two walkthrough examples, the material is assumed to be a homogeneous isotropic elastic with $E=1$, $\nu=0$. In the third example, elastic fibrous unit cell microstructure is considered; in the fourth example, inelastic fibrous unit cell microstructure is considered. In the first walkthrough example, the plate is meshed with tetrahedral elements of a size equal to one, whereas in the second, third, and fourth examples, a structured hexahedral mesh is considered.

6.3.1 First Walkthrough MDS-Macro Example

6.3.1.1 Geometry

The geometry of the model is defined in the Gmsh environment. To run Gmsh, first click on the "Geometry" button in the MDS-Macro GUI main screen (Figure 6.51) and provide the

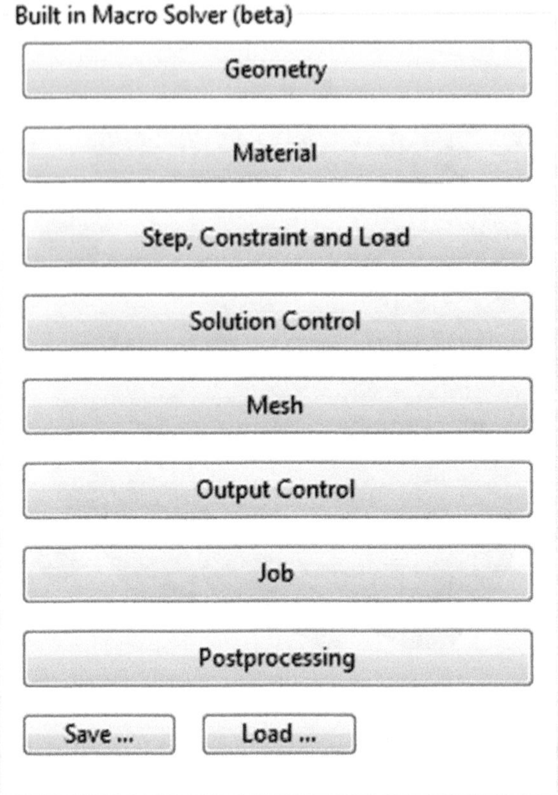

Figure 6.51 The major components of the MDS-Macro solver

model name in the pop-up window as shown in Figure 6.53. Select the "Open in Gmsh ..." button to open the Gmsh environment.

The Gmsh canvas window and the control window will pop up as shown in Figure 6.54.

Gmsh employs the boundary representation (BRep) approach to describe the model geometry. Models are created using a bottom-up approach by successively defining points, oriented lines (such as line segments, circles, ellipses, and splines), oriented surfaces (such as plane surfaces, ruled surfaces, and triangulated surfaces), and volumes.

There are two approaches to defining geometric entities:

Approach 1: Use the menu provided in the control window. When you click on a menu item, a submenu will show up. For instance, to add a new point, follow the following sequence: Elementary entities → Add → New → Point. Then define the X, Y, and Z coordinates in the "Contextual Geometry Definitions" window as shown in Figure 6.55. Click on the "Add" button to confirm adding the point.

To define the higher dimensional entities (lines, surface, and volume), select predefined entities at a lower dimension in the canvas window, and then follow the on-screen guidance as shown in Figure 6.56.

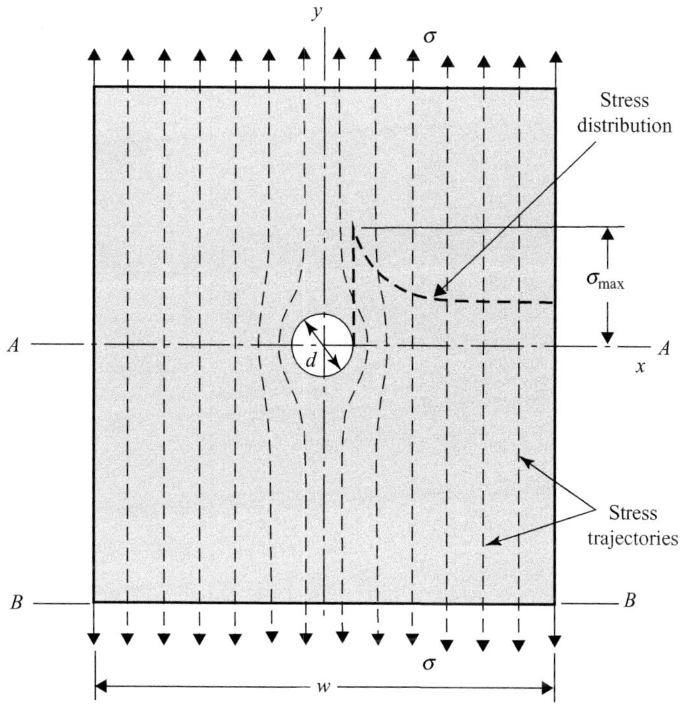

Figure 6.52 The classic plate with a hole problem

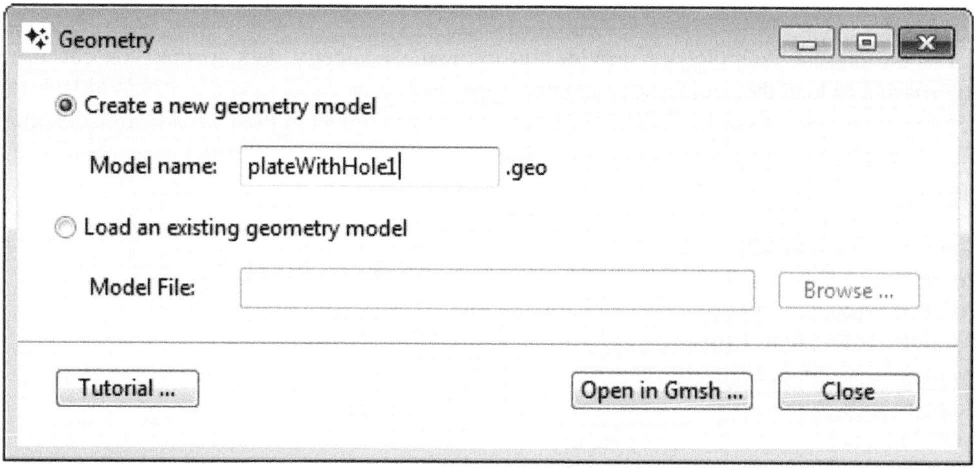

Figure 6.53 The first step in defining geometry

Approach 2: Directly write or edit the geometry data file using the Gmsh data file (for details see Gmsh Help documentation). To access the data file, click on the "Geometry" button to go back to the top level, and then click on the "Edit" button to pop up the data file as shown in Figure 6.57. After editing the data file, save the changes and click on the "Reload" menu to see the changes in the canvas window.

(a) (b)

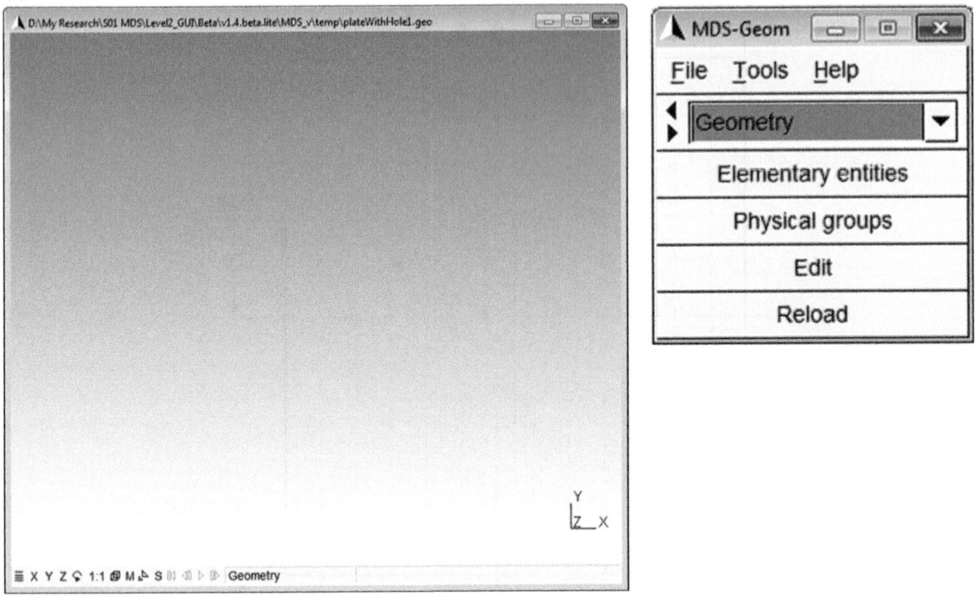

Figure 6.54 Gmsh environment: canvas window (a); control window (b)

Note that Gmsh will continue to write the geometry script from the end of the .geo file. You should leave one empty line at the end of the file so that the geometry commands will not be on the same line.

For the present example, we provide the following geometry data so that you can use Approach 2 to load the geometry all at once. Copy and paste the data to the plateWithHole1.geo file, save, and reload the Gmsh. The final geometry is shown in Figure 6.58. Note that the "Extrude" feature is used to generate the volume (see Gmsh Help for details). Keep the Gmsh open and go to the next step.

```
Point(1) = {0, 0, 0, 1.0};
Point(2) = {2.5, 0, 0, 1.0};
Point(3) = {20, 0, 0, 1.0};
Point(4) = {20, 20, 0, 1.0};
Point(5) = {0, 20, 0, 1.0};
Point(6) = {0, 2.5, 0, 1.0};
Line(1) = {6, 5};
Line(2) = {2, 3};
Line(3) = {5, 4};
Line(4) = {3, 4};
Circle(5) = {6, 1, 2};
Line Loop(6) = {1, 3, -4, -2, -5};
Plane Surface(7) = {6};
Extrude {0, 0, 5} { Surface{7}; }
```

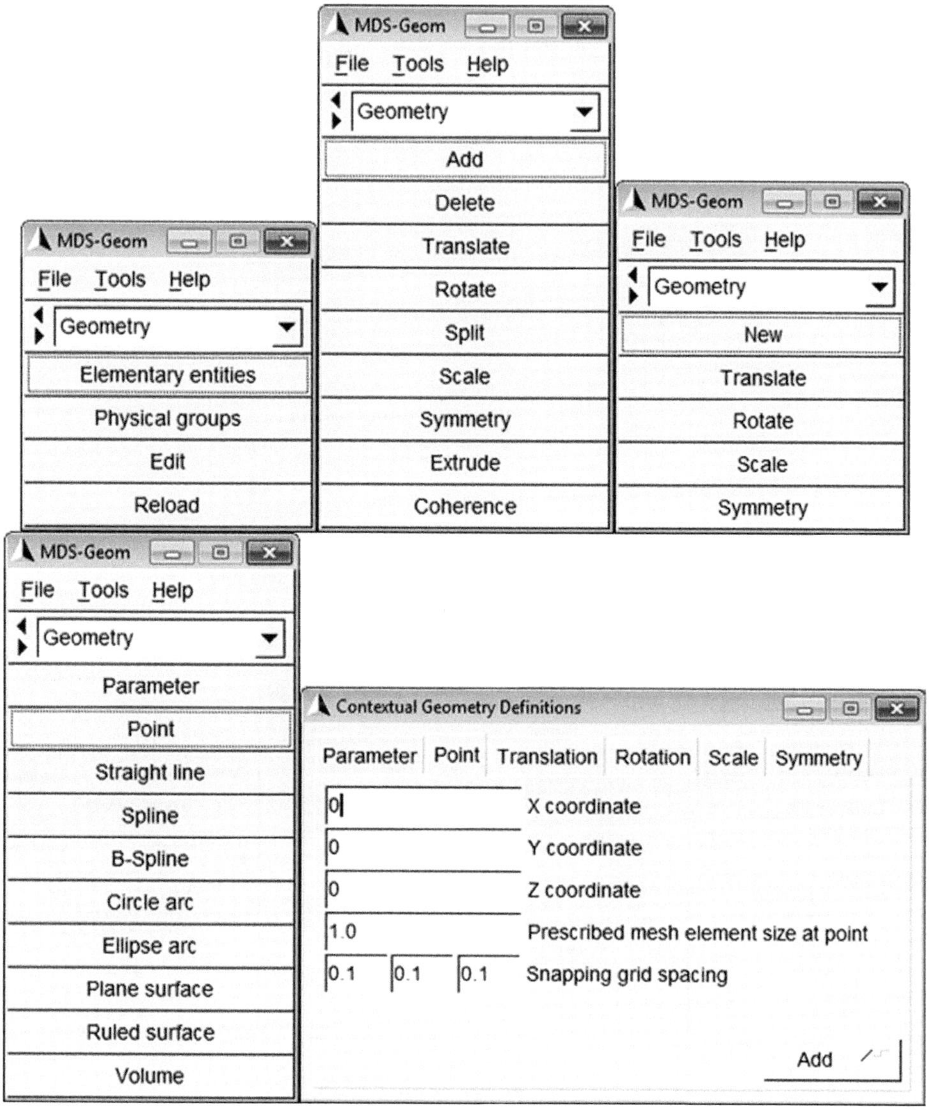

Figure 6.55 Gmsh menu path for adding a point

6.3.1.2 Material

Click on the "Material" button in the MDS-Macro GUI main screen to pop up the "Material properties" window as shown in Figure 6.59.

In the "Material properties" window, perform the following steps:

1. Create an isotropic material with $E=1$, $\nu=0$.
   Click the "Create …" button in the material list and define the material name, type, and parameters as shown in Figure 6.60. Then "Confirm" your input.

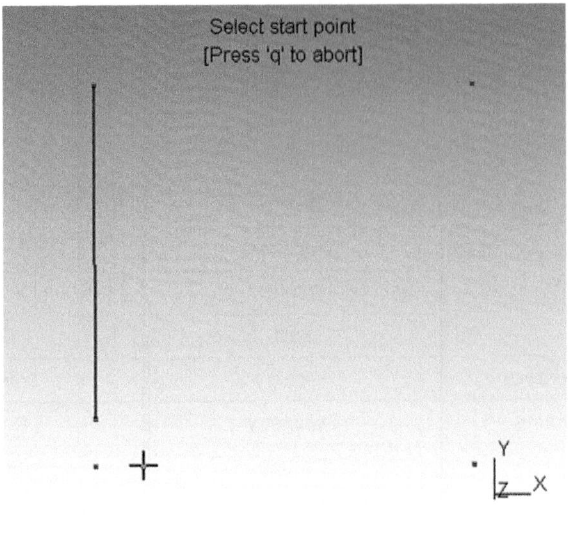

Figure 6.56 Adding lines in the Gmsh canvas window

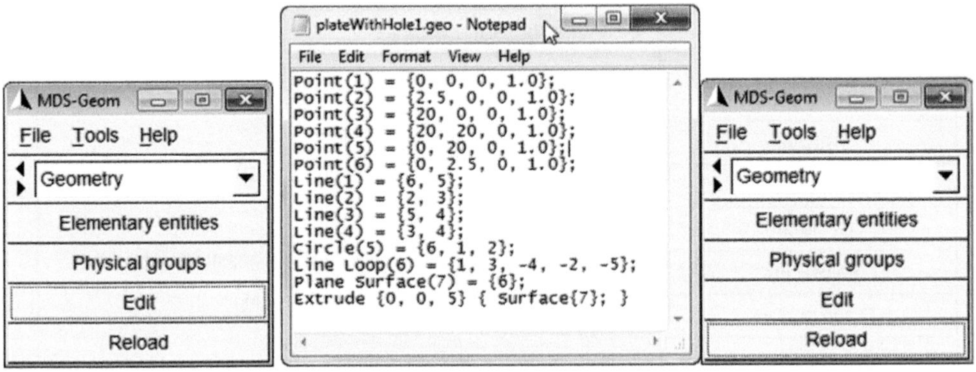

Figure 6.57 Editing the Gmsh data file directly

2. Define the volume group.
In MDS-Macro, the material, loading, and constraint attributes are defined on the geometric model through the so-called physical groups (physical point, physical line, physical surface, and physical volume). You need to create such a volume group and associate it with the material.
   Click the "Create ..." button in the volume group list and the "Physical Group Definition" window will pop up. Provide the name as the "wholeVolume." Click on the "Start" button and go to the Gmsh canvas window to select the yellow sphere, which represents the volume in Gmsh. The selected entity will be highlighted in red as shown in Figure 6.61. Go back to the definition window and click on the "End" button

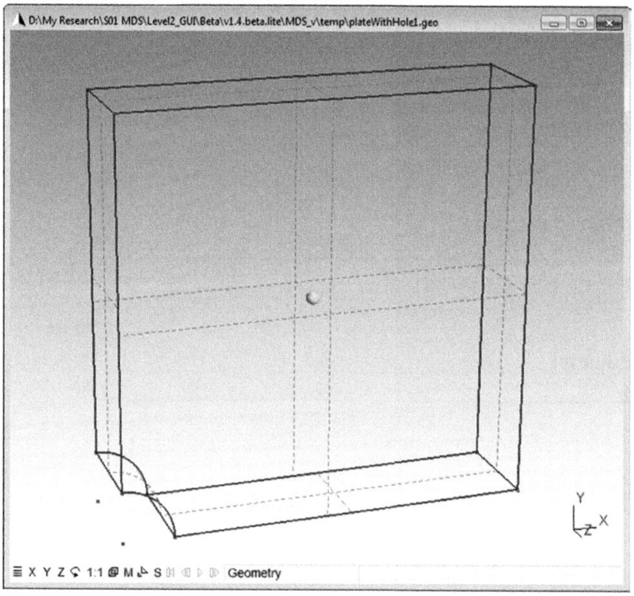

Figure 6.58 The final geometry of the plate with a hole problem

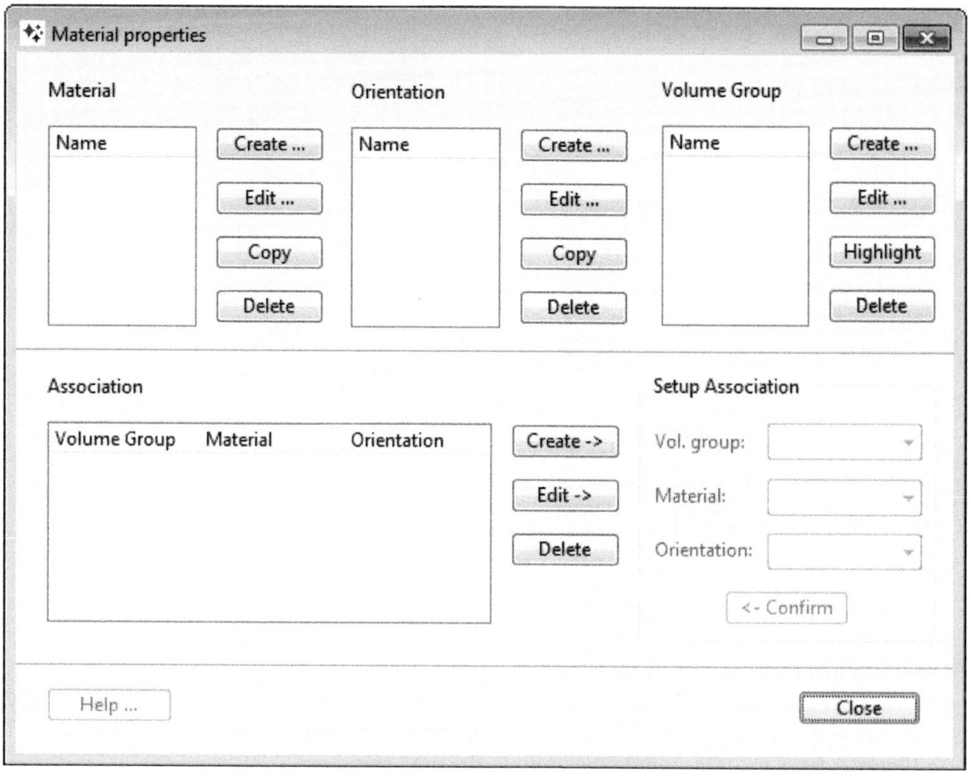

Figure 6.59 "Material properties" window

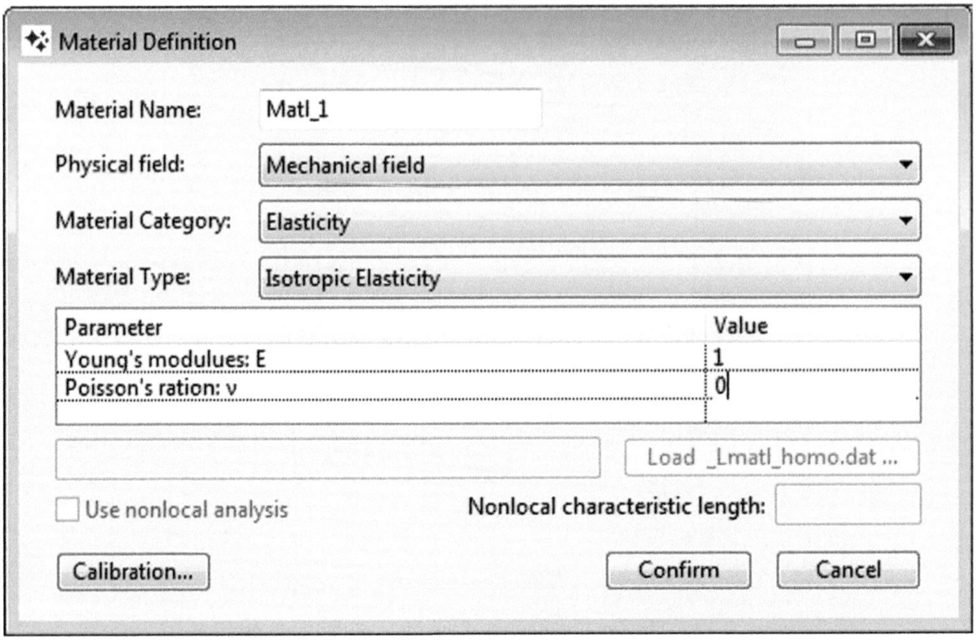

Figure 6.60 "Material Definition" window

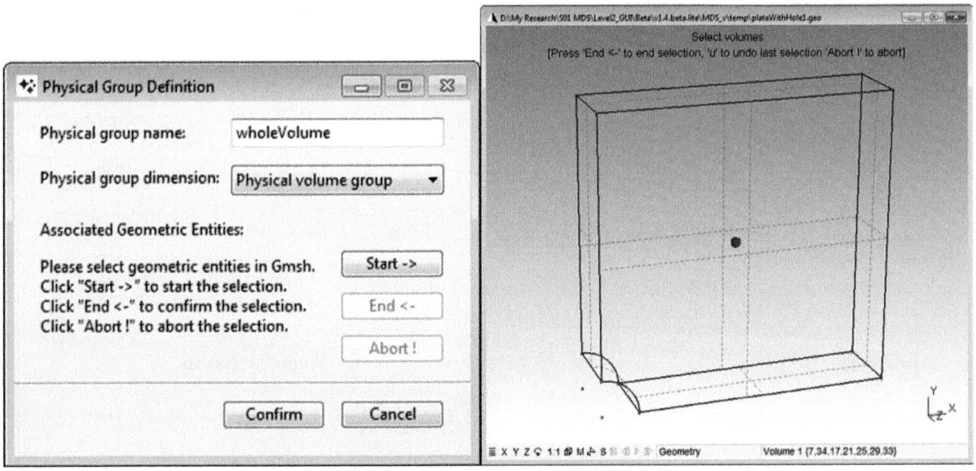

Figure 6.61 Physical group definition

to confirm the entity selection. Click the "Confirm" button to complete the volume group definition.

3. Associate the material with the volume group.
   Go back to the "Material properties" window and click the "Create →" button in the association list. Proceed by clicking the "← Confirm" button. The final window is shown in Figure 6.62. Close this window and go to the next step.

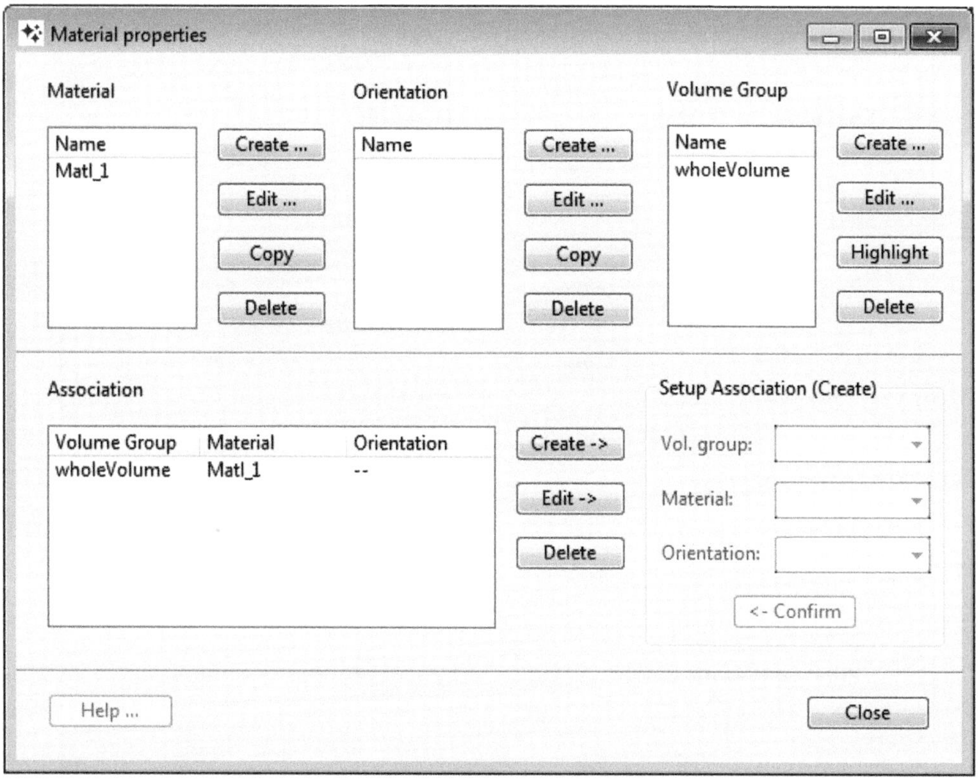

Figure 6.62 Final "Material properties" window

6.3.1.3 Step, Constraint, and Load

Click the "Step, Constraint, and Load" button in the MDS-Macro GUI main window to define the analysis step and boundary conditions. The corresponding window will pop up as shown in Figure 6.63.

In this window, perform the following steps:

1. Define the static linear step.
   Keep the step type selection as "Static, general." Click the "Create …" button in the step list and define the step with one increment as shown in Figure 6.64. "Confirm" your input.
2. Define four physical surfaces for the definition of boundary conditions and loading.
   Define three surfaces at X = 0, Y = 0, and Z = 0 planes for symmetric boundary conditions, and define one surface at the Y = 20 plane for surface traction. Here we only demonstrate the procedure for the X = 0 plane. Click the "Create …" button in the physical group list. As shown in Figure 6.65, name the physical group as "Surface_X_0." Then select "Physical surface group" as the dimension and click the "Start →" button to select the surface in the Gmsh canvas window as shown in Figure 6.66. Go back to the physical surface definition window and click on the "End" button to confirm the entity selection. Click "Confirm" to complete the surface group definition. Perform similar steps for the other three surfaces.

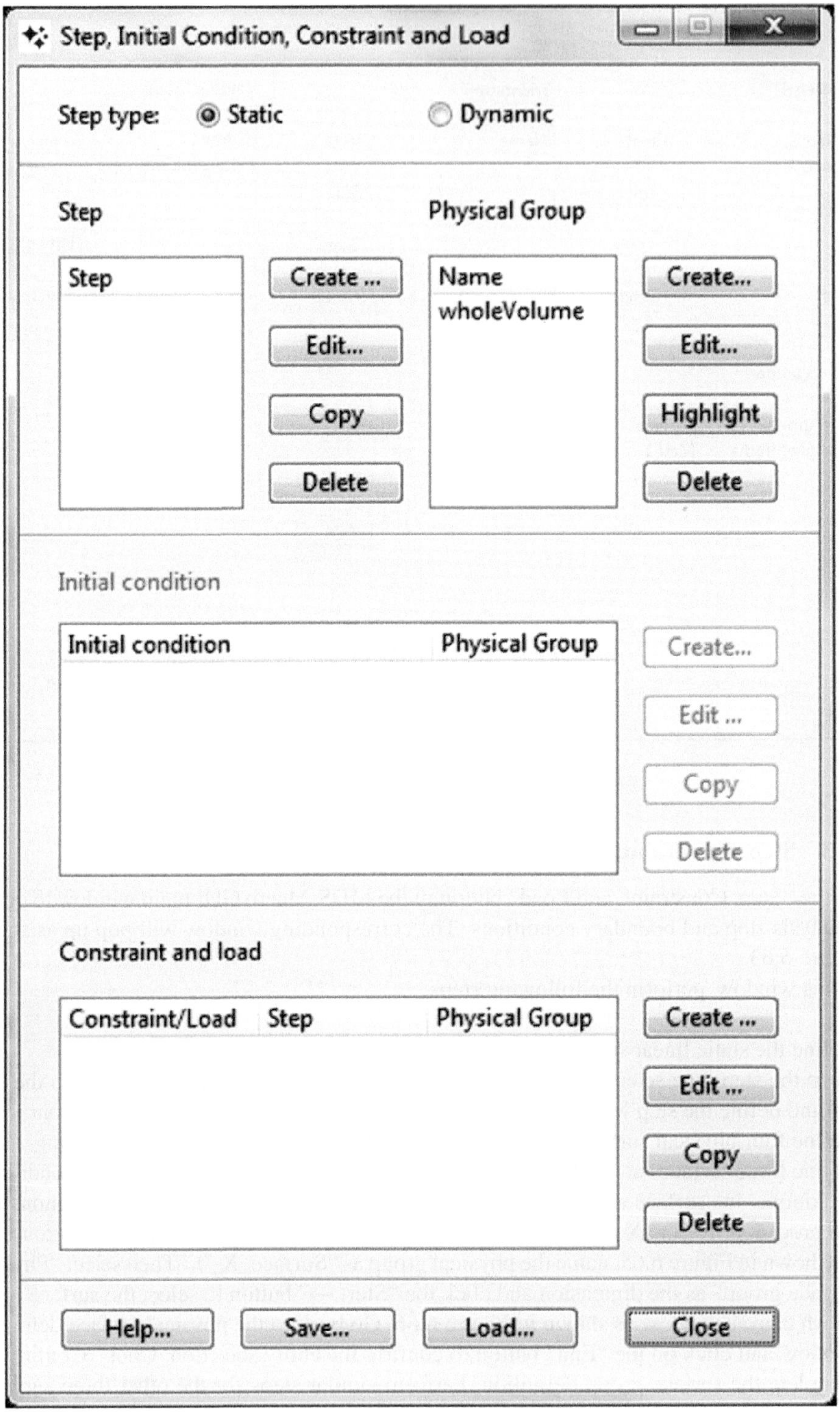

Figure 6.63 "Step, Initial Condition, Constraint, and Load" window

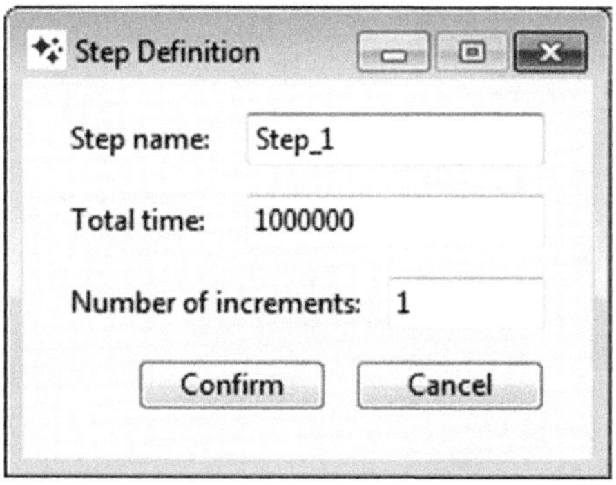

Figure 6.64 “Step Definition” window

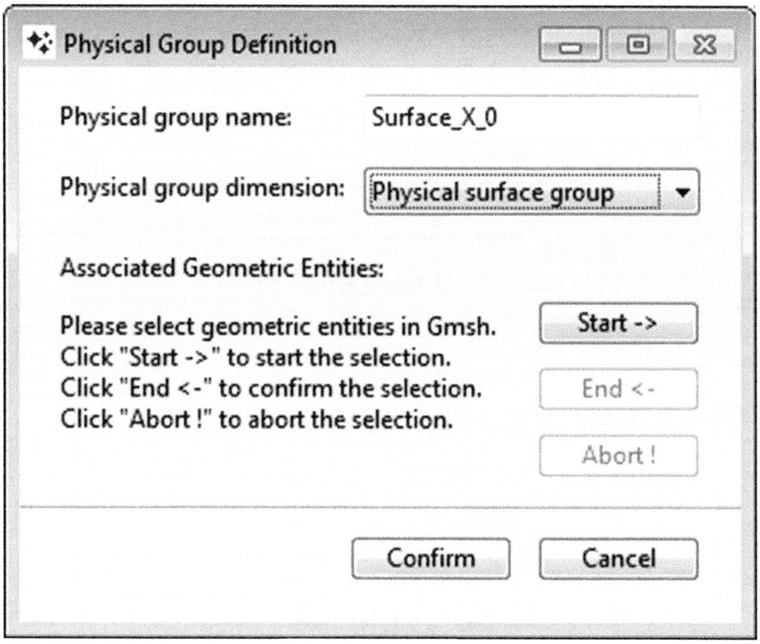

Figure 6.65 Physical surface definition

Finally, you should have the physical group list as shown in Figure 6.67. Note that for traction loading, “Surface_Y_20” is defined in the Y = 20 plane.

3. Define the symmetry boundary conditions and loading condition.
Define a zero displacement boundary condition along each of the symmetry planes. For the X = 0 plane, perform the following sequence. Click the “Create …” button in the constraint and load list. As shown in Figure 6.68, make sure the type is “Displacement constraint.” Set only $u_x = 0$ as active, and select “Surface_X_0” as the associated physical group. “Confirm”

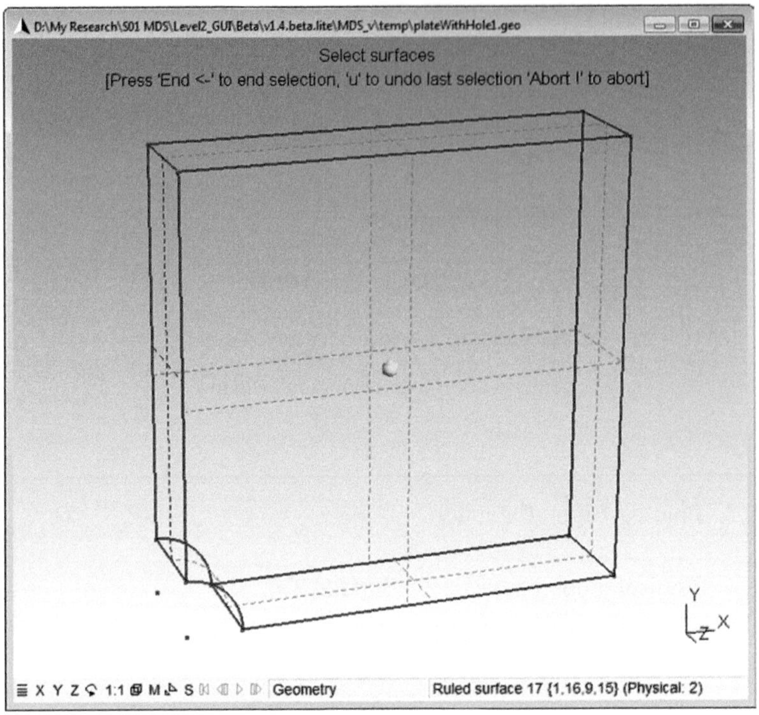

Figure 6.66 Physical surface selection

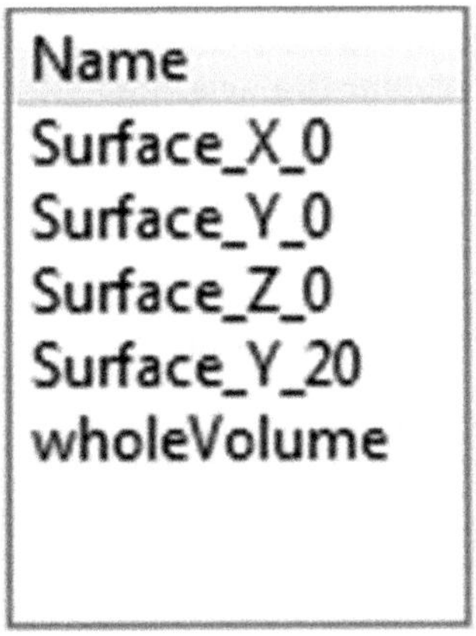

Figure 6.67 Final physical group list

your input. Perform similar steps to set up $u_y = 0$ for "Surface_Y_0" and $u_z = 0$ for "Surface_Z_0." For the surface traction at the Y = 20 plane, provide the information requested in Figure 6.69. "Confirm" your input.

The final "Step, Initial Condition, Constraint, and Load" window is shown in Figure 6.70. Close this window and go to the next step.

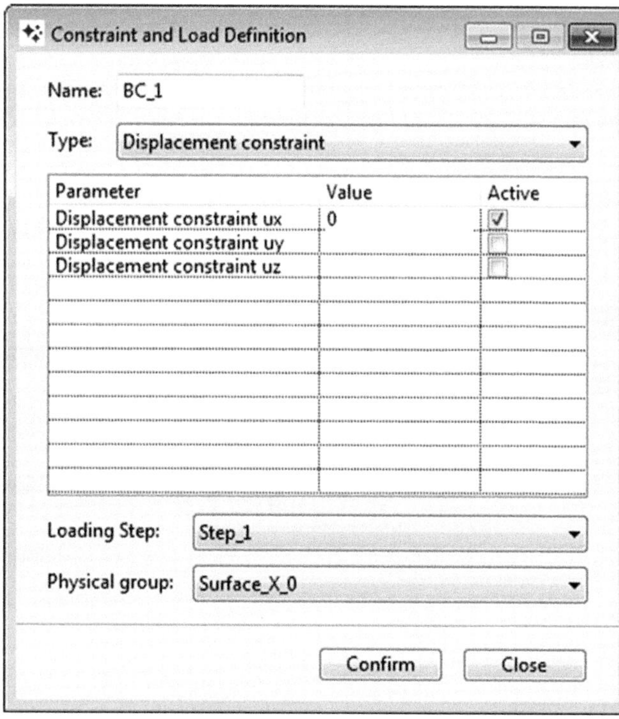

Figure 6.68 Symmetric boundary condition for Surface_X_0

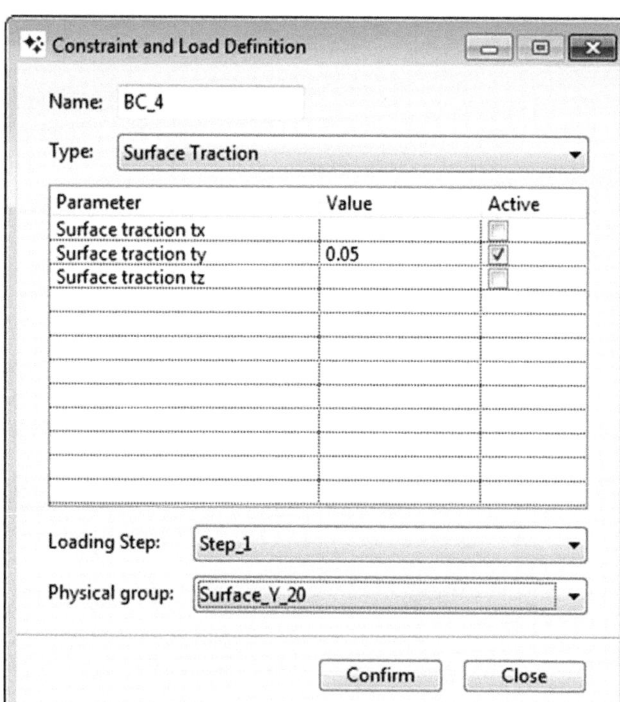

Figure 6.69 Loading condition for Surface_Y_20

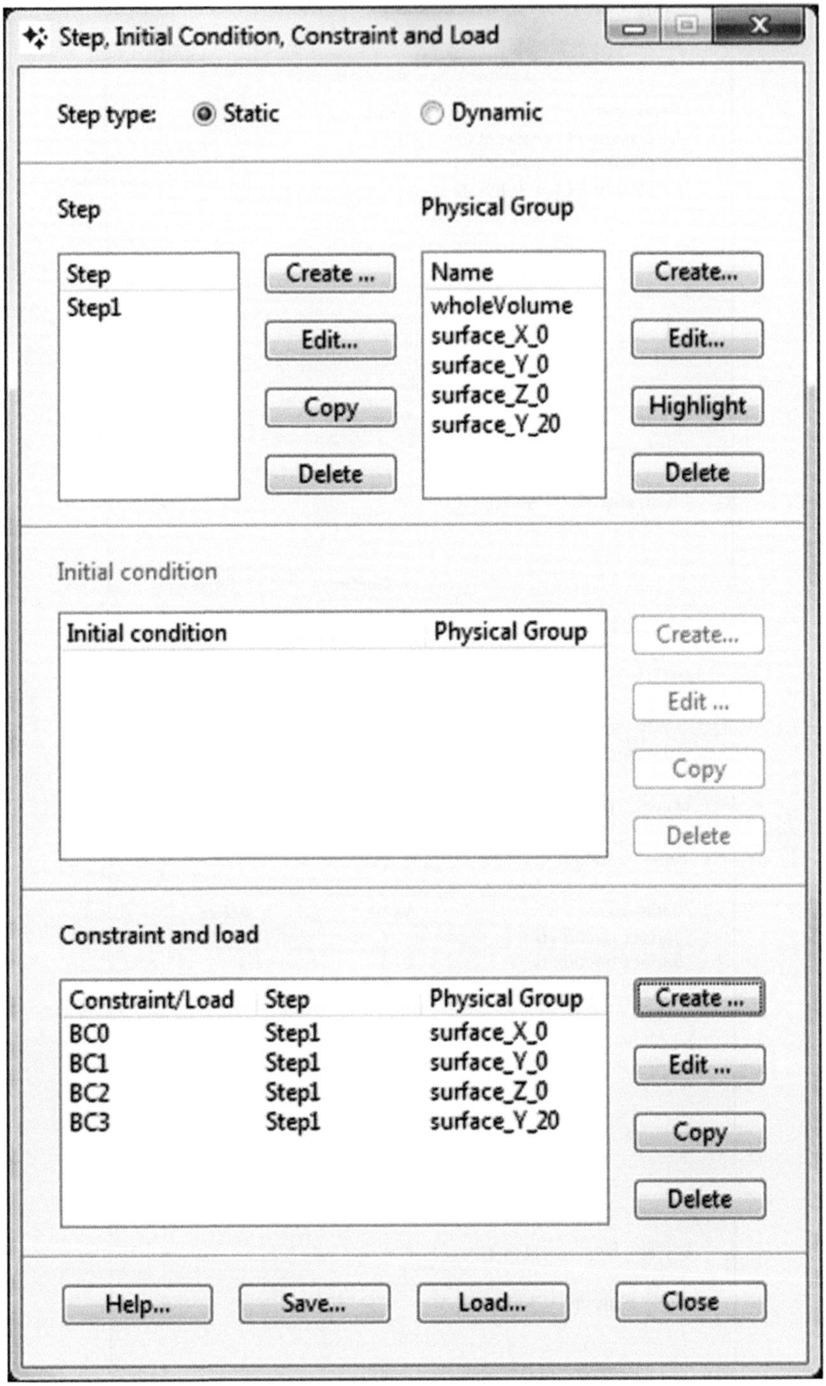

Figure 6.70 Final "Step, Initial Condition, Constraint, and Load" window

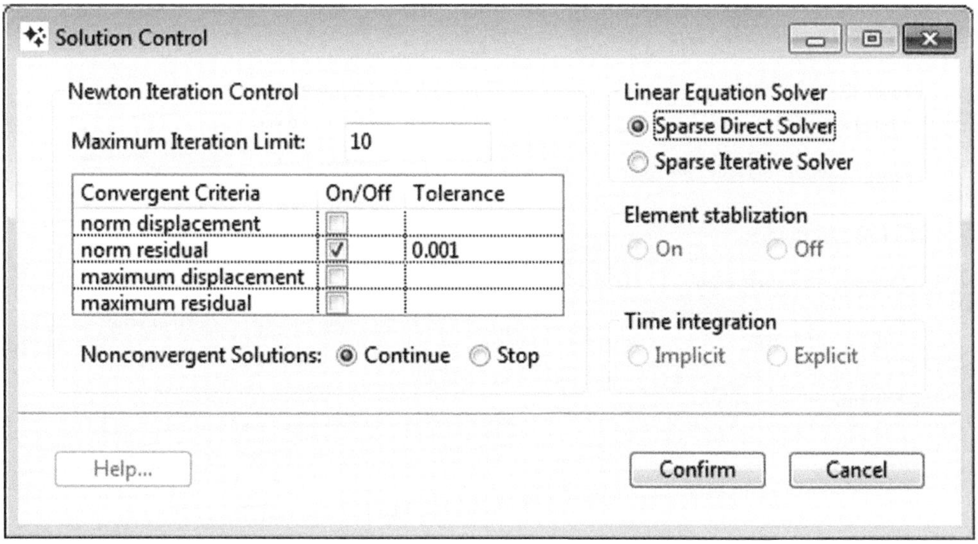

Figure 6.71 "Solution Control" window

6.3.1.4 Solution Control

Click the "Solution Control" button in the MDS-Macro GUI main screen to pop up the window shown in Figure 6.71. For this step, keep all the default settings. "Confirm" your input. Go to the next step.

6.3.1.5 Mesh

Click the "Mesh" button in the MDS-Macro GUI main screen to pop up the corresponding window as shown in Figure 6.72. Set up the mesh density control by defining the "Global – Approximate Element Size" to be equal to 1. Keep the rest of the default settings, and click the "Mesh" button to get the final mesh in the Gmsh canvas window as shown in Figure 6.73. Close the mesh input window and go to the next step.

6.3.1.6 Output Control

Click the "Output Control" button in the MDS-Macro GUI main screen to pop up the corresponding window, as shown in Figure 6.74. To request a certain output, define the groups in both the temporal domain (single time point or a series of continuous time points, that is, history) and the spatial domain (element group or node group). Then define the data and/or visualization (graphic output) requests for the chosen time and spatial group.

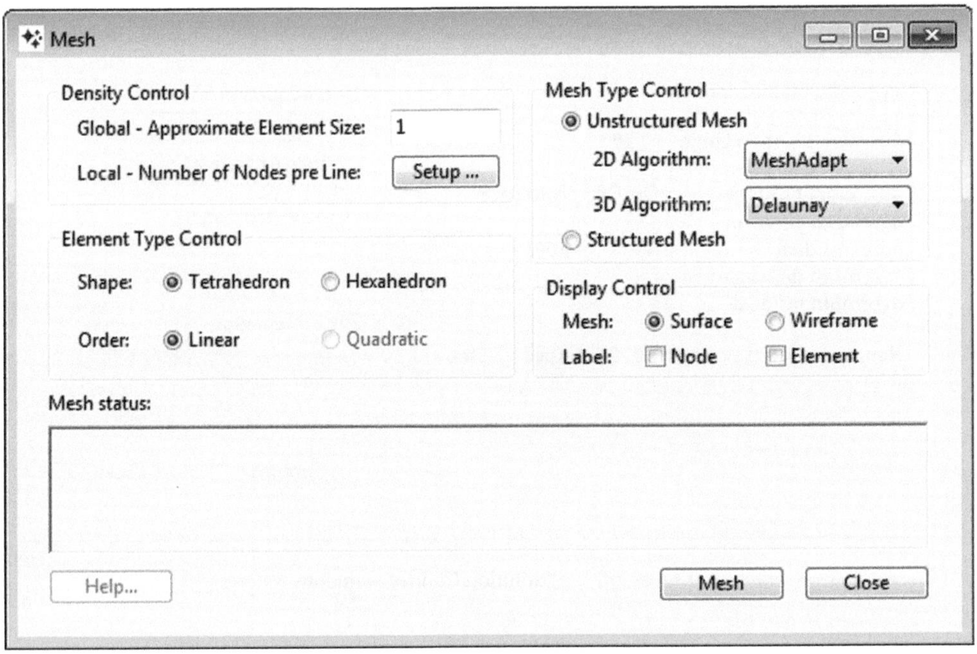

Figure 6.72 "Mesh" window

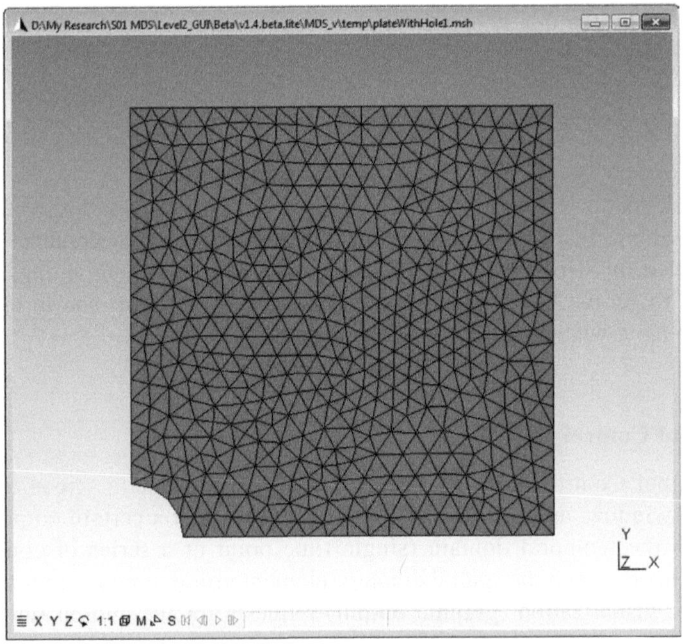

Figure 6.73 The resulting mesh in the Gmsh canvas window

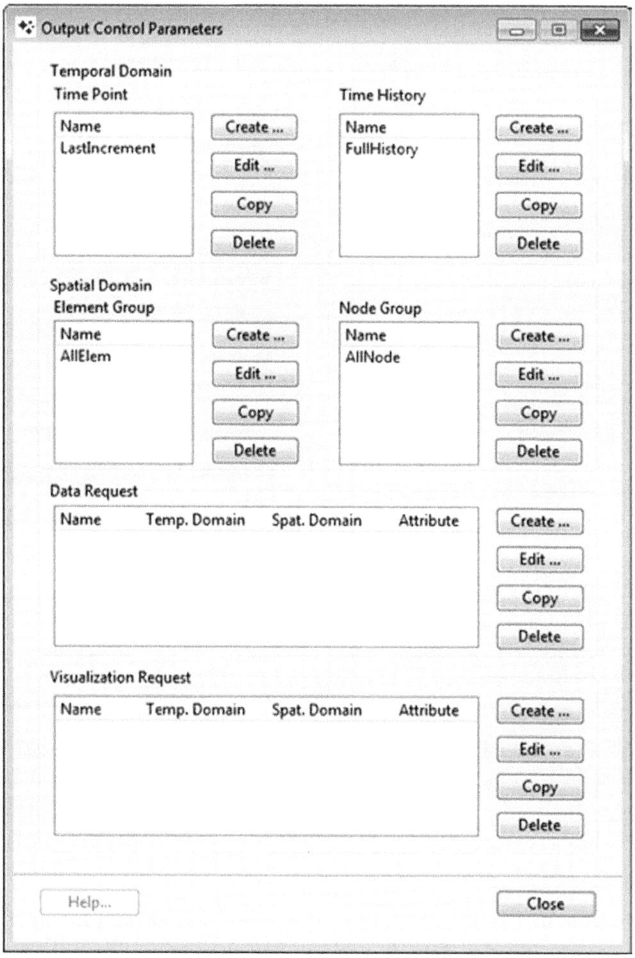

Figure 6.74 "Output Control Parameters" window

For each group type, MDS has a default group attribute. In the present example, we will use the default "LastIncrement" temporal group and the "AllElem" spatial group to output the data and visualize the stress component σ_{yy} for all elements at the end of the analysis.

To request a specific data output, click the "Create …" button in the data request list. Check the stress component S22 to output at the "LastIncrement" for "AllElem" as shown in Figure 6.75. "Confirm" your input.

Perform a similar step for the "Visualization Request" shown in Figure 6.75. "Confirm" your input. The final output control window is shown in Figure 6.76. Close this window. This completes the definition of all attributes of the problem.

In the next step, generate the final job input file and then submit it.

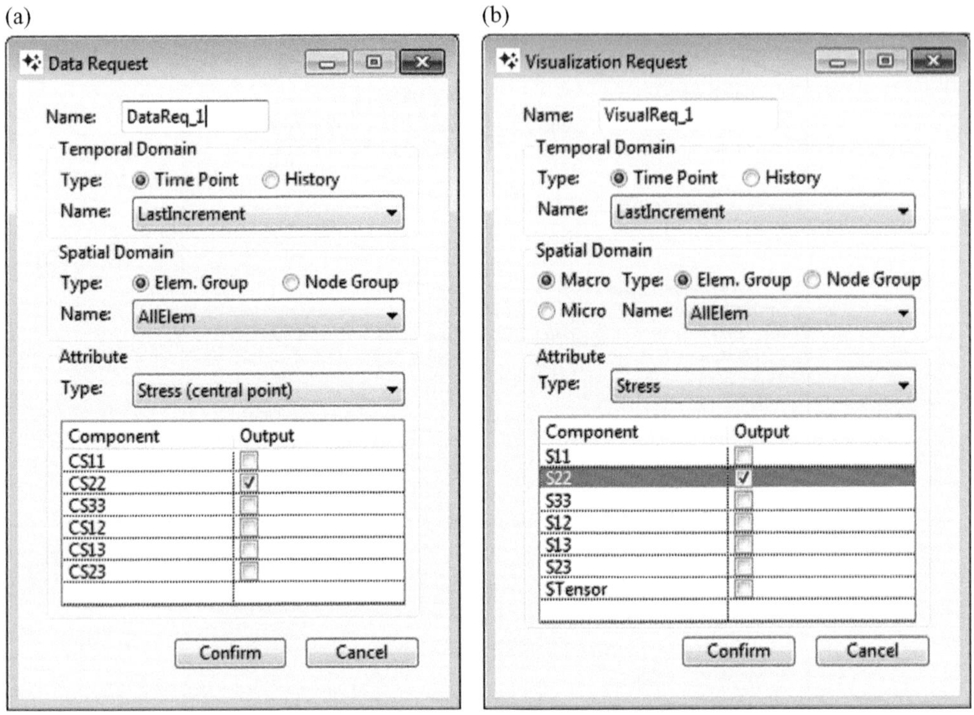

Figure 6.75 (a) "Data Request" window and (b) "Visualization Request" window

6.3.1.7 Job

Click the "Job" button in the MDS-Macro GUI main screen to pop up the job window. "Create …" a new job using the default settings and then "Confirm" the job. The final job window is shown in Figure 6.77.

Select "Job_1" and click the "Submit job" button. The "Backend Program" window will pop up as shown in Figure 6.78. Close this window when the execution of the job is completed.

Go to the next step to review the resulting data.

6.3.1.8 Postprocessing

Click the "Postprocessing" button in the MDS-Macro GUI main screen to pop up the "Macro postprocessing" window as shown in Figure 6.79. The stress S22 value in the element centroid (denoted as CS22 in Figure 6.79) for all elements is shown in the list on the right. Click the "View visualization" button to visualize the stress component in Gmsh.

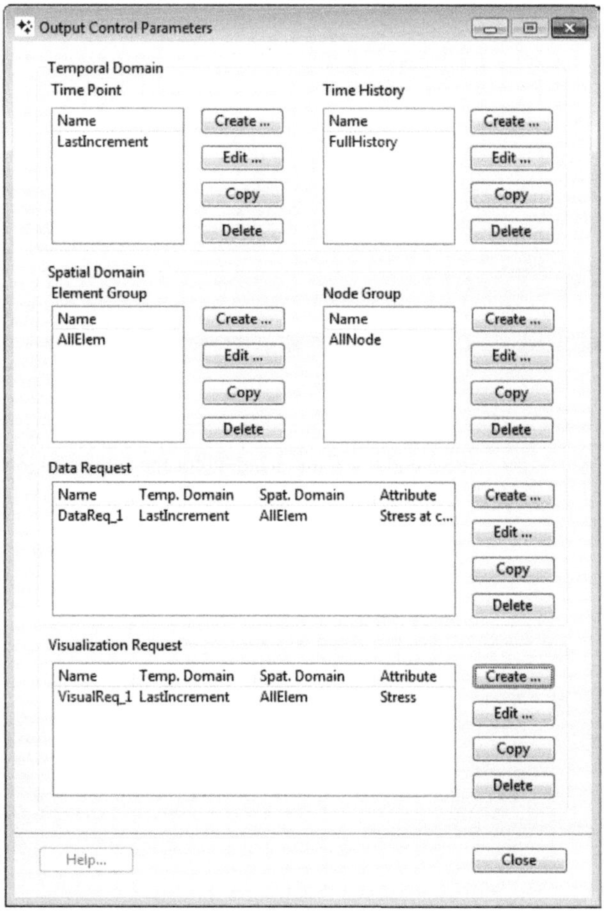

Figure 6.76 Final "Output Control Parameters" window

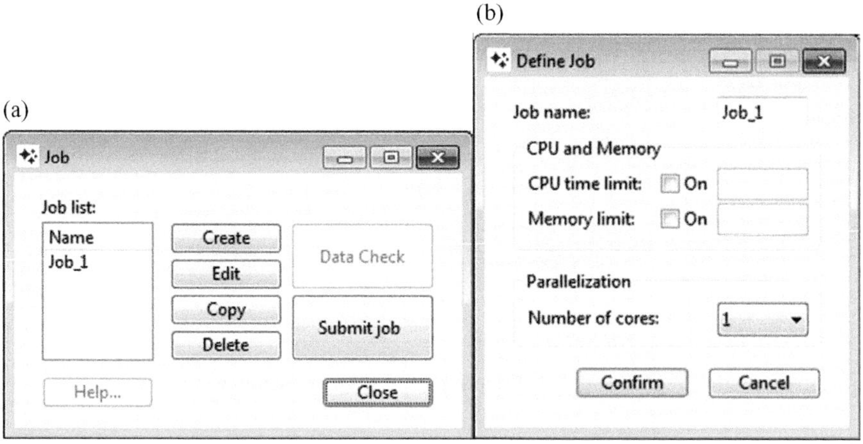

Figure 6.77 (a) "Job" window and (b)"Define Job" window

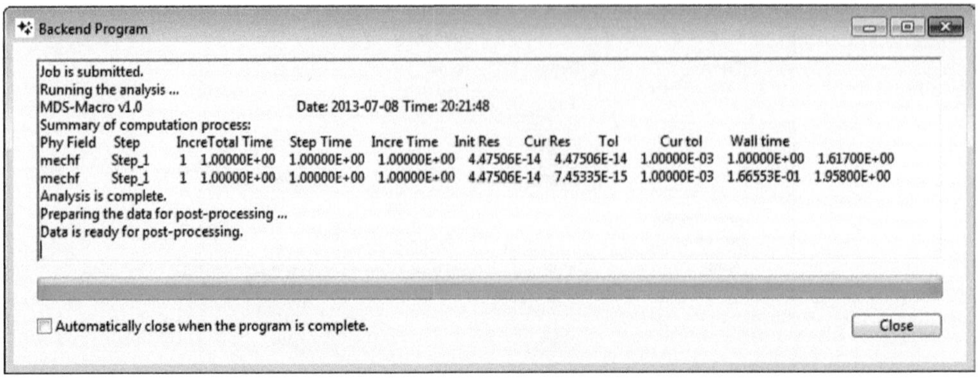

Figure 6.78 "Backend Program" window for the submitted job

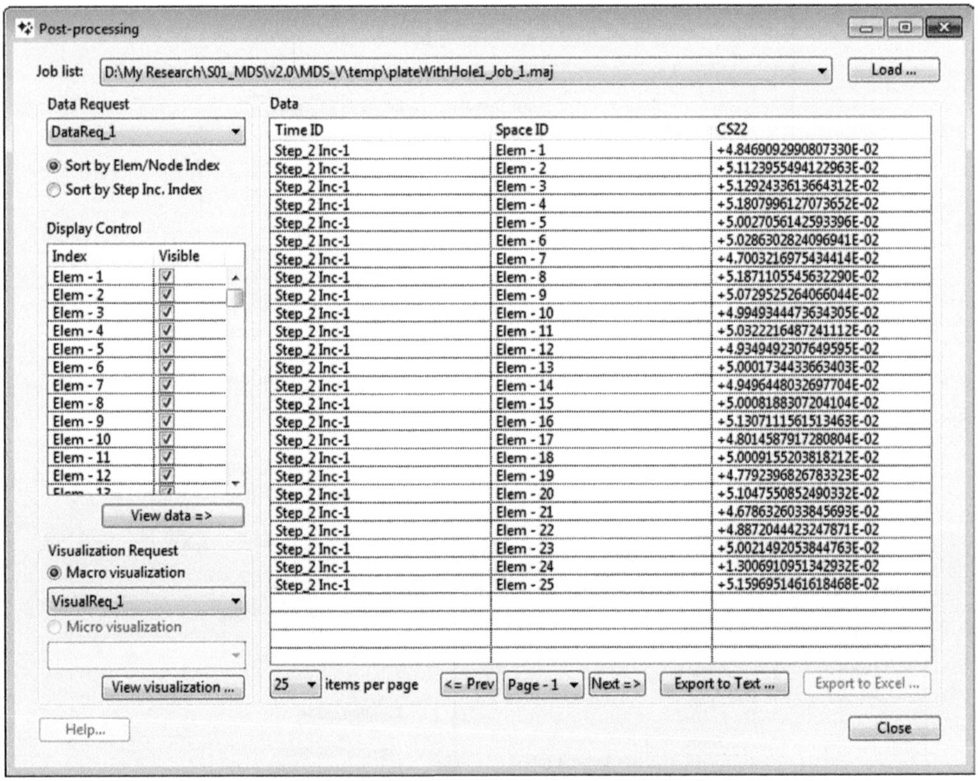

Figure 6.79 "Macro postprocessing" window

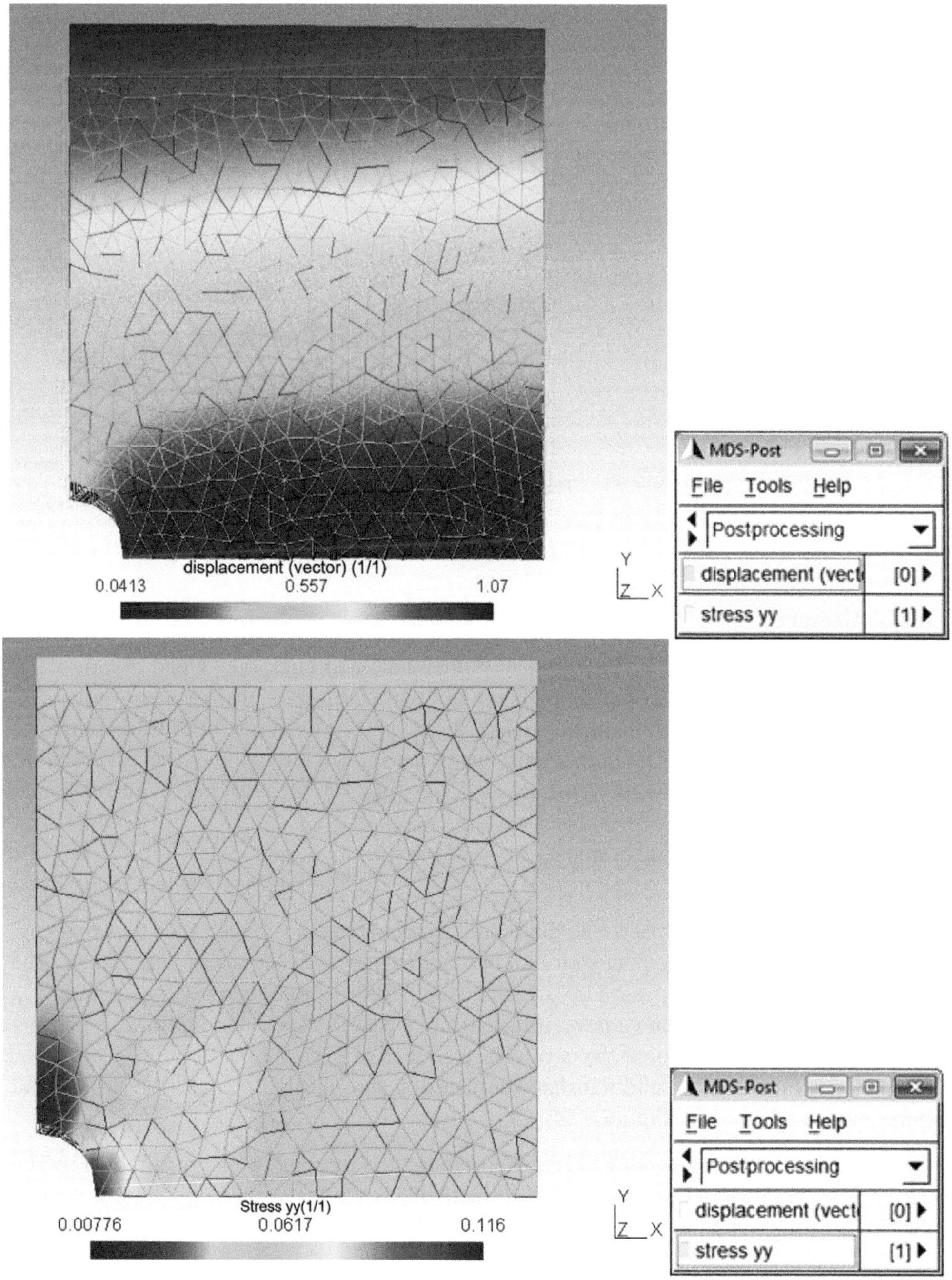

Figure 6.80 Postprocessing visualization Gmsh window

An independent Gmsh environment will pop up with two views, one for the default displacement and one for the user-requested stress component S22. Select "Postprocessing" in the Gmsh window and choose one of the views as shown in Figure 6.80.

This completes the tutorial for the first macro walkthrough example.

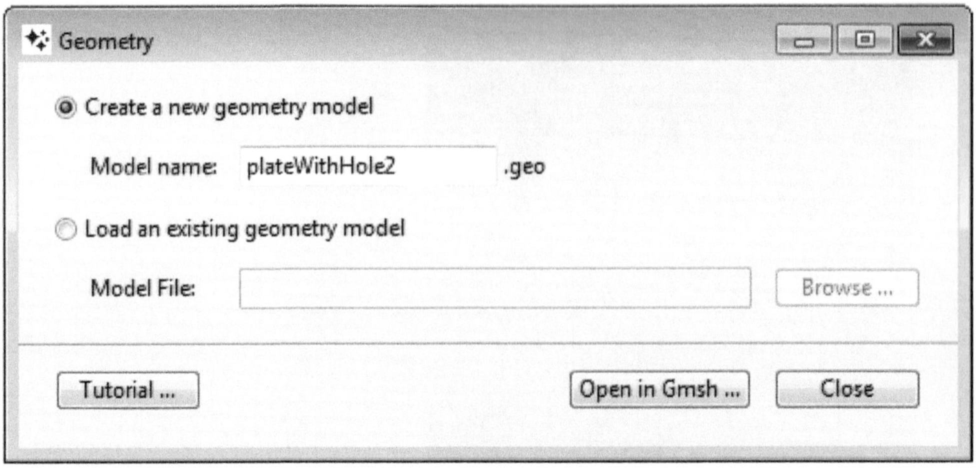

Figure 6.81 The first step of defining geometry

6.3.2 Second Walkthrough MDS-Macro Example

6.3.2.1 Geometry

Create a new geometric model named "plateWithHole2" and click the "Open in Gmsh ..." button to open the Gmsh environment as shown in Figure 6.81.

To construct a structured finite element mesh, partition the model into hexahedral domains. In the present example, the plate with a hole can be partitioned into two hexahedral domains. The plate consists of two pentagonal surfaces at Z=0 and Z=5. We will first divide the pentagons into two quadrilaterals.

As in the first walkthrough example, we will consider two approaches for creating the partitioned model:

Approach 1: Following Approach 1 in the first example, create a surface of Z=0. A diagonal line is added to subdivide the pentagon into two quadrilaterals as shown in Figure 6.82. The surface is defined as a "Ruled surface," which is required for structured mesh generation.

Choose "Translate" to create a new surface at Z=5. Select Elementary entities → Add → Translate → Surface, and choose the two surface entities (Figure 6.83).

Then connect the original and translated surfaces by straight lines. Create the surface and volume entities as shown in Figure 6.84.

Approach 2: You can directly copy the geometry data file given below and then reload it. The final geometry is given in Figure 6.85. Recall that you should leave one empty line at the end of file so that the geometry commands will not be on the same line.

```
Point(1) = {0, 0, 0, 1.0};
Point(2) = {2.5, 0, 0, 1.0};
Point(3) = {20, 0, 0, 1.0};
Point(4) = {20, 20, 0, 1.0};
Point(5) = {0, 20, 0, 1.0};
```

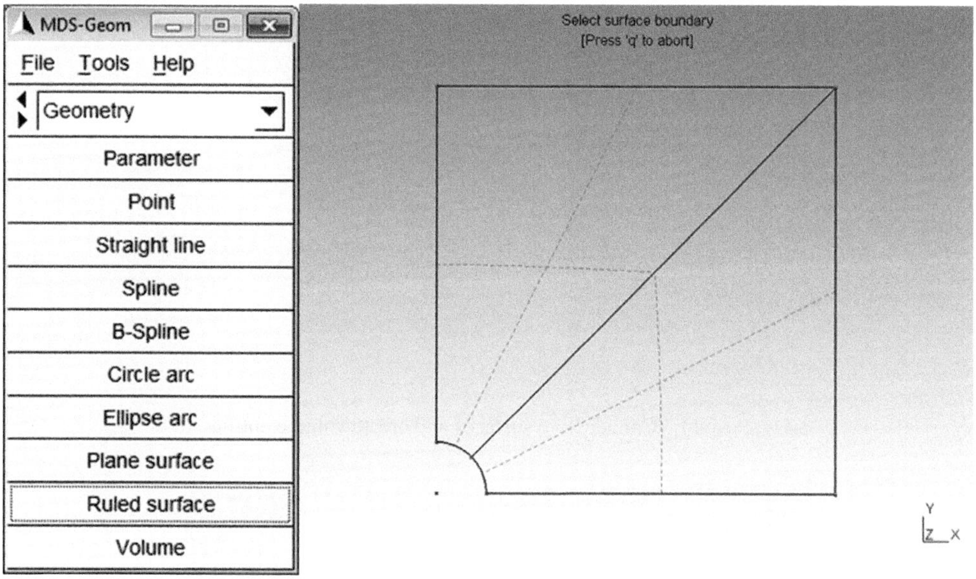

Figure 6.82 Defining the surface at $Z=0$

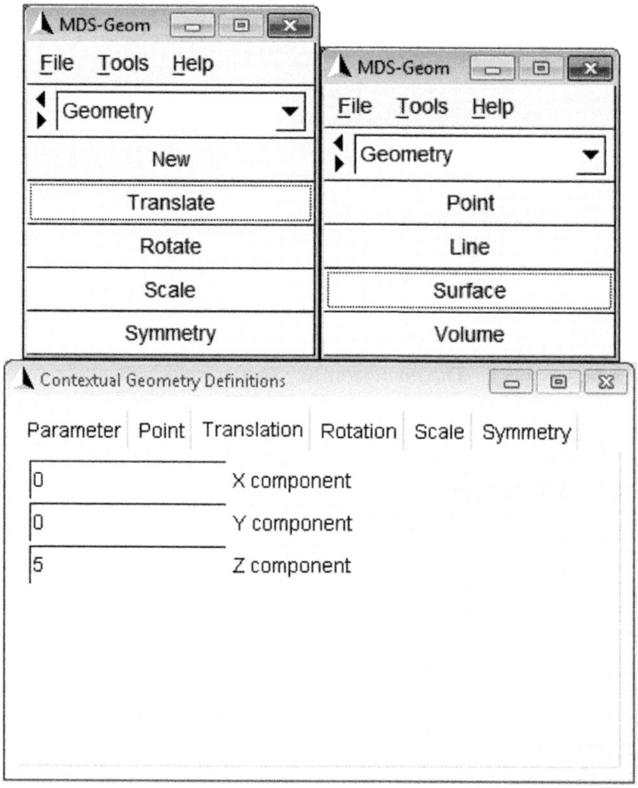

Figure 6.83 Gmsh menu path for translation

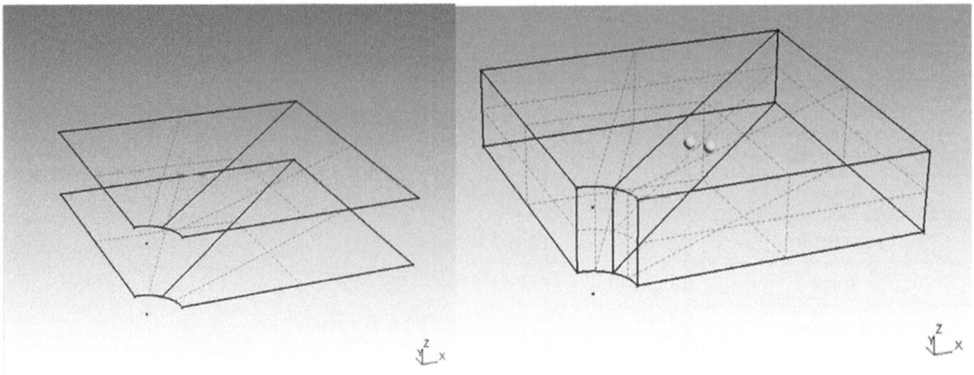

Figure 6.84 Connect two surfaces and create volume entities

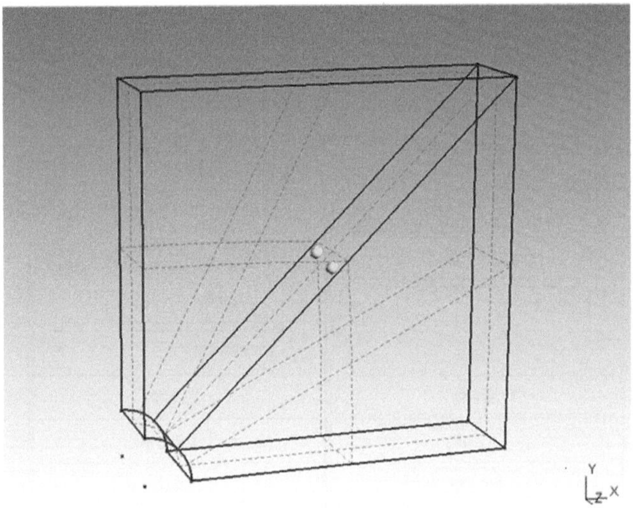

Figure 6.85 The final geometry for the plate with a hole

```
Point(6) = {0, 2.5, 0, 1.0};
Point(7) = {1.7678, 1.7678, 0, 1.0};
Line(1) = {4, 7};
Line(2) = {4, 3};
Line(3) = {3, 2};
Line(4) = {4, 5};
Line(5) = {5, 6};
Circle(6) = {6, 1, 7};
Circle(7) = {2, 1, 7};
Line Loop(8) = {5, 6, -1, 4};
Ruled Surface(9) = {8};
```

```
Line Loop(10) = {3, 7, –1, 2};
Ruled Surface(11) = {10};
Translate {0, 0, 5} {
  Duplicata { Surface{9, 11}; }
}
Line(22) = {9, 6};
Line(23) = {14, 7};
Line(24) = {26, 2};
Line(25) = {8, 5};
Line(26) = {18, 4};
Line(27) = {25, 3};
Line Loop(28) = {13, 22, –5, –25};
Ruled Surface(29) = {28};
Line Loop(30) = {14, 23, –6, –22};
Ruled Surface(31) = {30};
Line Loop(32) = {19, 23, –7, –24};
Ruled Surface(33) = {32};
Line Loop(34) = {18, 24, –3, –27};
Ruled Surface(35) = {34};
Line Loop(36) = {21, 27, –2, –26};
Ruled Surface(37) = {36};
Line Loop(38) = {15, 26, 1, –23};
Ruled Surface(39) = {38};
Line Loop(40) = {16, 25, –4, –26};
Ruled Surface(41) = {40};
Surface Loop(42) = {17, 35, 33, 11, 37, 39};
Volume(43) = {42};
Surface Loop(44) = {41, 12, 29, 31, 9, 39};
Volume(45) = {44};
```

6.3.2.2 Material

We use the same setting as in the first walkthrough example. The two volume entities will be assigned isotropic elastic material properties with $E=1$, $\nu=0$. In the "Material properties" window, perform the following steps:

1. Create an isotropic material with.
   Repeat the material definition as in the first walkthrough example and "Confirm" your input.
2. Define the volume group.
   Click the "Create …" button along the volume group list and the "Physical Group Definition" window will pop up. Name this volume group as "wholeVolume." Since there are two volume partitions, the "wholeVolume" group will contain the two partitions.
3. Associate the material with the volume group.
   Go back to the "Material properties" window and assign the volume group "wholeVolume" to "Mat_1." The final stage of this material step is shown in Figure 6.86.

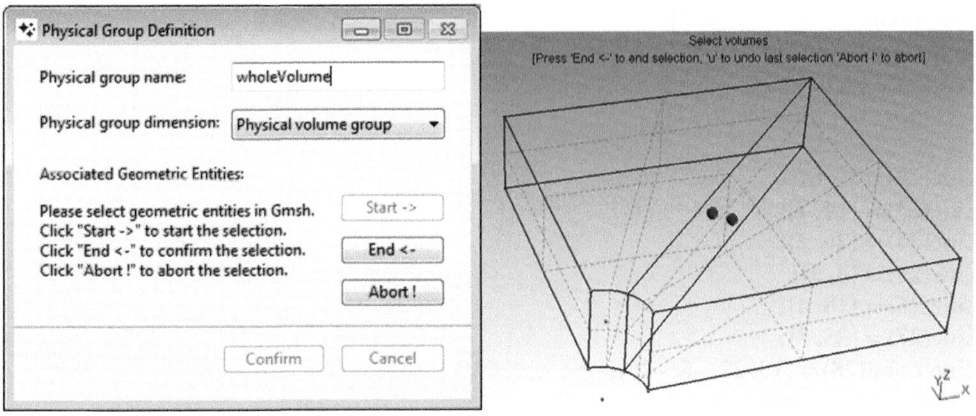

Figure 6.86 Physical group definition

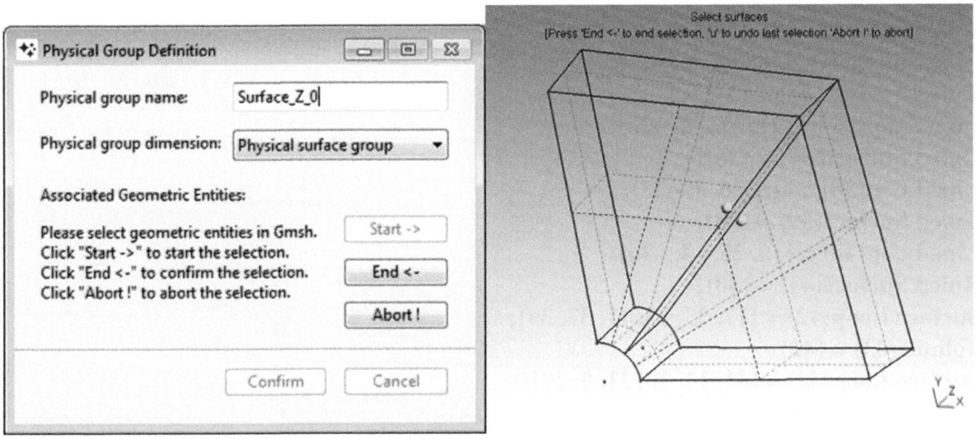

Figure 6.87 Create the physical surface and select surface entities at Z = 0

6.3.2.3 Step, Constraint, and Load

Consider the same settings for step, constraint, and load as in the first walkthrough example.

In the "Step, Constraint, and Load" window, perform the following steps:

1. Define a static linear step.
   As in the first walkthrough example, keep the job type as "static, general" and create one step with one increment.
2. For boundary conditions and loading definitions, define four physical surfaces: three surfaces at the X = 0, Y = 0, and Z = 0 planes for the symmetric boundary condition and one surface at the Y = 20 plane for surface traction. For illustration, consider the surface group at Z = 0. Since the geometric model is partitioned into two domains, the surface group at Z = 0 has two surface entities as shown in Figure 6.87.
   Continue to define the other surface groups.

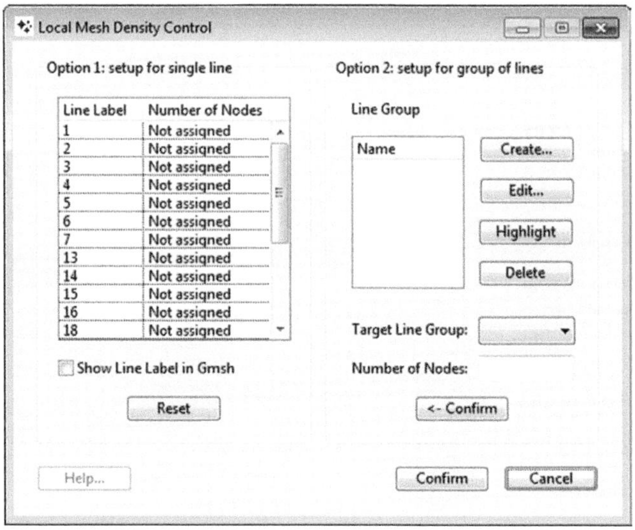

Figure 6.88 "Local Mesh Density Control" window

3. Define symmetric boundary conditions and loading.
   Repeat the same settings as in the first walkthrough example. For symmetry boundary conditions, select the boundary condition type as "Displacement in Global Coordinate," and set only $u_x = 0$ as active, associated with surface group "Surface_X_0." Then continue to set up $u_y = 0$ for "Surface_Y_0" and $u_z = 0$ for "Surface_Z_0."

For the surface traction at the Y = 20 plane, set up everything as shown in Figure 6.69. "Confirm" your input. The final "Step, Initial Condition, Constraint, and Load" window is shown in Figure 6.70.

6.3.2.4 Solution Control

Keep all default solution control parameters as in the first walkthrough example shown in Figure 6.71.

6.3.2.5 Mesh

Click the "Mesh" button in the MDS-Macro GUI main screen and the mesh control window will pop up as shown in Figure 6.72. Select a structured hexahedron mesh, which requires assigning a local number of nodes for every line. Click "Setup …" to set up the number of nodes per line. The control window is shown in Figure 6.88.

There are two options for setting the number of nodes for each line. Option 1 allows you to manually input the number of nodes for each line, as shown in Figure 6.89. You can check the "Show Line Label in Gmsh" to find the line entity label in the Gmsh canvas window.

Option 2 provides an alternative way to assign the same number of nodes for each line group. This option is more convenient for structured meshes, since it involves the same number

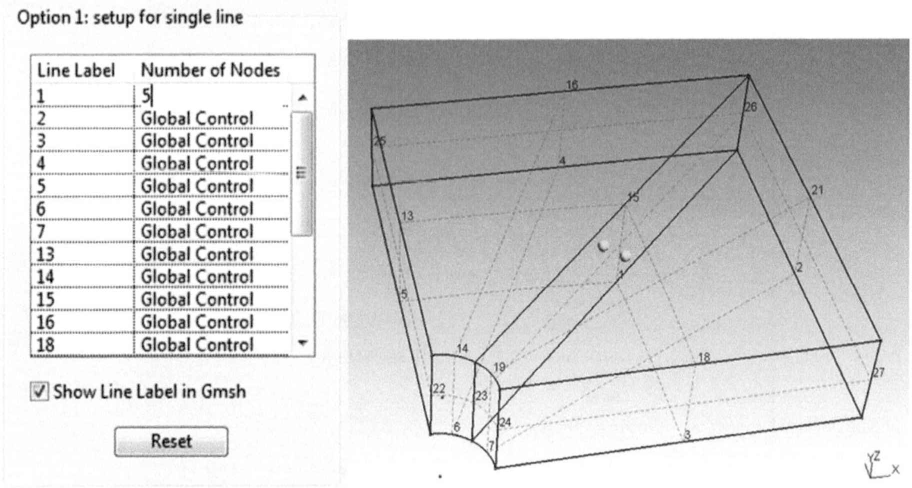

Figure 6.89 Setting the number of nodes for a single line entity

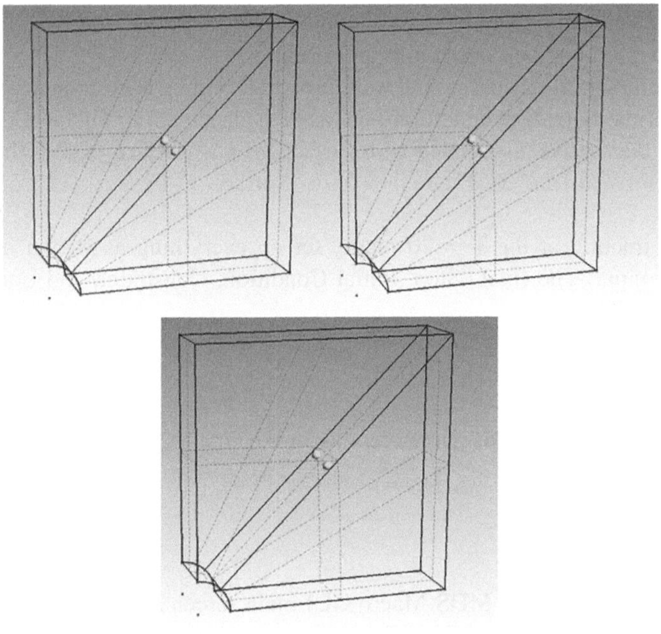

Figure 6.90 Line groups which have the same number of nodes

of nodes for a large number of lines. In the present example, we will define three line groups as shown in Figure 6.90, each having the same number of nodes.

For illustration, consider the first line group. First, click the "Create ..." button in the line group list. Input the name as "Line1." Click "Start →" to select the line entities in red as shown in Figure 6.91. Once again, click "End ←" to end the selection. "Confirm" to go back to the "Local Mesh Density Control" window.

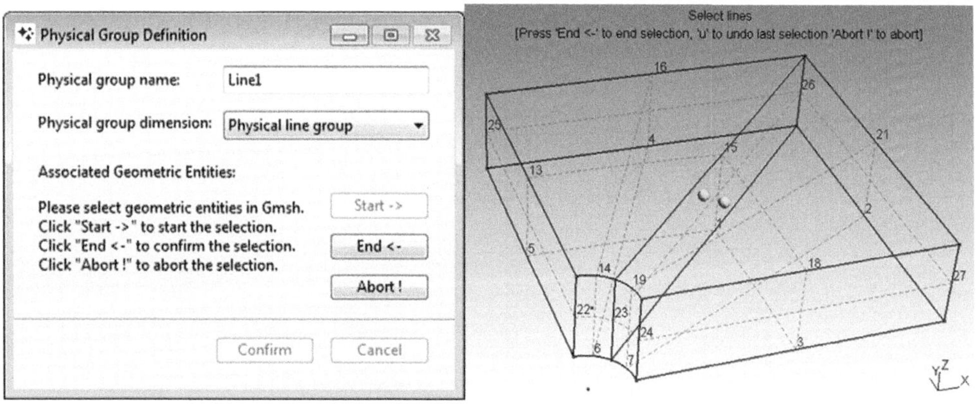

Figure 6.91 Create the line group and select line entities in Gmsh canvas window

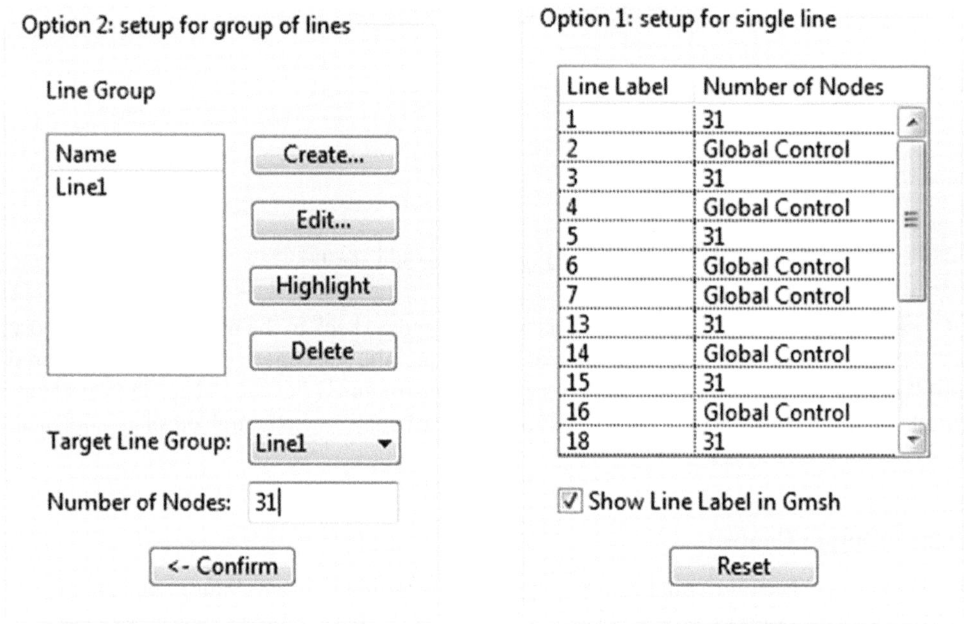

Figure 6.92 Assign the number of nodes for the line group

Once this line group is created, you can find it in the "Target Line Group" pop-up menu. Input 31 in the "Number of Nodes" and "← Confirm" it. You can see that the "Number of Nodes" list refreshes as shown in Figure 6.92. Since this line group contains line labels 1, 3, 5, 13, 15, and 18, these six lines share the same number of nodes. Since stress concentration occurs at line labels 3 and 18, you should set relatively dense mesh along this line group.

Repeat the above process to create the line group "Line2," which includes line labels 2, 4, 6, 7, 14, 16, 19, and 21, and assign the number of nodes as 11. Finally, create the line group "Line3," which contains line labels 22, 23, 24, 25, 26, and 27, and assign the number of nodes as 6. The final local number of nodes list is shown in Figure 6.93.

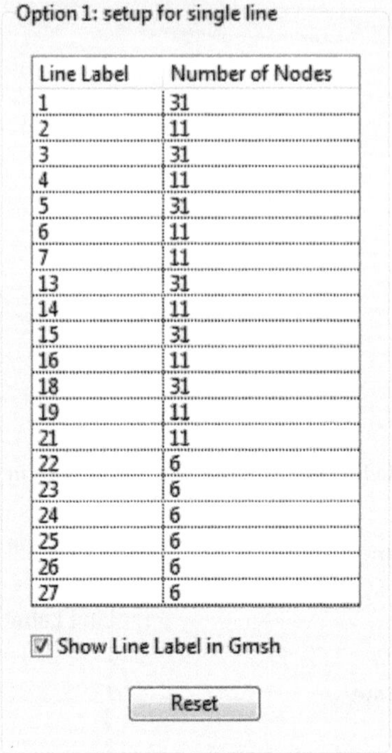

Line Label	Number of Nodes
1	31
2	11
3	31
4	11
5	31
6	11
7	11
13	31
14	11
15	31
16	11
18	31
19	11
21	11
22	6
23	6
24	6
25	6
26	6
27	6

Figure 6.93 Local number of nodes list

Close the "Local Mesh Density Control" window and go back to the mesh window. Choose the "Hexahedron" in "Element Type Control." Consequently, you can see that the "Unstructured Mesh" option is automatically disabled since the unstructured hexahedral mesh is not supported. Finally, click on the "Mesh" button. The structured hexahedral mesh is shown in Figure 6.94.

6.3.2.6 Output Control

Define the same data and visualization output request as in the first walkthrough example.

6.3.2.7 Job

Create and submit the job by following the same procedure described in the first walkthrough example. Once the job is completed, proceed to the postprocessing stage.

6.3.2.8 Postprocessing

Click "Postprocessing" and the stress component S22 value in the element centroid (CS22) for all elements is shown in the right column of Figure 6.95. The displacement and the stress S22 field are visualized by clicking on the "View visualization" button as shown in Figure 6.96.

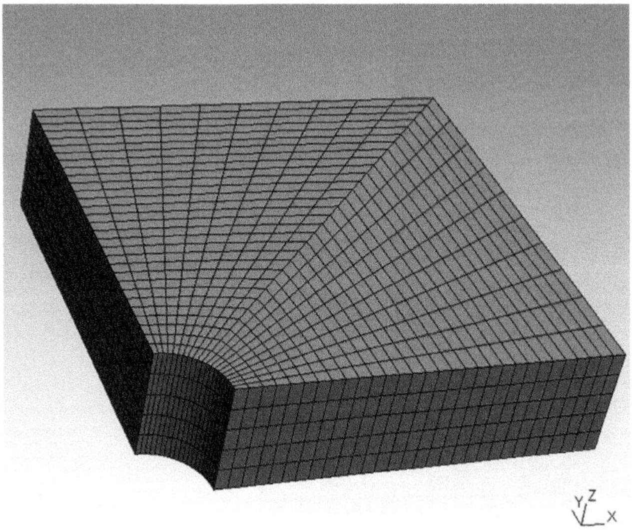

Figure 6.94 The resulting mesh at the Gmsh canvas window

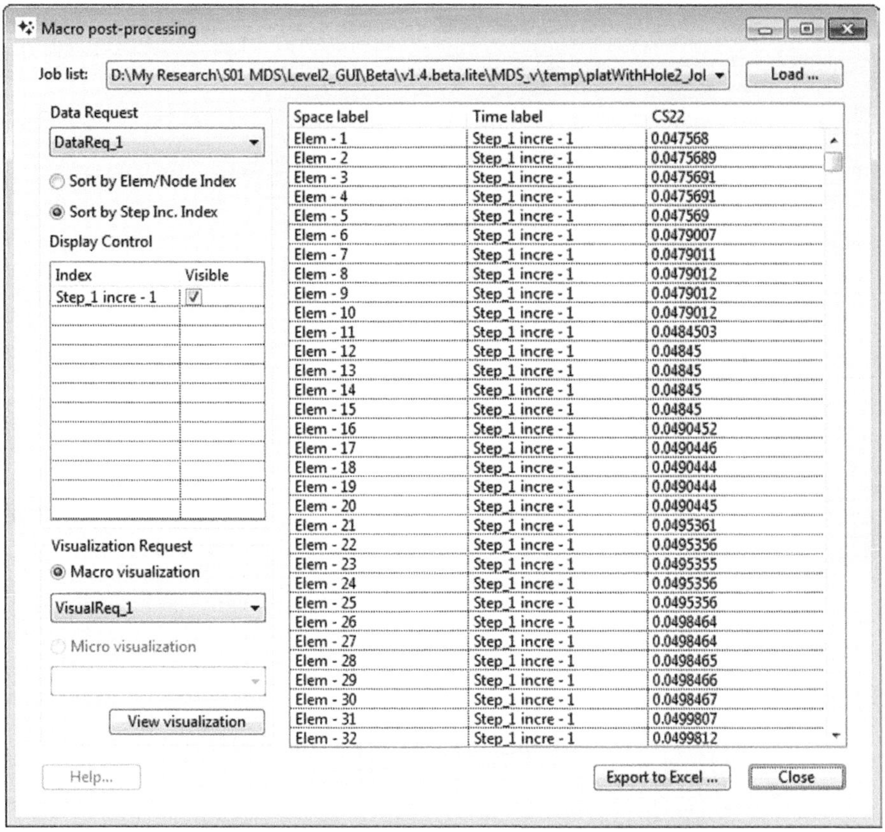

Figure 6.95 "Macro postprocessing" window

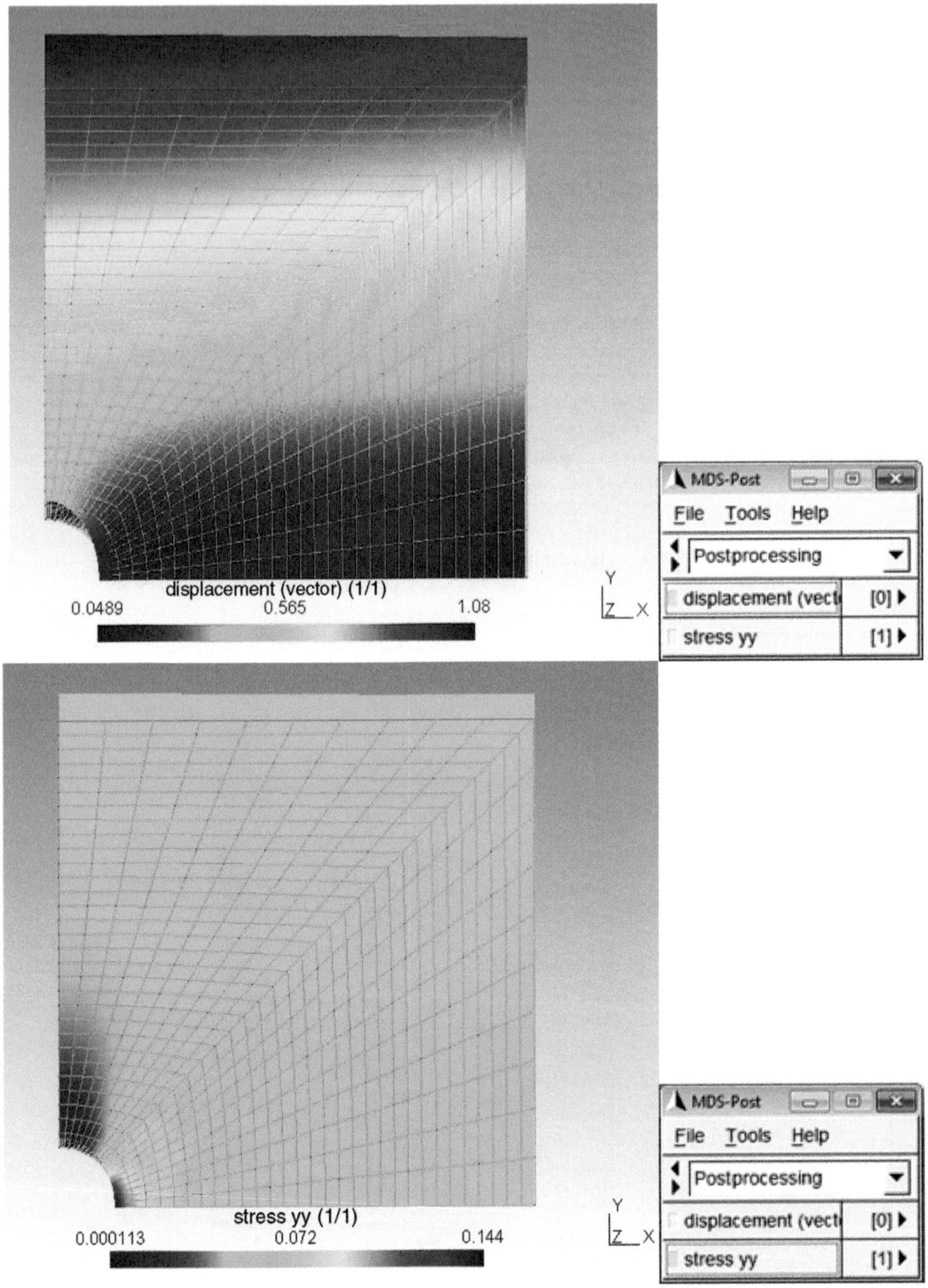

Figure 6.96 Postprocessing visualization Gmsh window

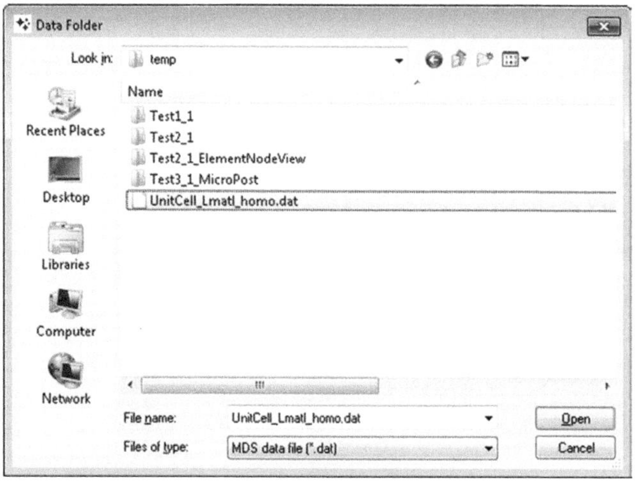

Figure 6.97 "Data Folder" window

6.3.3 *Third Walkthrough Example*

The third walkthrough example is concerned with a linear two-scale problem. At the microscale, a fibrous composite unit cell is considered with both fibers and matrix phases being isotropic elastic. A linear homogenization is performed, followed by assigning the homogenized material properties to the macro model. The macro problem is the same as in the second walkthrough example, and the unit cell model is the same as considered in section 6.2.3. Finally, the postprocessing is carried out at both the macro- and microscales.

6.3.3.1 Microscale Problem

First, select the "Micro Scale" label in the main MDS-Lite window in Figure 6.2. Carry out Steps 1 and 2 as indicated in the first walkthrough example in section 6.2.3.

6.3.3.2 Macroscale Problem

Create the geometry of the plate with a hole as in the second walkthrough example in section 6.3.2. For the material definition, begin with the process outlined in Figure 6.86 to create the volume physical group named "wholeVolume." Then click on the "Create ..." button to create a new material. To use the homogenized material properties of the fibrous composite unit cell, select the "Linear Multiscale" model and click the "Browse ..." button. Find the file named "UnitCell_Lmatl_homo.dat," which is generated in Step 2 of the unit cell problem as shown in Figure 6.97. Load the homogenized material properties as shown in Figure 6.98. "Confirm" to load this material.

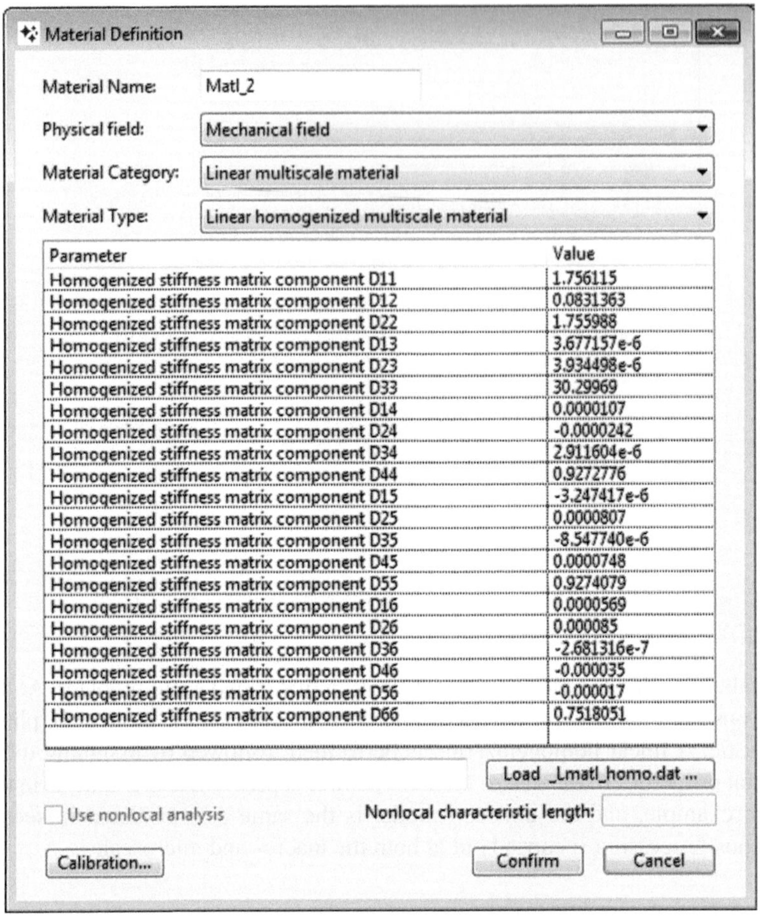

Figure 6.98 Linear multiscale material properties

Since the composite is not an isotropic material, the orientation has to be defined. By selecting the "Create ..." button in the orientation listbox, the window shown in Figure 6.99 pops up. At the microscale, the fiber is along the Z direction, whereas at the macroscale, the fiber is oriented in the loading (Y) direction. The transformation matrix between the two coordinate systems is depicted in Figure 6.99. "Confirm" to end the orientation definition.

Finally, set the association in Figure 6.100 and "Confirm" to end the setup.

Consider the step, constraint, load, solution control parameters, structured hexahedral mesh, and output request as in the second walkthrough example. In the output request window, create a visualization request, and select the Micro for the "Spatial domain". Choose element 1455, which is located at the tip of the hole. Create and submit the job the same way as in that example. Once the job is completed, proceed to the postprocessing step.

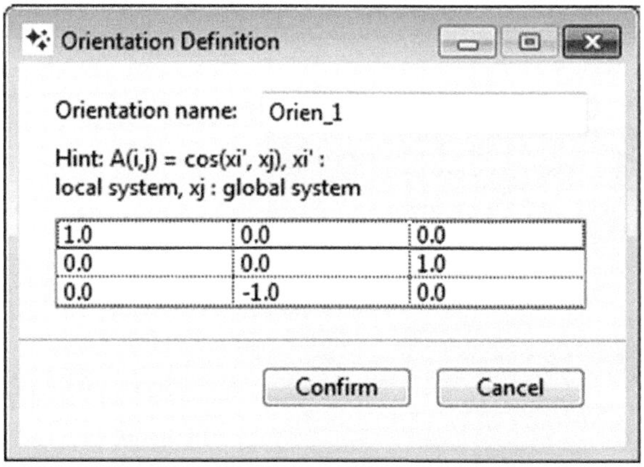

Figure 6.99 "Orientation Definition" window

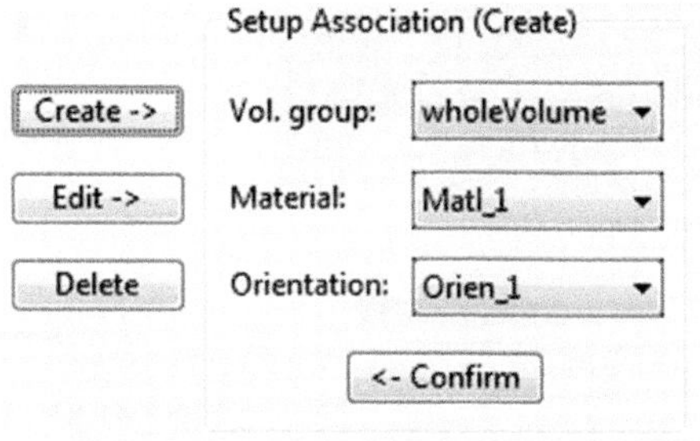

Figure 6.100 Setup association with orientation

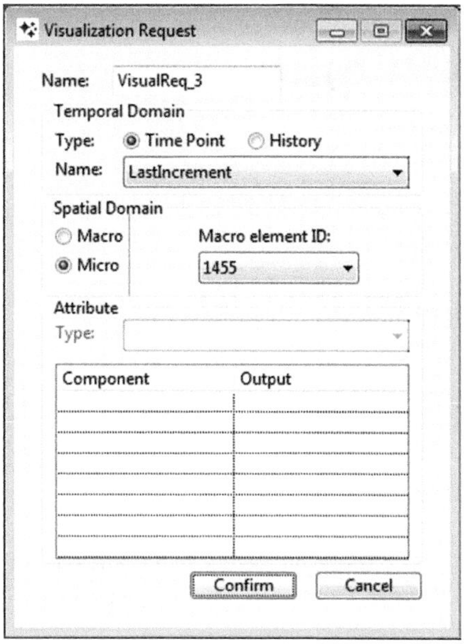

Figure 6.101 Create microscale visualization request

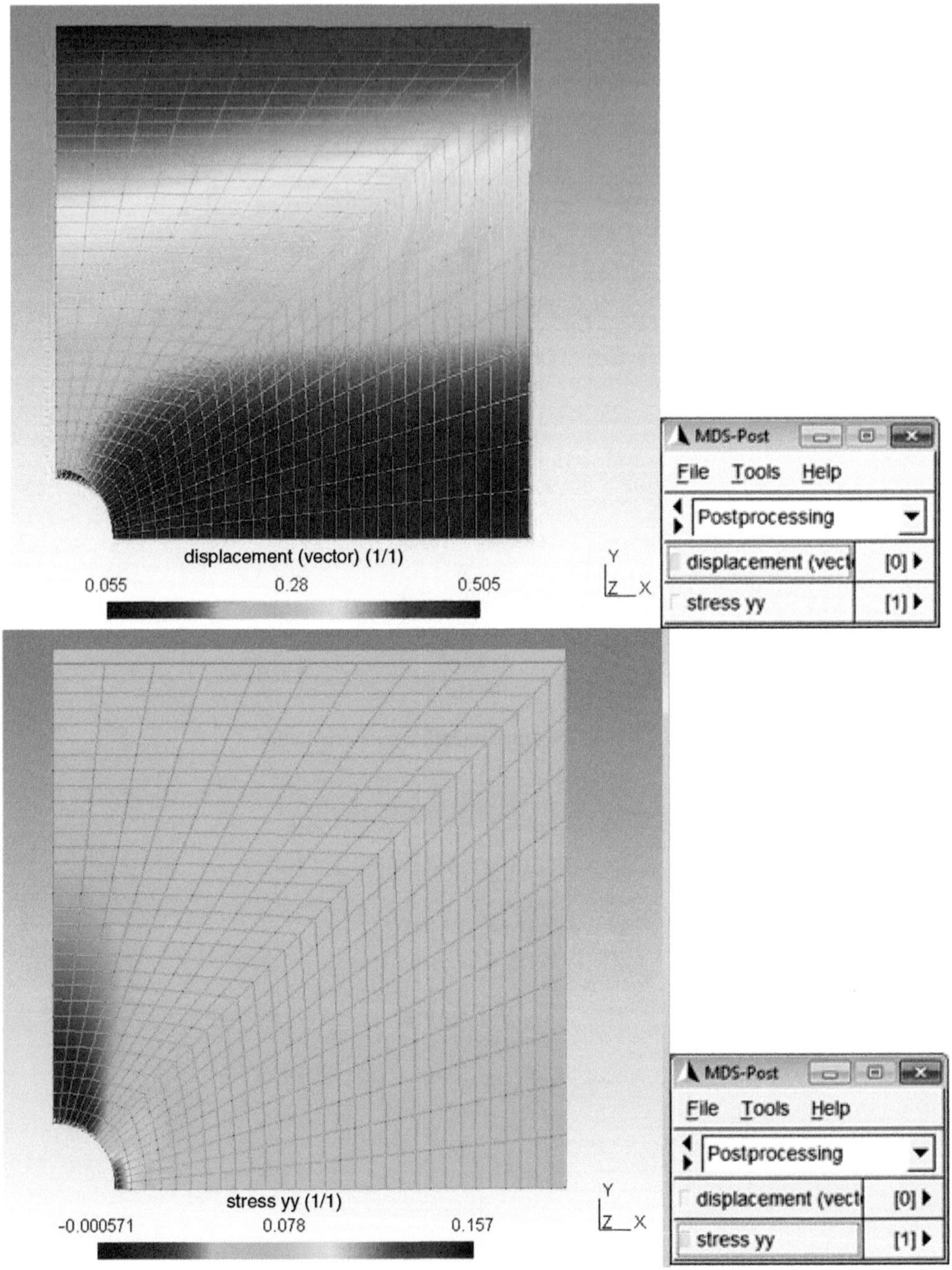

Figure 6.102 Macro displacement and stress distribution

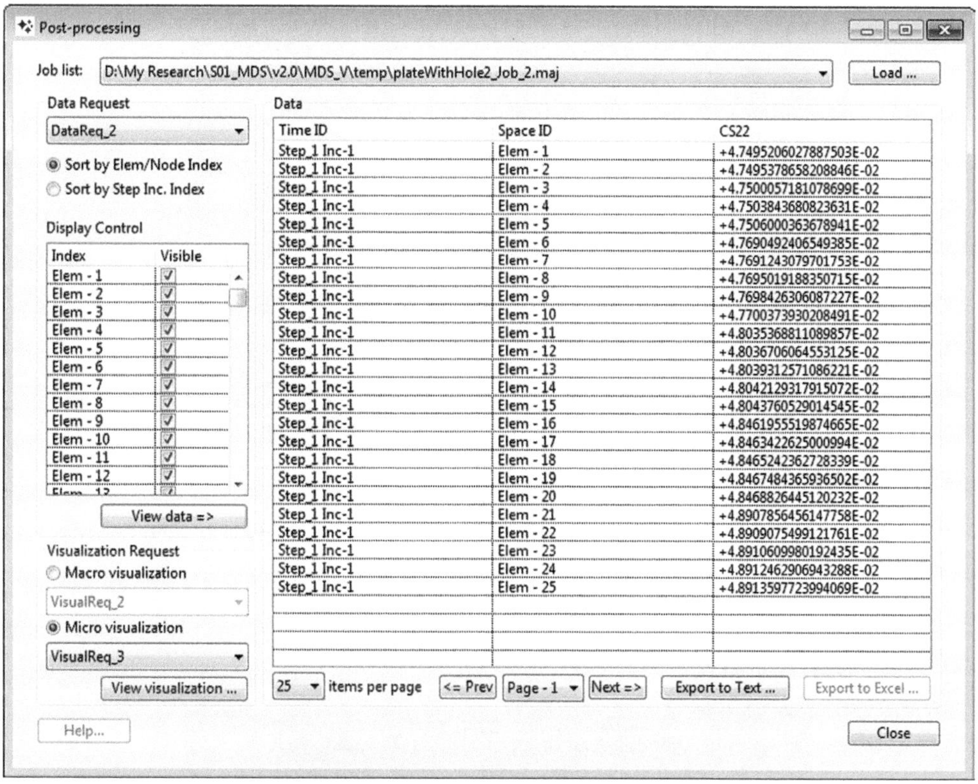

Figure 6.103 Micro visualization in the "Macro Postprocessing" window in MDS-Lite

For postprocessing at the macroscale, click "Postprocessing" under the "Macro Scale" label in the main MDS-Lite window. For visualization of the macro stress depicted in Figure 6.102, click "View visualization" as shown in Figure 6.95. Since the homogenized material is anisotropic, you can see that the stress distribution is considerably different from that of the isotropic material.

For postprocessing at the microscale, click the "Macro postprocessing" button in the main MDS-Lite window. This will take you to the "Macro postprocessing" window where you can see the list of elements to postprocess on the lower left, as shown in Figure 6.103.

In the "Micro visualization" list, you can see all of the macro elements that have homogenized material properties.

From the macro element list, select an element to postprocess at the microscale. Then click on "View visualization" to view the stress distribution at the microscale. Select the macro element #1455 located at the tip of the hole (Figure 6.104) where the macro stress is maximal, as shown in Figure 6.105.

You can see that the σ_{yy} in the fiber phase is roughly 2.5 times higher than the maximum macro stress. This is because most of the tensile load is carried out by the fibers.

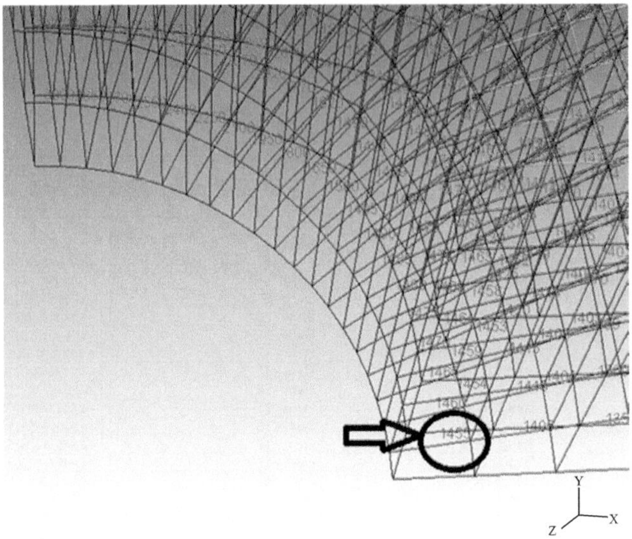

Figure 6.104 Postprocessing at the microscale

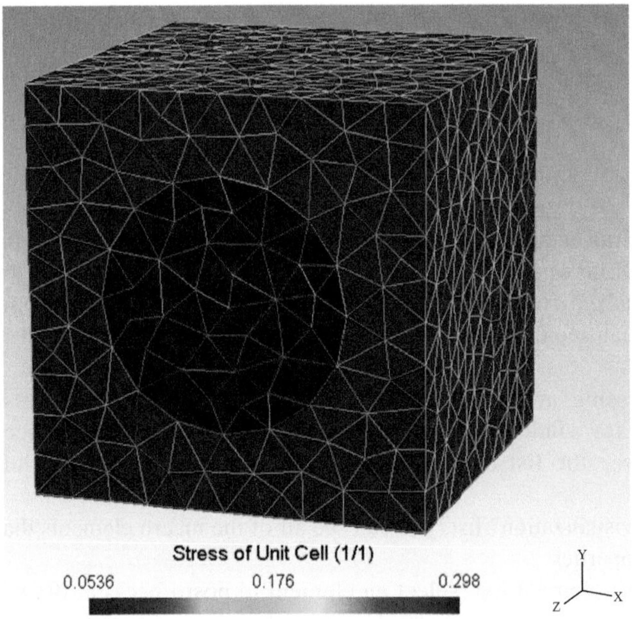

Figure 6.105 Unit cell stress distribution in the macro element #1455

6.3.4 Fourth Walkthrough Example

Consider a fibrous composite unit cell with a fiber volume fraction equal to 50% where fibers are aligned in the Z direction. In this example, we will consider an inelastic composite where the overall inelastic behavior of the material is characterized by the uniaxial tension in the X, Y, and Z directions and pure shear in the XY, XZ, and YZ directions. The reference solution (depicted in Figure 6.106) is obtained either from experimental data or from direct numerical simulation of the unit cell model. Note that since fibers are assumed to remain elastic throughout the loading history, uniaxial tension in the Z direction shows a monotonic increase in stress.

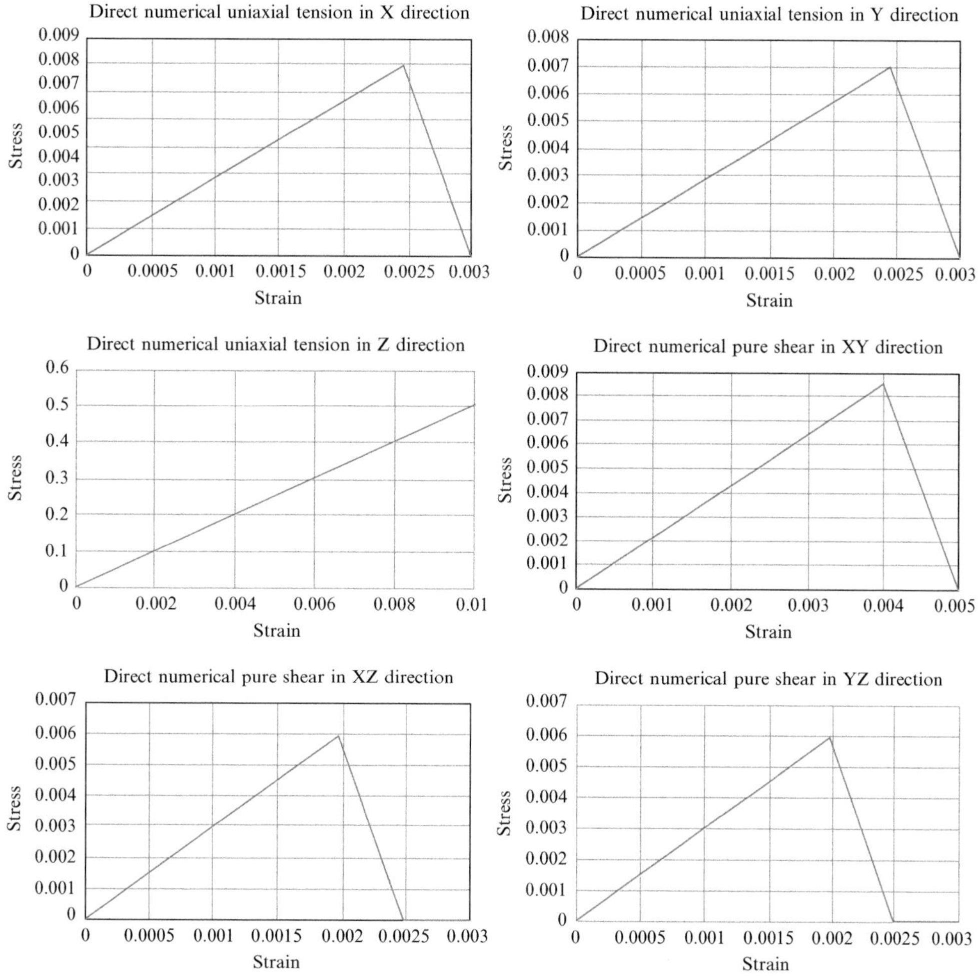

Figure 6.106 Reference solution for the uniaxial tension in X, Y, and Z directions and pure shear in XY, XZ and YZ directions

The overall inelastic anisotropic material properties are calibrated based on the reference solution described below.

6.3.4.1 Geometry

Following the same procedure as in the second walkthrough example, create the plate with a hole model depicted in Figure 6.85.

6.3.4.2 Material

Click “Material” in the main window. This will take you to the “Material Definition” window. Choose the “Nonlinear Multiscale” material category and click on the “Anisotropic Damage” type as shown in Figure 6.107. For the formulation details of the anisotropic damage model, see [7].

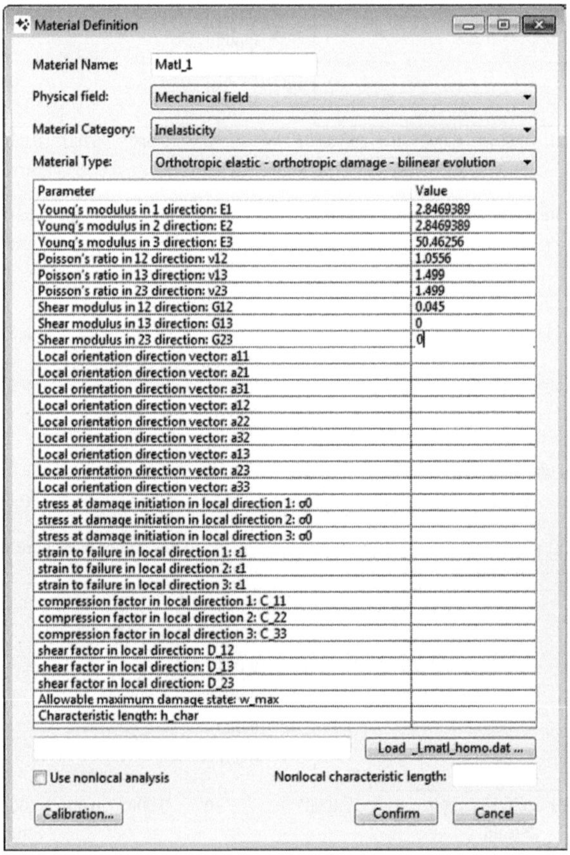

Figure 6.107 Anisotropic damage model

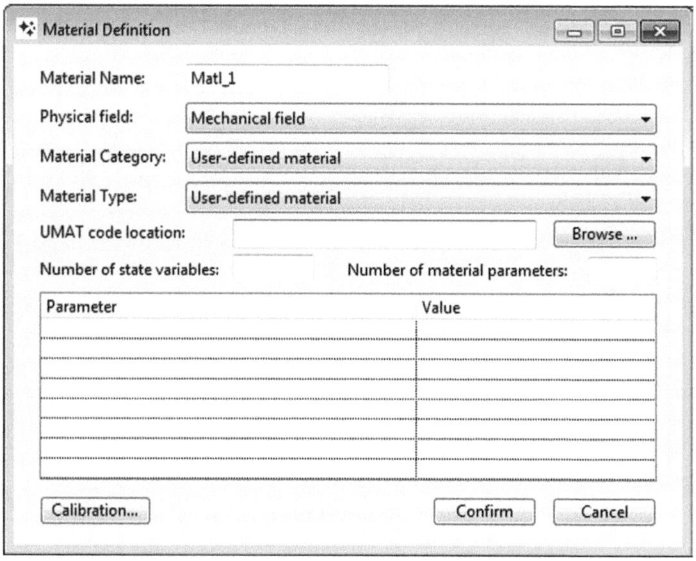

Figure 6.108 The interface of user-defined material

Remark 6.2 In MDS-Lite, you can specify a user-defined material (UMAT), which has to be written in Fortran 90 format. For instructions, consult the MDS website [6]. Once the UMAT has been defined, select the UMAT file using the "Browse …" button. Determine the number of state (internal) variables and material parameters that the UMAT calls for. Figure 6.108 depicts an example of the user-defined material, which requires 15 state (internal) variables and 12 material parameters.

Choose the homogenized elastic properties from the Lmatl_homo data file, which was defined in Step 2 of the unit cell problem, as shown in Figure 6.109.

You can rename any parameter, as shown in Figure 6.110. Set the characteristic length to zero to indicate that you will be performing a local analysis in the calibration phase. The internal variables will be calibrated to the reference solution. Click the "Calibration …" button at the lower left corner of the "Material Definition" window to invoke the calibration window.

On the left side of the "Material calibration" window, you can see all the parameter names and values. Additional columns include active tag, lower and upper bound values, and Master ID, which denotes the master–slave relation. The goal of parameter calibration is to identify optimal parameter values that minimize error between the simulation and the reference solution (see section 4.10 in Chapter 4).

In the following, we will calibrate nine internal variables, including stress at the damage initiation (strength) in the X, Y, and Z directions, strain at the fully damaged state in the X, Y, and Z directions, and shear strain upwind factors in the XY, YZ and XZ directions (see [7] for definition of model parameters). The compression factor is set to zero, denoting identical behavior in tension and compression.

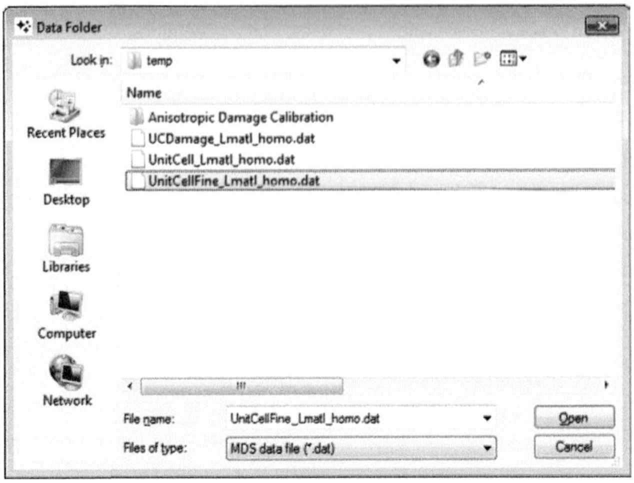

Figure 6.109 Browse Lmatl_homo data

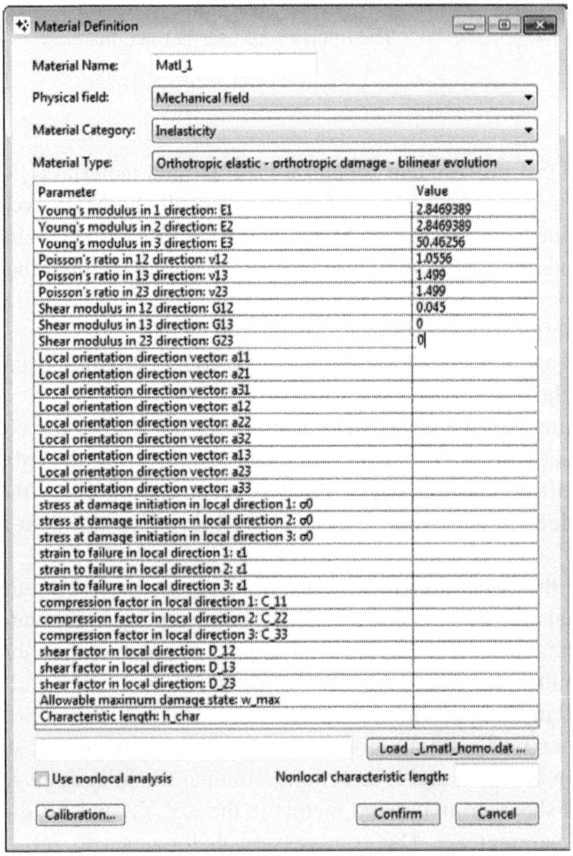

Figure 6.110 Linear model parameters of an anisotropic damage law

The nine internal variables are not calibrated simultaneously, but rather in three steps. In Step 1, given the reference solution in the uniaxial tension in the X and Y directions and XY shear, calibrate the five relevant parameters to the deformation in the XY plane. These include stress at the damage initiation, strain at the fully damaged state in the X and Y directions, and the shear strain upwind factor in the XY direction. In Step 2, given the reference solution in tension in the Z direction and XZ and YZ shear, calibrate the remaining four internal variables related to the deformation in the Z direction. These include stress at damage initiation, strain at the fully damaged state in the Z direction, and shear strain upwind factors in the XZ and YZ directions. Finally, calibrate all the internal variables simultaneously, given the complete set of the reference solution.

Note that the stress at damage initiation and the strain at the fully damaged state in the X and Y directions are identical due to transverse isotropy. This condition is imposed in the Master ID column, which shows that the value of the stress at damage initiation in the Y direction is equal to that in column 10, that is, stress at damage initiation in the X direction. Likewise, the value of the strain at the fully damaged state in the Y direction is equal to that in column 11, that is, strain at the fully damaged state in the Y direction. Load the three tests corresponding to the Step 1 calibration process in the window on the right as shown in Figure 6.111.

Click the "Run …" button to start the Step 1 calibration. The calibration process is monitored by the residual between the reference solution and the simulation in L_2 and max norms. You can modify the calibration control parameters, such as the convergence tolerance.

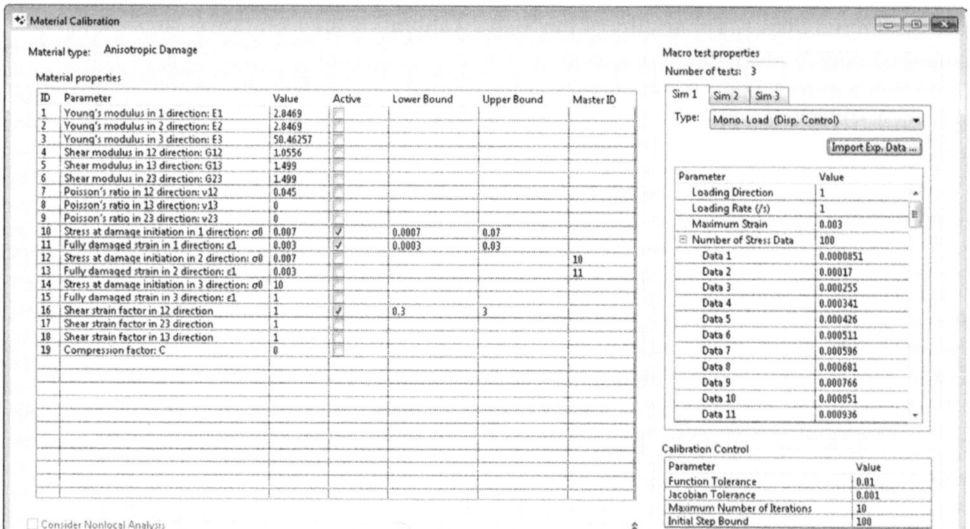

Figure 6.111 The first step calibration

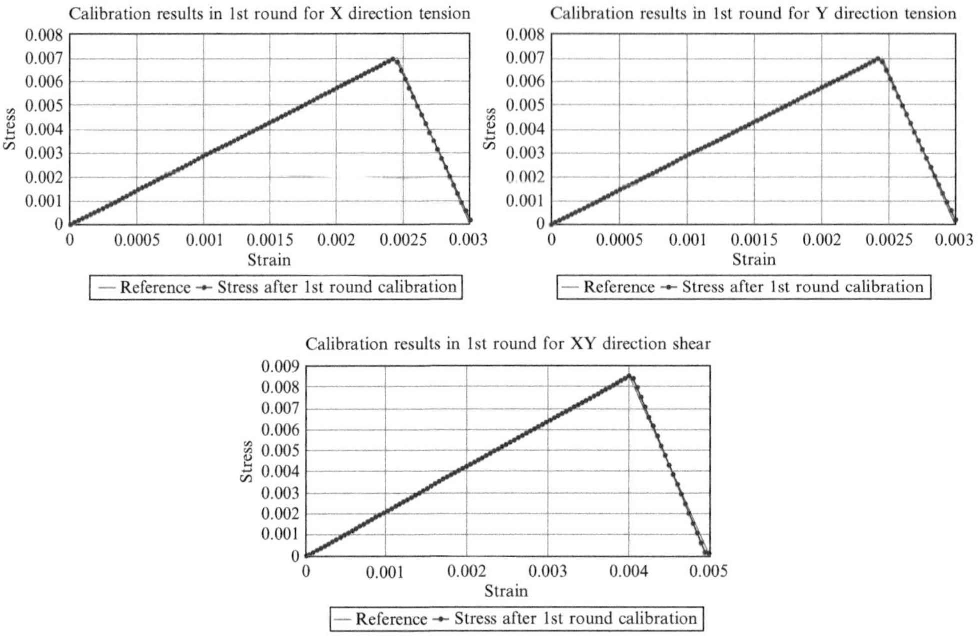

Figure 6.112 Comparison of the reference solution and simulation results after Step 1 calibration

Once the calibration ends, you can view the calibrated internal variables. If the solution has not converged to the specified tolerance, you can use the nonconverged results as a new initial guess and rerun Step 1. The results of the Step 1 calibration are as follows: stress at damage initiation in the X and Y directions equal to 0.006967252, strain at fully damaged state in the X and Y directions equal to 0.003006816, and shear strain upwind factor in the XY direction equal to 1.647929. Figure 6.112 depicts the comparison between the reference and simulation results.

Continue to the Step 2 calibration. The initial values and bounds are shown in Figure 6.113.

Click "Run ..." to initiate the Step 2 calibration. The results of the Step 2 calibration are as follows: stress at damage initiation in the Z direction equal to 9.997869, strain at the fully damaged state in the Z direction equal to 1.000128, and shear strain upwind factor in the YZ direction (same as the XZ direction) equal to 0.5775614. Figure 6.114 depicts the comparison between the reference and simulation results.

Finally, make all the internal variables active and perform a simultaneous calibration against the following six tests. You can choose a rather small tolerance since the initial values are close to the optimal. The final internal variables are:

Stress at damage initiation in the X and Y directions = 0.006948091
Strain at the fully damaged state in the X and Y directions = 0.003014976
Stress at damage initiation in the Z direction = 9.99977
Strain at the fully damaged state in the Z direction = 1.000014

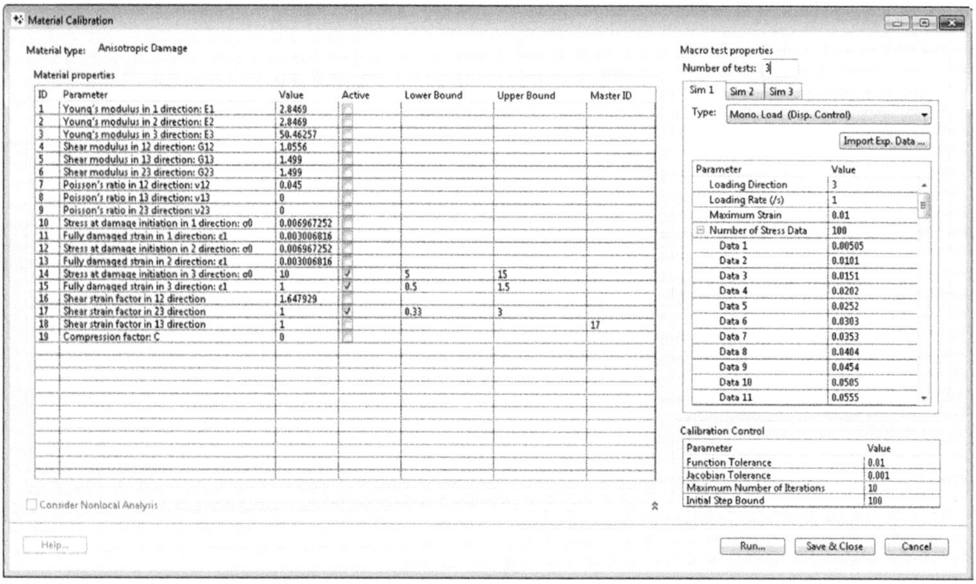

Figure 6.113 The second step calibration

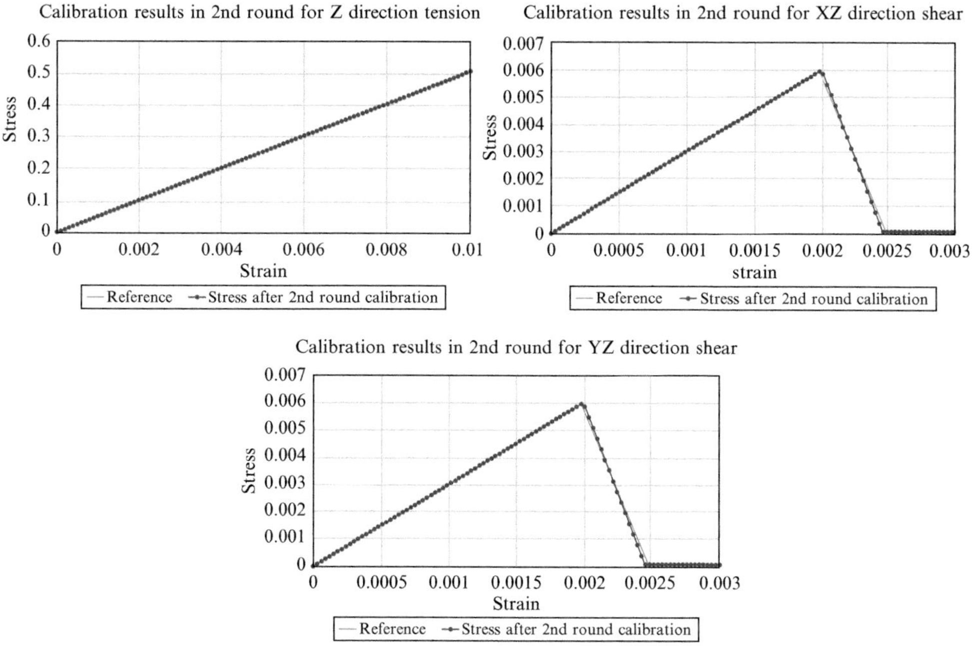

Figure 6.114 Comparison of the reference solution and simulation results after Step 2 calibration

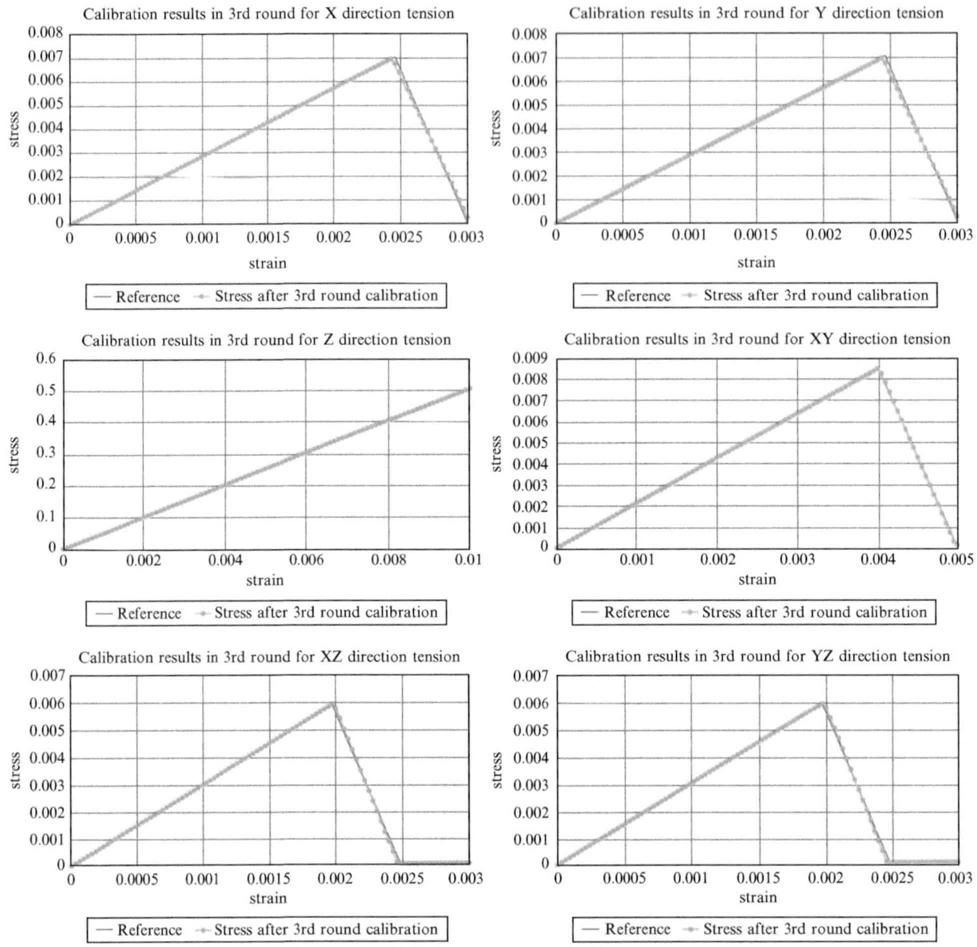

Figure 6.115 Comparison of the reference solution and simulation results after Step 3 calibration

Shear strain upwind factor in the XY direction = 1.647907
Shear strain upwind factor in the XZ and YZ directions = 0.5779028

Figure 6.115 depicts the comparison between the reference and simulation results.

To complete the material definition, it is necessary to define a characteristic material length due to strain softening and localization (see section 4 in Chapter 4). Choose the characteristic length as 0.025, which is approximately equal to the physical size of the unit cell. Note that the coarse-scale element size is approximately 0.05, that is, twice as large as the characteristic material size. To study the influence of the characteristic size, we will also consider a characteristic length equal to 0.25, that is, 10 times larger, as shown in Figure 6.116.

Follow the steps in the second walkthrough example to create the orientation matrix and physical volume groups, and then set up the association with physical groups, orientation matrix, and material.

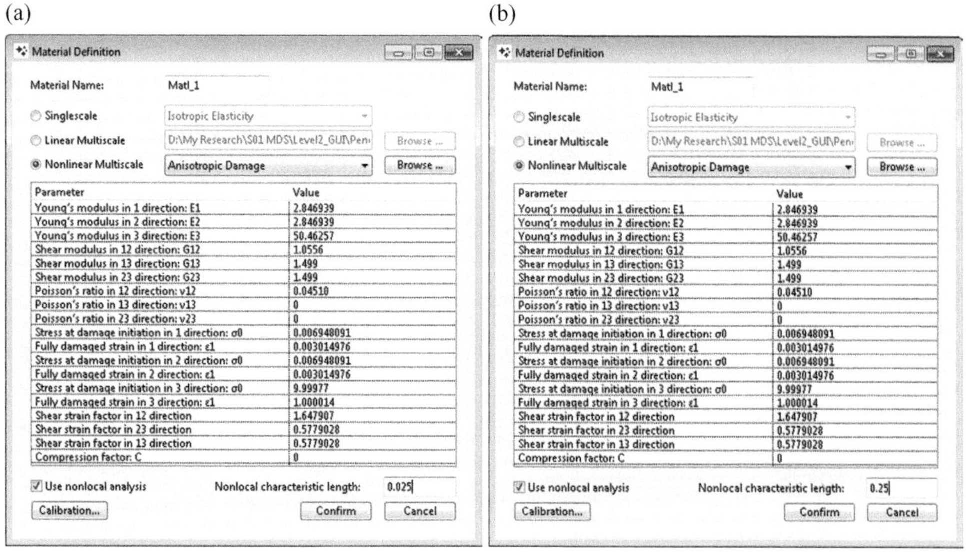

Figure 6.116 Material parameters for characteristic length equal to 0.025 (a) and 0.25 (b)

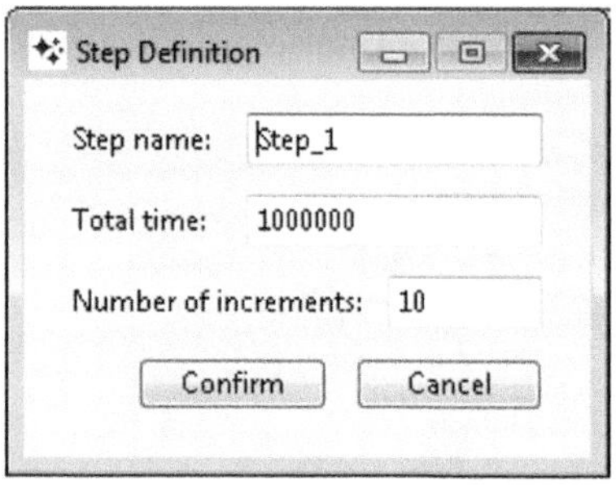

Figure 6.117 Step with multiple increments

6.3.4.3 Step, Initial Condition, Constraint, and Load

Create the initial step with 10 increments (Figure 6.117). For nonlinear analysis, we may use multiple increments to resolve damage evolution.

Keep the symmetry boundary condition as in the second walkthrough example for Surface_X_0, Surface_Y_0, and Surface_Z_0. For Surface_Y_20, use the displacement constraint as shown in Figure 6.118. Set the maximum deflection of Surface_Y_20 to 0.02.

6.3.4.4 Solution Control

Keep all the solution control parameters at the default values.

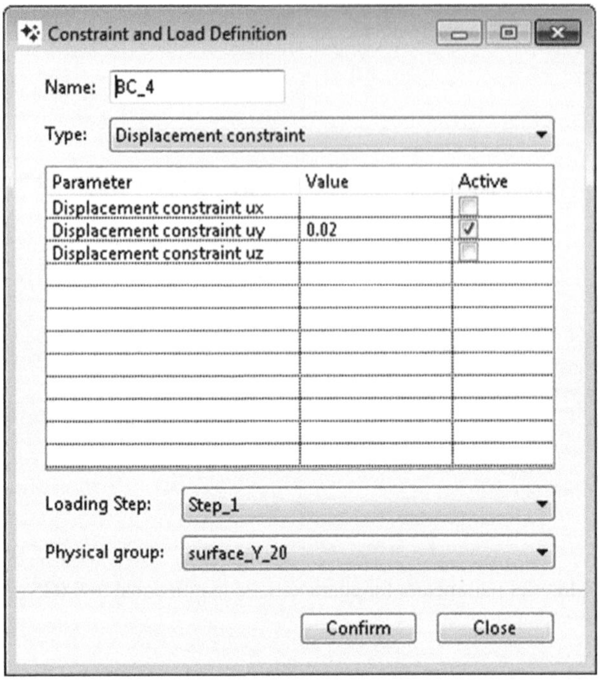

Figure 6.118 Displacement constraint

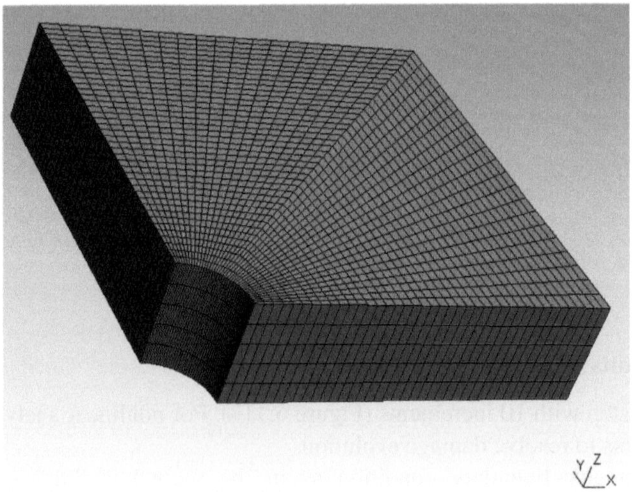

Figure 6.119 Structured hexahedron mesh for the fourth walkthrough example

6.3.4.5 Mesh

Use a structured hexahedral mesh. Assign the local number of nodes to the physical line groups, Line1, Line2, and Line3, which are shown in Figure 6.90 as 41, 21, and 6, respectively. The resulting mesh is shown in Figure 6.119.

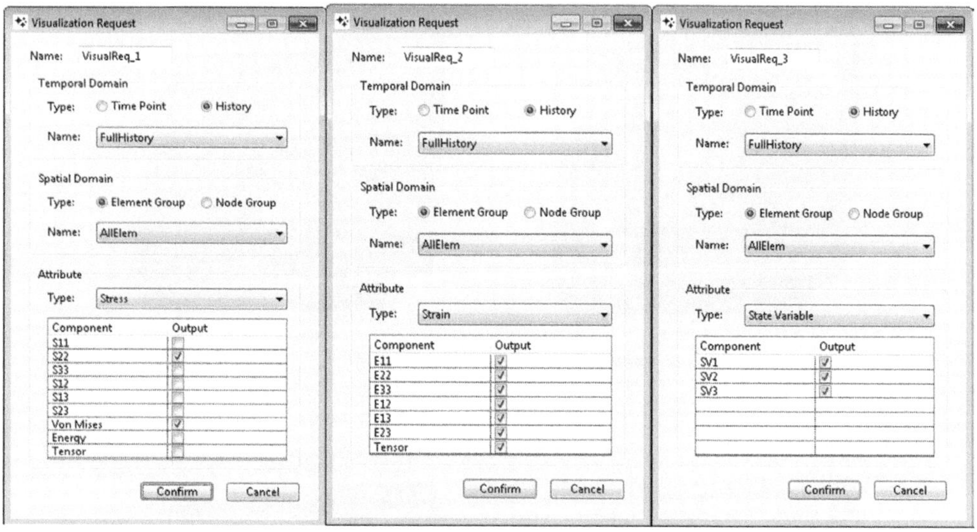

Figure 6.120 Visualization requests

6.3.4.6 Output Control

We will request three visualization outputs: stress, strain, and internal variables (damage parameters). The attributes of the visualization request are given in Figure 6.120.

6.3.4.7 Job

Create and submit the job. Once the job is completed, proceed to the postprocessing.

6.3.4.8 Postprocessing

The damage parameters are computed in the principal strain directions [7], with the largest damage parameter corresponding to the largest principal strain. In the following, we output the largest damage parameter, which is the best descriptor of physical damage. We define 10 increments in one step and require all visualizations to be based on full history. The maximum damage in frames 1 to 5 is zero, since all elements remain elastic up to that stage. In frame 6, the region near the hole starts to damage. Consequently, damage propagates along nearly a straight line (Figure 6.121).

Repeat the simulation using a size 10 times larger than the characteristic length, as shown in Figure 6.116.

The nonlocal theory forces localization to be wider, which makes the response considerably stiffer. Increase the maximum prescribed displacement on Surface_Y_20 to 0.03 and change the number of increments to 15, as shown in Figure 6.122.

Keep the same mesh and output control parameters. Create and submit the job. Proceed to postprocessing to visualize the maximum damage parameter. In frames 1 to 9, the plate

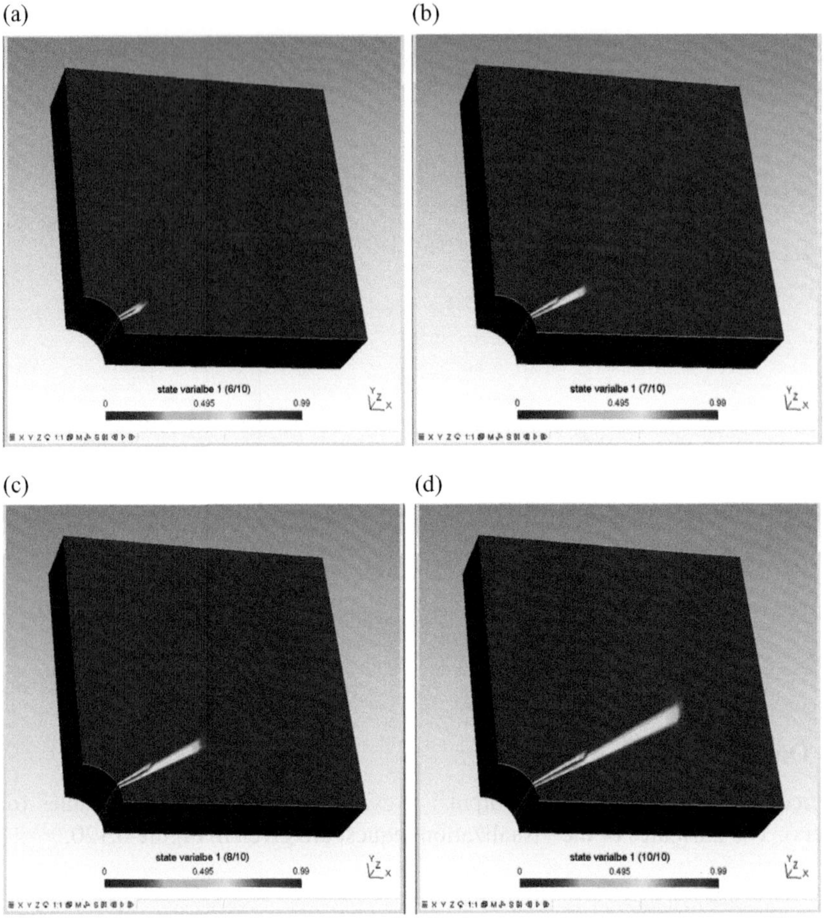

Figure 6.121 Maximum damage parameter in frames (a) 6, (b) 7, (c) 8, and (d) 10

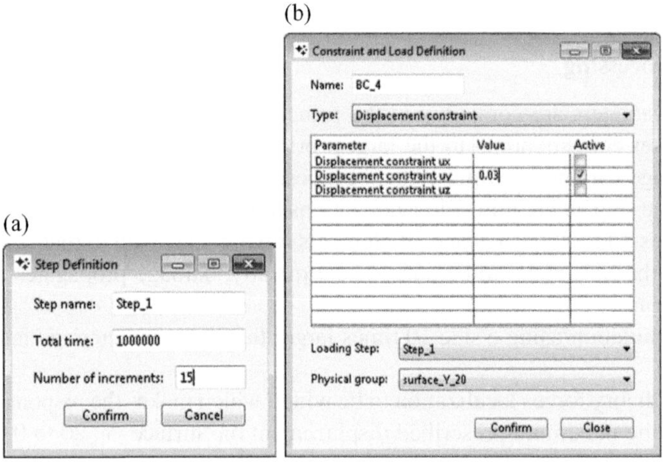

Figure 6.122 Defining number of increments (a) and maximum prescribed displacement value (b)

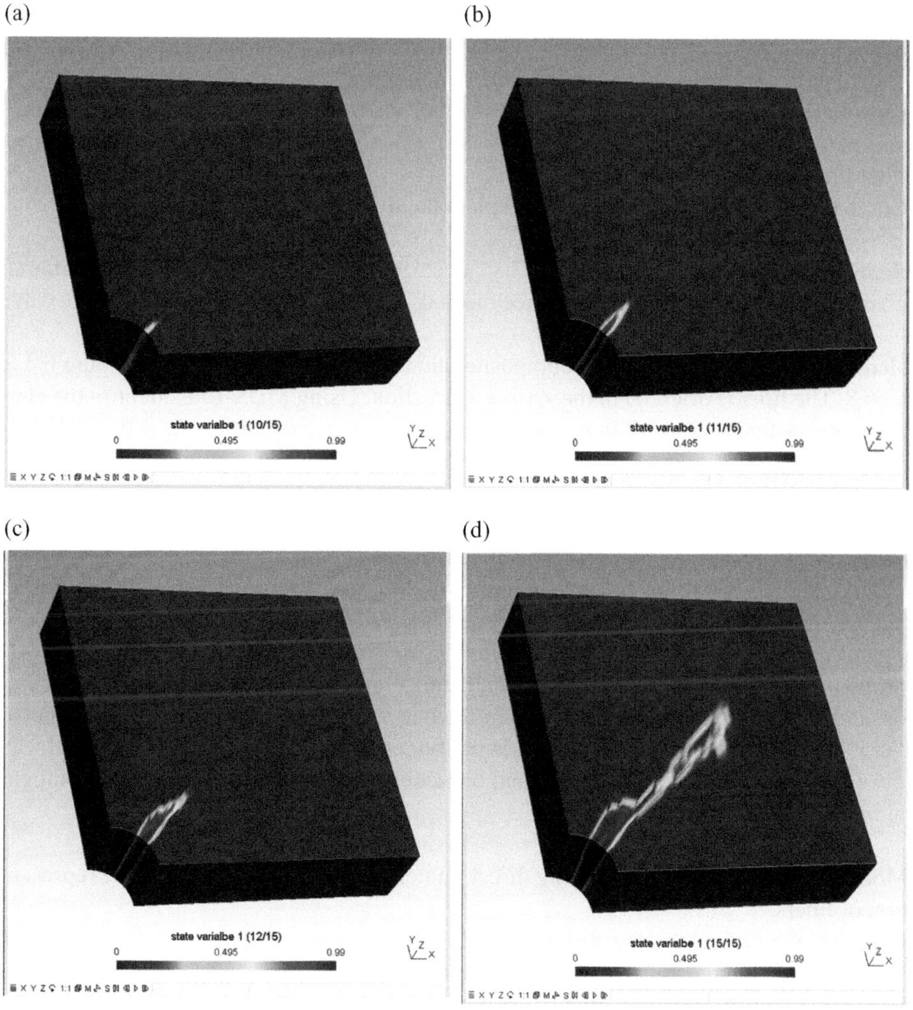

Figure 6.123 Maximum damage parameter in frames (a) 10, (b) 11, (c) 12, and (d) 15

remains elastic. Damage evolution starts in frame 10, which is delayed in comparison with the smaller characteristic size considered earlier (Figure 6.123). Compared with Figure 6.121, the localization width is considerably larger.

Problems

Problem 6.1 Using MDS-Macro, construct the fibrous composite unit cell problem defined in Figure 6.4 and Figure 6.8.

(a) Apply the unit displacement in the fiber direction while constraining the opposite unit cell face in the fiber direction. Compute the overall Young's modulus in the fiber direction.

(b) Subject the unit cell to tensile loading in the fiber direction by applying a traction load on the opposite faces of the unit cell. Constrain six degrees of freedom to eliminate rigid body motion. Compute the overall Young's modulus in the fiber direction.
(c) Consider a parametric fibrous unit cell model in MDS-Lite with the same volume fraction and material properties of microphases as in (a). Compute the Young's modulus in the fiber direction.
(d) Compare the above three results and explain the difference.

Problem 6.2 Repeat Problem 6.1 for a 30% spherical inclusion unit cell problem. Is the difference in the Young's modulus obtained by the three methods in (a)–(c) more pronounced? If so, why?

Problem 6.3 Consider the fibrous composite unit cell problem as defined in Figure 6.4 and Figure 6.8. The fiber is oriented in the Z (or x_3) direction. Using MDS-Lite, identify the elastic phase properties that will result in the following macroscopic properties: $E_{33}=10, E_{11}=E_{22}=1$, $G_{13}=G_{23}=1.5, G_{12}=1$. Assume that the matrix is isotropic, with fibers being either isotropic or transverse isotropic if necessary.

Problem 6.4 Consider the plate problem discussed in section 6.3 and the fibrous composite unit cell as defined in Figure 6.4 and Figure 6.8. Note that only the volume fraction is prescribed, whereas the physical size of the unit cell is not defined. This is because the first-order homogenization theory considered in Chapter 2 is invariant to the unit cell size. The goal of this problem is to assess the accuracy of homogenization for large unit cells.

Consider a unit cell size equal to 1/10th of the half width of the plate. Due to the relatively large unit cell size, the coarse-scale strain is no longer constant in the unit cell that is located in close proximity to the tip of the hole, and consequently, the results obtained by homogenization may not be accurate.

(a) Model the quarter of the plate using direct numerical simulation with discrete representation of fibers.
(b) Compare the critical (at the tip of the hole) stress results obtained by direct numerical simulation with the results obtained by homogenization.
(c) Repeat (a) and (b), but this time use a concurrent multiscale approach where a discrete representation of fibers is used in the vicinity of the tip of the hole, whereas elsewhere, homogenized properties are used.

Problem 6.5 Consider the 3D random chopped fiber composite in the MDS-Lite library with a volume fraction of fiber equal to 15% and matrix and fiber properties as indicated in Figure 6.8. Assume the fiber length to be equal to 1 and the fiber diameter to be cylindrical. The random chopped fiber composite in MDS-Lite is constructed using a hierarchical random sequential adsorption (HRSA) algorithm. Since the fibers are placed randomly, a different fiber topology is created each time. Determine the minimal unit cell size that will create macroscopically isotropic properties.

(a) Consider a fiber aspect ratio of 5 (that is, a diameter equal to 0.2) and vary the unit cell size from 1.5 to 2.5. For each unit cell size, construct a unit cell 30 times in MDS-Lite. Compute the average and standard deviation of Young's modulus in three directions.

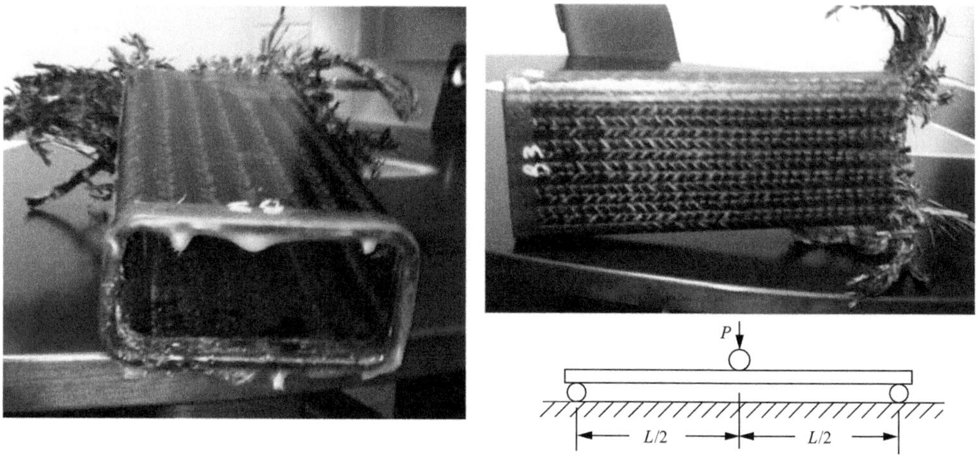

Figure 6.124 Composite beam. Length: $L = 1850$ mm. Outer cross section: 90 mm × 50 mm. Outer radius of curvature: 12 mm. Shell thickness: 4 mm. Fiber Young's modulus: $E_f = 21$ kN mm^{-2}. Matrix Young's modulus: $E_m = 2.5$ kN mm^{-2}. Poisson's ratio = 0.3

(b) Repeat (a) for a fiber aspect ratio of 15.
(c) What do you think is an optimal unit cell size?

Problem 6.6 Consider a composite beam as shown in Figure 6.124. Find the optimal composite microstructure that results in the lowest maximal von Mises stress and the lowest deflection. Consider four microstructures: plane weave, five harness weave, fibrous composite microstructure, and short fiber microstructure. Consider the fiber volume fraction of 30% for all microstructures. Make sure that the simulation results converge by considering increasingly refined finite element meshes of the composite beam.

References

[1] http://www.firehole.com.
[2] http://www.grc.nasa.gov/www/StructuresMaterials/MLP/software/mac-gmc/index.html.
[3] http://www.compositesanalysis.com.
[4] http://www.e-xstream.com/en/digimat.
[5] http://www.alphastarcorp.com.
[6] http://www.multiscale.biz.
[7] J. Fish, Y. Liu, V. Filonova, N. Hu, Zi. Yuan, Z. Yuan. A Phenomenological Regularized Multiscale Model International Journal for Multiscale Computational Engineering, 2013 (in press).
[8] http://www.geuz.org/gmsh/.
[9] Bailakanavar, M., Liu, Y., Fish, J. and Yuan, Z. Multiscale design systems automated modeling of random inclusion composites. International Journal for Multiscale Computational Engineering 2013 (in press).

Index

Practical Multiscaling, First Edition. Jacob Fish.

Printed and bound by CPI Group (UK) Ltd, Croydon, CR0 4YY

16/04/2025

14658389-0001